NAYFEH—Perturbation Methods
NAYFEH and MOOK—Nonlinear Oscillations
ODEN and REDDY—An Introduction to the Mathematical Theory of Finite Elements
PAGE—Topological Uniform Structures
PASSMAN—The Algebraic Structure of Group Rings
PRENTER—Splines and Variational Methods
RIBENBOIM—Algebraic Numbers
RICHTMYER and MORTON—Difference Methods for Initial-Value Problems, 2nd Edition
RIVLIN—The Chebyshev Polynomials
RUDIN—Fourier Analysis on Groups
SAMELSON—An Introduction to Linear Algebra
SIEGEL—Topics in Complex Function Theory
- Volume 1—Elliptic Functions and Uniformization Theory
- Volume 2—Automorphic Functions and Abelian Integrals
- Volume 3—Abelian Functions and Modular Functions of Several Variables

STAKGOLD—Green's Functions and Boundary Value Problems
STOKER—Differential Geometry
STOKER—Nonlinear Vibrations in Mechanical and Electrical Systems
STOKER—Water Waves
WHITHAM—Linear and Nonlinear Waves
WOUK—A Course of Applied Functional Analysis

A COURSE OF
APPLIED FUNCTIONAL ANALYSIS

ARTHUR WOUK
University of Alberta

A WILEY-INTERSCIENCE PUBLICATION

JOHN WILEY & SONS New York • Chichester • Brisbane • Toronto

CONTENTS

NOTATIONS

Conventions

C is the complex field.

If z is any complex number, then $\Re z$ denotes the real part of z, $\Im z$ the imaginary part of z.

R is the real field.

If x is an element of a set S, we write $x \in S$.

A collection C of elements $x \in S$ is called a subset of S, and we write $C \subset S$, or else $S \supset C$. If $C \neq S$, then C is a *proper* subset of S. If x is *not* an element of S, then we write $x \notin S$.

The empty set is denoted by $\varnothing$.

If S is a set on the real line, then $\inf S =$ *infimum* S is the greatest lower bound of S, while $\sup S =$ *supremum* S is the least upper bound of S.

On a set S let there be a proposition $P(x)$ which is either true or false for each $x \in S$. The set of x for which $P(x)$ is true is denoted by $\{x | P(x)\}$. The set of $x \in C \subset S$ for which $P(x)$ is true is denoted by $\{x \in C | P(x)\}$.

The *union* of two sets S and T is written $S \cup T$.

The *intersection* of two sets is written $S \cap T$.

An ordered pair of elements x and y is denoted by $\{x,y\}$.

The *product set* of two sets X, Y, the set of all ordered pairs $\{x,y\}$, where $x \in X$ and $y \in Y$, is denoted by $X \otimes Y$.

$\{x^n\}$ is the sequence $\{x^0, x^1, x^2, \ldots\}$. If $\{x^n\}$ is a sequence on the real line then $\limsup x^n$ is the largest real number x for which there is a subsequence $\{y^n\}$ with $x = \lim y^n$ and $\liminf x^n = -\limsup -x^n$.

$x \uparrow y$ means $x^1 \leqslant x^2 \leqslant x^3 \leqslant \ldots \leqslant y$ and $x^n \to y$.

$x^n \downarrow y$ means $-x^n \uparrow -y$.

Kronecker's delta: If i, j are any two integers from a set of integers I, then δ_i^j denotes the number 1 when $i = j$, otherwise the number 0.

We write $f = o(g)$ if the ratio $f/g \to 0$ as some common argument approaches a limit. We write $f = 0(g)$ if the ratio f/g is bounded as some common argument approaches a limit.

Symbols

In the following list of symbols, we give the symbol, a descriptive name or phrase, and a page reference for an explanation.

$\mathcal{C}$	the set of scalars	86
A^{-1}	inverse operator	88
$A^{1/2}$	positive square root of an operator A	218
A^T	transpose of A	93
$A \leqslant B$	partial ordering of operators	213
$A + B$	sum of two sets	342
$A \oplus B$	direct sum of two sets	342
$(\cdot)$	almost everywhere	43
$B(x, r)$	ball about x of radius r	21
$B\{x^1, x^2\}$	bilinear form	275
∂S	boundary of a set S	75
$BV[0, 1]$	space of functions of bounded variation	35
c_o	a sequence space	5
c	a sequence space	5
co A	convex hull of a set A	342
co f	convex hull of a functional	393
cl f	closed lower semicontinuous hull	393–394
$\overline{\text{co}}\ A$	closed convex hull of a set A	342
codim M	codimension of M	345
$C[a, b]$	space of continuous functions	7
$C[a, b)$	space of continuous functions	7
$C^1[a, b]$	space of continuously differentiable functions	7
$C^p[a, b]$	pth order differentiable functions	7
$\rightarrow$	convergence	19
$(\rightarrow)$	convergence almost everywhere	43
$\dim X$	dimension of a space X	87
$dF(x; h)$	Fréchet differential	267
$d_x F(x, y; h)$	partial differential	274
$d^+ f(x; h)$	directional derivative	335
$d^+(x; h)$	directional derivative	357
$d^-(x; h)$	directional derivative	357
D_T	domain of an operator T	20
$DF(x; h)$	linear Gâteaux differential	264
$DF(x)h$	linear Gâteaux differential	264

$L_p^m(0,1)$	a function space	334
$L_\infty^m(0,1)$	a function space	379
$L(f;\alpha)$	level set	353
$L_\Omega(f;\alpha)$	level set	353
$L_\Omega(f;x^0)$	level set	353
$\hat{L}_\Omega(\phi;z)$	level set	404
$\mathfrak{L}(x,u')$	Lagrangian	400
m	a sequence space	5
$\mu(E)$	measure of a set E	57–58
$N(A)$	null space of A	88
0	zero operator	105
$\rightarrow$	operator strong convergence	111
$\Rightarrow$	operator uniform convergence	111
$P\sigma(T)$	point spectrum of T	225
$\boldsymbol{P}$	value of primal problem	395
$\boldsymbol{P}^*$	value of dual problem	395
$\pi_{x'}$	hyperplane	345
$P(M,x)$	pseudotangent cone	384
$\overline{\mathbf{R}}$	the extended real field	352
$\mathbf{R}^n$	real n-space	4
$\mathbf{R}^{m-}$	negative orthant in R^{m-}	332
R_T	range of the operator T	20
R_λ	resolvent of T	225
$\rho(T)$	resolvent set of T	225
$\rho(x,y)$	a metric	3
$\rho_p(x,y)$	a metric	8
$\mathbf{S}\{x^1,x^2,\ldots,x^n\}$	span of $x^1,x^2,\ldots,x^n$	87
$[S]$	smallest linear manifold containing S	177
$\{S\}$	closure of $[S]$	177
$\overline{S}$	closure of a set S	75
S^0	annihilator of a set S in X	95
${}^0S$	annihilator of a set S in X'	95
sup	supremum = least upper bound	5
$\sigma(T)$	spectrum of T	225
$\Vert T\Vert$	norm of a bounded linear operator	98–99
T'	conjugate of an operator T	206
T^*	(Hilbert) adjoint of an operator T	210
T_0	Tangent manifold	325–326
$T(M,x)$	tangent cone	384
T^I	pseudoinverse of an operator T	375
$T^{-1}(x)$	preimage of x	372
$T_a^b(f)$	total variation of f	35

U	unitary operator	211
ucl g	upper closure	399
usc g	upper semicontinuous hull	399
$[u,v]$	inner product in the energy space	242
□u□	norm in the energy space	242
$\rightharpoonup$	weak convergence	198
$\overset{*}{\rightharpoonup}$	weak* convergence	199
$[x]$	coset	109
$\|x\|$	norm of a vector x	96
$\|x'\|$	norm of a bounded linear functional x'	190
x^f	linear functional	89–90
$x^\wedge$	extremals of x	367–373
$\langle x, x^f \rangle$	value of linear functional x^f at x	89–90
$x^f(x)$	value of linear functional x^f at x	89–90
x'	bounded linear functional	190
(x,y)	inner product of vectors x and y	170
$x \perp y$	x is orthogonal to y	177
X^f	algebraic conjugate of X	90
X^{ff}	second algebraic conjugate of X	92
X'	normed conjugate of X	190
X''	second normed conjugate of X	193
$[X \to Y]$	space of bounded linear operators	105–106
$[X]$	space of bounded linear operators	105–106
X/L	factor space	109
$X \ominus L$	orthogonal complement of L	186
$X \otimes Y$	product space	110
$W^1 \perp W^2$	the sets W^1, W^2 are orthogonal	177

A COURSE OF
APPLIED FUNCTIONAL ANALYSIS

1

METRIC SPACES AND FIXED POINTS

1.0 INTRODUCTION

Through the nineteenth century the basic entity of interest in applications, and hence in mathematical analysis, was the function, which by the end of that time had come to be considered among mathematicians as a single valued mapping or correspondence between two sets of objects, a domain D and a range R, both subsets of some Euclidean space: $f\colon D \to R$. Analysis consisted of the study of the local properties of such functions: invertibility (implicit function theorem); local approximation by polynomials (Taylor's theorem with remainder, interpolation methods, etc.), or by trigonometric functions (Fourier series); smoothness (differentiability, integrability), and so on. The methods of this type of analysis lead to the proliferation of special functions in attempts to discuss the solutions of differential equations that were not solvable in terms of the so-called elementary functions. The final state of this analysis may be found in the customary texts on "methods." About the methods and special functions one can say only that each "solves" the unique set of problems that generated it. However, many problems easily formulated in the same language were "unsolvable": for example, the singular Sturm-Liouville problem: $(py')' + (q+\lambda)y = 0$ on the half line $0 \leqslant x < \infty$, or else on the real axis $-\infty < x < \infty$.

Under the leadership of D. Hilbert, a group of analysts broke partly free from the limitations of the function in developing theories of integral equations and of equations in an infinite number of variables, establishing a new viewpoint in which certain global properties of functions were involved. In essence a global property may be

$$\int_0^\infty |f(x)|^2 \, ds < \infty$$

or

$$\sup_{-\infty<x<\infty} |f(x)| < \infty.$$

The analysts found that a new entity of primary interest was discovered, the functional as it came to be called, which stands in a relation to functions as functions stand in relation to points of a Euclidean space; the analysis of such functionals naturally came to be called *functional analysis*. The domains D of such functionals are themselves sets of functions. The range R, however, can be: sets of functions

$$z(x)=\int_a^x f[t,y(t)]\,dt \qquad y(x)\to z(x)$$

or numbers

$$A=\int_a^b y(x)\alpha(x)\,dx \qquad y(x)\to A$$

or infinite sequences

$$a_n=\int_0^\pi y(x)\sin nx\,dx \qquad y(x)\to\{a_n\}.$$

It thus became necessary to initiate a study of the various properties of such domains and ranges to permit analysis of how the functionals of interest behaved on them.

Several different points of view can be taken about the "spaces" with which we deal. Certain questions call for a domain D with a linear algebraic structure: that is if $x,y\in D$, and α,β are any complex numbers then $\alpha x+\beta y\in D$. Others call only for a distance concept to be formulated for elements of the domain D: for instance, $d(f,g)=\sup_{a\leqslant x\leqslant b}|f(x)-g(x)|$. Still other questions require a size for elements of the domain and range, generalizing the notion of the Euclidean length of a vector.

The basic questions of interest in applications, in a loose description, are of two sorts. The first widely occurring question is that of inversion: given an element y of the range R and a functional f, find those elements x of the domain D "satisfying" $f(x)=y$. The second is that of optimization: given a functional f from D to the real numbers, find those elements $x\in D$ such that the functional f "satisfies" $f(x)=\sup_{y\in D}f(y)$. In order to give precise meanings to these questions, we first study in some detail one type of space, the metric space.

1.1 METRIC SPACES

The concept of Euclidean distance, which is central to the analysis of such local properties of functions as continuity, may be generalized to an arbitrary set by the following:

DEFINITION 1.1.1. An abstract set X (whose elements $x,y,z,\ldots$ are called *points*) is called a *semimetric* space if there is a function ρ on $X\otimes X$ to the real numbers R such that, for any elements x,y,z,

1. $\rho(x,y)=0$ if $x=y$;
2. $\rho(x,y)\leqslant\rho(x,z)+\rho(y,z)$.

A semimetric space in which $\rho(x,y)>0$ when $x\neq y$ is called a *metric space*. The function ρ is called a *semimetric* or a *metric*. Sometimes a semimetric space is called a *quasi-metric space* or a *pseudometric space*.

Some customary properties of a distance function are established in the following:

THEOREM 1.1.1. *A semimetric ρ satisfies*

1. $\rho(x,y)\geqslant 0$
2. $\rho(x,y)=\rho(y,x)$.

PROOF. $\rho(x,x)\leqslant\rho(x,z)+\rho(x,z)$. Therefore $0\leqslant 2\rho(x,z)$ and $0\leqslant\rho(x,z)$. Since $\rho(x,y)\leqslant\rho(x,x)+\rho(y,x)$, we see that $\rho(x,y)\leqslant\rho(y,x)$. Also, $\rho(y,x)\leqslant\rho(y,y)+\rho(x,y)$, so that $\rho(y,x)\leqslant\rho(x,y)$, and hence $\rho(x,y)=\rho(y,x)$. ■

Thus we see that a *metric* possesses the following properties:

1. $\rho(x,y)=0$ if and only if $x=y$;
2. $\rho(x,y)=\rho(y,x)$ (symmetry);
3. $\rho(x,z)\leqslant\rho(x,y)+\rho(y,z)$ (triangle inequality).

When the function ρ is clearly understood, or is irrelevant to the discussion we will use X to denote the space X as well as the collection of elements originally described. Otherwise we must describe the pair (X,ρ) as the space.

Some examples of pairs (X,ρ) that satisfy our requirements are given below. Some of these are of interest only to the curious, others are of vital importance.

Example 1. The set X is arbitrary while the function ρ is taken to be

$$\rho(x,y)=0 \qquad \text{if } x=y,$$
$$\rho(x,y)=1 \qquad \text{if } x\neq y.$$

This is a metric space (why?) but of no use in analysis.

Example 2. The set X is $\mathbf{R}^n$, the collection of ordered sets of n real numbers $x=(x_1,x_2,\ldots,x_n)$ and the function $\rho(x,y)$ is

$$\rho(x,y)=\left[\sum_{i=1}^{k}(x_i-y_i)^2\right]^{1/2}.$$

Using the Cauchy inequality (Exercise 1.1.1) we prove the triangle inequality as follows:

$$\begin{aligned}\sum_{i=1}^{k}(x_i-z_i)^2&=\sum_{i=1}^{k}(x_i-y_i+y_i-z_i)^2\\&=\sum_{i=1}^{k}(x_i-y_i)^2+2\sum_{i=1}^{k}(x_i-y_i)(y_i-z_i)+\sum_{i=1}^{k}(y_i-z_i)^2\\&\leqslant\sum_{i=1}^{k}(x_i-y_i)^2+2\left[\sum_{i=1}^{k}(x_i-y_i)^2\sum_{i=1}^{k}(y_i-z_i)^2\right]^{1/2}\\&\qquad+\sum_{i=1}^{k}(y_i-z_i)^2\\&=\left\{\left[\sum_{i=1}^{k}(x_i-y_i)^2\right]^{1/2}+\left[\sum_{i=1}^{k}(y_i-z_i)^2\right]^{1/2}\right\}^2.\end{aligned}$$

Hence

$$[\rho(x,z)]^2\leqslant[\rho(x,y)+\rho(y,z)]^2.$$

If $k<n$, ρ is a semimetric; if $k=n$, ρ is a metric.

Example 3. The set X is $\mathbf{R}$, the set of real numbers, and

$$\rho(x,y)=|x-y|.$$

This is the case $n=k=1$ of Example 2, that is $\mathbf{R}=\mathbf{R}^1$.

Example 4. The set X is the same as in Example 2, but

$$\rho(x,y)=\max_{i=1,\ldots,k}|x_i-y_i|.$$

Again if $k<n$, ρ is a semimetric while if $k=n$, ρ is a metric. The proof that ρ satisfies properties 1, 2, and 3 is left as an exercise.

Example 5. The set X is l_2, the elements of which are the infinite sequences of complex numbers $x=\{x_1,x_2,x_3,\ldots\}$ such that $\sum_{i=1}^{\infty}|x_i|^2<\infty$. From the complex form of the Cauchy inequality, we see that $x\in X, y\in X$ implies

$$\left(\sum_{i=1}^{k}|x_iy_i|\right)^2\leqslant\sum_{i=1}^{k}|x_i|^2\sum_{i=1}^{k}|y_i|^2,\qquad k=1,2,3,\ldots$$

which can be used to prove that the function

$$\rho(x,y)=\left(\sum_{i=1}^{\infty}|x_i-y_i|^2\right)^{1/2}$$

is finite for any two elements of X. The proof of the triangle inequality is left as an exercise. Here ρ is easily seen to be a metric.

DEFINITION 1.1.2. If S is any set of real numbers, then by $\sup S$ ($\inf S$) is meant a number C such that if $s\in S$, then $s\leqslant C$ ($s\geqslant C$) while if $s\leqslant C'$ ($s\geqslant C'$) for all $s\in S$, then $C\leqslant C'$ ($C\geqslant C'$).

Example 6. X is m, the set of all infinite sequences of complex numbers $m=\{m_1,m_2,m_3,\ldots\}$ that are bounded: that is, for each m there is a real number B such that $|m_i|\leqslant B$, $i=1,2,3,\ldots$. The function $\rho(m,n)=\sup_i|m_i-n_i|$.

Example 7. X is c, the set of all convergent infinite sequences of complex numbers $m=\{m_1,m_2,m_3,\ldots\}$. The metric ρ is the same as in m. c is a metric space which is a proper subspace of m.

Example 8. X is c_0, the collection of all null sequences of complex numbers, $x=\{x_1,x_2,x_3,\ldots\}$ where $\lim_{i\to\infty}x_i=0$. The function ρ is the same as in Example 6. This is a proper subspace of c.

Example 9. X is the collection of all sequences of complex numbers $x=\{x_1,x_2,x_3,\ldots\}$ with the function ρ given by

$$\rho(x,y)=\sup_i \frac{|x_i-y_i|}{1+|x_i-y_i|}.$$

(That ρ is a metric follows easily from Exercise 1.1.7.)

Example 10. Another metric for X in Example 9 is

$$\rho(x,y)=\sum_{i=1}^{\infty}\frac{1}{2^i}\frac{|x_i-y_i|}{1+|x_i-y_i|}.$$

The series always converges since the ith term is dominated by 2^{-i}. The proof that ρ is a metric follows from the identity established in Exercise 1.1.7 and is left as an exercise for the reader.

Example 11. X is $\mathbf{E}^n$, the collection of ordered sets of n complex numbers $x=(x_1,\ldots,x_n)$ while

$$\rho_p(x,y)=\left[\sum_{i=1}^{n}|x_i-y_i|^p\right]^{1/p}$$

for some fixed $p\geqslant 1$. That ρ is a metric follows from the Minkowski inequality (Exercise 1.1.10). Example 2 is the real subspace of the case $p=2$.

Example 12. X is l_p, where $1\leqslant p<\infty$, the collection of sequences $x=\{x_1,x_2,x_3,\ldots\}$ of complex numbers satisfying

$$\sum_{i=1}^{\infty}|x_i|^p<\infty.$$

The function ρ is given by

$$\rho_p(x,y)=\left(\sum_{i=1}^{\infty}|x_i-y_i|^p\right)^{1/p}.$$

The Minkowski inequality may be used to show that $\rho_p(x,y)<\infty$ for all x,y and that the triangle inequality holds.

The remaining examples of metric spaces are function spaces, that is spaces in which the set X consists of functions on some domain in a

Euclidean space. For simplicity at this point, the space will be taken to be the real line and the domain some interval $a \leqslant t \leqslant b$ on that line. (Continuous analogs of the Cauchy, Hölder, and Minkowski inequalities, used below, are established in the exercises.)

Example 13. X is $C[a,b]$, the collection of all continuous functions from an interval $a \leqslant t \leqslant b$ to the real (or complex) numbers, with metric

$$\rho(x,y) = \max_{a \leqslant t \leqslant b} |x(t) - y(t)|.$$

Example 14. Another choice of metric for X in Example 13 is

$$\rho_p(x,y) = \left(\int_a^b |x(t) - y(t)|^p \, dt \right)^{1/p}.$$

The integral is taken in the sense of Riemann. The Schwarz inequality may be used to show that this is a metric for $p=2$. For $p \neq 2$, the Hölder and Minkowski inequalities are used.

Example 15. Yet another choice of metric for the continuous functions on $a \leqslant t \leqslant b$ is

$$\rho(x,y) = \int_a^b \frac{|x(t) - y(t)|}{1 + |x(t) - y(t)|} \, dt.$$

Example 16. Let X be $C[a,b)$, the continuous functions on the interval $a \leqslant t < b$, and choose

$$\rho(x,y) = \sup_{a \leqslant t < b} |x(t) - y(t)|.$$

X is a metric space. Every function in $C[a,b]$ corresponds to a unique function in $C[a,b)$, but there exist functions in $C[a,b)$ not in $C[a,b]$. One is $x(t) = \sin[(b-t)^{-1}]$. Every function in $C[a,b]$ is uniformly continuous (Heine-Borel theorem). This is not true in $C[a,b)$.

Example 17. Let X be $C^p[a,b]$, the functions with continuous derivatives of all orders not greater than p, and choose

$$\rho(x,y) = \max_{0 \leqslant j \leqslant p} \max_{a \leqslant t \leqslant b} \left| \frac{d^j x}{dt^j} - \frac{d^j y}{dt^j} \right|.$$

(Here the zero order derivative means the function.)

Example 18. Another choice of metric for X in Example 17 is

$$\rho_p(x,y)=\sum_{j=0}^{p}\left(\int_a^b\left|\frac{d^jx}{dt^j}-\frac{d^jy}{dt^j}\right|^p dt\right)^{1/p}$$

EXERCISES TO SECTION 1.1

1. (Cauchy inequality) If $a_1,\ldots,a_n,b_1,\ldots,b_n$ are real numbers then

$$\left(\sum_{i=1}^{n}a_ib_i\right)^2\leqslant\sum_{i=1}^{n}a_i^2\sum_{i=1}^{n}b_i^2$$

(Hint: the quadratic function of λ

$$\sum_{i=1}^{n}(a_i-\lambda b_i)^2=\sum_{i=1}^{n}(a_i^2-2\lambda a_ib_i+\lambda^2b_i^2)$$

is nonnegative, but has at most one real root.
Then the discriminant

$$4\left(\sum_{i=1}^{n}a_ib_i\right)^2-4\sum_{i=1}^{n}a_i^2\sum_{i=1}^{n}b_i^2\leqslant 0$$

with equality if and only if $a_i-\lambda b_i=0$, $i=1,\ldots,n$.)

2. (Complex form of the Cauchy inequality) If $a_1,a_2,\ldots,a_n,b_1,b_2,\ldots,b_n$ are complex numbers then

$$\left|\sum_{i=1}^{n}a_ib_i\right|^2\leqslant\left(\sum_{i=1}^{n}|a_i|^2\right)\left(\sum_{i=1}^{n}|b_i|^2\right)$$

with equality if and only if for some complex λ, $a_i=\lambda b_i$, $i=1,\ldots,n$.

3. In l_2, prove that $\rho(x,y)<\infty$ for every x,y.
Hint: Using the Cauchy inequality show that

$$\left(\sum_{i=1}^{k}|y_i-x_i|^2\right)^{1/2}\leqslant\left(\sum_{i=1}^{k}|y_i|^2\right)^{1/2}+\sum_{i=1}^{k}(|x_i|^2)^{1/2}$$

and then let $k\to\infty$. Also prove the triangle inequality. (Hint: Let $x_k=b_k-a_k, y_k=c_k-b_k$.)

4. Show that m is a metric space.

5. Show that c is a metric space.
6. Show that c_o is a metric space.
7. Prove that for any complex numbers A, B

$$\frac{|A+B|}{1+|A+B|} \leqslant \frac{|A|}{1+|A|} + \frac{|B|}{1+|B|}.$$

Use this fact to show that ρ, in Example 9, is a metric. (Hint: the function $x(1+x)^{-1}$ is strictly monotone increasing on $x \geqslant 0$, so that $|A+B| \leqslant |A|+|B|$ implies

$$\frac{|A+B|}{1+|A+B|} \leqslant \frac{|A|+|B|}{1+|A|+|B|}.)$$

8. If p, q are real numbers $\geqslant 1$ satisfying $1/p+1/q=1$, then

$$|AB| \leqslant \frac{|A|^p}{p} + \frac{|B|}{q}$$

for any complex numbers A, B, with equality if and only if $|A|=|B|$. (Hint: Show that the curve $y=x^{1/p}$ is concave down for $x>0$, so that the tangent to the curve always lies above the curve. This implies that the curve lies below the tangent at $x=1$, so that $x^{1/p} \leqslant x/p+1-1/p = x/p+1/q$. Then set $x=|A|^p/|B|^q$.)
9. (Hölder's inequality) For any complex numbers $x_1, \ldots, x_k, y_1, \ldots, y_k$, and any $p \geqslant 1$, $1/p+1/q=1$

$$\left|\sum_{i=1}^{k} x_i y_i\right| \leqslant \left(\sum_{i=1}^{k} |x_i|^p\right)^{1/p} \left(\sum_{i=1}^{k} |y_i|^q\right)^{1/q}.$$

(Hint: If $x_i' = x_i/(\Sigma_{i=1}^k |x_i|^p)^{1/p}, y_i' = y_i/(\Sigma_{i=1}^k |y_i|^q)^{1/q}$, $i=1,\ldots,k$, then

$$\sum_{i=1}^{k} |x_i' y_i'| \leqslant \frac{1}{p} + \frac{1}{q} = 1.$$

For $p=q=2$ this is the Cauchy inequality.) When does equality hold?
10. (Minkowski's inequality) For any complex numbers $x_1, \ldots, x_k$, $y_1, \ldots, y_k$, and any $p \geqslant 1$

$$\left(\sum_{i=1}^{k} |x_i+y_i|^p\right)^{1/p} \leqslant \left(\sum_{i=1}^{k} |x_i|^p\right)^{1/p} + \left(\sum_{i=1}^{k} |y_i|^p\right)^{1/p}.$$

(Hint: Since $(|A|+|B|)^p=(|A|+|B|)^{p-1}|A|+(|A|+|B|)^{p-1}|B|$ we see that

$$\sum_{i=1}^{k}(|x_i|+|y_i|)^p=\sum_{i=1}^{k}(|x_i|+|y_i|)^{p-1}|x_i|+\sum_{i=1}^{k}(|x_i|+|y_i|)^{p-1}|y_i|.$$

Applying the Hölder inequality and $(p-1)q=p$ leads after some manipulation to the desired result.) When does equality hold?

11. Show that l_p is a metric space.
12. Show that $C[a,b]$ is a metric space.
13. (Schwarz inequality) If f and g are continuous functions on $a \leqslant t \leqslant b$ then

$$\left|\int_a^b f(t)g(t)\,dt\right|^2 \leqslant \int_a^b |f(t)|^2\,dt \int_a^b |g(t)|^2\,dt.$$

(This was first stated by Buniakowsky but is most widely known as the Schwarz inequality.)

14. If f and g are continuous functions on $a \leqslant t \leqslant b$ and $p \geqslant 1, 1/p+1/q=1$, then

$$\left|\int_a^b f(t)g(t)\,dt\right| \leqslant \left(\int_a^b |f(t)|^p\,dt\right)^{1/p}\left(\int_a^b |g(t)|^q\,dt\right)^{1/q}$$

and

$$\left(\int_a^b |f(t)+g(t)|^p\,dt\right)^{1/p} \leqslant \left(\int_a^b |f(t)|^p\,dt\right)^{1/p}+\left(\int_a^b |g(t)|^p\,dt\right)^{1/p}.$$

(These are also known as the Hölder and Minkowski inequalities, but the extension to integrals is due to F. Riesz.) When does equality hold?

15. Show that the space of Example 14 is a metric space.
16. Show that the space of Example 15 is a metric space.
17. In Example 17, why is

$$\max_{a \leqslant t \leqslant b}\left|\frac{d^p x}{dt^p}-\frac{d^p y}{dt^p}\right|$$

not a metric?

18. In Example 18, could we choose as metric

$$\sum_{j=0}^{p-1}\max_{a \leqslant t \leqslant b}\left|\frac{d^j x}{dt^j}-\frac{d^j y}{dt^j}\right|+\int_a^b\left|\frac{d^p x}{dt^p}-\frac{d^p y}{dt^p}\right|dt?$$

1.2 FROM PROBLEMS TO FUNCTIONAL EQUATIONS

The natural statement of problems is rarely in the form of a functional equation, that is, a mapping between spaces of the sort described in Section 1.1. Below we give examples of some types of functional equations that arise in practice. This is not to be thought of as an exhaustive catalog, but the selection of examples is pertinent to the types of functional analysis that are considered below. The references can and should be used for a fuller understanding of the problem, its nonmathematical assumptions, and the simplifications involved in the derivation of the mathematical formulation. Our interest lies in the *form* of the resulting mathematical statement. (It may be worthwhile to note that there will be little emphasis on the parts of functional analysis that are related to the spectral theorem, and no example of this type is included.)

A. Optimal Economic Growth

We consider the question of "optimal economic growth in an aggregative closed economy" (see for instance Intriligator [1], 398–408).

Let $c(t)$ = consumption per worker,
$i(t)$ = investment per worker,
$k(t)$ = capital per worker,
$f(k)$ = output per worker,
μ = rate of depreciation of capital,
$\eta(t)$ = rate of growth of the labor force,
$U(c)$ = utility of consumption,
δ = discount rate for future utility,
W = welfare.

The basic equation of economic growth is a differential equation:

$$f(k(t)) = c(t) + [\mu + \eta(t)]k(t) + \frac{dk(t)}{dt}.$$

A planner choses a history $c(t)$ of consumption per worker on some time interval $t_0 \leqslant t \leqslant t_1$, which is subject to some constraint such as

$$0 \leqslant c(t) \leqslant f(k(t)), \qquad t_0 \leqslant t \leqslant t_1,$$

if, for instance, it is desirable that $i(t) = f(k(t)) - c(t)$ always be nonnegative. In order to make such a choice, the planner must have an economic objective, say the maximization of standard of living as measured by W.

Suppose that there exists a utility function $U(c)$ which is a monotone increasing function of c, and a discount rate δ for future utility, with total welfare W for a given strategy [choice of $c(t)$] given by

$$W=\int_{t_0}^{t_1} e^{-\delta(t-t_0)}U(c(t))\,dt.$$

We also must require that at $t=t_1$ the remaining capital is not all exhausted so that life may continue: in addition to knowing $k(t_0)=k_0$ we must require $k(t_1)\geqslant k_1$, some minimal acceptable capital for a second time period $t\geqslant t_1$. Then *optimal economic growth* means finding that $c(t)$ in some class C such that

$$W=\sup\left\{\int_{t_0}^{t_1} e^{-\delta(t-t_0)}U(c(t))\,dt \,|\, c(t)\in C\right\}$$

subject to the set of constraints

$$\begin{cases} \dfrac{dk}{dt}+[\mu+\eta(t)]k+c(t)=f(k), \\ k(t_0)=k_0, \\ k(t_1)\geqslant k_1, \\ 0\leqslant c\leqslant f(k). \end{cases}$$

The class C, for instance, might be required to contain only nondecreasing functions.

This is in the form of an optimal control problem and leads to two problems in functional analysis. W is a functional of $c(t)$ whose maximum is sought subject to constraints. In addition there is the problem of the existence of solutions to the set of constraints. If we rewrite the basic equation of economic growth as

$$k(t)=k_0+\int_{t_0}^{t}\{f(k(s))-c(s)-[\mu+\eta(s)]k(s)\}\,ds,$$

and define a *functional* $F(x;c)=y$ from functions $x(t)$ on $t_0\leqslant t\leqslant t_1$ to functions $y(t)$ on $t_0\leqslant t\leqslant t_1$ by

$$F(x;c)=k_0+\int_{t_0}^{t}\{f(x(s))-c(s)-[\mu+\eta(R)]x(s)\}\,ds,$$

then the existence of a solution corresponds to the existence of a *fixed point* k of F, that is, a k such that $k=F(k;c)$. The initial condition $k(t_0)=k_0$ will be satisfied automatically. The condition $k(t_1) \geqslant k_i$ will not be satisfied automatically for all c's and further analysis is necessary to specify C as the set of those c's such that both $0 \leqslant c(t) \leqslant f(s(t))$ and $k(t_1) \geqslant k_1$.

B. A Simple Queue

Suppose that there is a single service facility (say one barber) with N possible slots in the queue for waiting service (say N seats in the barber shop). If we assume that the probability of arrival at the queue in a time interval dt is λdt, while the probability of completion of service in a time interval dt is μdt, where P_n is the probability that the queue at time t is of length n (the queue is in state n), then it is easy to see, by counting state transitions, that for $t \geqslant t_0$

$$\begin{cases} \dfrac{dP_0}{dt} = -\lambda P_0 + \mu P_1 \\ \dfrac{dP_n}{dt} = \mu P_{n+1} - (\lambda + \mu) P_n + \lambda P_{n-1}, \qquad n = 1, \ldots, N-1 \\ \dfrac{dP_N}{dt} = \lambda P_{N-1} - \mu P_N; \end{cases}$$

we assume that when the queue is full (all N chairs are occupied) then no further arrivals join the queue, while all other arrivals are serviced immediately or else join the queue (see Morse [1], 14–16). The behavior of the queue is governed by a linear system of ordinary differential equations. The properties of the time behavior of the queue are those of the solution, if any, of the system of differential equations, here linear. The problem of the existence of solutions is formulated in terms of functionals in Section C below.

C. Systems of Differential Equations

Let x be a point of $\mathbf{R}^n$, the n-dimensional real Euclidean space; that is the set of all ordered sets of n real numbers $x = (x_1, \ldots, x_n)$. Let x' denote componentwise differentiation of $x(t) = (x_1(t), \ldots, x_n(t))$, where the $x_i(t)$ are differentiable functions. Then the general system of n first order differential equations has an initial value problem which can be written as

$$\begin{cases} x' = f(t, x), \\ x(t_0) = x_0, \end{cases}$$

where the second equation specifies the initial state of the system. Here

$f=(f_1,\ldots,f_n)$ is a set of n functions of the $n+1$ variables $t,x_1,\ldots,x_n$. If we let the integral of such a vector valued function be the vector of integrals of its components, it is easily seen that the above system is equivalent to the system

$$x(t)=x_0+\int_{t_0}^{t} f(t,x(t))\,dt.$$

We define a *functional* $F(y)$ defined on $t_0 \leqslant t \leqslant t$, for all vector valued functions $y(t)=(y_1(t),\ldots,y_n(t))$, $t_0 \leqslant t \leqslant t$, in some set (the values y lie in some region R of $\mathbf{R}^n$ for all t on $t_0 \leqslant t \leqslant t_1$) as

$$F(y)(t)=x_0+\int_{t_0}^{t} f(t,y(t))\,dt.$$

Then under suitable assumptions F maps the function $y=y(t)$ into some other vector valued function $z=z(t)$:

$$z=F(y).$$

The existence of solutions of the system of differential equations is equivalent to the existence of a *fixed point* of the mapping $F:y\rightarrow z$: that is of an x which is mapped into itself, so that $x=F(x)$.

D. Kinetics of Nuclear Reactors

The simplest models of the time dependent behavior of nuclear reactors are contained in the reactor kinetic equations (Greenspan, Keller, Okrent [1], 444–448). These assume a homogeneous reactor with all neutrons belonging to a single energy group (known sometimes as a point reactor, exhibiting no geometric structure). Assuming a *source* of neutrons of density $S(t)$, a total *neutron density n*, a number (usually six) of *emitters of delayed neutrons* of density C_i, and a *reactivity k*, simplified balancing of emission and absorption of neutrons leads to the system of equations

$$\begin{cases} \dfrac{dn}{dt}=\left[\dfrac{k-\sum_{i=1}^{6}\beta_i}{l}\right]n+\sum_{i=1}^{6}\lambda_i C_i+S, \\[2ex] \dfrac{dC_i}{dt}=\left(\dfrac{\beta_i}{l}\right)n-\lambda_i C_i, \qquad i=1,\ldots,6, \end{cases}$$

where $\beta_1,\ldots,\beta_6,\lambda_1,\ldots,\lambda_6,l$ are physical constants of the reactor.

If k is assumed to be a constant, this is a linear system of differential equations.

Another physical assumption about k, in a *thermal reactor*, makes k vary with the temperature T of the reactor according to the equations

$$\begin{cases} k = f(T), \\ \dfrac{dT}{dt} = \alpha_1 n - \alpha_2 - \alpha_3(T - T_0), \end{cases}$$

where $\alpha_1, \alpha_2, \alpha_3$, are physical constants and T_0 is the ambient temperature. This is a nonlinear system of differential equations.

In a third model (Hetrick [1], 161) the reactivity is assumed to vary according to the past history of the neutron density of the reactor in the form

$$k(t) = \int_{-\infty}^{t} h(t-s)[n(s) - n_0]\, ds$$

where $h(t)$ is a weighting function for past history of power output and n_0 is the neutron density at steady power output and the equilibrium temperature. The forms of $h(t)$ that are used include $\alpha e^{-\beta t}$ (or even $\alpha e^{-\beta(t-\tau)}$) for certain physical constants α, β (and a possible time delay τ), for the effect of past history. The system of equations is now a system of integro-differential equations, possibly involving a time delay. Devices similar to those used in Section C should be used by the reader to define the functional whose fixed point will be the solution of the kinetic equations in this case. This is an example of a *functional differential equation*.

E. Renewal Processes

Consider a telephone switching exchange that receives calls at times $t_1, t_2, t_3, \ldots$ where $0 < t_1 < t_2 < t_3 < \cdots$. It is often assumed that the times between successive calls, $T_1 = t_1, T_2 = t_2 - t_1, T_3 = t_3 - t_2, \ldots$, are independent identically distributed random variables. Estimating $N(T)$, the number of calls received in an interval, is of obvious importance to the designers of the exchange. If $P_n(T)$ is the probability that $N(T) = n$, then the mean value of $N(T), m(T)$, satisfies (Parzen [1], 170–172)

$$m(T) = \int_0^T [1 + m(T-s)] f(s)\, ds, \qquad t \geq 0,$$

where f is the probability density function for the interarrival times

$T_1, T_2, T_3, \ldots$. This is an equation of the form

$$g(t) = h(t) + \int_0^t g(t-s) f(s)\, ds, \qquad t \geqslant 0,$$

an integral equation known as the *renewal equation*.

This is a *functional equation*. The *functional* $F(g)$:

$$F(g)(t) = h(t) + \int_0^t g(t-s) f(s)\, ds,$$

maps a function g into another function, both defined for $t \geqslant 0$. The desired function g is again a fixed point of the mapping or functional F.

F. Boundary Value Problems

A pendulum with no damping forces, subject to a force f, moves so that the angle of deviation ϕ of the pendulum from the vertical satisfies

$$\frac{d^2\phi(t)}{dt^2} + \alpha^2 \sin\phi(t) = f(t)$$

where α is a physical constant of the pendulum. If $f(t) = \beta \sin \pi t$, an odd periodic function of period 2, it is of physical interest to show the existence of a solution ϕ which is also odd and periodic of period 2 (Tricomi [1], 213–216). It turns out that it is sufficient to solve the differential equation subject to the periodic boundary conditions $\phi(0) = \phi(1) = 0$.

Associated with a differential equation of the form

$$\frac{d^2y}{dx^2} = g(x), \qquad y(0) = y(1),$$

there is a Green's function (Tricomi [1], 116–117)

$$T(x,s) = \begin{cases} x(1-s), & 0 \leqslant x \leqslant s \\ s(1-x), & s \leqslant x \leqslant 1 \end{cases}$$

such that the solution $y(x)$ can be written as

$$y(x) = -\int_0^1 T(x,s) g(x)\, ds.$$

(This may be verified by direct calculation.) Then the pendulum problem could be rewritten as

$$\phi(t) = -\int_0^1 T(t,s)\left[f(s) - \alpha^2 \sin\phi(s)\right] ds,$$

if the solution ϕ were known to exist. But direct calculation shows that if this integral equation has a solution ϕ, it is also a solution of the differential equation. Thus we need only consider the functional $F(\phi)$ defined by

$$F(\phi)(t) = -\int_0^1 T(t,s)\left[f(s) - \alpha^2 \sin\phi(s)\right] ds$$

and show that it has a fixed point $\phi = F(\phi)$.

Other physical situations (Tricomi [1], 1–3) lead to integral equations of the form

$$\phi(x) = \lambda \int_0^1 K(x,s)\phi(s)\, ds, \qquad 0 \leqslant x \leqslant 1$$

where λ is an unspecified parameter. This is a functional equation of the form $\phi = \lambda F(\phi)$. A λ for which this equation has a nonzero solution ϕ is called an *eigenvalue*, and the functional equation $F(\phi) = (1/\lambda)\phi$ is called an *eigenvalue problem*.

1.3 EQUIVALENCE OF FUNCTIONAL EQUATIONS

In the examples of Section 1.2 we have generated three types of functional equations. If S, T, R denote functionals or mappings from a set X to itself, these can be abstracted in the following form:

1. The set X contains a zero element 0 and we seek an $x \in X$ such that $S(x) = 0$, that is, a *root* of the functional S.
2. We seek a *fixed point* x of the mapping T; that is, an x such that $T(x) = x$.
3. We seek an *eigenvalue* λ and *eigenvector* x of the mapping R; that is, a pair λ and x, where x is not the zero element such that $R(x) = \lambda x$.

We note here that these three formulations are essentially equivalent. That is, given any one form of the problems, we can (usually) transform it into the problem of a fixed point.

Given the problem of finding a root x of S, we can always define a new mapping T by $T(x) = x - S(x)$. Clearly, x is a root of S if and only if it is a fixed point of T. The same relation, given T, defines S such that x is a fixed point of T if and only if x is a root of S. The question of eigenvalue problems is more subtle. If X is a set of real elements, then we introduce a new space $X \otimes \mathbf{R}$ whose elements are pairs $(x,a), x \in X, a \in \mathbf{R}$, where $\mathbf{R}$ is the set of real numbers. We suppose that we can define a mapping $G: X \rightarrow \mathbf{R}$ such that G maps the zero element of X into zero, and maps nonzero elements of X into nonzero elements of $\mathbf{R}$. G is usually taken to be homogeneous so that $G(\alpha x) = \alpha^k x$ for some integer k. Define S by

$$S((x,a)) = [R(x) - ax, G(x) - 1].$$

Then finding a root of S corresponds to the solution of the eigenvalue problem for R. As an illustration, consider the eigenvalue problem $y'' = \lambda y$; $y(0) = y(\pi) = D$. Here one may take G as

$$G(x) = \int_0^\pi |x(t)|^2 \, dt$$

or

$$G(x) = \max_{0 \leqslant t \leqslant \pi} |x(t)|$$

to equal effect in removing the zero solution of the differential equation from contention.

1.4 COMPUTING FIXED POINTS BY SUCCESSIVE APPROXIMATIONS

Successive approximation or functional iteration is one of the oldest methods for solving a functional equation $x = T(x)$, used originally to find roots of functions of one variable and to improve the accuracy of approximate solutions of differential equations. The program is to set up a sequence of approximate solutions starting with some initial guess x^0, and to let

$$x^{n+1} = T(x^n), \qquad n = 0, 1, 2, \ldots .$$

If at each stage the newly found approximant x^{n+1} is again within the domain on which T can act, it makes sense to ask whether in some sense $\lim_{n\to\infty} x^n$ exists in X, and if $T(\lim_{n\to\infty} x^n)$ makes sense, does it happen that $\lim_{n\to\infty} T(x^n) = T(\lim_{n\to\infty} x^n)$? There are really three questions of interest:

1. What is a suitable definition of a limit in a given metric space?
2. When does a given mapping T commute with a given limiting process in a given metric space?
3. When does x^n "converge" to a solution of $x = T(x)$?

Consider question 1. If ρ is a metric, that is, a distance measure, then $\lim_{n\to\infty} x^n = x$ should mean $\rho(x^n,x)\to 0$, $n\to\infty$. A sequence $\{x^n\}$ should be convergent if $\rho(x^n,x^m)\to 0$, $n,m\to\infty$. (Note: not $\rho(x^n,x^{n+1})\to 0$, $n\to\infty$. For instance, if $X=\mathbf{R}$ and $x^n = 1+1/2+1/3+\cdots+1/n$, then $|x^{n+1}-x^n| = 1/(n+1)\to 0$, $n\to\infty$ while $x^n\to\infty, n\to\infty$).

If we refer to the spaces of Section 1.1 we see that in Example 1, there are too few convergent sequences for anything to be accomplished. If we consider Examples 2, 4, and 11, we can find many convergent sequences. But in different metrics, convergence may have different meanings. In the present case this is not so. To see this, observe that in $\mathbf{R}^n$ (or $\mathbf{E}^n$), with $\rho(x,y) = \max_i |x_i - y_i|$, $\rho(x^n,x^m)\to 0$, $n,m\to\infty$ implies that each coordinate $\{x_i^n\}$ is a real (or complex) Cauchy convergent sequence. Since for the real (or complex) numbers, every Cauchy convergent sequence has a real (or complex) limit, this implies that any sequence in $\mathbf{R}^n$ (or $\mathbf{E}^n$) that is convergent for this $\rho(x,y)$ has a limit in $\mathbf{R}^n$ (or $\mathbf{E}^n$). Further, for all $p \geqslant 1$

$$\max_{1\leqslant i\leqslant n} |x_i - y_i| \leqslant \left(\sum_{i=1}^{n} |x_i - y_i|^p \right)^{1/p} \leqslant n^{1/p} \max_{1\leqslant i\leqslant n} |x_i - y_i|.$$

Thus all three of the metrics on $\mathbf{R}^n$ (or $\mathbf{E}^n$) have the property that a convergent sequence has a limit. Let us give this property a name.

DEFINITION 1.4.1. A sequence $\{x^n\}$ in X is called a *Cauchy sequence* if $\rho(x^n,x^m)\to 0$, $n,m\to\infty$.

DEFINITION 1.4.2. A sequence $\{x^n\}$ in X has a *limit* x in X if $\rho(x^n,x)\to 0$, $n\to\infty$. (We will write $x^n\to x$, $n\to\infty$, or even $x^n\to x$ if no misunderstanding is possible.)

DEFINITION 1.4.3. A metric space X is *complete*, if every Cauchy sequence in X has a limit in X.

Above we have shown that the three metric spaces of examples 2, 4, and 11 are complete and that the limit is independent of the metric. This is not generally true.

An important four point relation is the following:

LEMMA 1.4.1. $|\rho(x,y)-\rho(z,u)| \leqslant \rho(x,z)+\rho(y,u)$

PROOF. The lemma is a consequence of the inequalities

$$\rho(x,y) \leqslant \rho(x,z)+\rho(z,y) \leqslant \rho(x,z)+\rho(z,u)+\rho(u,y)$$

which imply

$$\rho(x,y)-\rho(z,u) \leqslant \rho(x,z)+\rho(y,u);$$

similarly we show that

$$\rho(z,u)-\rho(x,y) \leqslant \rho(x,z)+\rho(y,u)$$

which implies the lemma. ■

The continuity of $\rho(x,y)$ on $X \otimes X$ now follows:

LEMMA 1.4.2. *If* $x^n \to x$, $y^n \to y$, $n \to \infty$, *then* $\rho(x^n, y^n) \to \rho(x,y)$, $n \to \infty$.

PROOF. From Lemma 1.4.1 we see that

$$|\rho(x^n,y^n)-\rho(x,y)| \leqslant \rho(x^n,x)+\rho(y^n,y) \to 0, \qquad n \to \infty \quad ■$$

That $C[a,b]$ is complete in the metric

$$\rho(x,y)=\max_{a \leqslant t \leqslant b} |x(t)-y(t)|$$

is the well-known theorem of Seidel: a uniformly convergent sequence of continuous functions on a finite interval has a continuous limit to which it converges uniformly.

However, $C[a,b]$ is *not* complete in the metric

$$\rho(x,y)=\left(\int_a^b |x(t)-y(t)|^2\,dt\right)^{1/2},$$

for there exist sequences of continuous functions $y^n(t)$ such that y^n does not converge to a continuous function yet $\rho(y^n,0) \to 0$, $n \to \infty$. (Why?)

Consider now question 2. It is really a question about the continuity of the mapping T as a function of x. We make some definitions.

DEFINITION 1.4.4. Let T map X to Z. By D_T, the *domain* of T is meant the set of $x \in X$ at which T is defined. R_T, the *range* of T, is the set of $z \in Z$ of the form $z = Tx$, $x \in D_T$.

DEFINITION 1.4.5. Given $x \in X$, a metric space. By the *open (closed) ball of radius r about x*, $B(x,r)$ is meant the set of $y \in X$ such that

$$\rho(x,y) < r \quad (\rho(x,y) \leqslant r).$$

DEFINITION 1.4.6. Given $x \in D_T$. Let $B(x,r) \cap D_T$ contain points other than x for all sufficiently small r. The mapping T is said to be *continuous* at x if given any $\varepsilon > 0$ there exists a $\delta = \delta(\varepsilon) > 0$ such that $\rho(T(x), T(y)) < \varepsilon$ for all $y \in B(x,\delta) \cap D_T$. We write $\rho(T(x), T(y)) \to 0$ as $y \to x$.

Consider now question 3. The fundamental theorem on the method of successive approximations is the following:

THEOREM 1.4.1. (Banach) *Let $T: X \to X$ be defined on a complete metric space X with $D_T = X$. Let α satisfy $0 < \alpha < 1$ and $\rho(T(x), T(y)) \leqslant \alpha\rho(x,y)$ for all $x,y \in X$. Then T has a unique fixed point x^*. The successive approximations $x^{n+1} = T(x^n)$ converge uniformly to x^* on any sphere $\rho(x^0, x^*) \leqslant \delta$.*

PROOF. Since $D_T = X$, all x^n are defined, $n = 0, 1, 2, \ldots$, and

$$\rho(x^{n+1}, x^n) = \rho(T(x^n), T(x^{n-1})) \leqslant \alpha\rho(x^n, x^{n-1}).$$

Repeating this n times gives

$$\rho(x^{n+1}, x^n) \leqslant \alpha^n \rho(x^1, x^0).$$

Now if $m < n$

$$\begin{aligned}
\rho(x^m, x^n) &\leqslant \rho(x^m, x^{m+1}) + \rho(x^{m+1}, x^n) \\
&\leqslant \rho(x^m, x^{m+1}) + \rho(x^{m+1}, x^{m+2}) + \cdots + \rho(x^{n-1}, x^n) \\
&\leqslant (\alpha^m + \alpha^{m+1} + \cdots + \alpha^{n-1})\rho(x^1, x^0) \\
&\leqslant \alpha^m(1 + \alpha + \cdots + \alpha^{n-1-m})\rho(x^1, x^0) \\
&\leqslant \alpha^m(1-\alpha)^{-1}\rho(x^1, x^0).
\end{aligned}$$

If $n, m \to \infty$, then $\rho(x^m, x^n) \to 0$. If X is complete, $\{x^n\}$ converges, say to x^*. Then $T(x^*) = x^*$: for $\rho(x^{n+1}, T(x^*)) = \rho(T(x^n), T(x^*)) \leqslant \alpha\rho(x^n, x^*) \to 0$, $n \to \infty$. Thus $0 \leqslant \rho(x^*, T(x^*)) \leqslant \rho(x^*, x^{n+1}) + \alpha\rho(x^*, x^{n+1}) \to 0$, $n \to \infty$. So $x^* = T(x^*)$ and

$$\rho(x^{n+1}, x^*) \leqslant \alpha^{n+1}\rho(x^0, x^*).$$

If $\rho(x^0, x^*) \leqslant \delta$, then $\rho(x^{n+1}, x^*) \to 0$ uniformly with respect to x^0. The fixed point is unique for if $T(x^*) = x^*$, $T(x^{**}) = x^{**}$, then $0 \leqslant \rho(x^*, x^{**}) = \rho(T(x^*), T(x^{**})) \leqslant \alpha\rho(x^*, x^{**})$ where $\alpha < 1$. If $\rho(x^*, x^{**}) \neq 0$, this implies $1 \leqslant \alpha$, a contradiction. ■

The fundamental error estimate implied by the proof above is

$$\rho(x^n, x^*) \leqslant \alpha^n \delta$$

whenever

$$\rho(x^0, x^*) \leqslant \delta.$$

Even more useful is the estimate

$$\rho(x^n, x^*) \leqslant \left(\frac{\alpha^n}{1-\alpha}\right)\rho(x^1 - x^0)$$

which follows from

$$\rho(x^n, x^p) \leqslant \left(\frac{\alpha^n}{1-\alpha}\right)\rho(x^1 - x^0),$$

for all $n < p < \infty$, by letting $p \to \infty$.

A mapping T which satisfies

$$\rho(T(x), T(y)) \leqslant \alpha\rho(x_1 y), \qquad 0 < \alpha < 1$$

is often called a *contraction mapping* or a *contraction*. Theorem 1.4.1 is called the Banach fixed point theorem, also the *contraction mapping principle*.

EXERCISES TO SECTION 1.4

1. Exhibit a sequence of functions $y^n \in C[0, 1]$ such that

$$\int_0^1 |y^n(t)|^2 \, dt \to 0, \; n \to \infty$$

   while y^n does not converge.
2. In a semimetric space, the proof of Theorem 1.4.1 can still be formally carried through, if the definitions are appropriately made (See Collatz [1], pp. 208–210.) Carry out the details.

3. Even though the continuity of T was in question, it is not a part of the assumptions of Theorem 1.4.1. Show that a contraction mapping is continuous.
4. Is m, the metric space of Example 6, complete?
5. c is a proper subspace of m. Show that c is complete.
6. c_o is a proper subspace of c. Show that c_o is complete.
7. The spaces of Examples 9 and 10 are complete. Show this.
8. The space of Example 15 is not complete—show this by considering the sequence of functions

$$x^n(t)=\begin{cases} 0, & a \leqslant t \leqslant b-\dfrac{1}{n} \\ n(t-b)+1, & b-\dfrac{1}{n} \leqslant t \leqslant b. \end{cases}$$

9. Show that the space $C[a,b)$ of Example 16 is complete. (Hint: for every c, $a<c<b$, the space $C[a,c]$ is complete. Any function in $C[a,b)$ is in each of $C[a,c]$, $a<c<b$. Also,

$$\sup_{a \leqslant t \leqslant c} |x(t)-y(t)| \leqslant \sup_{a \leqslant t < b} |x(t)-y(t)|.\Big)$$

1.5 SOME APPLICATIONS OF THE CONTRACTION MAPPING PRINCIPLE

1. Let $f(x)$ be a function on $[a,b]$ which is Lipschitz continuous there with constant $k<1$; that is, for all t,s on the interval, $|f(t)-f(s)| \leqslant k|t-s|$. Suppose that $f:[a,b]\to[a,b]$. Then $f(t^*)=t^*$ at one and only one point $t^* \in [a,b]$. For any starting value $t^0 \in [a,b]$ the sequence $\{t^n\}$ defined by $t^{n+1}=f(t^n)$ converges to t^* and $|t^n-t^*| \leqslant k^n(b-a)$.
2. Let $X=\mathbf{E}^n$ and let $T(x)=Ax+b$ where $A=(a_{ij})$ is an $n\times n$ matrix and $b\in\mathbf{E}^n$. Then the functional equation $x=T(x)=Ax+b$ is equivalent to a matrix equation $Mx=b$, with $M=I-A$, where M is a $n\times n$ matrix and I is the $n\times n$ identity matrix. We need only rewrite this as

$$x=x-Mx+b$$

or

$$x=(I-M)x+b=Ax+b.$$

When is T a contraction? The answer depends upon the choice of the metric:

(a)Let $\rho(x,y)=\max_i|x_i-y_i|$, $x,y\in\mathbf{E}^n$. Then

$$\begin{aligned}\rho(Ax+b,Ay+b) &= \max_{1\leqslant i\leqslant n}|[A(x-y)]_i|\\ &= \max_{1\leqslant i\leqslant n}\left|\sum_{j=1}^{n} a_{ij}(x_j-y_j)\right|\\ &\leqslant \max_{1\leqslant i\leqslant n}\sum_{j=1}^{n}|a_{ij}|\,|x_j-y_j|\\ &\leqslant \max_{1\leqslant i\leqslant n}\sum_{j=1}^{n}|a_{ij}|\max_{1\leqslant j\leqslant n}|x_j-y_j|\\ &\leqslant\left(\max_{1\leqslant i\leqslant n}\sum_{j=1}^{n}|a_{ij}|\right)\rho(x,y).\end{aligned}$$

Hence T is a contraction if

$$\alpha=\max_{1\leqslant i\leqslant n}\left(\sum_{j=1}^{n}|a_{ij}|\right)<1.$$

(b)Let $\rho(x,y)=\Sigma_{i=1}^{n}|x_i-y_i|$. Then

$$\begin{aligned}\rho(Ax+b,Ay+b)| &= \sum_{i=1}^{n}\left|\sum_{j=1}^{n} a_{ij}(x_j-y_j)\right|\\ &\leqslant \sum_{i=1}^{n}\sum_{j=1}^{n}|a_{ij}|\,|x_j-y_j|\\ &\leqslant\left(\max_{1\leqslant j\leqslant n}\sum_{i=1}^{n}|a_{ij}|\right)\sum_{j=1}^{n}|x_j-y_j|.\end{aligned}$$

Hence T is a contraction in this metric if

$$\alpha=\max_{1\leqslant j\leqslant n}\left(\sum_{i=1}^{n}|a_{ij}|\right)<1.$$

(c)If $\rho(x,y)=(\sum_{i=1}^{n}|x_i-y_i|^2)^{1/2}$, then

$$[\rho(Ax+b,Ay+b)]^2=\sum_{i=1}^{n}\left[\sum_{j=1}^{n}a_{ij}(x_j-y_j)\right]^2$$

$$\leqslant \sum_{i=1}^{n}\sum_{j=1}^{n}(a_{ij})^2\sum_{j=1}^{n}(x_j-y_j)^2$$

by the Cauchy inequality. Thus

$$\rho(Ax+b,Ay+b)\leqslant\left\{\left[\sum_{i=1}^{n}\sum_{j=1}^{n}(a_{ij})^2\right]^{1/2}\right\}\rho(x,y)$$

and T is a contraction if

$$\alpha=\sum_{i=1}^{n}\sum_{j=1}^{n}(a_{ij})^2<1.$$

The choice of metric determines the condition that insures that T is a contraction mapping.

3. *The Fredholm nonhomogeneous integral equation of the second kind*:

$$x(t)=\lambda\int_a^b k(t,s,x(s))\,ds+b(t).$$

If we work in $C[a,b]$ with $b\in C[a,b]$ and k continuous on $[a,b]\otimes[a,b]\otimes\mathbf{R}$, while

$$|k(t,s,x')-k(t,s,x'')|\leqslant M|x'-x''|,\qquad \begin{cases} a\leqslant s\leqslant b,\\ a\leqslant t\leqslant b.\end{cases}$$

then

$$T(x)(t)=\lambda\int_a^b k(t,s,x(s))\,ds+b(t)$$

is a contraction for $0<\lambda<1/M(b-a)$ with the choice of metric

$$\rho(x,y)=\sup_{a\leqslant t\leqslant b}|x(t)-y(t)|.$$

This follows from the calculation

$$\rho(T(x),T(y)) \leqslant \lambda \sup_{a \leqslant t \leqslant b} \int_a^b |k(t,s,x(s)) - k(t,s,y(s))|\,ds$$
$$\leqslant \lambda M(b-a)\rho(x,y).$$

With slight modifications, this formal analysis proves the existence of solutions to many classical problems. Thus if

$$\begin{cases} \lambda = 1, \\ K(t,s,x) = \begin{cases} f(s,x), & t \geqslant s, \\ 0, & t < s, \end{cases} \\ b(t) = x^0. \end{cases}$$

then we are solving the ordinary differential equation $dx/ds = f(s,x)$ for the initial condition $x(a) = x^0$. $T(x)$ still maps continuous functions into continuous functions. The existence of the solution is guaranteed for the interval $M|t-a| < 1$. The successive approximations in this case are usually called the Picard iterates and the method is called the Picard method. (It was first used, however, by Cauchy.) The form of the iteration is usually written as

$$x^{n+1}(t) = \int_a^t f(s, x^n(s))\,ds + x^0, \qquad n = 0, 1, 2, \ldots.$$

Sometimes $f(s,x)$ is only defined on some rectangle $[a,b] \otimes [c,d]$ with $x^0 \in [c,d]$. Then if $|f| \leqslant M'$ there, and $x^n(t)$ takes values only in $[c,d]$, we have the inequality

$$|x^{n+1}(t) - x^0| \leqslant \int_a^t |f(s, x^n(s))|\,ds$$
$$\leqslant M'|t-a|,$$

which guarantees that $x^{n+1}(t)$ is in the domain of f for $M'|t-a| \leqslant \min(|d-x^0|, |c-x^0|)$. Then the convergence of the Picard iterations to the unique solution $x(t)$ is guaranteed only on the interval $|t-a| < \min(|d-x^0|/M', |c-x^0|/M', 1/M) = h$. However, by an obvious process of continuation the solution can be extended until it reaches the boundary of the domain of f.

Note that we do not need to use all of $C[a,b]$ in X. It would be adequate to consider the subspace with $x(a) = x^0$.

If $k(t,s,x)=k(t,s)x$ we are solving the nonhomogeneous linear Fredholm equation of the second kind.

If $k(t,s,x)=0$, $s>t$, we are solving the inhomogeneous Volterra integral equation.

In order for the mapping T to map $C[a,b]$ into $C[a,b]$, some conditions may be necessary on k and $b(t)$. These vary from case to case. Thus, we require $b(t)\in C[a,b]$. In the case of $dx/ds=f(s,x)$ it is sufficient for $f(s,x(s))$ to be integrable for every $x(s)\in C[a,b]$. In the case of the Fredholm equation we usually demand $k(t,s,x)$ to be continuous on $a\leqslant s\leqslant b, a\leqslant t\leqslant b$, while in the Volterra case we need only $k(t,s,x)$ continuous on $a\leqslant s\leqslant t\leqslant b$. In the Fredholm case we obtain existence for all sufficiently small λ, while in the Volterra case we obtain existence for all λ for sufficiently small intervals $b-a$.

One last remark: this proof establishes the convergence of the Neumann series for the linear integral equation. In fact our proof of convergence coincides with the customary proof there if we write it out extensively. Suppose

$$x(t)=\lambda\int_a^b K(t,s)x(s)\,ds+b(t).$$

The successive approximations for $x^0=0$ are

$$x^1(t)=\lambda\int_a^b K(t,s)b(s)\,ds+b(t),$$

$$x^2(t)=\lambda^2\int_a^b\int_a^b K(t,s)K(s,u)b(u)\,du\,ds+\lambda\int_a^b K(t,s)b(s)\,ds+b(t)$$

$$=\lambda^2\int_a^b K^2(t,s)b(s)\,ds+\lambda\int_a^b K(t,s)b(s)\,ds+b(t)$$

where we write

$$K^{i+1}(t,s)=\int_a^b K^i(t,u)K(u,s)\,du,\qquad i=1,2,3,\ldots.$$

In fact

$$x^n(t)=\sum_{i=1}^n\lambda^i\left(\int_a^b K^i(t,s)b(s)\,ds\right)+b(t)$$

and by the Banach fixed point theorem x^n converges uniformly to the solution of the integral equation given by the Neumann series

$$x(t)=\sum_{i=1}^{\infty}\lambda^i\int_a^b K^i(t,s)b(s)\,ds+b(t)$$

for $\lambda<1/M(b-a)$.

EXERCISES TO SECTION 1.5

1. Show that the solution $x(t)$ obtained by the Picard method for an ordinary differential equation on $|t-a|<h$ has a unique limit $x(a+h)$ as $|t-a|\uparrow h$. From this it follows that the Picard method can be reapplied at $a+h$ to extend the solution to a domain $t>a+h$.
2. Let X be $\mathbf{E}^n$ with M an $m\times n$ matrix, $b\in\mathbf{E}^n$. Consider the equation $Mx=b$. If $m_{ii}\neq 0$, $i=1,\ldots,n$, this system can be rewritten as

   $$x_i=\sum_{j\neq i}\frac{(-m_{ij})}{m_{ii}}x_j+\frac{b_i}{m_{ii}}$$

   or as

   $$x=Vx+d$$

   where V is an $n\times n$ matrix with $v_{ii}=0$, $i=1,\ldots,n$; $v_{ij}=-m_{ij}/m_{ii}$; $i\neq j$ and $d_i=b_i/m_{ii}$, $i=1,\ldots,n$. Show that any of the three conditions below imply the convergence of the iteration.

   $$x^{n+1}=Vx^n+d, \qquad n=1,2,\ldots,$$

   to the solution of $Mx=b$.
   (a) $\Sigma_{i=1}^n\Sigma_{j=1}^n|m_{ij}/m_{ii}|^2<1$;
   (b) $\Sigma_{j=1}^n|m_{ij}|<2|m_{ii}|$, $i=1,\ldots,n$;
   (c) $\max_{1\leqslant j\leqslant n}\Sigma_{i=1}^n|m_{ij}/m_{ii}|<2$.
3. Let $k(s,t,u)$ be continuous on $0\leqslant s,t\leqslant 1$ for each u and

   $$|k(s,t,u_1)-k(s,t,u_2)|\leqslant L|u_1-u_2|$$

   for all $|u_1|,|u_2|\leqslant U$, where U is independent of (s,t). Then for all sufficiently small λ there exists exactly one solution $\bar{u}(s)$ of

   $$u(s)=\lambda\int_0^1 k(s,t,u(t))\,dt$$

satisfying $|\bar{u}(t)| \leqslant U$. The method of successive approximation converges to $\bar{u}$ for every u_0 with $|u_0(s)| \leqslant U$.

4. The application of the contraction mapping principle to matrix equations in sequence spaces is more complex than in the case $X = \mathbf{E}^n$ treated in the text. Restrict attention to the spaces m, l_1, l_2 and find conditions on an infinite matrix A such that $T(x) = Ax + b$ is a contraction.

1.6 INADEQUACY OF THE RIEMANN INTEGRAL FOR ITERATIVE PROCESSES

For application of the Banach fixed point theorem (1.4.1), in addition to knowing that T is a contraction, it is necessary to know which spaces are complete. We have seen so far that the spaces $\mathbf{R}^n, \mathbf{E}^n$ are complete for all the proposed metrics and for all $n = 1, 2, 3, \ldots$. The space $C[a,b]$, $C[a,b)$, and also the spaces $C(a,b]$ and $C(a,b)$ are complete with the metric of Example 16. In other metrics involving integrals not even $C[a,b]$ turns out to be complete. The problem is seemingly only a technical one. In the theory of the Riemann integral we are unable to make the interchange

$$\int_a^b \lim_{n\to\infty} x^n(t)\,dt = \lim_{n\to\infty} \int_a^b x^n(t)\,dt$$

for $x^n \in C[a,b]$, where $[a,b]$ is a finite interval, except by using uniform convergence or else Osgood's theorem, which requires that $\lim_{n\to\infty} x^n(t) = x(t)$ where $x(t)$ is piecewise continuous, and that there is a constant M such that $|x^n(t)| \leqslant M$ for all n, t. It is our purpose in this section to show that there are natural and important situations in which the *sup* metric is inappropriate, and piecewise continuity is too strong a condition on the limit function. We really need to find a wider class of integrable functions and a broader definition of integral that is compatible with the desired interchange of order of limit and integral.

Let $\{x^n\}$ be a Cauchy sequence in l_2. (For ease of analysis we will use the real space l_2; the complex case is essentially the same.) Then

$$\sum_{i=1}^{\infty} (x_i^n)^2 < \infty$$

while

$$\sum_{i=1}^{\infty} (x_i^n - x_i^m)^2 \to 0, n, m \to \infty.$$

Since

$$|x_i^n - x_i^m| \leqslant \rho(x^n, x^m) \to 0, n, m \to \infty$$

each coordinate $\{x_i^n\}$ is a Cauchy sequence in **R**, so that, by the completeness of **R**, there is an x_i with $x_i^n \to x_i, n \to \infty$. Let x denote the sequence of limits $x_1, x_2, x_3, \ldots$. For any $M > 0$,

$$\sum_{i=1}^{M} (x_i^n - x_i^m)^2 \leqslant \sum_{i=1}^{\infty} (x_i^n - x_i^m)^2.$$

For any $\epsilon > 0$, there is N_ϵ such that

$$\sum_{i=1}^{\infty} (x_i^n - x_i^m)^2 < \epsilon, \qquad n, m \geqslant N_\epsilon.$$

If we let $n \to \infty$, fixing m, we conclude that

$$\sum_{i=1}^{M} (x_i - x_i^m)^2 < \epsilon, \qquad m \geqslant N_\epsilon, \text{ all } M.$$

Since this is true for all M, it follows that

$$\sum_{i=1}^{\infty} (x_i - x_i^m)^2 < \epsilon, \qquad m \geqslant N_\epsilon.$$

But for any integer P,

$$\sum_{i=1}^{P} (x_i)^2 = \sum_{i=1}^{P} (x_i - x_i^m + x_i^m)^2$$

$$\leqslant \sum_{i=1}^{P} (x_i - x_i^m)^2 + 2\sum_{i=1}^{P} |x_i^m(x_i - x_i^m)| + \sum_{i=1}^{P} (x_i^m)^2.$$

The Cauchy inequality implies that

$$\sum_{i=1}^{P} |(x_i^m)(x_i - x_i^m)| \leqslant \left\{ \left[\sum_{i=1}^{P} (x_i^m)^2 \right] \left[\sum_{i=1}^{P} (x_i - x_i^m)^2 \right] \right\}^{1/2}.$$

Hence for all m, P,

$$\sum_{i=1}^{P} (x_i)^2 \leqslant \epsilon + 2\rho(x^m, 0)\epsilon^{1/2} + \rho^2(x^m, 0) < \infty.$$

Hence $x \in l_2$, and, for any $\epsilon > 0$, there is N_ϵ such that

$$\rho(x, x^m) < \epsilon, \qquad m \geqslant N_\epsilon,$$

which implies $\rho(x, x^m) \to 0$, $m \to \infty$. Thus *every Cauchy sequence* in l_2 *has a limit in* l_2. That is, l_2 *is a complete metric space.*

In many physical applications, and other situations in which periodic phenomena occur, it is important to consider the space of functions that are Riemann integrable on $[-\pi, \pi]$, with mean value zero and (again for technical convenience only) even there, so that $f(-t) = f(t)$. It is known from the theory of Fourier series that there corresponds to each such f a cosine series of the form

$$\tfrac{1}{2} a_0 + a_1 \cos t + a_2 \cos 2t + a_3 \cos 3t + \cdots$$

where the sequence $a_0, a_1, a_n, \ldots$ is defined by Euler's formulas:

$$a_n = \frac{1}{\pi} \int_{-\pi}^{\pi} f(t) \cos nt \, dt, \qquad n = 0, 1, 2, \ldots.$$

When f is of mean value zero, $a_0 = 0$. The integrals exist since the product of a bounded continuous function and a Riemann integrable function is again Riemann integrable. A fundamental theorem of the theory of Fourier series, Parseval's theorem, asserts that if $[f(t)]^2$ and $f(t)$ are both Riemann integrable, then

$$\sum_{n=1}^{\infty} (a_n)^2 = \frac{1}{\pi} \int_{\pi}^{\pi} [f(t)]^2 dt.$$

Since

$$\int_{-\pi}^{\pi} [f(t) - g(t)] \cos nt \, dt = \int_{-\pi}^{\pi} f(t) \cos nt \, dt - \int_{-\pi}^{\pi} g(t) \cos nt \, dt,$$

we can associate a metric with the set of series of Fourier coefficients of Riemann integrable functions of mean zero,

$$\rho(a, b) = \left[\sum_{n=1}^{\infty} (a_n - b_n)^2 \right]^{1/2}$$

and a metric with the functions themselves

$$\rho(f, g) = \left(\frac{1}{\pi} \int_{-\pi}^{\pi} [f(t) - g(t)]^2 dt \right)^{1/2}.$$

The Fourier coefficients lie in the complete metric space l_2.

Suppose now that we are given a fixed point problem $f=F(f)$, where F maps continuous functions into continuous functions. (The latter condition is typical of the problems we have stated.) Then if f corresponds to the element $a\in l_2$, a satisfies $a=\hat{F}(a)$ for some mapping $\hat{F}$. The correspondence between f and a is 1:1; that is, if f corresponds to a and g corresponds to b then $\rho(f,g)=\rho(a,b)$ so $f=g$ if, and only if $a=b$, when f,g are continuous. Also if $x^{n+1}=F(x^n)$ and $a^{n+1}=\hat{F}(a^n)$ are corresponding iterations, $\rho(a^n,a^m)=\rho(f^n,f^m)$ and F is a contraction if and only if $\hat{F}$ is a contraction with the same value of α in both cases. (Why?) If $\hat{F}$ is a contraction, then $a^n\to a^*=\hat{F}(a^*), n\to\infty$ for a unique a^*. *But there is no guarantee that the corresponding sequence of continuous functions converges to a continuous function in the metric adopted.*

As one example, let

$$x^n(t)=\begin{cases}0 & \dfrac{1}{n}\leqslant|t|\leqslant\pi\\ \left[-n\left(|t|-\dfrac{1}{n}\right)\right]^{1/2} & 0\leqslant|t|\leqslant\dfrac{1}{n}.\end{cases}$$

Then

$$\int_{-\pi}^{\pi}[x^n(t)-0]^2\,dt=\frac{1}{n}\to 0,\qquad n\to\infty$$

but on $[-\pi,\pi]$

$$x^n(t)\to\begin{cases}1 & t=0,\\ 0 & t\neq 0.\end{cases}$$

This sequence converges to a function that differs from a continuous function at only one point so the case may seem weak. Consider next a more serious difficulty:

Let the nth function be given by

$$x^n(t)=\begin{cases}-1 & -\pi\leqslant t\leqslant-\dfrac{1}{n}\\ nt & -\dfrac{1}{n}\leqslant t\leqslant\dfrac{1}{n}\\ +1 & \dfrac{1}{n}\leqslant t\leqslant\pi.\end{cases}$$

Then

$$\int_{-\pi}^{\pi} [x^n(t) - x^*(t)]^2 \, dt \to 0, \qquad n \to \infty$$

where

$$x^*(t) = \begin{cases} -1 & -\pi \leqslant t < 0 \\ 0 & t = 0 \\ 1 & 0 < t \leqslant \pi. \end{cases}$$

Now $x^*(t)$ is not continuous. Is it conceivable that there is a continuous function $x^{**}(t)$ such that $x^n(t) \to x^{**}(t)$? Suppose this were so; then

$$\int_{-\pi}^{\pi} [x^{**}(t) - x^*(t)]^2 \, dt$$

$$= \lim_{n \to \infty} \int_{-\pi}^{\pi} [x^n(t)\,dt - x^*(t)]^2 \, dt = 0.$$

Hence

$$\int_{-\pi}^{0} [x^{**}(t) + 1]^2 \, dt = 0.$$

The only continuous function that satisfies this is $x^{**}(t) = -1$, $-\pi \leqslant t \leqslant 0$. Similarly $x^{**}(t) = +1$, $0 \leqslant t \leqslant \pi$. This violates the assumption that x^{**} is continuous.

This would suggest that we should consider the piecewise continuous functions as acceptable functions. But in Exercise 1.6.1 it will be seen that this is not good enough. Using straightforward techniques, it is possible to find examples in which the discontinuities of the limit function are "dense" on the interval, and the limit is not integrable; something more than piecewise continuity is needed. The notion of integral must be extended so that the function space itself becomes a complete space, if this is possible, so that to every sequence a there corresponds a unique integrable function f whose square is integrable.

Remark 1. The analysis above is somewhat disingenuous. We have not exhibited a problem of the type specified for which the limit process works for the Fourier coefficients, but not for the functions. The latter examples exist but are too complex for easy comprehension. We will settle here for a plausibility argument.

Remark 2. Another extension of the notion of integral, the Riemann-Steiltjes integral, leads to a complete metric space. See Exercises 1.6.3 and 1.6.4.

EXERCISES TO SECTION 1.6

1. We work on the interval $[0,1]$ for convenience. The function $x \in C[0,1]$ can be extended to be continuous for all t by setting

$$x(t)=\begin{cases} x(0) & -\infty<t\leqslant 0 \\ x(t) & 0\leqslant t\leqslant 1 \\ x(1) & 1\leqslant t<\infty. \end{cases}$$

Form a sequence $\{t_n\}$ on $[0,1]$ by setting $t_0=0, t_1=1, t_2=2^{-1}, t_3=2^{-2}, t_4=3\times 2^{-2}, t_5=2^{-3}, t_6=3\times 2^{-3}, t_7=5\times 2^{-3}, t_8=7\times 2^{-3},\ldots$; at each stage form the next group of t's by striking from the set of points $j\times 2^{-k}$, $j=0,1,2,\ldots,2^k$, all those that already have been put in the sequence $\{t_n\}$. Let $x_j(t)=x(t-t_j)$. This has the effect of shifting the extended function to the right by a distance t_j. *Show that* between any two points t', t'' of $[0,1]$ there will lie at least one point of the sequence $\{t_j\}$. Now consider

$$y^n(t)=\sum_{j=0}^{\infty} 2^{-j}x_j^n(t)$$

where $x^n(t)$ is the function

$$x^n(t)=\begin{cases} -1 & 0\leqslant t\leqslant \frac{1}{2}-\frac{1}{n} \\ n\left(t-\frac{1}{2}\right) & \frac{1}{2}-\frac{1}{n}\leqslant t\leqslant \frac{1}{2}+\frac{1}{n} \\ +1 & \frac{1}{2}+\frac{1}{n}\leqslant t\leqslant 1. \end{cases}$$

Show that the infinite series converges uniformly to $y^n(t)$, which is therefore a continuous function on $[0,1]$.
Now

$$y^n(t)=\sum_{j=1}^{k-1} 2^{-j}x_j^n(t)+2^k x_k^n(t)+\sum_{j=k+1}^{\infty} 2^{-j}x_j^n(t) \tag{1}$$

Show that $\lim_{n\to\infty} y^n(t)=y(t)$ exists at every point of $[0,1]$ and that $y(t)$ is not continuous at any of the points t_k, $k=0,1,2,\ldots$. (Hint: in eq. 1, each of the sums and its limit is continuous at t_k while the term

$2^{-k}x_k^n(t)$ has as its limit a function that takes the value -2^{-k} for $t<t_k$, 0 for $t=t_k$, and 2^{-k} for $t>t_k$. Thus the set of discontinuities of $y(t)$ contains all the points t_k, $k=0,1,2,\ldots$.) Show also that

$$\int_0^1 (y^n(t)-y^m(t))^2\,dt \to 0, \qquad n,m\to\infty.$$

2. (Bounded variation) Let $-\infty<a<b<\infty$. Let $a=t_0<t_1<t_2<\cdots<t_n=b$ be any finite decomposition of the interval $[a,b]$. For any complex valued function f set

$$\sum_{ab}(f)=\sum_{i=1}^{n}|f(t_i)-f(t_{i-1})|.$$

Let

$$T_a^b(f)=\sup_{\{t_i\}}\sum_{ab}(f),$$

where the supremum is over all finite decompositions of $[a,b]$. $T_a^b(f)$ is called the total variation of f. If $T_a^b(f)<\infty$, f is called "of bounded variation." Show

(a) If $f(t)$ is monotone, $f(t)$ is of bounded variation.
(b) If $f,g\in BV([a,b])$ (i.e., are of bounded variation on $[a,b]$) then $\alpha f+\beta g\in BV([a,b])$ for any constants α,β.
(c) $T_a^b(f)=T_a^c(f)+T_c^b(f)$, $a\leqslant c\leqslant b$.
(d) Show that $T_a^t(f)$ is a monotone increasing function of t.
(e) Show that $f(t)-T_a^t(f)$ is a monotone decreasing function of t.
(f) A function is of bounded variation if, and only if, it is the difference of two monotone functions.
(g) Show that any function of bounded variation is differentiable almost everywhere. (Hint: *Lebesgue's theorem*: *Every monotonic function $f(x)$ possesses a finite derivative almost everywhere*; see, for example, Riesz-Nagy [1], 5–6).

3. *Riemann-Stieltjes integrals*: Let $\alpha(t)\in BV([a,b])$. For all partitions of $[a,b]$ and points ξ_k consider the set of numbers

$$\sum=\sum_1^n f(\xi_k)[\alpha(t_k)-\alpha(t_{k-1})], \qquad t_{k-1}\leqslant\xi_k\leqslant t_k.$$

If $\lim \Sigma$ exists as $\max_k |t_k - t_{k-1}| \to 0$, we call the limit

$$\int_a^b f(t)\, d\alpha(t).$$

(a) Show that if $f(t) \in C^0[a,b]$, then $\int_a^b f(t)\, d\alpha(t)$ exists.
(b) So does $\int_a^b \alpha(t)\, df(t)$ (with the obvious definition).
(c) $\int_a^b f(t)\, d\alpha(t) + \int_a^b \alpha(t)\, df(t) = f(t)\alpha(t)|_a^b$.
(d) The space $BV([a,b])$ is a complete metric space if we set

$$\rho(f,g) = |f(a) - g(a)| + T_a^b(f-g).$$

2
INTEGRATION THEORY

2.0 INTRODUCTION

Our starting point for the definition of an integral is the desire to have it defined for as wide a class of functions as possible, to have it commute with limiting processes that are weaker than uniform convergence, or the conditions of Osgood's theorem, and yet for it to coincide with the Riemann integral for all Riemann integrable functions. This suggests that it might be possible to define an integral for a suitable class of elementary functions that coincides with the Riemann integral on that class and then to extend the definition to cover those cases in which both

$$\lim_{n\to\infty} \int_a^b f^n(t)\,dt$$

and

$$\lim_{n\to\infty} f^n(t)$$

exist, in a suitable sense, so that commutation is preserved. The crux of the question is the "suitable sense" for the limit. We follow F. Riesz in the development of this idea to the extent that it provides the tools needed in application. The integral obtained coincides with the ordinary Lebesgue integral. The proof of this fact is left to supplementary reading. Our procedure differs from that of Lebesgue in focusing directly on the limiting process, instead of on the usual method of defining a measure that carries the limiting processes implicitly, and then proving the commutation property for sequences of functions. (We have in mind, however, the final powerful theorems of Lebesgue on the possibility of the interchange of integral and limit, and would be hard pressed to define the suitable limit process if we did not know in advance the final form of the theorems to be proved.) Having extended the integral once, the question arises: Can we extend it once more? When we answer this question negatively, we have our integral, provided that the class of integrable functions is wide enough.

Here the process will be carried out on the real line $-\infty<t<\infty$; it is left as an exercise to extend the notion to integrals over $\mathbf{R}^n$ and $\mathbf{E}^n$. In addition, the possibility of an integral with respect to a measure other than ordinary length will also be left as an exercise.

2.1 STEP FUNCTIONS AND THEIR INTEGRALS

Our starting point is the fact that the area of a rectangle is the product of the base and the altitude. By the *support* of a function $f(t)$ is meant the set of points t at which $f(t)\neq 0$. When the support of f is a finite set of nonoverlapping intervals on each of which f is a constant, the Riemann integral assigns a number that is the sum of the (signed) areas of the rectangles formed from the graph of the function and the t axis in the obvious way. We formalize this notion in the following definitions:

DEFINITION 2.1.1. Let $-\infty\leqslant a<b\leqslant\infty$. Let $\{I_k\}$ be a finite collection of finite open intervals $a_k<t<b_k$ with $a\leqslant a_k<b_k\leqslant b$, no two of which overlap. A function $\phi(x)$ which satisfies

$$\phi(x)=\begin{cases} c_k, & x\in I_k, \\ 0, & x\notin \bigcup\limits_k I_k, \end{cases}$$

for some real constants c_k is called a *step function*. The class of all such step functions is denoted by S_0.

We associate with each interval I_k its length $|I_k|=b_k-a_k$.

DEFINITION 2.1.2. To each $\phi\in S_0$ we associate a number called the integral of ϕ by

$$\int_a^b \phi(t)\,dt=\sum_k c_k|I_k|.$$

The integral on the class S_0 obviously has the following properties:

1. If $\phi\geqslant 0$, then $\int_a^b\phi(t)\,dt\geqslant 0$.
2. If $\phi\in S_0$ and α is a real number, then $\alpha\phi\in S_0$ and

$$\int_a^b \alpha\phi(t)\,dt=\alpha\int_a^b\phi(t)\,dt.$$

3. If ϕ', ϕ'' are in S_0, then so is $\phi' + \phi''$ and

$$\int_a^b [\phi'(t) + \phi''(t)]\,dt = \int_a^b \phi'(t)\,dt + \int_a^b \phi''(t)\,dt.$$

4. If ϕ', ϕ'' are in S_0 and $\phi' \leqslant \phi''$, then

$$\int_a^b \phi'(t)\,dt \leqslant \int_a^b \phi''(t)\,dt.$$

This integral may be thought of as a *functional* whose domain is S_0 and whose range is R, the real line.

DEFINITION 2.1.3. By an *open covering* of a set E of points of R is meant a collection of open intervals $\{J_\alpha\}$, possibly infinite in number, such that each point $x \in E$ is contained in at least one J_α of the collection. If the index α runs through the integers, and if

$$\epsilon = \sum_\alpha |J_\alpha| < \infty$$

we call the covering an ϵ-cover. If for each value of $\epsilon > 0$ there is an ϵ-cover of E, E is called a *zero set*. A property that holds at all points of $[a,b]$ except for a zero set E is said to hold *almost everywhere*.

As an example of a zero set we consider the set of t in the interval $[0,1]$ whose coordinate is a rational number. To see that this is a zero set, we first write each rational number as a fraction p/q in lowest terms, so that the integers p and q have no common divisor. For each q, $q = 1,2,3,\ldots$, we write, in increasing order, the rationals (in lowest terms) with denominator q. Then we write out the sequence $0, 1, \frac{1}{2}, \frac{1}{3}, \frac{2}{3}, \frac{1}{4}, \frac{3}{4}, \frac{1}{5}, \frac{2}{5}, \frac{3}{5}, \frac{4}{5}, \frac{1}{6}, \frac{5}{6}, \ldots$. Every rational occurs once and only once in this enumeration. Let r_k denote the kth element of this sequence, $k = 0,1,2,\ldots$. Cover r_k by the interval $(r_k - 2^{-(k+2)}\epsilon, r_k + 2^{-(k+2)}\epsilon)$. This is an open interval J_k of length $2^{-(k+1)}\epsilon$. Then using the formula for the sum of a geometric progression we see that

$$\sum_{k=1}^{\infty} |J_k| = \epsilon \sum_{k=0}^{\infty} 2^{-(k+1)} = \frac{1}{2}\,\frac{1}{1-1/2}\,\epsilon = \epsilon.$$

Thus for every $\epsilon > 0$, there is an ϵ-cover of the rationals.

DEFINITION 2.1.4. A sequence of functions $\{f^n\}$ is said to be *increasing almost everywhere* if whenever $m > n$, $f^n(t) \leqslant f^m(t)$ almost everywhere. If the sequence $\{-f^n\}$ is increasing almost everywhere we say that $\{f^n\}$ is *decreasing almost everywhere*.

The following two lemmas will justify the extension of the integral to a wider class of functions:

LEMMA 2.1.1. *If* $\{\phi^n\}$ is a *sequence of step functions* $\phi^n \in S_0$ *that decrease almost everywhere to zero, then*

$$\int_a^b \phi^n(t)\,dt \to 0, \qquad n \to \infty.$$

LEMMA 2.1.2. *If* $\{\phi^n\}$ *is a sequence of step functions* $\phi^n \in S_0$ *that increase almost everywhere and if there exists a real number* $A < \infty$ *such that*

$$\int_a^b \phi^n(t)\,dt \leqslant A, \qquad n = 1, 2, 3, \ldots,$$

then $\phi^n(t)$ *tends to a finite limit almost everywhere.*

Considering Lemma 2.1.2 and property (4) of the integral, it is clear that

$$\lim_{n\to\infty} \int_a^b \phi^n(t)\,dt$$

exists while

$$\lim_{n\to\infty} \phi^n(t) = \phi(t)$$

exists almost everywhere. It makes sense to make the following:

DEFINITION 2.1.5. If $\{\phi^n\}$ is a sequence of step functions increasing almost everywhere to a function $\phi(t)$, we define an integral of ϕ by

$$\int_a^b \phi(t)\,dt = \lim_{n\to\infty} \int_a^b \phi^n(t)\,dt.$$

In order for this to make sense we must show that

$$\int_a^b \phi(t)\,dt$$

is independent of the particular defining sequence $\{\phi^n\}$. Also this definition makes ordinary sense only if the sequence $\{\phi^n\}$ satisfies the condition that the integrals of the sequence are uniformly bounded; we will sometimes use it in the extended sense that if the right side of the definition has value $+\infty$, so does the left side. The class of functions ϕ that are the limit of sequences in S_0 increasing almost everywhere we call S_1.

The class S_1 contains functions unbounded above, but all the members of the set are clearly bounded below. This asymmetry is resolved by extending S_1 to a set $S_2=\{h=f-g|f\in S_1,g\in S_1\}$ which closes S_1 algebraically. On S_1 we define an integral via

$$\int_a^b h(t)\,dt=\int_a^b f(t)\,dt-\int_a^b g(t)\,dt.$$

This definition will turn out to be independent of the representation $h=f-g$ and to possess no further extension along the lines of Lemma 2.1.2 and Definition 2.1.5. This will complete the extension of the functional on S_0.

EXERCISES TO SECTION 2.1

1. Show that a finite set of points is always a zero set.
2. If E^1,E^2 are zero sets, show that $E^1\cup E^2$ is also a zero set.
3. If the sequence $E^1,E^2,E^3,\ldots$ is a sequence of zero sets, then show that $\bigcup_k E^k$ is also a zero set. (Hint: cover E^k by a cover of size $2^{-k}\epsilon$.)
4. Prove properties 1, 2, 3, 4 of the integral on S_0.
5. Show that if ϕ,ψ are in S_0, so are the functions $\max(\phi,\psi)$, $\min(\phi,\psi)$.
6. The function on $[0,1]$

$$f(t)=\begin{cases}1, & \text{if } t \text{ is irrational}\\ 0, & \text{if } t \text{ is rational}\end{cases}$$

is in S_1. Show this.

2.2 PROOFS OF THE FUNDAMENTAL LEMMAS

PROOF OF LEMMA 2.1.1. Since the sequence $\{\phi^n\}$ decreases almost everywhere to zero, $\phi^1(t)\geqslant\phi^n(t)\geqslant 0$ for all n. There is some interval $[a',b']$ that contains the support of ϕ^1, and $\phi^n(t)=0$, $n=1,2,3,\ldots$ if $t\notin[a',b']$. The set of points E^0 at which $\lim_{n\to\infty}\phi^n(t)\neq 0$ lies in $[a',b']$ and is a zero set. Further, each step function ϕ^n is continuous except at a finite set of points E^n in $[a',b']$ and a finite set of points is always a zero set. Then

$$A = \bigcup_{k=0}^{\infty} E^k$$

is also a zero set in $[a',b']$. Given any $\epsilon > 0$ we can find a sequence of open intervals $\mathfrak{A}_0$ of total length ϵ that covers A.

If $t \notin A$, then $\phi^n(t) \to 0$, $n \to \infty$. Thus for each such t there is an $N = N(t,\epsilon)$ such that $\phi^n(t) < \epsilon$ whenever $n \geqslant N$. Since ϕ^n is constant and continuous at $t \notin A$, there is an open interval $I_{n,t}$ about t on which $\phi^n(t) < \epsilon$, $n \geqslant N$. Each function ϕ^n has a finite number of such open intervals. We order these open intervals from left to right for a fixed n, and then by increasing n. Thus we get a second sequence of open intervals, $\mathfrak{A}_1$, and each point $t \in [a',b']$ is covered by at least one open interval of either $\mathfrak{A}_0$ or $\mathfrak{A}_1$. We now apply the Heine-Borel theorem which implies that there is a finite set of intervals in $\mathfrak{A}_0 \cup \mathfrak{A}_1$ that covers $[a',b']$, say $u^1, u^2, \ldots u^k$ in $\mathfrak{A}_0$ and $v^1, v^2, \ldots, v^m$ in $\mathfrak{A}_1$. Clearly

$$\sum_{j=1}^{k} |u^j| < \epsilon.$$

If $t \in v^i$, for any $i = 1, \ldots, m$, $0 \leqslant \phi^n(t) \leqslant \phi^1(t) \leqslant \max_{[a',b']} \phi^1(t) = M$. But each v^i comes from some ϕ^{n_i} with $\phi^{n_i}(t) \leqslant \epsilon$, $t \in v^i$. Let $\eta = \max(n_1, n_2, \ldots, n_m)$. Then $\phi^n(t) < \epsilon$ if $n \geqslant \eta$ for $t \in v^1 \cup v^2 \cup \ldots \cup v^m$. Then

$$0 \leqslant \int_a^b \phi^n(t)\,dt \leqslant \epsilon M + \epsilon(b' - a'), \qquad n \geqslant \eta.$$

So

$$0 \leqslant \limsup_{n \to \infty} \int_a^b \phi^n(t)\,dt \leqslant \epsilon\left[M + (b' - a')\right].$$

Since this estimate holds for all $\epsilon > 0$, we must have

$$0 \leqslant \limsup_{n \to \infty} \int_a^b \phi^n(t)\,dt \leqslant 0$$

so that

$$\lim_{n \to \infty} \int_a^b \phi^n(t)\,dt = 0. \qquad \blacksquare$$

PROOF OF LEMMA 2.1.2. The sequence $\{\phi^n\}$ increases almost everywhere, as does the sequence $\{\phi^n-\phi^1\}$, while $\phi^n-\phi^1\geqslant 0$ for all n. Thus, we may assume that the original sequence satisfies $\phi^n\geqslant 0$. For each fixed t either $\lim_{n\to\infty}\phi^n(t)$ is finite or else $\phi^n(t)\to\infty$, $n\to\infty$. Let D_0 be the set of $t\in[a,b]$ at which $\phi^n(t)\to\infty$, $n\to\infty$. Choosing any $\epsilon>0$, we consider $D_\epsilon=\{t|\phi^n(t)>A/\epsilon$ for all $n\geqslant$ some $N\}$. The definition of D_ϵ is justified by the increasing property of the sequence $\{\phi^n\}$. Then $D_0\subset D_\epsilon$. If we define $D_{\epsilon,n}=\{t|\phi^n(t)>A/\epsilon\}$, then it is immediate that

$$D_\epsilon=\bigcup_{n=1}^{\infty} D_{\epsilon,n}.$$

Each $D_{\epsilon,n}$ is a finite collection of nonoverlapping intervals whose total length $|D_{\epsilon,n}|$ must satisfy

$$|D_{\epsilon,n}|A/\epsilon<\int_a^b\phi^n(t)\,dt\leqslant A.$$

We need merely define another step function ψ^n which coincides with ϕ^n on $D_{\epsilon,n}$ and is zero elsewhere and apply property 4 of the integral. Thus $|D_{\epsilon,n}|<\epsilon$ for all n.

The sequence $\{D_{\epsilon,n}\}$ satisfies $D_{\epsilon,n}\subset D_{\epsilon,n+1}$ for all n since the sequence $\{\phi^n\}$ is increasing. Each point $t\in D_\epsilon$ is contained in some open interval $O_{n,t}\subset D_{\epsilon,n}$ and the open intervals $O_{n,t}$ satisfy $O_{n,t}\subset O_{n+1,t}$ for all n. An increasing sequence of open intervals has end points that form monotone sequences on the interval $[a,b]$; they converge to an open interval. Thus the sets $D_{\epsilon,n}$ converge to D_ϵ which is a collection of open intervals of total length not greater than ϵ. Hence D_0 is a zero set. ■

2.3 THE INTEGRAL IN S_1

The first question that arises with regard to the definition of the integral is the following: is the integral a property of the defining sequence $\{\phi^n\}$ or is it a property only of the limit function ϕ? The answer is a consequence of Lemma 2.3.1 below. Before stating the lemma we introduce some notations that will ease the problem of writing. If a sequence of functions $\{f^n\}$ converges almost everywhere to a function f we write $f^n(\longrightarrow)f$, $n\to\infty$. If the sequence $\{f^n\}$ is increasing almost everywhere (decreasing almost everywhere) we write $f^n(\uparrow)f$, $n\to\infty$ ($f^n(\downarrow)f$, $n\to\infty$). If f and g are functions on the same domain and $f\leqslant g$ ($f\geqslant g$) except on some zero set, then we write $f(\leqslant)g$ ($f(\geqslant)g$); if $f=g$ almost everywhere, we write $f(=)g$.

LEMMA 2.3.2. *On* $[a,b]$ *let* $\{\phi^n\},\{\psi^n\}$, *be sequences in* S_0 *such that* $\phi^n(\uparrow)f$, $n\to\infty$, $\psi^n(\uparrow)g$, $n\to\infty$, *where* f *and* g *are permitted to take the value* $+\infty$. *Let*

$$f(\leqslant)g.$$

Then

$$\lim_{n\to\infty}\int_a^b \phi^n(t)\,dt \leqslant \lim_{n\to\infty}\int_a^b \psi^n(t)\,dt,$$

where the limits are permitted the value $+\infty$.

PROOF. Given a function $h(t)$, let $h(t)^+$ be defined by $h(t)^+ = \max[h(t),0]$. For a fixed m, define the sequence

$$\eta^n(t) = [\phi^m(t) - \psi^n(t)]^+.$$

Then by our assumption, $\eta^n(t)(\downarrow)0$, $n\to\infty$ and from Lemma 2.1.1 we conclude

$$\int_a^b \eta^n(t)\,dt \to 0, \qquad n\to\infty.$$

Since $\phi^m - \psi^n \leqslant \eta^n$, we see that

$$\int_a^b [\phi^m(t) - \psi^n(t)]\,dt \leqslant \int_a^b \eta^n(t)\,dt.$$

But ϕ^n and ψ^n are in S_0 so properties 2 and 3 allow us to conclude

$$\int_a^b \phi^m(t)\,dt \leqslant \int_a^b \psi^n(t)\,dt + \int_a^b \eta^n(t)\,dt.$$

Letting $n\to\infty$ yields

$$\int_a^b \phi^m(t)\,dt \leqslant \lim_{n\to\infty}\int_a^b \psi^n(t)\,dt;$$

then letting $m\to\infty$ shows the truth of the lemma. ■

If we apply the lemma to the case $f(\geqslant)g$ we would conclude

$$\lim_{n\to\infty}\int_a^b \phi^n(t)\,dt \geqslant \lim_{n\to\infty}\int_a^b \psi^n(t)\,dt.$$

In the case $f(=)g$ we conclude that the *integral is independent of the defining sequence in* S_1. In addition if f,g, are in S_1 and $f(\leqslant)g$ we can rewrite the conclusion of the lemma as

$$\int_a^b f(t)\,dt \leqslant \int_a^b g(t)\,dt$$

where $+\infty$ is permitted as the value on both sides of the inequality, or else on the right side alone.

We note also that in S_1, the integral possesses the following four properties, analogous to those in S_0:

1. If $f(\geqslant)0$, then $\int_a^b f(t)\,dt \geqslant 0$.
2. If f,g are in S_1, so is $f+g$ and
$$\int_a^b [f(t)+g(t)]\,dt = \int_a^b f(t)\,dt + \int_a^b g(t)\,dt.$$
3. If f is in S_1 and $\alpha \geqslant 0$ then $\alpha f \in S_1$ and
$$\int_a^b \alpha f(t)\,dt = \alpha \int_a^b f(t)\,dt.$$
4. If f,g are in S_1 and $f(\leqslant)g$, then
$$\int_a^b f(t)\,dt \leqslant \int_a^b g(t)\,dt.$$

The asymmetric nature of property 3 is inherited from the increasing property of the sequence of step functions used in the definition of the integral on S_1. For this reason we extend the class of function S_1 to $S_2 = \{h \mid h = f - g, f, g \in S_1\}$.

EXERCISES TO SECTION 2.3

1. Prove properties 1, 2, 3, and 4 in S_1.
2. Show that if f,g are in S_1, so are the functions $\max(f,g)$ and $\min(f,g)$.

2.4 THE INTEGRAL IN S_2

We start with a proof of property 2 of the integral in S_1.

LEMMA 2.4.1. *If f,g are in S_1, then $f+g$ is in S_1 and*

$$\int_a^b [f(t)+g(t)]\,dt = \int_a^b f(t)\,dt + \int_a^b g(t)\,dt.$$

PROOF. If f,g are in S_1, then there exist sequences $\{\phi^n\},\{\psi^n\}$ in S_0, as in Lemma 2.1.2; then $(\phi^n+\psi^n)(\uparrow)(f+g)$, $n\to\infty$ while

$$\int_a^b [\phi^n(t)+\psi^n(t)]\,dt=\int_a^b \phi^n(t)\,dt+\int_a^b \psi^n(t)\,dt.$$

We can take limits as $n\to\infty$ and conclude the truth of the lemma. ■

We are really interested in differences, not sums; about these we prove:

LEMMA 2.4.2. *If f_1,f_2,g_1,g_2, are in S_1 and*

$$f_1-f_2(=)g_1-g_2$$

then

$$\int_a^b f_1(t)\,dt-\int_a^b f_2(t)\,dt=\int_a^b g_1(t)\,dt-\int_a^b g_2(t)\,dt.$$

PROOF. $f_1-f_2(=)g_1-g_2$ if and only if $f_1+g_2(=)f_2+g_2$. We now apply Lemma 2.4.1. ■

Thus we can make the following:

DEFINITION 2.4.1. If $h\in S_2$, so $h=f-g$ with $f,g\in S_1$, then set

$$\int_a^b h(t)\,dt=\int_a^b f(t)\,dt-\int_a^b g(t)\,dt.$$

Lemma 2.4.2 asserts that the value of the integral depends only upon the function h, not upon the functions f,g of the representation of $h=f-g$. For S_2 we now have the following four properties of the integral:

1. If $h(\geqslant)0$, then $\int_a^b h(t)\,dt\geqslant 0$.
2. If h,k are in S_2, so is $h+k$ and

$$\int_a^b [h(t)+k(t)]\,dt=\int_a^b h(t)\,dt+\int_a^b k(t)\,dt.$$

3. If h is in S_2 and α is real then $\alpha h\in S_2$ and

$$\int_a^b \alpha h(t)\,dt=\alpha\int_a^b h(t)\,dt.$$

4. If h,k are in S_2 and $h(\leqslant)k$ then

$$\int_a^b h(t)\,dt \leqslant \int_a^b k(t)\,dt.$$

The class S_2 is algebraically closed. In addition we have the following properties

5. If $h \in S_2$, then $|h| \in S_2$.
6. If $h \in S_2$, then $h^+ \in S_2$.
7. If $h \in S_2$, then $h^- = (h - h^+) \in S_2$.
8. If $h \in S_2$ on $[a,b]$ and on $[b,c]$, then $h \in S_2$ on $[a,c]$ and

$$\int_a^c h(t)\,dt = \int_a^b h(t)\,dt + \int_b^c h(t)\,dt.$$

Our only tool for showing that a function is in S_1 or S_2 involves convergence of increasing sequences. This is both awkward and misleading, as we see from the following characterization of S_2:

THEOREM 2.4.1. *If $h \in S_2$ then there exists a sequence $\{\phi^n\}$ in S_0 such that $\phi^n(\longrightarrow)h, n\to\infty$ and*

$$\int_a^b |\phi^n(t) - h(t)|\,dt \to 0, \qquad n\to\infty.$$

PROOF. We represent $h = f - g$ with f,g in S_1. Then there exist sequences $\{\psi_1^n\}, \{\psi_2^n\}$ in S_0 such that $\psi_1^n(\uparrow)f$, $n\to\infty$ and $\psi_2^n(\uparrow)g$, $n\to\infty$, while

$$\int_a^b \psi_1^n(t)\,dt \to \int_a^b f(t)\,dt, \qquad n\to\infty$$

and

$$\int_a^b \psi_2^n(t)\,dt \to \int_a^b g(t)\,dt, \qquad n\to\infty.$$

Then $(\psi_1^n - \psi_2^n)(\to)h$, $n\to\infty$, and

$$\int_a^b |\psi_1^n(t) - \psi_2^n(t) - h(t)|\,dt \leqslant \int_a^b [\,|f(t) - \psi_1^n(t)| + |g(t) - \psi_2^n(t)|\,]\,dt$$

$$= \int_a^b f(t)\,dt - \int_a^b \psi_1^n(t)\,dt + \int_a^b g(t)\,dt - \int_a^b \psi_2^n(t)\,dt$$

$$\to 0, \qquad n\to\infty. \qquad \blacksquare$$

With the aid of this theorem we can relate our integral to the Riemann integral. We recall first the fact that *on a finite interval* $[a,b]$ *a function* f *is Riemann integrable if and only if* f *is bounded and the discontinuities of* f *form a zero set* (compare, e.g., Franklin [1], 254). Then we can establish the following:

THEOREM 2.4.2. *If* f *is Riemann integrable on a finite interval* $[a,b]$, *then* $f \in S_2$ *and its integral coincides with its Riemann integral there.*

PROOF. Decompose $[a,b]$ into $k = 2^n$ equal subintervals by $k+1$ points $a = a_1 < a_2 < \cdots < a_k < a_{k+1} = b$. On each subinterval $[a_j, a_{j+1}]$ define a function $\phi^n \in S_0$ by

$$\phi^n = \inf\{ f(t) | t \in [a_j, a_{j+1}] \}.$$

Then $\phi^n(\uparrow) f$, $n \to \infty$, that is, except at the discontinuities of f. We also define a step function ψ^n by

$$\psi^n(t) = \inf\{ -f(t) | t \in [a_j, a_{j+1}] \}.$$

Then $-\psi^n(\downarrow) f$, $n \to \infty$. Also

$$\phi^1 \leqslant \phi^2 \leqslant \phi^3 \leqslant \cdots \leqslant -\psi^3 \leqslant -\psi^2$$

so

$$\int_a^b \phi^n(t)\, dt \leqslant \int_a^b [-\psi^m(t)]\, dt.$$

Letting $n \to \infty, m \to \infty$, the limits exist and

$$\lim_{n \to \infty} \int_a^b \phi^n(t)\, dt \leqslant \lim_{m \to \infty} \int_a^b [-\psi^m(t)]\, dt.$$

But

$$\lim_{n \to \infty} \int_a^b \phi^n(t)\, dt = \int_a^b f(t)\, dt$$

and

$$\lim_{n \to \infty} \int_a^b [-\psi^n(t)]\, dt = \int_a^b f(t)\, dt.$$

On the other hand, $\int_a^b \phi^n(t)\, dt$ and $\int_a^b [-\psi^n(t)]\, dt$ are respectively a

Darboux lower sum and a Darboux upper sum for the Riemann integral and hence both converge to the Riemann integral as $n \to \infty$. ■

We have actually shown that a function f which is Riemann integrable is in S_1, as is $-f$.

EXERCISES TO SECTION 2.4

1. Prove properties 1, 2, 3, and 4 in S_2.
2. Show that if $h=f-g$ then $|h|=\max(f,g)-\min(f,g)$.
3. Show that if $h=f-g$ then

$$h^{+}=\max(f,g)-g=f-\min(f,g).$$

4. Let $h^{-}=h-h^{+}$, that is, $h^{-}=h$ when $h \leqslant 0$; show that

$$h^{-}=\max(f,g)-f=g-\min(f,g).$$

5. Prove properties 5, 6, 7, and 8.
6. Show that

$$f(t)=\begin{cases} t^{-p} & 0<t\leqslant 1,\\ 1 & t=0,\end{cases}$$

is in S_1 on $[0,1]$ for $0<p\leqslant 1$. (This function is not Riemann integrable.) Show that $-f \notin S_1$.

2.5 CLOSURE OF S_1 AND S_2

Next we study the following question: If instead of sequences of step functions in S_0, we take sequences of functions in S_1 or S_2 with the same property of increase, and with uniformly bounded integrals, do they yield new functions for which an integral can be defined? The answer is no. In providing the answer we develop the basic theorems on passage to the limit in an integral.

THEOREM 2.5.1. *(Closure of S_1) Let $\{f^n\}$* be *a sequence of functions in S_1 such that for all n*

$$f^n(t)(\leqslant)f^m(t), \qquad m\geqslant n,$$

and for some finite A,

$$\int_a^b f^n(t)\,dt \leqslant A.$$

Then there exists a function $f \in S_1$ *such that*

$$f^n(t)(\uparrow)f(t), \qquad n \to \infty$$

and

$$\int_a^b f^n(t)\,dt \longrightarrow \int_a^b f(t)\,dt, \qquad n \to \infty.$$

PROOF. To each $f^n \in S_1$, we associate a sequence $\{\phi_n^m\}$ in S_0 such that $\phi_n^m(t)(\uparrow)f^n(t), m \to \infty$ and

$$\int_a^b \phi_n^m(t)\,dt \ \uparrow \ \int_a^b f(t)\,dt, \qquad m \to \infty.$$

We arrange a tableau

$$\begin{array}{l}
\phi_1^1 \leqslant \phi_1^2 \leqslant \phi_1^3 \leqslant \cdots \leqslant \phi_1^n \leqslant \cdots (\leqslant) f^1 \\
\phi_2^1 \leqslant \phi_2^2 \leqslant \phi_2^3 \leqslant \cdots \leqslant \phi_2^n \leqslant \cdots (\leqslant) f^2 \\
\vdots \\
\phi_n^1 \leqslant \phi_n^2 \leqslant \phi_n^3 \leqslant \cdots \leqslant \phi_n^n \leqslant \cdots (\leqslant) f^n \\
\vdots
\end{array}$$

For each n, we define a new member of S_0, φ^n as

$$\varphi^n(t) = \max_{i,j \leqslant n} \phi_i^j(t);$$

that is, we take the maximum of the functions in the upper left $n \times n$ section of the tableau. Because f^n is itself an increasing sequence, $\varphi^n(\leqslant)f^n$ for all n. Also

$$\int_a^b f^n(t)\,dt$$

is an increasing sequence of numbers bounded above by A, so it has a limit. Then we have the inequality

$$\int_a^b \varphi^n(t)\,dt \leqslant \int_a^b f^n(t)\,dt \leqslant A.$$

Thus $\{\varphi^n\}$ satisfies the condition of Lemma 2.1.2, and there exists a function $f \in S_1$ such that $\varphi^n(\uparrow)f, n \to \infty$ and

$$\int_a^b \varphi^n(t)\,dt \quad \to \int_a^b f(t)\,dt \leqslant A.$$

By the definition of $\{\varphi^n\}$, $\phi_n^m \leqslant \varphi^m$ as soon as $m \geqslant n$; letting $m \to \infty$ shows that $f^n(\leqslant)f$. The combined inequality $\varphi^n(\leqslant)f^n(\leqslant)f$ shows that the increasing sequence $f^n(\uparrow)f, n \to \infty$ and the integrated form

$$\int_a^b \varphi^n(t)\,dt \leqslant \int_a^b f^n(t)\,dt \leqslant \int_a^b f(t)\,dt \leqslant A$$

shows that

$$\lim_{n \to \infty} \int_a^b f^n(t)\,dt = \int_a^b f(t)\,dt. \quad \blacksquare$$

In S_2 we have a similar development as follows:

THEOREM 2.5.2. *(Closure of S_2) Let $\{h^n\}$ be a sequence of functions in S_2 such that for all n*

$$h^n(t)(\leqslant)h^m(t), \qquad m \geqslant n,$$

and for some finite A

$$\int_a^b h^n(t)\,dt \leqslant A.$$

Then there exists a function $h \in S_2$ such that

$$h^n(t)(\uparrow)h(t), \qquad n \to \infty,$$

and

$$\int_a^b h^n(t)\,dt \longrightarrow \int_a^b h(t)\,dt, \qquad n \to \infty.$$

PROOF. We rewrite the sequence h^n as

$$h^n = h^1 + k^1 + k^2 + \cdots + k^{n-1}$$

where $k^n(t) = h^{n+1}(t) - h^n(t)$; thus $k^n(t)(\geqslant)0$. Then

$$\begin{aligned}\sum_{j=1}^{n} \int_a^b k^j(t)\,dt &= \int_a^b \sum_{j=1}^{n} k^j(t)\,dt \\ &\leqslant A + \int_a^b -h^1(t)\,dt \\ &\leqslant A + \int_a^b |h^1(t)|\,dt \\ &\leqslant B < \infty.\end{aligned}$$

Thus

$$\sum_{j=1}^{\infty} \int_a^b k^j(t)\,dt = \lim_{n\to\infty} \int_a^b \left(\sum_{j=1}^{n} k^j(t) \right) dt < \infty.$$

Since $k^n \in S_2$, $k^n = f^n - g^n$ where f^n, g^n are in S_1 and there exists a sequence $\{\psi_n^m\}$ in S_0 such that $\psi_n^m(\uparrow)g^n, m\to\infty$ and

$$\int_a^b \psi_n^m(t)\,dt \to \int_a^b g^n(t)\,dt, \qquad m\to\infty.$$

We find an $M = M(n)$ such that

$$0 < \int_a^b g^n(t)\,dt - \int_a^b \psi_n^M(t)\,dt < 2^{-n}.$$

Then we can rewrite $k^n = f^n - g^n = (f^n - \psi_n^M) - (g^n - \psi_n^M) = \bar{f}^n - \bar{g}^n$. Now the difference of a function in S_1 and a function in S_0 is easily seen to be in S_1, so $\bar{f}^n, \bar{g}^n$ are in S_1, while

$$\begin{aligned}\sum_{j=1}^{n} \int_a^b \bar{g}^j(t)\,dt &= \int_a^b \sum_{j=1}^{n} \bar{g}^j(t)\,dt \\ &\leqslant 2, \qquad n = 1, 2, 3 \ldots .\end{aligned}$$

In addition $\bar{g}^n(\geqslant)0$, so $k^n(\geqslant)0$ implies $\bar{f}^n(\geqslant)0$. Then, for all n,

$$\begin{aligned}\int_a^b \sum_{j=1}^{n} \bar{f}^j(t)\,dt &= \sum_{j=1}^{n} \int_a^b \bar{f}^j(t)\,dt \\ &= \sum_{j=1}^{n} \left(\int_a^b k^j(t)\,dt + \int_a^b \bar{g}^j(t)\,dt \right) \\ &\leqslant B + 2.\end{aligned}$$

Since $\Sigma_{j=1}^n \bar{f}^j(t)$ is an increasing sequence, we conclude from Theorem 2.5.1 the existence of an $\bar{f} \in S_1$ such that $\left(\Sigma_{j=1}^n f^j(t)\right)(\uparrow)\bar{f}(t), n \to \infty$ and

$$\int_a^b \sum_{j=1}^n f^j(t)\,dt \to \int_a^b \bar{f}(t)\,dt, \qquad n \to \infty.$$

The same situation holds for $\bar{g}^n$, so that there is a $\bar{g}(t)$ $(=)$ $\lim_{n\to\infty} \Sigma_{j=1}^n \bar{g}^j(t)$, and the term by term integration of the series is valid. Then $\bar{f} - \bar{g} \in S_2$, so $h = h^1 + \bar{f} - \bar{g} \in S_2$, $h^n(\uparrow)h, n \to \infty$, and

$$\int_a^b h^n(t)\,dt \to \int_a^b h(t)\,dt, \qquad n \to \infty. \ \blacksquare$$

COROLLARY 1. *Every series $\Sigma_{n=1}^\infty k^n(t)$ of integrable functions for which*

$$\sum_{n=1}^\infty \int_a^b |k^n(t)|\,dt < \infty$$

converges almost everywhere to an integrable function and

$$\sum_{n=1}^\infty \int_a^b k^n(t)\,dt = \int_a^b \sum_{n=1}^\infty k^n(t)\,dt.$$

The proof follows from the decomposition $k^n = (k^n)^+ + (k^n)^-$, and an inspection of the proof of the theorem.

COROLLARY 2. *If $h \in S_2$ and*

$$\int_a^b |h(t)|\,dt = 0$$

then $h(t)(=)0$.

COROLLARY 3. *Let $\{h^n\}$ be a sequence of functions in S_2, with $h^n(\leqslant)h^m$, $m \geqslant n$, such that $h^n(\uparrow)h, n \to \infty$, where $h \in S_2$. Then*

$$\lim_{n\to\infty} \int_a^b h^n(t)\,dt = \int_a^b h(t)\,dt.$$

Two major results, Fatou's lemma and Lebesgue's theorem on dominated convergence, provide the final results on the integration of sequences that we will use.

LEMMA 2.5.1. *Let $\{f^n\}$ be a sequence of functions in S_2 where each $f^n(\geqslant)0$. Then*

$$\int_a^b \liminf_{n\to\infty} f^n(t)\,dt \leqslant \liminf_{n\to\infty} \int_a^b f^n(t)\,dt.$$

Here we allow either both sides, or the right side of the inequality to take the value $+\infty$.

PROOF. Let $\{f^n\}\subset S_2$, $f^n(\geqslant)0$. Let

$$h_n^m(t) = \inf_{n\leqslant i\leqslant m} f^i(t).$$

Then $0(\leqslant)h_n^m \leqslant f^n$, while $h_n^m(\geqslant)h_n^{m+1}$, $m=n+1, n+2, \ldots$. Then $-h_n^m\in S_2$ for all m and $|-h_n^m|(\leqslant)f^n$ for all m. Therefore, since $-h_n^m$ is an increasing sequence, we apply Theorem 2.5.1 to assert that $\lim_{m\to 0} -h_n^m(=)-h_n$ exists in S_2. Then $h_n=\inf_{m\geqslant n} f^m\in S_2$, and

$$\int_a^b h_n(x)\,dx \leqslant \int_a^b f^m(x)\,dx, \qquad m\geqslant n.$$

Therefore

$$\int_a^b h_n(x)\,dx \leqslant \liminf_{m\to\infty} \int_a^b f^m(x)\,dx.$$

Also $h_n \leqslant h_{n+1} \leqslant \ldots$ and

$$\int_a^b h_n(x)\,dx \leqslant \liminf_{m\to\infty} \int_a^b f^m(x)\,dx.$$

If the right side is finite, this implies that $\lim_{n\to\infty} h_n$ exists, hence that

$$\int_a^b \liminf_{n\to\infty} f^n(x)\,dx \leqslant \liminf_{n\to\infty} \int_a^b f^n(x)\,dx.$$

If the right side is infinite, so

$$\liminf_{n\to\infty} \int_a^b f^n(x)\,dx = \infty,$$

and

$$\int_a^b h_n(x)\,dx \to +\infty, \qquad n\to\infty,$$

then

$$\int_a^b \liminf_{n\to\infty} f^n(x)\,dx = +\infty. \quad \blacksquare$$

Two alternate forms of Fatou's lemma are corollaries.

COROLLARY 1. *(Fatou's lemma) Let* $\{f^n\} \subset S_2$, $f^n(\geqslant)0$, *while* $f^n(t)$ $(\longrightarrow) f(t), n\to\infty$ *and for some finite* M

$$\int_a^b f^n(t)\,dt \leqslant M.$$

Then $f \in S_2$ *and*

$$\int_a^b f(t)\,dt \leqslant \liminf_{n\to\infty} \int_a^b f^n(t)\,dt.$$

PROOF. In this case, both sides of the inequality of Lemma 2.5.1 are finite, while $\liminf_{n\to\infty} h_n(t)(=)f(t)$. $\blacksquare$

COROLLARY 2. *(Fatou's lemma) Let* $g, \{f^n\} \subset S_2$ *with* $g(\leqslant)f^n$. *Then*

$$\int_a^b \liminf_{n\to\infty} f^n(t)\,dt \leqslant \liminf_{n\to\infty} \int_a^b f^n(t)\,dt.$$

Again, infinite values are permitted as before. Here g plays the same role as 0 in the proof of Lemma 2.5.1.

THEOREM 2.5.3. *(Lebesgue's theorem on dominated convergence) Let* $\{f^n\} \subset S_2$, $f^n(t)(\longrightarrow) f(t), n\to\infty$ *and for some* $g \in S_2$

$$|f^n(t)|(\leqslant)g(t).$$

Then $f \in S_2$ *and*

$$\lim_{n\to\infty} \int_a^b f^n(t)\,dt = \int_a^b f(t)\,dt.$$

PROOF. In addition to the sequences h_n^m, h_n we define sequences g_n^m, g_n via

$$g_n^m(t) = \sup_{n\leqslant i\leqslant m} f^i(t)$$

and

$$g_n(t) = \sup_{i \geqslant n} f^i(t).$$

Then $g_n(t)(\downarrow)f(t), n \to \infty$ as before, and the function $g(t)$ provides a constant

$$A = \int_a^b g(t)\,dt < \infty,$$

which bounds, for all n, both

$$\left|\int_a^b g_n(t)\,dt\right|$$

and

$$\left|\int_a^b h(t)\,dt\right|.$$

This implies $f \in S_2$ and that both

$$\int_a^b h_n(t)\,dt \quad \to \quad \int_a^b f(t)\,dt, \qquad n \to \infty$$

and

$$\int_a^b g_n(t)\,dt \quad \to \quad \int_a^b f(t)\,dt, \qquad n \to \infty.$$

But for all n

$$\int_a^b h_n(t)\,dt \leqslant \int_a^b f^n(t)\,dt \leqslant \int_a^b g_n(t)\,dt,$$

which implies the truth of the theorem when $n \to \infty$. ∎

COROLLARY 1. *If* $\{f^n\} \subset S_2$, $f^n(\longrightarrow)f, n \to \infty$ *and*

$$|f(t)|(\leqslant)g(t)$$

where $g \in S_2$, *then* $f \in S_2$.

PROOF. Let

$$\bar{f}^n(t) = \begin{cases} g(t), & \text{if } f(t) > g(t), \\ f^n(t), & \text{if } |f^n(t)| \leqslant g(t), \\ -g(t), & \text{if } f(t) < -g(t). \end{cases}$$

Then $\bar{f}^n(\longrightarrow)f, n\rightarrow\infty$ and we can apply the Lebesgue theorem on dominated convergence. ■

COROLLARY 2. *If* $|f(t)|(\leqslant)g(t)$ *with* f *continuous almost everywhere and* $g\in S_2$, *then* $f\in S_2$.

PROOF. Approximate the function $y=x$ on $[a,b]$, by a step function ϕ^n of 2^n equal steps. If both a and b are finite, this is trivial. If a is finite and b is infinite, do this on the interval $[a,n]$ as soon as $n>a$. If a is infinite and b is finite do this on $[-n,b]$ as soon as $-n<b$. If both a,b are infinite, do this on $[-n,n]$. Then $f^n(t)=f[\phi^n(t)](\longrightarrow)f(t), n\rightarrow\infty$ while $|f^n(t)|(\leqslant)g(t)$. Apply Theorem 2.5.3. ■

2.6 SUPPLEMENTS ON INTEGRATION THEORY

At this point the theory of the integral is sufficiently well developed for most applications. However, in order to connect the theory developed above with other theories of integration, and to make other notations and terminologies accessible to the reader, we provide here a series of remarks that make these connections and provide a few further results which we will use without proof.

Remark 1. (Measure theory) Every function in S_2 is a limit almost everywhere of step functions, but the converse is not true without some further conditions, as our various theorems on the interchange of limit and integral show. We extend S_2 slightly to a larger class of functions in the following.

DEFINITION 2.6.1. A function f for which there exists a sequence $\{\phi^n\}\in S_0$ such that $\phi^n(\longrightarrow)f, n\rightarrow\infty$ is called *measurable.*

Theorem 2.4.1 asserts that *every function in* S_2 *is measurable*. Corollary 1 to Theorem 2.5.3 may be read as: *a measurable function f which is dominated by a function* $g\in S_2$: $|f(t)|(\leqslant)g(t)$ *is itself in* S_2. Some further properties of measurable functions are established in the exercises.

DEFINITION 2.6.2. To any set $E\subset[a,b]$, we assign a *characteristic function* Φ_E where

$$\Phi_E(t)=\begin{cases}1, & t\in E,\\ 0, & t\notin E.\end{cases}$$

If the function Φ_E is *measurable*, the set E is also said to be *measurable* and of measure.

$$\mu(E)=\begin{cases}\int_a^b \Phi_E(t)\,dt, & \Phi_E\in S_2,\\ \infty, & \Phi_E\notin S_2.\end{cases}$$

In this terminology, Exercise 2.1.6 can be restated as saying that *the characteristic function of the irrational numbers on* [0, 1] *is measurable* or that *the irrationals on* [0, 1] *are a measurable set*. (In fact they are of measure 1.)

Remark 2. (The fundamental theorems of the integral calculus) These concern the relation between integral and derivative, integration by parts, and the change of variable in the integral. In the integral theory just developed it is possible to prove (e.g., Riesz-Nagy [1], 47–8):

THEOREM 2.6.1. (*Fundamental theorem of the calculus*) *If for each* $x, a<x<b, f$ *is in* S_2 *on the interval* $[a,x]$, *then*

$$F(x)=\int_a^x f(t)\,dt$$

is differentiable almost everywhere and

$$F'(t)(=)f(t).$$

We subsume a possible converse to this theorem in the following definition.

DEFINITION 2.6.3. A function $F(t)$ that is differentiable almost everywhere and that satisfies

$$F(x)=\int_a^x F'(t)\,dt$$

is said to be *absolutely continuous.*

Normally this definition is a theorem and absolute continuity is defined differently. (e.g., Riesz-Nagy [1], 50–54.) To see that this definition is not vacuous, consider the function h defined by $h(t)=\Sigma_1^n 2^{-j}$, $1-1/n \leqslant t < 1-1/(n+1)$, $n=1,2,3,\ldots$. Then $dh/dt(=)0$ so h is not the integral of its derivative. A function such as h all of whose variation occurs on a zero set is called *purely singular*.

THEOREM 2.6.2. (*Integration by parts*) *Let* $-\infty<a<b<\infty$. *If* $F(t)$ *and* $G(t)$ *are absolutely continuous with derivatives* $f(t), g(t)$ *respectively, then*

$$\int_a^b F(t)g(t)\,dt+\int_a^b f(t)G(t)\,dt=F(b)G(b)-F(a)G(a).$$

THEOREM 2.6.3. (*Change of variables in integration*) *Let $s(t)$ be absolutely continuous and monotone nondecreasing on $a \leqslant t \leqslant b$ and let $f \in S_2$ on $\alpha = s(a) \leqslant s \leqslant s(b) = \beta$. Then $f(s(t))s'(t) \in S_2$ on $[a,b]$ and*

$$\int_\alpha^\beta f(s)\,ds = \int_a^b f(s(t))s'(t)\,dt.$$

Both theorems are clearly true for step functions $f(t), g(t)$. The proofs proceed in a direct manner from that fact.

Remark 3. (Lebesgue-Stieltjes integrals) In the definition of the integral we took the "size" of an interval I to be its length. In the definition of the integral of a step function we can replace this by a "size function" $\mu(I)$ as follows: Let $\sigma(t)$ be a monotone nondecreasing function on $a \leqslant t \leqslant b$ and let $\lim_{h \to 0} \sigma(t - h^2) = \sigma(t)$ on $a < t \leqslant b$. Let $\mu([\alpha, \beta)) = \sigma(\beta) - \sigma(\alpha)$ for any interval $[\alpha, \beta)$ in $[a,b]$. Call $\{O'\}$ the collection of all open intervals whose end points are continuity points of $\sigma(t)$. (The discontinuity points are a zero set; see e.g., Franklin [1], 255–256.) Define a μ-zero set to be one possessing for each $\epsilon > 0$ a cover $\{I_k\} \subset \{O'\}$ with

$$\sum_{k=1}^{\infty} \mu(I_k) < \epsilon.$$

The theory developed above can be repeated to define a new class S_2 with integrals

$$\int_a^b f(t)\,d\mu$$

with respect to μ. This assigns to all the measurable sets E a measure $\mu(E)$ by

$$\mu(E) = \begin{cases} \displaystyle\int_a^b \Phi_E(t)\,d\mu, & \text{if } \Phi_E \in S_2, \\ \infty, & \text{if } \Phi_E \notin S_2. \end{cases}$$

This integral is the Lebesgue-Stieltjes integral. Only if $\sigma(t)$ is absolutely continuous does it reduce to an ordinary Lebesgue integral

$$\int_a^b f(t)\sigma'(t)\,dt.$$

Remark 4. (The integral on $\mathbf{R}^n$) We sketch the idea of multiple integration for $n = 2$; the higher extensions are no more difficult. In $\mathbf{R}^2$ we take, as the equivalent of the one-dimensional interval, the two-dimensional rectangle $I_t \otimes I_s$, where I_t, I_s are finite intervals on the real line $\mathbf{R}$. A step is the characteristic function of $I_t \otimes I_s$. If f_k is the characteristic function of the kth rectangle of a set of n nonoverlapping rectangles, then a step function is, for some real $c_1, \ldots, c_n$,

$$\phi(t,s) = \sum_{i=1}^{n} c_i f_i(t,s).$$

Its integral is

$$\int_{\mathbf{R}^2} \phi(t,s)\,d(ts) = \sum_{i=1}^{n} c_i \int_{\mathbf{R}^2} f_i(t,s)\,d(ts)$$

where $\int_{\mathbf{R}^2} f_i(t,s)\,d(ts)$ is the area of the ith rectangle.

Proceeding as above we define an integral on $\mathbf{R}^2$. If E is a measurable set in $\mathbf{R}^2$, then we can define an integral over E by considering, for any f, the function f_E defined by

$$f_E(t,s) = \begin{cases} f(t,s), & (t,s) \in E, \\ 0, & (t,s) \notin E, \end{cases}$$

and setting

$$\int_E f(t,s)\,d(ts) = \int_{\mathbf{R}^2} f_E(t,s)\,d(ts).$$

The fundamental new theorem of interest for multiple integrals is the following.

THEOREM 2.6.4. (*Fubini's theorem*) *Let $f(t,s)$ be in S_2 on a set $E = E_t \otimes E_s$. Then*

1. *For almost all s in E_s, $f(t,s)$ is in S_2 on E_t;*
2. *The function $F(s) = \int_{E_t} f(t,s)\,dt$ is in S_2 on E_s;*
3. $\int_E f(t,s)\,d(ts) = \int_{E_s} F(s)\,ds = \int_{E_s}\left[\int_{E_t} f(t,s)\,dt\right] ds.$

The proof follows directly from the easily established validity of the result for step functions $\phi(t,s)$.

The result is also valid when $t \in \mathbf{R}^p$, $s \in \mathbf{R}^q$ with no formal change in the statement of the theorem.

A partial converse to Fubini's theorem is the following.

THEOREM 2.6.5. *If $f(t,s)$ is a measurable function on $E = E_t \otimes E_s$, where E, E_t, E_s are measurable sets, such that*

$$\int_{E_s} |f(t,s)|\,ds$$

exists almost everywhere on E_t and is itself integrable on E_t, that is

$$\int_{E_t}\left[\int_{E_s} |f(t,s)|\,ds\right] dt$$

exists, then $f(t,s)$ is integrable on E and

$$\int_E f(t,s)\,d(ts) = \int_{E_t}\left[\int_{E_s} f(t,s)\,ds\right]dt$$
$$= \int_{E_s}\left[\int_{E_t} f(t,s)\,dt\right]ds.$$

Remark 5. (Integral of functions to other ranges) The integral for complex valued functions $f(t) = u(t) + i\,v(t)$ is defined in an obvious fashion as

$$\int_E f(t)\,dt = \int_E u(t)\,dt + i\int_E v(t)\,dt.$$

If the range of f is in $\mathbf{E}^n$, so f is a vector valued function, the integral is defined by means of the integrals of the components. These definitions hold for t in $\mathbf{E}^p$.

Remark 6. How far have we come? Two theorems quoted next give an interesting view of measurability and convergence almost everywhere of measurable functions.

THEOREM 2.6.6. (*Lusin's theorem*) *Let f be a measurable function defined on a measurable set E, with $|f(t)|(<)\infty$, $t \in E$. Then for any $\epsilon > 0$ there exists a continuous function g on E such that*

$$\mu\{t | f(t) \neq g(t)\} < \epsilon.$$

THEOREM 2.6.7. (*Egoroff's theorem*) *Let E be a measurable set of finite measure. Let $\{f^n\}$ be a sequence of measurable functions that converge almost everywhere on E. Then for every $\epsilon > 0$ there is a measurable subset of E, E_ϵ such that $\mu(E - E_\epsilon) < \epsilon$, and the sequence f^n converges uniformly on E_ϵ.*

EXERCISES TO SECTION 2.6

1. Show that the function t^{-1} is measurable but not integrable on $[0,1]$.
2. Show that a constant nonzero function is measurable but not integrable on $[0,\infty]$ and on $(-\infty,\infty)$.
3. Show that if f,g are measurable, then so are $\alpha f + \beta g, f \circ g$, $\max(f,g)$, $\min(f,g)$ and even $1/f$ if $f(\neq)0$.
4. Show that if a sequence of measurable functions $\{f^n\}$ converges almost everywhere to a function f, then f is measurable. (Hint: given

$h \in S_2$, $h > 0$ for all t, then form the related sequence

$$g^n = \frac{hf^n}{h + |f^n|}$$

which is both measurable and dominated by h, so integrable. Passage to the limit yields an integrable function g and

$$f = \frac{hg}{h - |g|}.)$$

5. Show that the measure of the irrationals on [0, 1] is 1.
   (Hint: $\mathbf{\Phi}_E(=)1,\ 0 \leqslant t \leqslant 1$.)
6. Show that the rationals on [0, 1] are measurable with measure zero.
7. Show that if E_1, E_2 are measurable, then so are $E_1 \cup E_2, E_1 \cap E_2$.
8. Show that $E_1 \subset E_2$ implies $\mu(E_1) \leqslant \mu(E_2)$ for measurable E_1, E_2.
9. In Lebesgue's theory a function f is said to be measurable if for every c the set $\{t | f(t) < c\}$ is measurable. By considering the function

$$f_c(t) = \begin{cases} f(t), & f(t) \geqslant c, \\ c, & f(t) < c, \end{cases}$$

   and the limit as $d \downarrow 0$ of

$$\frac{f_c(t) - f_{c-d}(t)}{d}$$

   show that this yields the characteristic function of $\{t | f(t) < c\}$, which is measurable as a consequence of Exercise 2.6.4.
10. Verify Theorem 2.6.2 when f, g are step functions and Theorem 2.6.3 when f is a step function. (For outlines of the rest of the proofs see, for example, Riesz-Nagy [1], 54–56.)
11. Let f, g be in S_2 on a measurable set E. For some $M > \infty$ let $|f(t)|(\leqslant)M$. Show that $fg \in S_2$ on E and

$$\left|\int_E [f(t)g(t)]\,dt\right| \leqslant M \int_E |g(t)|\,dt.$$

12. Let $E' \subset E$, both measurable. Then $f \in S_2$ on E implies

$$\int_{E'} |f(t)|\,dt \leqslant \int_E |f(t)|\,dt.$$

   (Hint: $|\mathbf{\Phi}_{E'}| \leqslant 1$.)

13. Starting with the definition of step functions appropriate to two dimensions define S_2 on $(-\infty < t, s < \infty)$. Prove the term by term integration theorems for double integrals. (These are all immediate translations to $\mathbf{R}^2$ of the proofs given in the text for $\mathbf{R}^1$.) Prove Fubini's theorem.

3

COMPLETE SPACES OF INTEGRABLE FUNCTIONS

3.0 INTRODUCTION

We now have at our command more powerful theorems concerning the interchange of limit and integral. We will see below that these are sufficient to form complete metric spaces of integrable functions that resolve the anomalies discussed in Section 1.6.

We will work with complex valued functions whose domain of definition is a given measurable set E in a finite dimensional Euclidean space of some fixed dimension, not otherwise specified. Two functions that differ only on a set of measure zero will be considered as the same function. We will use the integration theory developed in Chapter 2 as if it had been developed for these functions. It will be clear that the domain of definition of the functions plays little or no role in the arguments. We will only discuss the case of the Lebesgue integral, remarking at this point that we could as easily utilize the Lebesgue-Stieltjes integral.

3.1 THE SPACES $L_p(E)$

Metric spaces can be formed from the measurable functions on E as follows:

DEFINITION 3.1.1. Choose p, $1 \leqslant p < \infty$. The set of all measurable functions f on E such that $|f|^p$ is integrable:

$$\int_E |f(t)|^p \, dt < \infty,$$

with metric

$$\rho(f,g) = \left[\int_E |f(t) - g(t)|^p \, dt \right]^{1/p},$$

is called the space $L_p(E)$.

In the present discussion, E is held fixed and we will usually write only L_p, with E understood. In order to justify and use the definition, we need to extend the Schwarz, Hölder, and Minkowski inequalities.

LEMMA 3.1.1. *Let* $1<p<\infty$ and $1/p+1/q=1$. *Let* $f\in L_p, g\in L_q$. *Then* $f\bar{g}\in L_1$, *and*

$$\left|\int_E f(t)\bar{g}(t)\,dt\right| \leqslant \left[\int_E |f(t)|^p\,dt\right]^{1/p}\left[\int_E |g(t)|^q\,dt\right]^{1/q}$$

PROOF. Since $|f|^p$ and $|g|^q$ are in S_2, there exist sequences $\{\phi^n\}$ $\{\psi^n\}$ such that $\phi^n(\longrightarrow)|f|^p, n\to\infty$, $\psi^n(\longrightarrow)|g|^q, n\to\infty$,

$$\int_E |\phi^n(t)-|f(t)|^p|\,dt\to, \qquad n\to\infty,$$

and

$$\int_E |\psi^n(t)-|g(t)|^q|\,dt\to 0, \qquad n\to\infty.$$

Now $|f|^p(\geqslant)0, |g|^q(\geqslant)0$, so the modified sequences $\{\hat{\phi}^n\}$ $\{\hat{\psi}^n\}$, defined by

$$0\leqslant\hat{\phi}^n(t)=\max[0,\phi^n(t)]$$

and

$$0\leqslant\hat{\psi}^n(t)=\max[0,\psi^n(t)]$$

satisfy the same relations. Then

$$(\hat{\phi}^n)^{1/p}(\hat{\psi}^n)^{1/q}(\longrightarrow)|f\bar{g}|, \qquad n\to\infty.$$

From Exercise 1.1.8 we see that

$$|f\bar{g}|(\leqslant)(1/p)|f|^p+(1/q)|g|^q.$$

The right side of this inequality is in S_2, while $f\bar{g}$ is measurable, so from Corollary 1 to Theorem 2.5.3, Lebesgue's theorem on dominated convergence, we conclude that $f\bar{g}\in S_2$. If we can set

$$A=\frac{|f|}{\left(\int_E |f(t)|^p\,dt\right)^{1/p}}$$

and

$$B=\frac{|g|}{\left(\int_E |g(t)|^q\,dt\right)^{1/q}}$$

in Exercise 1.1.8 and integrate both sides we find

$$\frac{\int_E |f(t)\bar{g}(t)|\,dt}{\left[\int_E |f(t)|^p\,dt\right]^{1/p}\left[\int_E |g(t)|^q\,dt\right]^{1/q}} \leqslant \frac{1}{p}+\frac{1}{q}=1,$$

which establishes the Hölder inequality (and for $p=2$ the Schwarz inequality.)

If either $f(=)0$ or $g(=)0$, the inequality is trivial. ■

LEMMA 3.1.2. *If f,g are in L_p, $1 \leqslant p < \infty$, then $f+g \in L_p$ and*

$$\left(\int_E |f(t)+g(t)|^p\,dt\right)^{1/p} \leqslant \left(\int_E |f(t)|^p\,dt\right)^{1/p} + \left(\int_E |g(t)|^p\,dt\right)^{1/p}$$

PROOF. $|f+g|^p$ is measurable and $|f+g|^p \leqslant (|f|+|g|)^p \leqslant [2\max(|f|,|g|)]^p$. But the set $F=\{t\,|\,|f|-|g|>0\}$ is measurable, as is $E-F=\{t\,|\,|f|-|g|\leqslant 0\}$. Let $m(t)=\max(|f(t)|,|g(t)|)$. Then on F, $m^p=|f|^p$ so m^p is integrable there, while on $E-F$, $m^p=|g|^p$ and is integrable there. Then m^p is integrable on E, and so $|f+g|^p$ is integrable there.

If $p=1$ there is nothing left to prove. If $1<p<\infty$, then

$$\begin{aligned}|f+g|^p &\leqslant |f||f+g|^{p-1}+|g||f+g|^{p-1}\\ &\leqslant (|f|+|g|)|f+g|^{p-1}.\end{aligned}$$

Now $|f+g|^{p-1} \in L_q$ where $q=p/(p-1)$, while both $|f|,|g| \in L_q$. Using the Hölder inequality shows that

$$\int_E |f(t)|\,|f(t)+g(t)|^{p-1}\,dt \leqslant \left(\int_E |f(t)|^p\,dt\right)^{1/p}\left(\int_E |f(t)+g(t)|^p\,dt\right)^{1/q}.$$

A similar inequality for g can be derived and then

$$\int_E |f(t)+g(t)|^p\,dt \leqslant \left[\left(\int_E |f(t)|^p\,dt\right)^{1/p} + \left(\int_E |g(t)|^p\,dt\right)^{1/p}\right] \times \left(\int_E |f(t)+g(t)|^p\,dt\right)^{1/q}.$$

Since $1-1/q=1/p$, this proves the Minkowski inequality. ■

The second lemma shows that $\rho(f,g)$ is well defined for any two functions $f,g \in L_p$, on replacing g by $-g$. If we replace f by $f-h$, g by $h-g$ for any $h \in L_p$ the triangle inequality is established. Corollary 2 to Theorem 2.5.2 shows that ρ is a metric.

The following analysis will lead to a justification of the definition of a space to be known as L_∞. If the set E is of infinite measure, first truncate it in each Euclidean coordinate to the range $[-n,n]$. The truncated set of finite measure will also be denoted by E for convenience. Choose a p, $1 \leqslant p < \infty$.

DEFINITION 3.1.2. For each $f \in L_p$ we define a pth mean by

$$M_p(f;E) = \left[\frac{1}{\mu(E)} \int_E |f(t)|^p\,dt\right]^{1/p}.$$

LEMMA 3.1.3. *If $f \in L_p$, then $f \in L_r, 1 \leqslant r \leqslant p$ and $M_r(f;E)$ is monotone increasing in r.*

PROOF. If $1 \leqslant r \leqslant p$, then $|f|^r(\leqslant)\max(1,|f|^p)$ which is integrable on sets of finite measure, while $|f|^r$ is clearly also measurable, so integrable. In the Hölder inequality put $|f|^r$ for f, 1 for g, and note that for $1 \leqslant r \leqslant s \leqslant p$

$$\int_E |f(t)|^r\,dt \leqslant \left(\int_E [|f(t)|^r]^{s/r}\,dt\right)^{r/k} \left(\int_E 1\,dt\right)^{(1-r/s)}$$
$$\leqslant \mu(E)^{(1-r/s)} \left(\int_E |f(t)|^s\,dt\right)^{r/s}$$

or $M_r(f;E) \leqslant M_s(f;E)$, with strict inequality unless $f(=)1$. ■

LEMMA 3.1.4. *If f is measurable and $|f|(\leqslant)M$ for some constant $M < \infty$, then $f \in L_p$ for all $p, 1 \leqslant p < \infty$ and $M_p(f;E) \leqslant M$.*

PROOF. Since a constant M^p is integrable whenever E is of finite measure, $|f|^p$ is always integrable (why?) while

$$\int_E |f(t)|^p \, dt \leqslant \int_E M^p \, dt = M^p \mu(E). \quad \blacksquare$$

We conclude from these lemmas the central fact:

COROLLARY 1. *If f is measurable* and $|f|(\leqslant)M$, *then*

$$\lim_{p\to\infty} M_p(f;E) \leqslant M.$$

DEFINITION 3.1.3. A measurable function f corresponding to which there exists a constant $\mathfrak{F}$ such that

$$\mu\{t|f(t) > \mathfrak{F}\} = 0$$

while for any $\epsilon > 0$

$$\mu\{t|f(t) > \mathfrak{F} - \epsilon\} > 0$$

is said to have an *essential upper bound* $\mathfrak{F}$, and we write $\mathfrak{F} = \operatorname{ess\,sup}_E f$. The E will be omitted if the context is clear.

The corollary can now be restated in the form:

$$\lim_{p\to\infty} M_p(f;E) \leqslant \operatorname*{ess\,sup}_E |f|$$

We will now show that this is in fact an equality. To see this choose $A < \operatorname{ess\,sup}|f|$ and let $\mathbf{\Phi}_A$ be the characteristic function of the set $\{t||f(t)| > A\}$. Since $\mathbf{\Phi}_A^p = \mathbf{\Phi}_A$ for all p,

$$\begin{aligned}
M_p(A\mathbf{\Phi}_A;E) &= AM_p(\mathbf{\Phi}_A;E) \\
&= A\frac{\mu\{t||f| > A\}}{\mu(E)^{1/p}} \\
&\leqslant M_p(f;E) \\
&\leqslant M_p(\text{ess sup}|f|\mathbf{\Phi}_E;E) \\
&= \text{ess sup}|f|\,\mu(E)^{1/p}.
\end{aligned}$$

Letting $p \to \infty$ provides the inequality

$$A \leqslant \lim_{p\to\infty} M_p(f;E) \leqslant \text{ess sup}|f|.$$

Since this is true for all $A < \text{ess sup}|f|$, we must in fact have the following:

COROLLARY 2. *If f is measurable and $|f|(\leqslant)M < \infty$ then*

$$\lim_{p\to\infty} M_p(f;E) = \operatorname*{ess\,sup}_E |f|.$$

We can now remove the restriction that E be of finite measure. If E_n denotes the restriction of an E of infinite measure as above, we can write this limit as

$$\lim_{p\to\infty}\left[\frac{1}{\mu(E_n)}\int_{E_n}|f(t)|^p\,dt\right]^{1/p} = \operatorname*{ess\,sup}_{E_n}|f|.$$

Then

$$\lim_{n\to\infty}\lim_{p\to\infty}\left[\frac{1}{\mu(E_n)}\int_{E_n}|f(t)|^p\,dt\right]^{1/p} = \operatorname*{ess\,sup}_{E}|f|.$$

DEFINITION 3.1.4. The set of all measurable functions f such that $\text{ess sup}_E |f| < \infty$ is called $L_\infty(E)$. The metric is taken as

$$\rho(f,g) = \operatorname*{ess\,sup}_{t\in E}|f(t)-g(t)|.$$

We note here some facts about the L_p spaces which we have proven:

1. *If E is of finite measure, then $L_p(E) \subset L_q(E), 1 \leqslant p \leqslant q \leqslant \infty$.*
2. *The functions $f \in L_\infty(E)$ are in $L_p(E')$ for all $p, 1 \leqslant p \leqslant \infty$, and for all sets E' of finite measure that are contained in E.*

EXERCISES TO SECTION 3.1

1. Let $E = F \cup G$, all measurable sets, with $F \cap G = \varnothing$, and let f be a function defined on E which is integrable on F and G. Show that f is integrable on E and

$$\int_E f(t)\,dt = \int_F f(t)\,dt + \int_G g(t)\,dt.$$

(Hint: $f = f\Phi_F + f\Phi_G$.)

2. When does equality hold in the Hölder and Minkowski inequalities?
3. If f is measurable, then so is f^α for all $\alpha \geqslant 0$, and if $\mu\{t|f=0\}=0$, also for all $\alpha<0$.
4. If E is of finite measure then any constant function is integrable on E. (Hint: write the constant as $c\Phi_E$.)
5. Extend the Minkowski inequality to $p=\infty$.
6. Show that the function ρ for L_∞ is a metric.
7. Show that the Hölder inequality is valid for $p=1$, $q=\infty$ in the form

$$\left|\int_E f(t)\bar{g}(t)\,dt\right| \leqslant \operatorname{ess\,sup}_E|g| \int_E |f(t)|\,dt.$$

8. Let $k(s,t)\in L_2(E)$ where E is $\mathbf{R}^2$, the two dimensional Euclidean space. Prove that for any $f\in L_2\ (-\infty<t<\infty)$

$$(Kf)(s)=\int_{-\infty}^{\infty} k(s,t)f(t)\,dt$$

is an element of $L_2\ (-\infty<s<\infty)$.
9. Choose $p, 0<p<\infty$ and let X_p be the space of all infinite sequences of complex numbers $\alpha=(\alpha_1,\alpha_2,\alpha_3,\ldots)$ such that

$$\sum_{i=1}^{\infty}|\alpha_i|^p<\infty.$$

Show that

$$\left(\sum_{i=1}^{n}|\alpha_i|^{p'}\right)^{1/p'} \leqslant \left(\sum_{i=1}^{n}|\alpha_i|^{p}\right)^{1/p}$$

for $0<p<p'$, so $X_p\subset X_{p'}, 0<p<p'$, and $l_p\subset l_{p'}, 1\leqslant p\leqslant p'<\infty$. (Hint: Hardy, Littlewood, Polya [1], 4.). Then show that

$$\lim_{p\to\infty}\left(\sum_{i=1}^{n}|\alpha_i|^{p}\right)^{1/p}$$

exists for all n and equals $\max_{1\leqslant i\leqslant n}|\alpha_i|$, so that

$$\lim_{n\to\infty}\lim_{p\to\infty}\left(\sum_{i=1}^{n}|\alpha_i|^{p}\right)^{1/p} = \sup_i|\alpha_i|.$$

We define $l_\infty=\{\alpha|\sup_i|\alpha_i|<\infty\}$. (Hint:

$$\left(\sum_{i=1}^{n}|\alpha_i|^p\right)^{1/p}=\gamma\left(\sum_{i=1}^{n}\frac{|\alpha_i|^p}{\gamma^p}\right)^{1/p}.$$

Set $\gamma=\max_{1\leqslant i\leqslant n}|\alpha_i|$, $\beta_i=\alpha_i/\gamma$. Then

$$\left(\sum_{i=1}^{n}|\beta_i|^p\right)^{1/p}\leqslant n^{1/p}\to 1, \qquad p\to\infty.$$

But

$$\left(\sum_{i=1}^{n}|\beta_i|^p\right)^{1/p}\geqslant 1.)$$

3.2 THE L_p SPACES ARE COMPLETE

If $\rho_p(f,g)$ denotes the metric in the space $L_p(E), 1\leqslant p\leqslant\infty$, we wish to show that to every (Cauchy) sequence in $L_p, \{f^n\}$, with $\lim_{n,m\to\infty}\rho_p(f^n,f^n)=0$, there corresponds an (essentially) unique function $f\in L_p(E)$ such that $\rho_p(f^n,f)\to 0, n\to\infty$.

DEFINITION 3.2.1. If a sequence $\{f^n\}$ and a function f in $L_p(E)$ satisfy $\rho_p(f^n,f)\to 0, n\to\infty$ we say that f^n *converges to f in the mean* $L_p(E)$.

The fundamental theorem here (for $p=2$ known as the Riesz-Fischer theorem) is the following:

THEOREM 3.2.1. *Let* $1\leqslant p\leqslant\infty$ *and* $\{f^n\}\in L_p(E)$. *Necessary and sufficient to the existence of an* $f\in L_p(E)$ *such that* f^n *converges to f in the mean* $L_p(E)$ *is*

$$\rho_p(f^n,f^m)\to 0, \qquad n,m\to\infty.$$

PROOF. Necessity follows directly from the triangle inequality:

$$\rho_p(f^n,f^m)\leqslant\rho_p(f^n,f)+\rho_p(f,f^m).$$

For the proof of sufficiency, first let $1<p<\infty$ and let E' denote any measurable subset of E of finite measure. On E', $L_r\subset L_p, 1\leqslant r\leqslant p$ so by the

Hölder inequality with $1/p+1/q=1$,

$$\int_{E'} |f^n(t)-f^m(t)||1|dt \leqslant \rho_p(f^n,f^m)\left(\int_{E'} 1\,dt\right)^{1/q}. \tag{1}$$

Now we choose a subsequence of $\{f^n\},\{f^{m_i}\}$ such that $\rho_p(f^n,f^{m_i})<2^{-i}$, $n \geqslant m_i$. Then

$$\sum_{k=1}^{\infty} \int_{E'} |f^{m_{k+1}}(t)-f^{m_k}(t)|dt \leqslant \mu(E')^{1/q} \sum_{k=1}^{\infty} 2^{-k}. \tag{2}$$

Corollary 1 to Theorem 2.5.2. (closure of S_2) asserts that

$$f^{m_1}(t)+\sum_{k=1}^{\infty}[f^{m_{k+1}}(t)-f^{m_k}(t)] = \lim_{k\to\infty} f^{m_k}(t)$$

converges almost everywhere to a function $f \in L_1(E')$. Since E' is arbitrary, f is defined on E. Also by the triangle inequality

$$\begin{aligned}\rho_p(f^{m_k},0) &\leqslant \rho_p(f^{m_1},0)+\rho_p(f^{m_k},f^{m_1})\\ &\leqslant \rho_p(f^{m_1},0)+\tfrac{1}{2},\end{aligned}$$

so that there is a constant $M<\infty$ with

$$\int_E |f^{m_k}(t)|^p\,dt \leqslant M, \qquad k=1,2,3,\ldots.$$

Now $|f^{m_k}(t)|^p(\longrightarrow)|f(t)|^p, k\to\infty$, and Fatou's lemma implies

$$\int_E \liminf_{k\to\infty} |f^{m_k}(t)|^p\,dt \leqslant M,$$

so $f \in L_p(E)$ and

$$\int_E |f(t)|^p\,dt \leqslant M.$$

Fix m_r, let $n \geqslant m_r$; the triangle inequality yields

$$\begin{aligned}\rho_p(f^{m_k},f^n) &\leqslant \rho_p(f^{m_k},f^{m_r})+\rho_p(f^{m_r},f^n)\\ &\leqslant 2^{-r+1}.\end{aligned}$$

Then Fatou's lemma provides

$$\int_E \liminf_{k\to\infty} |f^{m_k}(t)-f^n(t)|^p\,dt \leqslant 2^{-r+1}$$

or

$$\int_E |f(t)-f^n(t)|^p\,dt \leqslant 2^{-r+1}.$$

Thus for any r

$$\limsup_{n\to\infty} \int_E |f(t)-f^n(t)|^p\,dt \leqslant 2^{-r+1}$$

so $\rho_p(f,f^n)\to 0, n\to\infty$, and f^n converges to f in the mean $L_p(E)$.For the case $p=1, q=\infty$, we replace eq. 1 by

$$\int_{E'} |f^n(t)-f^m(t)|\,dt = \rho_1(f^n,f^m),$$

and eq. 2 by

$$\sum_{k=1}^{\infty} \int_{E'} |f^{m_{k+1}}(t)-f^{m_k}(t)|\,dt \leqslant \sum_{k=1}^{\infty} 2^{-k};$$

the rest of the proof remains valid.

In the case $p=\infty$, except for a set $S\subset E$ of measure zero, $|f^n(t)-f^m(t)| \leqslant \operatorname{ess\,sup}_E |f^n(t)-f^m(t)|$, for all n,m. (Why?). Hence on $E-S$, the sequence f^n converges to a function f. Since

$$|f^n(t)|(\leqslant)|f^n(t)-f^N(t)|+|f^N(t)|,$$

it follows that for any $\epsilon>0$ there exists $N=N(\epsilon)$ with $\rho_\infty(f^n,f^m)<\epsilon$, $n,m\geqslant N$ and

$$|f(t)|\leqslant \epsilon + |f^N(t)|$$

on $S-E$. Thus $f\in L_\infty(E-S)$. On S, set $f=0$. Then $|f^n(t)-f(t)|(\leqslant)\epsilon$ for $n\geqslant N$. Hence $\rho_\infty(f^n,f)\to 0, n\to\infty$.

In all cases, the uniqueness almost everywhere of the limit function follows from the triangle inequality. ■

COROLLARY 1. *If* $\{f^n\} \in L_p(E)$, $1 \leqslant p \leqslant \infty$, *is a Cauchy sequence, then there is a limit function* $f \subset L_p(E)$ *and a subsequence* $\{f^{n_k}\}$ *with* $f^{n_k}(\to)f$, $k \to \infty$.

In the case $p=2$, the space L_2 is the complete metric space in which Fourier analysis may be developed naturally.

EXERCISES TO SECTION 3.2

1. Show that if $f \in L_p(-\pi, \pi)$, $1 \leqslant p \leqslant \infty$, then

$$\int_{-\pi}^{\pi} f(t) \cos nt \, dt$$

and

$$\int_{-\pi}^{\pi} f(t) \sin nt \, dt$$

exist and are finite for $n=0,1,2,3,\ldots$. Another statement of this is the existence of

$$\int_{-\pi}^{\pi} f(t) e^{int} \, dt, \qquad n=0, \pm 1, \pm 2, \ldots.$$

2. If $f(t) \in L_1(-\infty, \infty)$ then show that

$$\int_{-\infty}^{\infty} f(t) e^{i\lambda t} \, dt$$

exists for all real λ. Does this hold for $f \in L_2(-\infty, \infty)$?

4

SEPARABILITY AND COMPACTNESS

4.0 INTRODUCTION

Some topological notions are needed to describe different types of metric spaces, notions that generalize the ideas from the theory of point sets to these more abstract spaces. We use the word topology very loosely. To a certain extent we mean only the geometric notions familiar from point sets on the real line and in $\mathbf{E}^n$. However at certain times we will speak of topologies for a space. In the latter case we shall be dealing only with the ideas connected with various types of convergence, such as uniform convergence or convergence in the mean L_p. No deeper meaning is attached to the word topology used here.

4.1 SEPARABLE SPACES

Open and closed sets in a metric space are determined by the following:

DEFINITION 4.1.1. A set S in a metric space X is *open* if for any $s \in S$ there is an open ball $B(s,r) \subset S$. By the *closure* $\bar{S}$ of a set S (not necessarily open) in X is meant the set of points $x \in X$ for which there exist Cauchy sequences $\{s^i\} \in S$ with $\rho(s^i, x) \to 0, i \to \infty$. A set S is *closed* if $S = \bar{S}$. By $\operatorname{int} S$, the *interior of* S is meant the largest open set contained in S. By the *boundary of* S, ∂S, is meant the intersection of $\bar{S}$ and $\overline{X-S}$.

THEOREM 4.1.1. *S is an open set if and only if $X-S$ is a closed set.*

PROOF. Suppose $X-S$ is closed. Suppose S is not open. Then there exists $p \in S$ such that $B(p,r)$ contains a point not in S, say $x^r \in X-S$, for all sufficiently small r. Then $\rho(x^r, x^{r'}) < 2\max(r,r') \to 0$, as $r, r' \to 0$. Hence $\{x^{r_k}\}$ is a Cauchy sequence for $r_k = 1/k$, k sufficiently large, and $\rho(p, x^{r_k}) \leqslant 1/k \to 0$, $k \to \infty$. Then $p \in X-S$, a contradiction. Hence S is open.

Conversely if S is open, let $\{x^k\} \in X - S$ be a Cauchy sequence converging to a point p. If $p \notin X - S, p \in S$, so $B(p,\epsilon) \subset S$ for some $\epsilon > 0$. Then $\rho(p,x) \geqslant \epsilon$ for all $x \in X - S$, which contradicts $\rho(p,x^k) \to 0$, $k \to \infty$. Hence $X - S = \overline{X - S}$. ■

DEFINITION 4.1.2. The set A is said to be *dense* in the set B if $A \subset B$ and $\bar{A} \supset B$. The set is said to be *nowhere dense* if $\operatorname{int} \bar{A} = \varnothing$.

DEFINITION 4.1.3. If there exists, in a metric space X, a set S which is denumerable and dense in X, then X is called *separable*.

The interest of separable spaces, in addition to the fact that many of the spaces we have encountered are separable, is the reduction of the complexity of much of analysis in them. That is, a property or a theorem whose arguments are elements of a separable space, and which holds under small perturbations of the arguments, need only be proven for some denumerable set whose closure is the whole space. Some examples of separable spaces are the following:

Example 1. The real line $\mathbf{R}$ is separable. The metric is $\rho(t',t'') = |t' - t''|$, and a dense denumerable set is the set of all rational numbers (Exercise 4.1.3).

Example 2. The space $C[a,b]$ of Example 13 of Section 1.1 is a separable space for any $a, b, -\infty < a < b < \infty$. This follows in three steps. First, by a change of independent variable we can always assume $a = 0, b = 1$. Then for any $x \in C[0,1]$ the Weirstrass approximation theorem asserts the existence, for any $\epsilon > 0$, of a polynomial $p(t) = \sum_{i=0}^{n} \alpha_i t^i$ such that $\rho(x,p) < \epsilon/2$; here $n = n(\epsilon, x)$. Lastly, if r_i denotes a complex number with rational real and imaginary parts such that $|\alpha_i - r_i| < \epsilon/2(n+1)$, $i = 0, \ldots, n$, then

$$\left| \sum_{i=0}^{n} \alpha_i t^i - \sum_{i=0}^{n} r_i t^i \right| < \frac{\epsilon}{2}, \qquad 0 \leqslant t \leqslant 1,$$

so that

$$\rho\left(x, \sum_{i=0}^{n} r_i t^i \right) < \epsilon.$$

Thus the polynomials with (complex) rational coefficients are dense in $C[0,1]$. But these are a denumerable set (Exercise 4.1.4).

Example 3. The real space l_p, $1 \leqslant p < \infty$ is separable. We define a dense denumerable set S as follows:

$$S = \{ r | r = (r_1, r_2, r_3, \ldots),\ r_i \text{ rational},\ r_i = 0, i \geqslant n = n(r) \}.$$

Then, by Exercise 4.1.4, for each fixed n the corresponding r's are denumerable, and by Exercise 4.1.3, S is denumerable. Given $x \in l_p$, $x = (\alpha_1, \alpha_2, \alpha_3, \ldots)$; for any $\epsilon > 0$ find m such that

$$\sum_{i=m+1}^{\infty} |\alpha_i|^p < \frac{\epsilon^p}{2^p}$$

and rational numbers r_i such that

$$|\alpha_i - r_i| < \frac{\epsilon^p}{(2m)^p}, \qquad i = 1, \ldots, m.$$

Then, if $r = (r_1, r_2, \ldots, r_m, 0, 0, 0, \ldots)$, $\rho(x, r) < \epsilon$. A similar argument works for the complex l_p space, $1 \leqslant p < \infty$.

Example 4. The space L_p, $1 \leqslant p < \infty$ is separable. To see this start with the step functions $\phi = \Sigma_k c_k f_k$, which are clearly dense in L_p. Here f_k is the characteristic function of some "rectangle" in $\mathbf{E}^n$ and c_k is a scalar. The support "rectangle" of f_k can be approximated arbitrarily closely by one with only rational numbers in the definition of the boundary; the c_k can also be arbitrarily closely approximated by rational numbers. Then we arrive at the set of "rational step functions" which are dense in $S_0 \subset L_p$ in the sense of uniform convergence, hence in the L_p metric. It now becomes an involved but straightforward matter to show that the "rational step functions" are denumerable.

The following example shows that some spaces are not separable.

Example 5. The space m of all infinite sequences $a = (\alpha_1, \alpha_2, \alpha_3, \ldots)$ with metric $\rho(a, b) = \sup_i |\alpha_i - \beta_i|$ is *not* separable. To see this, consider the subset $I \subset m$ of sequences of zeros and ones. I is a set in which $x \neq y$ implies $\rho(x, y) = 1$. On the other hand, suppose $S \subset m$ were a dense denumerable set $S = (s^1, s^2, s^3, \ldots)$. We can cover each s^i by the ball $B(s^i, \frac{1}{3})$ and $m \subset \cup_{i=1}^{\infty} B(s^i, \frac{1}{3})$. Then, since I is not denumerable (Exercise 4.1.5), two elements of I, x, y, must fall in the same ball. Hence we must have both $\rho(x, y) < \frac{2}{3}$ and $\rho(x, y) = 1$, which is impossible. By a quite similar argument we can show that the space L_∞ of all bounded measurable functions is not separable.

EXERCISES TO SECTION 4.1

1. Let S, S', S'' be sets in a metric space X. Show:
   (a) If S is open, then $S = \operatorname{int} S$.
   (b) If $S' \subset S''$, then $\bar{S}' \subset \bar{S}''$. If S' is strictly smaller than S'', must $\bar{S}'$ be strictly smaller than $\bar{S}''$?
   (c) $\overline{S' \cup S''} = \overline{S'} \cup \overline{S''}$
   (d) $\overline{S' \cap S''} \subset \bar{S}' \cap \bar{S}''$. Show by a counter example that equality need not hold.
   (e) $\partial S = \bar{S} - \operatorname{int} S = \overline{X - S} - \operatorname{int}(X - S)$
2. Let $\{S^n\}$ be a collection of closed sets in a metric space X. Then $\cap_{n=1}^{\infty} S^n$ is closed and $\cup_{n=1}^{k} S^n$ is closed for any $k < \infty$. Let $\{O^n\}$ be a collection of open sets in X; then $\cup_{n=1}^{\infty} O^n$ and $\cap_{n=1}^{k} O^n$ are open, $k < \infty$.
3. The rational numbers on the real line are denumerable. (Hint: show instead that if $E^n, n = 1, 2, 3, \ldots$ is a denumerable set, $E^n = \{x_1^n, x_2^n, x_3^n, \ldots\}$, then $\cup_{n=1}^{\infty} E^n$ is denumerable, by using the following tableau

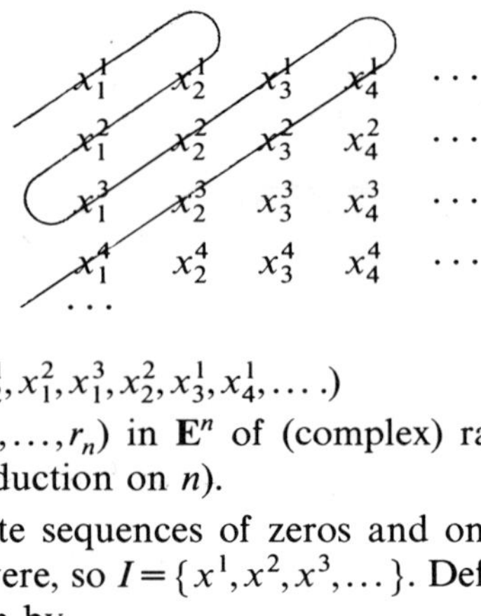

   and counting $x_1^1, x_2^1, x_1^2, x_1^3, x_2^2, x_3^1, x_4^1, \ldots$.)
4. The n-tuples $(r_1, r_2, \ldots, r_n)$ in $\mathbf{E}^n$ of (complex) rational r_i are denumerable. (Hint: use induction on n).
5. The set I of infinite sequences of zeros and ones is not denumerable. (Hint: suppose it were, so $I = \{x^1, x^2, x^3, \ldots\}$. Define a sequence y whose ith element is given by

$$y_i = \begin{cases} 0, & x_i^i = 1, \\ 1, & x_i^i = 0. \end{cases}$$
)

4.2 COMPACT SETS

On the real line a fundamental property of continuous functions is the fact that a continuous function on a closed bounded interval achieves its maximum (and minimum) there. In a more abstract setting, we wish to know the sets that can play the role of the closed bounded sets when we wish to assure that a continuous functional achieves its maximum (or minimum) there. This question is intrinsic to many problems of the existence of best approximations and of the existence of solutions to variational problems. The following definition of these sets is not obviously related to this idea.

DEFINITION 4.2.1. A subset S of a metric space X is said to be *compact* if every collection of open sets $\{O^\alpha\}$ such that $S \subset \cup_\alpha O^\alpha$ (an *open cover*) possesses a *finite subcover* (i.e. a finite collection $O^1, O^2, \ldots, O^N$ which is an open cover.)

This definition is often useful, but an alternate, sequential definition of compactness is contained in the following:

LEMMA 4.2.1. *A set S in a metric space X is compact if and only if every infinite subset of S contains a sequence that is convergent to a limit in S.*

PROOF. Suppose first that S is compact, but M is an infinite set with no subsequence convergent to a point of S. Then each $s \in S$ can be covered by an open ball $B(s,\epsilon)$ which contains at most one point of M (i.e. it may be that $s \in M \subset S$.) This provides an open cover for S, which has a finite subcover, $B^1, \ldots, B^N$. Since M can have at most one point in each such ball, M is finite, which is impossible.

Now suppose that every infinite subset $M \subset S$ contains a convergent subsequence with limit in S, but S is not compact. Choose $\epsilon > 0$; fix $s^1 \in S$. Choose s^2 such that $\rho(s^1, s^2) \geqslant \epsilon$. Choose s^3 so that $\rho(s^i, s^3) \geqslant \epsilon$, $i = 1,2$. Continue in this fashion to define s^n such that $\rho(s^i, s^n) \geqslant \epsilon$, $i = 1,2,\ldots,n-1$. This can be done for all n since S is not compact, and the balls of radius ϵ are an open cover which does not possess a finite subcover. The sequence $\{s^n\}$ must possess a subsequence $\{s^{n_k}\}$ that converges, so $\rho(s^{n_k}, s^{n_j}) < \epsilon$ if n_k, n_j are sufficiently large; this is a contradiction. ■

We can now extend the theorem on maximum values of continuous functions to metric spaces.

THEOREM 4.2.1. *Let S be a compact subset of a metric space X and $f(x)$ a continuous functional on S. Then*

1. *$f(x)$ is bounded on S;*
2. *$f(x)$ achieves its maximum and minimum on S.*

PROOF. Suppose f is not bounded on S. Then for any n there is an $s^n \in S$ with $f(s^n) \geqslant n$. Since the set S is compact, there is a subsequence such that $s^1, s^2, s^3, \ldots$ is convergent to some $s \in S$. Then since f is continuous, $f(s^n) - f(s^m) \to 0$, $n, m \to \infty$ and $f(s^n) \to f(s)$, $n \to \infty$. But a convergent sequence is bounded while $f(s^n) \to \infty$, $n \to \infty$. This is a contradiction; thus f is bounded from above. A similar proof shows that f is bounded from below.

Now let $M = \sup_{s \in S} f(s)$. For any $\epsilon > 0$ there is an $s \in S$ such that $M \geqslant f(s) > M - \epsilon$. If $\epsilon = 1/n$, call this s^n. Then $f(s^n) \to M$, $n \to \infty$. Now s^n has a convergent subsequence $\{s^{n_k}\}$ with $\rho(s^{n_k}, s) \to 0$, $n \to \infty$ and $s \in S$. By continuity, $f(s) = M$. A similar proof shows that f achieves its minimum. ■

Not all sets are compact. In $C[0,1]$, for instance, consider the functional

$$f(s) = \int_0^1 |s(t)|\, dt,$$

and the set S of all $s \in C[0,1]$ with $s(0) = 0, s(1) = 1$. If we note that for $s = t^n$

$$f(t^n) = \int_0^1 |t^n|\, dt = \frac{1}{n+1}$$

we see that

$$\inf_{s \in S} f(s) = 0.$$

But no $s \in S \subset C[0,1]$ has $f(s) \equiv 0$.

Continuity, demanded in Theorem 4.2.1, is more than is needed, and even more than is usually available to us. We make the following:

DEFINITION 4.2.2. Let X be a metric space. A functional $f(x)$ is said to be *upper semicontinuous at* x if $x^n \to x$, $n \to \infty$ implies

$$\limsup_{n \to \infty} f(x^n) \leqslant f(x).$$

It is said to be *lower semicontinuous* if

$$\liminf_{n \to \infty} f(x^n) \geqslant f(x).$$

COROLLARY 1. *An upper semicontinuous functional defined on a compact subset S of a metric space X is bounded above on S and achieves its maximum on S.*

COROLLARY 2. *A lower semicontinuous functional defined on a compact set S in a metric space X is bounded from below on S and achieves its minimum on S.*

The proofs are the observation that the inequality $\limsup_{n\to\infty} f(x^n) \leqslant f(x)$ is all that is needed in the proof of the preceding theorem to prove boundedness from above. In the second part of the proof we conclude

$$\limsup_{n\to\infty} f(s^{n_k}) = M \leqslant f(s)$$

while we know

$$f(s) \leqslant M.$$

for all s, whence $f(s) = M$. A similar argument works for the second corollary.

In the calculus of variations, the functionals involved are usually only lower semicontinuous, by virtue of Fatou's lemma, since, for a differentiable path $x = x(t)$, they are of the form

$$\int_a^b L(t, x, x')\, dt$$

for some integrable Lagrangian function L. Suppose that we deal with this functional

$$\mathcal{F}(x) = \int_0^1 L(t, x, x')\, dt.$$

Let $x \in S$, *a compact set of paths* in the plane with $x(0) = A, x(1) = B$. For any sequence $x^n \in S$ we have from Fatou's lemma

$$\int_0^1 \liminf_{n\to\infty} L(t, x^n, x^{n\prime})\, dt \leqslant \lim_{n\to\infty} \inf \int_0^1 L(t, x^n, x^{n\prime})\, dt.$$

If $x^n \to \bar{x}$, $n \to \infty$ then we have

$$\mathcal{F}(\bar{x}) \leqslant \liminf_{n\to\infty} \mathcal{F}(x^n).$$

Thus $\mathcal{F}$ is lower semicontinuous. If S is compact this implies that there is an M such that $\mathcal{F}(x) \geqslant M$, $x \in S$ and there is an $\bar{x} \in S$ such that

$$\mathcal{F}(\bar{x}) \leqslant \mathcal{F}(x), \qquad x \in S.$$

Frequently we construct a sequence $\{x^n\} \subset S$ by choosing a collection Q^n of subsets of S with $Q^n \subset Q^{n+1}$, on which we can actually minimize $\mathcal{F}$, and making $x^n \in Q^n$ satisfy

$$\mathcal{F}(x^n) = \inf_{x \in Q^n} \mathcal{F}(x).$$

Then $\{x^n\}$ satisfies $\mathcal{F}(x^n) \geqslant \mathcal{F}(x^{n+1})$ for all n; if it converges to some $\bar{x} \in S$, it is a minimizing sequence on $\cup_{n=1}^{\infty} Q^n$. (If S is separable, the construction of this sequence is particularly straightforward; for instance Q^n may be the trigonometric or ordinary polynomials of degree n).

If $\cup_n Q^n$ is dense in S we must actually get a local minimum of $\mathcal{F}$. If the functional $\mathcal{F}$ is convex (so has a unique local, hence global minimum) this is the global minimum.

For the determination of compact sets and the study of what are called compact operators we introduce a new definition.

DEFINITION 4.2.3. A set S in a metric space X is *precompact* if $\bar{S}$ is compact.

Then it is an obvious corollary of Lemma 4.2.1 that *S is precompact if and only if every infinite sequence in S contains a convergent subsequence.*

LEMMA 4.2.2. *Let S be a precompact subset of a metric space X and f a uniformly continuous functional on S. Then f is bounded on S.*

PROOF. Suppose f is not bounded. Then for any $n > 0$ there exists an $s^n \in S$, with $f(s^n) \geqslant n$. The set $\{s^n\}$ contains a convergent subsequence s^{n_i} with $f(s^{n_i}) > n_i$. On the other hand, by uniform continuity, for any $\epsilon > 0$ there is a $\delta(\epsilon)$ with

$$|f(s^*) - f(s^{**})| < \epsilon \text{ for } \rho(s^*, s^{**}) < \delta.$$

This makes the sequence $f(s^{n_i})$ bounded, which is an obvious contradiction. ∎

COROLLARY 1. *A precompact set is bounded.*

PROOF. $\rho(s,0)$ is a uniformly continuous functional on X, since $|\rho(s,0) - \rho(t,0)| \leqslant \rho(s,t) + \rho(0,0) = \rho(s,t)$. ∎

LEMMA 4.2.3. *A bounded set S in* $\mathbf{E}^n$ *is precompact.*

PROOF. Take any sequence in $S \subset \mathbf{E}^n$, $s^1, s^2, s^3, \ldots$. Since it is bounded, the components of s^i: $(\alpha_i^1, \ldots, \alpha_i^n)$ form bounded sequences on the real line.

Successive selection of subsequences yields a subsequence for which the components converge. Hence S is precompact. ■

This shows that necessary and sufficient for precompactness in $\mathbf{E}^n$ is boundedness, while necessary and sufficient for compactness in $\mathbf{E}^n$ is that the set be both closed and bounded (for in this case $S=\bar{S}$.)

LEMMA 4.2.4. *If S is a precompact set in a metric space X then $S\subset\overline{M}$ where M is denumerable.*

PROOF. Suppose S is not separable but precompact. Then for some n we could find an infinite sequence $s^1, s^2, s^3, \ldots$ in S with $\rho(s^i, s^j)\geqslant 1/n$, $i\neq j$. But for S precompact, the sequence $s^1, s^2, s^3, \ldots$ must have a convergent subsequence. This violates $\rho\geqslant 1/n$. Hence for each n we can find a finite set $s^1, \ldots, s^{N(n)}$ in S, and a finite cover of S by the balls $B(s^i, 1/n)$. The union of all these points is enumerable, and every point of S is a limit of points in this union. Hence S contains a dense denumerable set. ■

COROLLARY 1. *A precompact metric space is separable.*

We now collect some information about compact subsets of various function spaces.

A. Compact Subsets of $C[0,1]$

We use the metric

$$\rho(x,y)=\max_{0\leqslant t\leqslant 1}|x(t)-y(t)|.$$

A set $S\subset C[0,1]$ is *uniformly bounded* if

$$|\rho(x,0)|\leqslant M<\infty$$

for all $x\in S$. The *oscillation* of x, $\omega(x,\delta)$ is defined as

$$\omega(x,\delta)=\sup_{|t'-t''|\leqslant\delta}|x(t')-x(t'')|$$

for $t', t''\in[0,1]$.

If $\omega(x,\delta)<\epsilon$ for $\delta=\delta(\epsilon)$ for any ϵ and for all $x\in S$, then we call S an *equicontinuous* family. The fundamental theorem is the following:

THEOREM 4.2.2. (*Arzèla*) *$S\subset C[0,1]$ is precompact if and only if S is uniformly bounded and equicontinuous.*

(Proofs of this theorem are widely available. See for instance Franklin [1], 412–415, or Kolmogorov and Fomin [1], 54–55).

A sufficient condition for a family to be equicontinuous is that it be uniformly bounded and that there exist a positive constant K such that

$$|x(t') - x(t'')| \leqslant K|t' - t''|$$

for all $x \in S$ and all t', t'' in the domain of definition of the family. This *Lipschitz condition* is the usual form in which we encounter equicontinuity. Clearly $\omega(x, \delta) \leqslant K\delta < \epsilon$ if $\delta < \epsilon / K$.

B. Compact Sets in L_p.

For L_p-space, things are not so simple. For any $x \in S$ set

$$x_h(t) = \frac{1}{2h} \int_{t-h}^{t+h} x(s)\, ds$$

where we set $x = 0$ for $t \notin [0, 1]$. Let $\omega(x, h)$ be given by

$$\left(\int_0^1 |x(t) - x_h(t)|^p \, dt \right)^{1/p}.$$

THEOREM 4.2.3. (*Kolmogorov*) *A set* $S \subset L_p(0, 1)$ *is precompact if and only if*

1. $\left(\int_0^1 |x(t)|^p \, dt \right)^{1/p} \leqslant M < \infty$, *for all* $x \in S$ *and some* M.
2. *For any* $\epsilon > 0$, *there is a* $\delta = \delta(\epsilon)$ *such that* $\omega(x, h) < \epsilon$ *for* $|h| \leqslant \delta$ *for all* $x \in S$.

(For a proof see Lusternick and Sobelev, [1], 41–44). Sufficient for condition 2 is the following due to *F.* Riesz: for any $\epsilon > 0$ there is a $\delta = \delta(\epsilon)$ with

$$\left(\int_0^1 |x(t+h) - x(t)|^p \, dt \right)^{1/p} < \epsilon$$

for all $|h| \leqslant \delta(\epsilon)$.)

C. Compact Sets of Rectifiable Paths.

In the calculus of variations we want the following: Let Γ be any continuous curve in $\mathbf{E}^n$, parameterized by $\Gamma(t)$, $0 \leqslant t \leqslant 1$. Define a metric on such curves by

$$\rho(\Gamma_1, \Gamma_2) = \inf \max_t \rho_n(\Gamma_1(t), \Gamma_2(t))$$

where ρ_n is Euclidean distance and the infinum is taken over all different continuous parametrizations of the curves Γ_1, Γ_2. Then we have the following theorem.

THEOREM 4.2.4. (*Hilbert*) *The set of all rectifiable curves with fixed endpoints that lie in a bounded sphere in* $\mathbf{E}^n$ *and whose lengths are uniformly bounded is compact.*

(For a proof see, for instance, Kolmogorov and Fomin, [1], 66–70).

5

ALGEBRAIC STRUCTURE OF LINEAR VECTOR SPACES

5.0 INTRODUCTION

We are concerned here with those concepts that arise in spaces that are linearly closed: $f, g \in X$ implies $\alpha f + \beta g \in X$ for any α, β. We assume no metric structure and isolate for greater clarity aspects of the spaces that are almost 'formal' in character.

5.1 LINEAR VECTOR SPACES

The abstract notion of a *linear vector space* is founded on an algebraic structure. We consider a set X of elements $x, y, z \ldots$. On X we assume that there is a group operation $+$ that satisfies $x, y \in X$ implies $x + y \in X$, and

1. *Commutativity*: $x + y = y + x$;
2. *Associativity*: $(x + y) + z = x + (y + z)$;
3. *Identity*: there is an element $0 \in X$ with $x + 0 = x$ for all x;
4. *Inverse*: $x \in X$ implies that there is a $y \in X$ with $x + y = 0$.

X is called the set of *vectors*; the elements of X are called *vectors*.

In addition we suppose given a set of scalars $\mathcal{C}$, with elements $\alpha, \beta, \gamma, \ldots$, which possess both an operation $+$, satisfying rules of the same form as (1), (2), (3), and (4), and a multiplication: $\alpha, \beta \in \mathcal{C} \Rightarrow \alpha\beta \in \mathcal{C}$ satisfying

1′. *Commutativity*: $\alpha\beta = \beta\alpha$;
2′. *Associativity*: $(\alpha\beta)\gamma = \alpha(\beta\gamma)$;
3′. *Identity*: there is an element $1 \in \mathcal{C}$ with $\alpha 1 = \alpha$;
4′. *Inverse*: if $\alpha \neq 0$, then there is β with $\alpha\beta = 1$.

Customarily $\mathcal{Q}$ is either the complex or the real numbers and we speak of a complex or a real space.

Last, we assume given a law of composition of elements $\alpha \in \mathcal{Q}$, $x \in X$, which for emphasis we write $\alpha \circ x$, defined on all of $\mathcal{Q} \otimes X$ such that

5. *Distributivity*: $\alpha \circ (x+y) = \alpha \circ x + \alpha \circ y$
6. *Distributivity*: $(\alpha+\beta) \circ x = \alpha \circ x + \beta \circ x$
7. *Associativity*: $\alpha \circ (\beta \circ x) = (\alpha\beta) \circ x$
8. *Normalization*: $1 \circ x = x$

The triple $(X, \mathcal{Q}, \circ)$ we call a *linear vector space*. Except for the purposes of the special emphasis, we drop the awkward notation $\alpha \circ x$ and write αx. In a rather obvious manner, most of the spaces we have seen to this point are examples of linear vector spaces. We sketch the customary structure of a linear vector space next.

DEFINITION 5.1.1. A subset M of X that is closed under the operation of vector addition and scalar multiplication is called a *linear manifold*.

DEFINITION 5.1.2. Given $x^1, x^2, \ldots, x^k \in X$. If there exists a set of scalars $\alpha_1, \alpha_2, \ldots, \alpha_k$, not all zero such that the linear combination $\alpha_1 x^1 + \alpha_2 x^2 + \ldots + \alpha_k x^k = 0$, we call the x^i *linearly dependent*. If no such set of scalars α_i exists, then the x^i are called *linearly independent*. An infinite set of x's, $\{x^i\}$ is called *linearly independent* if no finite subset is linearly dependent.

DEFINITION 5.1.3. We define the *dimension* of a linear manifold M as the n for which there exists a set of n linearly independent vectors in M, while every set of $n+1$ vectors in M is linearly dependent. We write dim $M = n$. If no finite n is the dimension of M, we write dim $M = \infty$ and call M *infinite dimensional*.

DEFINITION 5.1.4. The *span* of a set of vectors $x^1, \ldots, x^n$, $\mathbf{S}\{x^1, \ldots, x^n\} = \{y \mid y = \Sigma_{i=1}^n \alpha_i x^i$ for some scalars $\alpha_i\}$. If a linear manifold $M \subset \mathbf{S}\{x^1, \ldots, x^n\}$, $x^1, \ldots, x^n$ is called a *basis* for M.

DEFINITION 5.1.5. If M, N are two linear manifolds in a linear vector space X, then their sum $M + N = \{x \mid x = m + n, m \in M, n \in N\}$.

EXERCISES TO SECTION 5.1

1. $\{x^i\}$ is linearly dependent if and only if at least one x^j equals a linear combination of a finite number of other x^i's.

2. If the linear manifold $M = \mathbf{S}\{x^1,\ldots,x^n\}$, while $y^1,\ldots,y^m \in M$ are linearly independent, then $m \leqslant n$. If $m < n$ then there exist $y^{m+1},\ldots,y^n \in M$ such that $M \subset \mathbf{S}\{y^1,\ldots,y^m,y^{m+1},\ldots,y^n\}$. If M, N are linear manifolds and $M \subset N \subset X$, then $\dim M \leqslant \dim N$.
3. Show that the sum of two linear manifolds is a linear manifold.
4. If X is the set of real valued functions $x(t)$ in one real variable, $-\infty < t < \infty$, $M = \{x(t)|x(t) = -x(-t)\}$ while $N = \{x(t)|x(t) = x(-t)\}$, then show that $X = M + N$.
5. Show that if M, N are linear manifolds in a linear vector space X, every $x \in M + N$ has a unique representation $x = m + n, m \in M, n \in N$, if and only if $M \cap N = \{0\}$.

5.2 MAPPINGS OF LINEAR VECTOR SPACES

Suppose that we are given two linear vector spaces $(X, \mathscr{C}, \circ)$ and $(Y, \mathscr{C}, \circ)$ over the same scalar field $\mathscr{C}$. A *mapping* from X to Y is a collection of ordered pairs $\{x,y\}, x \in X, y \in Y$ in which each x occurs at most once. We write $x \to y$. If, whenever the mapping is defined for x, x', x'', it satisfies

1. $x' \to y', x'' \to y''$ implies $x' + x'' \to y' + y''$, and
2. $x \to y$ implies $\alpha \circ x \to \alpha \circ y$,

the mapping is called *linear*. We can write $\{x, Ax\}$ or $Ax = y$ instead of $x \to y$. The $x \in X$ for which A is defined are called the *domain* D_A of A. The set of y's that occur as right-hand elements in the pairing $\{x,y\}$ is called the *range* R_A of A. If conditions 1 and 2 are not both satisfied we will write $A(x) = y$. In both cases the *mapping* A is also called an *operator*. We also write $A : X \to Y$ implying only $D_A \subset X, R_A \subset Y$. The set of x's in the pairing $\{x, 0\}$ is called the *null space* $N(A)$ of A.

If in the collection $\{\{x,y\}\}$ of ordered pairs, each y occurs only once, then A is one-to-one $(1:1)$ and there is defined an *inverse* mapping $y \to x$ which we write $\{A^{-1}y, y\}$ or $x = A^{-1}(y)$. We have the fundamental:

THEOREM 5.2.1. *If A is a linear mapping from X to Y, then A^{-1} exists if and only if*

$$Ax = 0 \textit{ implies } x = 0. \tag{1}$$

PROOF. If A^{-1} exists, then suppose $Ax = 0$. $A^{-1}Ax = A^{-1}0$. By property 2 of linear mappings $A^{-1}0 = 0$ while $A^{-1}Ax = x$. If instead eq. 1 holds, then $Ax' = Ax''$ implies $A(x' - x'') = 0$ so $x' = x''$, A is $1:1$, and A^{-1} exists. ■

Theorem 5.1.1 is the basic theorem on the solution of linear equations; it is the weak form of the

Alternative Theorem: *Either $Ax=y$ has a unique solution, or $Ax=0$ has a solution $x\neq 0$.*

DEFINITION 5.2.1. Two linear vector spaces X, Y over the same field of scalars $\mathcal{C}$ are said to be *isomorphic* if there exists a linear mapping $A:X\to Y$ with $D_A=X, R_A=Y$ which is $1:1$ (so that A^{-1} exists).

THEOREM 5.2.2. *Let $A:X\to Y$ be a linear mapping where* $\dim X=\dim Y<\infty$. *Then A^{-1} exists if and only if $R_A=Y$.*

PROOF. By Exercise 5.1.3, if A^{-1} exists, then $\dim R_A=\dim Y$, so $R_A=Y$, as linear manifolds. If $R_A=Y$, then $\dim Y=n<\infty$ means there exist n independent y^i's with $Y=\mathbf{S}\{y^1,\ldots,y^n\}$, while there is an x^i such that $Ax^i=y^i$, $i=1,\ldots,n$. If the x^i were dependent then

$$\sum_{i=1}^{n}\alpha_i x^i=0$$

implies

$$A\sum_{i=1}^{n}\alpha_i x^i=\sum_{i=1}^{n}\alpha_i Ax^i=\sum_{i=1}^{n}\alpha_i y^i=0. \tag{2}$$

Thus the x^i are independent. Then $D_A\supset\mathbf{S}\{x^1,\ldots,x^n\}$. As linear manifolds of dimension n, we must have $D_A=\mathbf{S}\{s^1,\ldots,x^n\}$. Then $Ax=0$ implies, for some $\alpha_1,\ldots,\alpha_n$, eq. 2. But the independence of $y^1,\ldots,y^n$ makes $\alpha_1=\cdots=\alpha_n=0$, so $x=\sum_{i=1}^{n}\alpha_i x^i=0$. Thus A^{-1} must exist. ■

Finite dimensionality was necessary: In M, the space of all sequences, the mapping $A:M\to M$ where $y_i=x_{i+1}$ has $D_A=M, R_A=M$, but A^{-1} does not exist.

THEOREM 5.2.2. *Any two linear vector spaces over the same field of scalers $\mathcal{C}$ of the same finite dimension n are isomorphic.*

PROOF. Choose a basis for each of X, Y and set $Ax^i=y^i$. Then $D_A=X$ and $R_A=Y$. Then A^{-1} exists for $Ax=0$ implies eq. 2 again for some $\{\alpha_i\}$; the independence of $y^1,\ldots,y^n$ implies $\alpha_i=0$, and $x=0$. ■

DEFINITION 5.2.2. Any $\mathcal{C}$ is itself a one-dimensional vector space over itself. Then any mapping of $(X,\mathcal{C},\circ)\to(\mathcal{C},\mathcal{C},\circ)$ is called a *functional*. If the mapping is linear, it is called a *linear functional*. We denote a linear functional on X by $\langle x,x^f\rangle$ for the linear functional $\langle\cdot,x^f\rangle$ evaluated at x, or else by $x^f(x)$.

If x_1^f and x_2^f are two linear functionals on X, then we can define a linear vector space of linear functionals on X, by the rules

$$\langle x,x_1^f+x_2^f\rangle=\langle x,x_1^f\rangle+\langle x,x_2^f\rangle$$

$$\langle x,\alpha x^f\rangle=\alpha\langle x,x^f\rangle.$$

The zero linear functional is the mapping 0^f with $\langle x,0^f\rangle=0$ or $0^f(x)=0$ for all $x\in X$.

DEFINITION 5.2.3. The linear vector space X^f of all linear functionals x^f on X, is called the *algebraic conjugate space of X*.

Example 1. Let $X=(\alpha_1,\ldots,\alpha_n)$, the n-tuples over $\mathscr{C}$. Choose as a basis $e^1,\ldots,e^n$ for X: $e^i=(\alpha_1,\ldots,\alpha_n)$, $\alpha_j=\delta_i^j$ (Kronecker's δ). Then any $x=\sum_{i=1}^n\alpha_ie^i$, and any $x^f\in X^f$ takes the form

$$\langle x,x^f\rangle=\langle\sum_{i=1}^n\alpha_ie^i,x^f\rangle=\sum_{i=1}^n\alpha_i\langle e^i,x^f\rangle=\sum_{i=1}^n\alpha_i\xi^i.$$

Thus any set of n numbers $\xi^i\in\mathscr{C}$ define a linear functional. The mapping of X^f into X via

$$Ax^f=(\xi_1,\ldots,\xi_n)$$

where $\xi_i=\xi^i$, $i=1,\ldots,n$ is a $1:1$ invertible operator from X^f to X, so that they are isomorphic.

Example 2. Let $X=l_2$. Then $x=(\alpha_1,\alpha_2,\ldots)$ with $\sum_{i=1}^\infty|\alpha_i|<\infty$, and $x^f=(\xi^1,\xi^2,\ldots)$ with $\sum_{i=1}^\infty|\xi^i|^2<\infty$ makes

$$\sum_{i=1}^\infty|\alpha_i\xi^i|\leqslant\left(\sum_{i=1}^\infty|\alpha_i|^2\right)^{1/2}\left(\sum_{i=1}^\infty|\xi^i|^2\right)^{1/2}<\infty.$$

Hence $\langle x,x^f\rangle=\sum_{i=1}^\infty\alpha_1\xi^i$ is a linear functional.

Example 3. Let $X=l_p$, $1<p<\infty$; then $x=(\alpha_1,\alpha_2,\ldots)$ and $x^f=(\xi^1,\xi^2,\ldots)$ where $1/p+1/q=1$, and

$$\sum_{i=1}^\infty|\xi^i|^{1/q}<\infty$$

implies

$$\left|\sum_{i=1}^{\infty} \alpha_i \xi^i\right| < \infty.$$

Hence

$$\langle x, x^f \rangle = \sum_{i=1}^{\infty} \alpha_i \xi^i$$

defines a linear functional. Thus $l_q \subset (l_p)^f$.

Example 4. Let $X = C[a,b], -\infty < a < b < \infty$. Then for any $\alpha(s)$ of bounded variation

$$\langle x, x^f \rangle = \int_a^b x(s)\, d\alpha(s)$$

exists. Thus $BV[a,b] \subset (C[a,b])^f$.

When X is the space of Example 1, $x \in X$ makes $x = \sum_{i=1}^n \alpha_i e^i$, and

$$\langle x, x^f \rangle = \sum_{i=1}^{n} \alpha_i \langle e^i, x^f \rangle.$$

Thus x^f is defined by its effect on $\{e^i\}$; two linear functionals x_1^f, x_2^f, with $\langle e^i, x_1^f \rangle = \langle e^i, x_2^f \rangle$, $i = 1, \ldots, n$ are the same functional. If we set $\langle e^i, x^f \rangle = \xi^i$, we find $\langle x, x^f \rangle = \sum_{i=1}^n \alpha_i \xi^i$ and X^f is isomorphic to X. The two bases in X and X^f such that

$$\langle x^i, x^{jf} \rangle = \delta_i^j$$

are called *dual bases*.

DEFINITION 5.2.4. A linear mapping P of a linear vector space X into itself is called a *projection* if $P^2 = P$.

EXERCISES TO SECTION 5.2

1. If A is linear then D_A, R_A are linear manifolds.
2. If A is linear and A^{-1} exists, A^{-1} is linear.
3. Let A be a linear operation, $\dim D_A = n$. Then $\dim R_A = m \leqslant n$. If A^{-1} exists, then $\dim R_A = n$. This holds also for $n = \infty$. As a corollary, show that $M \subset N \subset X, \dim M = \dim N = n < \infty$ implies $M = N$.

4. Let $Y=C[0,1]$. Let $X\subset Y, x\in X$ if x has continuous first and second derivatives on $[0,1]$ and $x(0)=x'(0)=0$. Let $p,q\in Y$ and let T be that operator from X to Y given by

$$Tx=x''+px'+qx.$$

Clearly $R_T\subset Y$. Is $R_T=Y$? Is T bounded? Does T^{-1} exist? (Show that $Tx=0$ implies $x=0$; $R_T=Y$.)
Let $p=q=0$, $Tx=x''$, $D_T=\{x|x''\in Y,\ x(0)=x(1)=0\}$. Then $Tx=y$ implies

$$x=\int_0^s\left[\int_0^u y(t)\,dt\right]du-s\int_0^1\left[\int_0^u y(t)\,dt\right]du.$$

Find a kernel $k(s,t)$ such that

$$x(s)=\int_0^1 k(s,t)y(t)\,dt.$$

5. Let $X=Y=C[0,\pi]$, the continuous complex valued functions of t. Let T be the operator with domain D_T consisting of functions x with $x''\in X, x(0)=x(\pi)=0$. Let $Tx=y$ mean

$$-x''(s)+\lambda x(s)=y(s)$$

where λ is complex. Let $\mu=(\lambda)^{1/2}$; show that

$$x(s)=\mu^{-1}\int_0^s y(t)\sinh\mu(t-s)\,dt+c_1e^{\mu s}+c_2e^{-\mu s}.$$

Find c_1,c_2 so that $x\in D_T$. Characterize those λ's for which T^{-1} exists and find $k(s,t)$ such that

$$x(s)=\int_0^\pi k(s,t)y(t)\,dt.$$

6. If $x^1,\ldots,x^n$ is a basis for X, then the dual set is a basis for X^f.
7. $(X^f)^f=X$ if X is a *finite* dimensional linear vector space.
8. Show that a projection P satisfies
   (a) $R_P=N(I-P)=\{x\in X|Px=x\}$;
   (b) $R_P+N(P)=X$;
   (c) $R_P\cap N(P)=\{0\}$.
9. Show that if M,N are linear manifolds in a linear vector space X with $M\cap N=\{0\}$ and $X=M+N$, then there is a unique projection P with

$R_P = M$ and $N(P) = N$. Such a pair of manifolds are called *algebraic complements*.

10. If X is a linear vector space whose dimension $n < \infty$, and $x^1, x^2, \ldots, x^n$ are a linearly independent set in X, then any $x = \alpha_1 x^1 + \cdots + \alpha_n x^n$ for some scalars $\alpha_1, \ldots, \alpha_n$. If $e_1, \ldots, e_n$ is any sequence of zeros and ones, then the mapping $x \to \alpha_1 e_1 x^1 + \cdots + \alpha_n e_n x^n$ is a projection. Find the algebraic complement of its range.

5.3 TRANSPOSE OF A LINEAR OPERATOR

Let X and Y be linear vector spaces and $A: X \to Y$ a linear operator from X to Y. Each of X, Y has a conjugate space X^f, Y^f. Consider the linear functional in Y^f, $\langle Ax, y^f \rangle_Y$ composed via A from X. Clearly it is a linear functional on X, so it is some element $\langle x, x^f \rangle_X$ in X^f. That is

$$\langle Ax, y^f \rangle_Y = \langle x, x^f \rangle_X$$

is an identity for all x in X. This defines a mapping from Y^f to X^f, which is called A^T, the *transpose of* A:

$$\langle Ax, y^f \rangle_Y = \langle x, A^T y^f \rangle_X.$$

Pictorially:

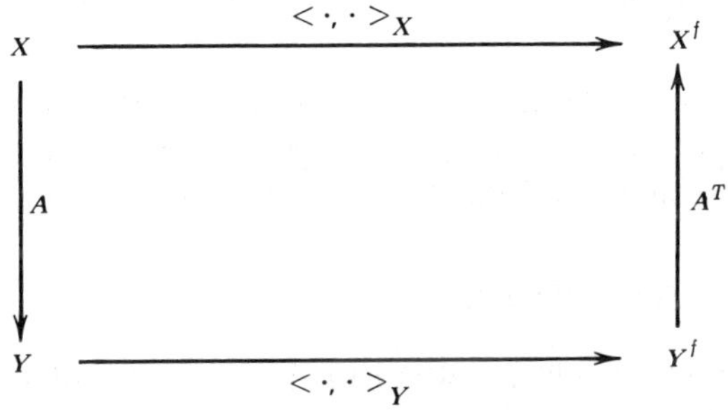

If X and Y are finite dimensional spaces, say $\dim X = n, \dim Y = m$, then the mapping A may be represented by a rectangular matrix, via

$$Ax^j = \sum_{i=1}^{m} \alpha_{ij} y^i, \qquad j = 1, \ldots, n$$

where $\{x^j\}, \{y^i\}$ are bases for X and Y respectively. Since specifying A on a basis specifies it completely, the $m \times n$ matrix (α_{ij}) together with the two basis $\{x^j\},\{y^j\}$ specify A completely. If we choose dual bases in X^f and Y^f, so that

$$\langle x^j, x^{fk}\rangle_X = \delta_j^k \qquad 1 \leqslant j,k \leqslant n,$$

$$\langle y^j, y^{fk}\rangle_y = \delta_j^k \qquad 1 \leqslant j,k \leqslant m,$$

then we see that

$$A^T y^{fk} = \sum_{p=1}^{n} \beta_{pk} x^{fp}, \qquad k=1,\ldots,m.$$

A^T is specified by (β_{ki}), an $n\times m$ matrix, for these bases. Now

$$\langle Ax^j, y^{fk}\rangle = \sum_{i=1}^{m} \alpha_{ij}\langle y^i, y^{fj}\rangle_Y = \alpha_{kj}$$

while

$$\langle x^j, A^T y^{fk}\rangle = \sum_{p=1}^{n} \beta_{pk}\langle x^j, x^{fp}\rangle_X = \beta_{jk}.$$

Thus

$$\alpha_{kj} = \beta_{jk}, \qquad j=1,\ldots,n, \qquad k=1,\ldots,m.$$

Thus our definition of transpose coincides with the customary one for matrices. Transpose is an *algebraic concept* related to the formal structure of a linear vector space and the space of linear functionals on it.

EXERCISES TO SECTION 5.3

1. The algebraic conjugate (or dual) of an n-dimensional vector space is n-dimensional.
2. If X is n-dimensional, $x\in X$, then $\langle x, x^f\rangle = 0$ for all $x^f \in X^f$ implies $x=0$.
3. (*Corollary*) Given X, X^f, X^{ff}, X finite dimensional, and

   $$\langle x, x^f\rangle_X \equiv \langle x^f, x^{ff}\rangle_{X^f}$$

   as x^f varies over X^f, set $Jx = x^{ff}$. Then J is a linear mapping and $Jx=0$ implies $x=0$, so J is invertible.

4. Let M be a linear manifold of dimension $m<n$ in linear space X, and suppose $x_0 \in X-M$. Then there exists an element $x_0^f \in X^f$ such that $x_0^f(x)=0$ if $x \in M$ and $x_0^f(x_0)=1$.
5. (*Definition*) Given a set $S \in X$. By S^0 is meant the set of $x^f \in X^f$ such that $\langle x, x^f \rangle = 0$ for all $x \in S$. (S^0 is the *annihilator* of S). If $S \in X^f$, by ${}^0S$ is meant the set of $x \in X$ such that $\langle x, x^f \rangle = 0$, for all $x^f \in S$. Let $A: X \to Y$ be a linear mapping. $N(A)$, the *null manifold* of A is the set $x \in X$ such that $Ax=0$. Show that

   (a) A^{-1} exists if and only if $N(A)=0$.
   (b) $R_A^0 = N(A^T)$.
   (c) $R_A = {}^0N(A^T)$.
   (d) $R_A = Y$ if and only if $(A^T)^{-1}$ exists.
   (e) $N(A) = {}^0R_{A^T}$.
   (f) If X is finite dimensional, $R_{A^T} = N(A)^0$.

6. If $R_{A^T} = X^f$, then A^{-1} exists.
7. If X is finite dimensional, and A^{-1} exists, then $R_{A^T} = X^f$.
8. If P is a projection of a finite dimensional linear vector space X into itself, we can choose a basis for X and a dual basis in X^f such that

$$P = \begin{bmatrix} 1 & & & & & & & & & \\ & 1 & & & & & & & & \\ & & 1 & & & & & & & \\ & & & \cdot & & & & & & \\ & & & & \cdot & & & & & \\ & & & & & \cdot & & & & \\ & & & & & & 1 & & & \\ & & & & & & & 0 & & \\ & & & & & & & & \cdot & \\ & & & & & & & & & \cdot \\ & & & & & & & & & & \cdot \\ & & & & & & & & & & & 0 \end{bmatrix}$$

   (Hint: Choose a basis for R_P and one for $N(P)$.)
9. If μ is a scalar, P a projection of a linear vector space, and I the identity, then show that, provided $\mu \neq 1$,

$$(I-\mu P)^{-1} = I + \left(\frac{\mu}{1-\mu}\right)P.$$

6

NORMED SPACES AND CONTINUOUS OPERATORS

6.0 INTRODUCTION

We now put together the algebraic structure and the metric structure to form the normed linear vector space. A "size" is attributed to an element x of the space X corresponding to its distance $\rho(x,0)$ from the origin. This can be done for those metrics that are compatible with the algebraic structure in the sense that $\rho(x,y)=\rho(x-y,0)$ for any $x,y \in X$. This stems from the practice in vector analysis of describing a point P of $\mathbf{E}^n$ by the vector from the origin to the point P and giving to the vector a length equal to the distance from 0 to P. In this context we study the continuous linear operators from one normed linear vector space to another.

6.1 NORMED SPACES AND CONTINUOUS OPERATORS

DEFINITION 6.1.1. Let $(X, \mathscr{C}, \circ)$ be a linear vector space. By a *norm* on X we mean a function $\|x\|: X \to \{t | 0 \leqslant t < \infty\}$ such that

1. $\|x+x'\| \leqslant \|x\| + \|x'\|$ for all $x, x' \in X$;
2. $\|\alpha x\| = |\alpha| \, \|x\|$ for all $\alpha \in \mathscr{C}$, $x \in X$,
3. $\|x\| = 0$ if and only if $x = 0$.

If property 3 is weakened to

3′. $x = 0$ only if $\|x\| = 0$,

we have a *pseudonorm*, or *seminorm*. A norm is a metric if we set $\rho(x^1, x^2) = \|x^1 - x^2\|$. The verification is immediate.

DEFINITION 6.1.2. Two normed linear vector spaces X, Y, with norms $\|\cdot\|_X, \|\cdot\|_Y$, respectively, are called *congruent* if there exists

1. A linear operator, $T, D_T = X, R_T = Y$, possessing an inverse, such that,
2. $\|Tx\|_Y \equiv \|x\|_X$, for all $x \in X$.

T is called an *isometric* operator.

We have already seen examples of this concept. Consider the even periodic functions of a common period $[0, 2\pi]$ of mean zero, with norm

$$\|f\| = \left(\frac{1}{\pi}\int_0^{2\pi} |f(t)|^2\, dt\right)^{1/2}$$

and the Fourier Sequences defined by

$$\left\{\alpha_1, \alpha_2, \alpha_3, \ldots \mid \sum_{i=1}^{\infty} |\alpha_i|^2 < \infty\right\}.$$

The mapping $Tf = \{\alpha_n\}$, where

$$\alpha_n = \frac{1}{\pi}\int_0^{2\pi} (\cos n\pi t) f(t)\, dt, \qquad n = 1, 2, 3, \ldots$$

is an isometry, by virtue of the Parseval theorem.

The notion of continuity is expressed in terms of the norm. Thus we call the mapping $A: X \to Y$ continuous at x^0 if

$$\|A(x) - A(x^0)\|_Y \to 0, \qquad \|x - x^0\|_X \to 0.$$

When the meaning is clear we will write $\|\cdot\| = \|\cdot\|_Y$, and $\|\cdot\| = \|\cdot\|_X$. For linear operators we have a fundamental theorem on such continuity:

THEOREM 6.1.1. *Let the linear operator T map X into Y. Let $X = D_T$. Then*

1. *If T is continuous at x^0, then it is continuous everywhere on X, and in fact is uniformly continuous on X.*
2. *T is continuous if and only if there exists a finite M such that*

$$\|Tx\|_Y < M\|x\|_X \tag{1}$$

for all $x \in X$.

We call an operator T satisfying eq. 1 a *bounded* operator, even if it is nonlinear.

PROOF. Let T be continuous at x^0 and linear. Then for any x, $x' \in X$,

$$Tx - Tx' = T(x + x^0 - x') - Tx^0.$$

Set $x^* = x + x^0 - x'$. Then $\|Tx - Tx'\| = \|Tx^* - Tx^0\| \to 0$, $\|x^* - x^0\| \to 0$. But $\|x^* - x^0\| = \|x - x'\|$. Hence $\|Tx - Tx'\| \to 0$, $\|x - x'\| \to 0$. For any $\epsilon > 0$, choose δ such that

$$\|Tx^* - Tx^0\| < \epsilon \text{ for } \|x^* - x^0\| < \delta.$$

Then $\|Tx - Tx'\| < \epsilon$ for $\|x - x'\| < \delta$ for all x, x' in X. Thus the continuity is uniform on X.

Now suppose T is bounded. Then

$$\|Tx\| \leqslant M \|x\|$$

implies T is continuous at $x^0 = 0$, and hence everywhere.

Conversely suppose T is continuous everywhere. Then T is continuous at zero. Given $\epsilon > 0$ there exists $\delta = \delta(\epsilon)$ such that

$$\|Tx\| \leqslant \epsilon \text{ if } \|x\| \leqslant \delta(\epsilon).$$

Let x^* be any element, $x^* \neq 0$. Put

$$z = \left(\frac{\delta(\epsilon)}{\|x^*\|} \right) x^*.$$

Then

$$\|z\| = \left(\frac{\delta(\epsilon)}{\|x^*\|} \right) \|x^*\| = \delta(\epsilon).$$

Hence $\|Tz\| \leqslant \epsilon$, so for all $x^* \in X$,

$$\|Tx^*\| \leqslant \frac{\epsilon}{\delta(\epsilon)} \|x^*\|.$$

Thus T is bounded. ■

DEFINITION 6.1.3. If T is a continuous linear operator from X to Y with $X = D_T$, we define its *bound* to be $\inf M$ for all M such that

$$\|Tx\| \leqslant M \|x\| \text{ for all } x \in X,$$

and write $\|T\|=\inf\{M \mid \|Tx\| \leqslant M\|x\|, \text{ all } x \in X\}$. $\|T\|$ is also called the *norm* of T for reasons that become apparent in the next section.

The fundamental theorem on invertibility of linear operators is the following:

THEOREM 6.1.2. *Let T be a linear operator from X to Y; $D_T=X$. The inverse mapping T^{-1} exists and is continuous on R_T if and only if there exists a constant $m>0$ such that*

$$\|Tx\| \geqslant m\|x\|, \qquad x \in X.$$

PROOF. If $\|Tx\| \geqslant m\|x\|$, then $Tx=0$ implies $x=0$. Hence the alternative theorem implies that T^{-1} exists with $D_{T^{-1}}=R_T$. For any $y \in R_T$ set $x=T^{-1}y$. Then

$$\|TT^{-1}y\| = \|y\| \geqslant m\|T^{-1}y\|.$$

Hence

$$\|T^{-1}y\| \leqslant m^{-1}\|y\|$$

for all $y \in R_T$. Hence T^{-1} is a bounded, linear, and continuous operator on R_T.

Conversely, if T^{-1} exists and is continuous on R_T it is bounded on R_T so

$$\|T^{-1}y\| \leqslant M\|y\|$$

for all $y \in R_T$. For any x, set $x=T^{-1}y$. Then $y=Tx$ so

$$\|T^{-1}y\| = \|T^{-1}Tx\| = \|x\| \leqslant M\|Tx\|.$$

Hence

$$\|Tx\| \geqslant M^{-1}\|x\|. \quad \blacksquare$$

If T, T^{-1} are both continuous, we call T *bicontinuous*.

When are two norms equivalent? We enlarge the question by the following definition.

DEFINITION 6.1.4. Two normed linear vector spaces X, Y are *homeomorphic* if there exists a 1 : 1, bicontinuous, linear mapping T with $D_T=X$, $R_T=Y$.

The mapping need not be isometric; we only ask that it preserve convergence:

COROLLARY 1. *X and Y are homeomorphic if and only if there exists a linear operator T with $X = D_T$, $Y = R_T$, and constants m, M such that*

$$0 < m\|x\| \leqslant \|Tx\| \leqslant M\|x\|, \qquad x \in X, x \neq 0.$$

COROLLARY 2. *The space X with norms $\|\cdot\|_1$ and $\|\cdot\|_2$ has the same convergent sequences in both norms if and only if there exist constants m, M with*

$$0 < m\|x\|_1 \leqslant \|x\|_2 \leqslant M\|x\|_1, \qquad x \in X, x \neq 0.$$

We consider two such norms as equivalent.

Some examples of homeomorphic spaces are the following: let X be $\mathbf{E}^n$, the set of n-tuples $(\alpha_1, \ldots, \alpha_n)$ over $\mathcal{C}$, with the three norms

$$\|x\|_\infty = \max(|\alpha_1|, \ldots, |\alpha_n|);$$

$$\|x\|_1 = \sum_{i=1}^{n} |\alpha_i|;$$

$$\|x\|_2 = \left(\sum_{i=1}^{n} |\alpha_i|^2 \right)^{1/2}.$$

Let T be a linear operator mapping X into X. We obtain three bounds for T. If we choose as basis in $\mathbf{E}^n$, $e^1, \ldots, e^n$, where $e_i^j = \delta_i^j$, we can represent T with respect to this basis. If $x = \sum_{i=1}^{n} \alpha_i e^i$, and

$$Tx = \sum_{i=1}^{n} \alpha_i Te^i = \sum_{i=1}^{n} \alpha_i \sum_{j=1}^{n} \beta_{ij} e^j$$

$$= \sum_{j=1}^{n} \left(\sum_{i=1}^{n} \alpha_i \beta_{ij} \right) e^j,$$

then T is represented by the matrix (β_{ij}). Then

$$\|Tx\|_\infty = \max_{1 \leqslant j \leqslant n} \left\{ \sum_{i=1}^{n} \alpha_i \beta_{ij} \middle| \max(|\alpha_1|, \ldots, |\alpha_n|) = 1 \right\}.$$

Thus $\|Tx\|_\infty \leqslant \max_{1 \leqslant j \leqslant n} \sum_{i=1}^{n} |\beta_{ij}|$. This value is actually achieved for a suitable choice of x (with $\alpha_j = \pm 1, j = 1, \ldots, n$) so

$$\|T\|_\infty = \max_{1 \leqslant j \leqslant n} \sum_{i=1}^{n} |\beta_{ij}|.$$

Now

$$\begin{aligned}
\|Tx\|_1 &= \sum_{j=1}^{n} \left| \sum_{i=1}^{n} \alpha_i \beta_{ij} \right| \\
&\leqslant \sum_{j=1}^{n} \sum_{i=1}^{n} |\alpha_i||\beta_{ij}| \\
&\leqslant \max_{1 \leqslant i \leqslant n} \left(\sum_{j=1}^{n} |\beta_{ij}| \right) \sum_{i=1}^{n} |\alpha_i|.
\end{aligned}$$

Thus

$$\|Tx\|_1 \leqslant \max_{1 \leqslant i \leqslant n} \sum_{j=1}^{n} |\beta_{ij}|, \qquad \|x\|_1 = 1.$$

But suppose

$$\max_{1 \leqslant i \leqslant n} \sum_{j=1}^{n} |\beta_{ij}| = \sum_{j=1}^{n} |\beta_{sj}|.$$

Then set $\alpha_s = 1, \alpha_i = 0,\ i \neq s$, so

$$\sum_{j=1}^{n} \left| \sum_{i=1}^{n} \alpha_i \beta_{ij} \right| = \sum_{j=1}^{n} |\beta_{sj}|.$$

Thus $\|Tx\|_1$ achieves its bound for this x, so

$$\|T\|_1 = \max_{1 \leqslant i \leqslant n} \sum_{j=1}^{n} |\beta_{ij}|.$$

Last, by the Cauchy inequality

$$\begin{aligned}
\|Tx\|_2^2 &= \sum_{j=1}^{n} \left| \sum_{i=1}^{n} \alpha_i \beta_{ij} \right|^2 \\
&\leqslant \sum_{j=1}^{n} \sum_{i=1}^{n} |\alpha_i|^2 \sum_{i=1}^{n} |\beta_{ij}|^2 \\
&\leqslant \sum_{i,j=1}^{n} |\beta_{ij}|^2, \qquad \|x\|_2 = 1.
\end{aligned}$$

Thus

$$\|T\|_2 \leqslant \left(\sum_{i,j=1}^{n} |\beta_{ij}|^2 \right)^{1/2}.$$

But here we note that we cannot remove the inequality. For instance, if T is the identity, $Tx = x$ for all x, then $\beta_{ij} = 0$ $i \neq j$, $\beta_{ii} = 1$, $i = 1, \ldots, n$ so $\|T\|_2 = 1$, but

$$\sum_{i,j=1}^{n} |\beta_{ij}|^2 = n.$$

The three norms are equivalent:

$$\|x\|_\infty \leqslant \|x\|_1 \leqslant n\|x\|_\infty$$

and

$$\|x\|_\infty \leqslant \|x\|_2 \leqslant n^{1/2}\|x\|_\infty.$$

The operator norms they induce are *not* identical.

EXERCISES TO SECTION 6.1

1. From the definition show that

$$\begin{aligned} \|T\| &= \sup_{\|x\| \leqslant 1} \|T\| \\ &= \sup_{\|x\| = 1} \|Tx\| \\ &= \sup_{x \neq 0} \frac{\|Tx\|}{\|x\|}. \end{aligned}$$

   In addition let S be any subset of X. Show that T is bounded if and only if $T(S) \subset Y$ is bounded whenever S is bounded in X.
2. A metric space whose elements are a linear vector space and whose metric satisfies $\rho(x,y) = \rho(x-y, 0)$, $\rho(\alpha x, 0) = |\alpha|\rho(x, 0)$ is a normed linear vector space with $\|x\| = \rho(x, 0)$.
3. Let r be a nonnegative measurable function on a measurable set E. Define, for any measurable function x on E such that $x^2 \in L_1(E)$,

$$\|x\|_{2,r} = \left[\int_E x^2(t) r(t)\, dt \right]^{1/2}.$$

Show that this is a norm on a space $L_2(E;r)$. If there exist constants m, M such that $0<m(\leqslant)r(t)(\leqslant)M<\infty$, show that $L_2(E;r)=L_2(E)$ and

$$0<m\|x\|_2\leqslant\|x\|_{2,r}\leqslant M\|x\|_2$$

where $\|\cdot\|_2$ is the norm obtained from $\rho_2(x,y)$ as in Exercise 6.1.2.

4. Let $E=\{t|0\leqslant t\leqslant 1\}$. Let $r(t)=t$, so that $M=1$, but no $m>0$ exists as in Exercise 6.1.3. Show that $L_2(E;r)\supset L_2(E)$, and in fact contains functions not in $L_2(E)$. (Hint: consider the function $t^{-2/3}$.)
5. Let $E=\{t|0\leqslant t\leqslant 1\}$. Let $r(t)=t^{-1}$, so that $m=1$ but no $M<\infty$ exists as in Exercise 6.1.3. Show that $L_2(E;r)$ is a proper subspace of $L_2(E)$.
6. Let $X=\{\pi(t)|\pi(t)=\sum_{i=0}^{n}\alpha_i t^i, \alpha_i\in\mathscr{C}\}$. Let $\|x\|=\max_i|\alpha_i|$. Show that this is a norm. Let D be the differentiation operator, so that

$$D\pi(t)=\pi'(t)=\sum_{i=1}^{n} i\alpha_i t^{i-1}.$$

Find $\|D\|$. Does D^{-1} exist?

7. The sequence space l_p, $1\leqslant p<\infty$ is a subspace of $l_\infty=m=$ space of infinite sequences with norm $\|x\|_\infty=\sup_i|x_i|$. Find the closure of l_1 in l_∞ (with respect to $\|\cdot\|_\infty$.) What about l_p, $1<p<\infty$?
8. Let $k(s,t,u)$ be continuous on the set $\{[0,1]\otimes[0,1]\otimes(-\infty,\infty)\}$ and satisfy

$$|k(s,t,u)|\leqslant M$$

there. The integral operator $T(x)=y$, where

$$y(s)=\int_0^1 k[s,t,x(t)]\,dt$$

is defined for all functions in $C[0,1]$. Show that the image of $\|x\|\leqslant N$ under T is a compact set in $C[0,1]$ for every finite N. (Hint: *use* Arzèla's theorem.)

9. Let $k(s,t)$ be a continuous function on $0\leqslant s$, $t\leqslant 1$. Let $x(t)\in L_2(0,1)$. The integral operator $T(x)=y$, where

$$y(s)=\int_0^1 k(s,t)x(t)\,dt$$

maps the unit sphere $\|x\|\leqslant 1$ in $L_2(0,1)$ into a compact subset of $L_2(0,1)$. Show this. (Hint: $R_T\subset C[0,1]$. $T\{x|\|x\|\leqslant 1\}$ is compact in

$C[0,1]$; use Arzèla's theorem. Hence $T\{x \mid \|x\| \leqslant 1\}$ is compact in $L_2(0,1)$.)

10. Consider the integral equation $x(s)=\int_0^1 k(s,t)x(t)\,dt+f(s)$ where $x,f \in C[0,1]$ and $k(s,t)$ is continuous on $0\leqslant s,t\leqslant 1$. Suppose k can be written as

$$k(s,t)=\sum_{i=1}^{n}\phi_i(s)\psi_i(t),$$

where the functions ϕ_i are linearly independent. Show that x can be written as

$$x(s)=\sum_{i=1}^{n}\alpha_i\phi_i(s)+f(s)$$

and determine the formulas for α_i. Show that the α's are the solution of an algebraic system of equations. (These are called *degenerate* kernels.)

11. On the real line define a mapping $u\to x$ by

$$x(t)=\begin{cases} a(t)u(t), & t\geqslant t_0,\\ 0, & t\leqslant t_0.\end{cases}$$

Let I be an interval. Find conditions on $a(t)$ such that the mapping $u\to x$:

(a) Maps $L_2(I)$ to $L_2(I)$,
(b) Maps $C(I)$ to $C(I)$.

Is the mapping linear? Is it bounded? (Prove the results). On $-\infty<t<\infty$ let

$$X=\left\{x(t)\,\Big|\,\sup_{-\infty<t<\infty}|x(t)|<\infty\right\}$$

with

$$\|x\|=\sup_{-\infty<t<\infty}|x(t)|.$$

Let $a(t)=t/(1+t^2)^{1/2}$. Determine the norm of the operator. What if $a(t)=t/(1+t^2)$?

12. In Exercise 1.5.4 we have actually done the work to establish conditions sufficient that an infinite matrix A be a bounded operator from m to m and from l_p to l_p, $1 \leqslant p < \infty$. Show this. Note again that the matrix $A=(a_{ij})=I$, where $a_{ii}=1$, $i=1,2,3,\ldots$ is bounded, but the sufficient condition found is not verified.
13. Let A be an infinite matrix such that, for some real number a and some integer k,

    (a) $|a_{ij}| \leqslant a$, $i,j=1,2,3,\ldots$;
    (b) $|a_{ij}|=0$ if $|i-j|>k$.

    Then A is a bounded operator from l_p to l_p, $1 \leqslant p \leqslant \infty$. Show this and find a bound for $\|A\|$.

6.2 LINEAR OPERATORS AS A NORMED LINEAR VECTOR SPACE

We fix our attention on two normed linear vector spaces X and Y over the same field of scalars $\mathscr{C}$ together with the set of all linear operators T from X to Y; this is itself a linear vector space in the following obvious way. If T', T'' are any two such linear operators, then by $\alpha T' + \beta T''$, $\alpha, \beta \in \mathscr{C}$ we mean the operator that maps $x \in X$ into $y \in Y$ where

$$y = \alpha T'x + \beta T''x.$$

The *zero operator* 0 maps every $x \in X$ to $0 \in Y$.

We restrict our attention to those T's for which our operator norm is defined:

$$\|T\| = \sup_{\|x\|=1} \|Tx\|,$$

that is, *the bounded linear operators from* X to Y; then we note the following:

1. $\|T\| = \sup_{\|x\|=1} \|Tx\| \geqslant 0$. If $\|T\|=0$, then $Tx=0$ for all x, so $T=0$. If $\|T\| \neq 0$, then there is an x with $Tx \neq 0$, so $\|T\| \geqslant 0, \|T\|=0$ if and only if $T=0$.
2. $\|\alpha T\| = \sup_{\|x\|=1} \|\alpha Tx\| = |\alpha| \sup_{\|x\|=1} \|Tx\| = |\alpha|\,\|T\|$.
3. $\|T'+T''\| = \sup_{\|x\|=1} \|T'x + T''x\| \leqslant \sup_{\|x\|=1} \|T'x\| + \sup_{\|x\|=1} \|T''x\| \leqslant \|T'\| + \|T''\|$.

Thus the norm makes the linear space of all bounded linear operators from X to Y a normed linear vector space. We denote this space by $[X \to Y]$; when $X = Y$ we write $[X]$ for $[X \to X]$.

DEFINITION 6.2.1. Let $T, \{T^n\} \subset [X \to Y]$. If $\|T^n - T\| \to 0$, $n \to \infty$, we speak of *norm convergence*; since $\|T^n - T\| \to 0$, $n \to \infty$ implies $\|T^n x - Tx\| \to 0$, $n \to \infty$, uniformly in x, we also speak of *convergence in the uniform topology*.

A complete normed linear vector space is usually called a *Banach space.*

THEOREM 6.2.1. *If Y is a Banach space, then $[X \to Y]$ is a Banach space.*

PROOF. Suppose Y is complete. Let $\{T^n\}$ be a norm convergent sequence in $[X \to Y]$. Then for any $\epsilon > 0$,

$$\|T^n - T^m\| < \epsilon \qquad n, m \geqslant N_\epsilon,$$

so

$$\|T^n x - T^m x\| < \epsilon \|x\|, \qquad n, m \geqslant N_\epsilon,$$

for all $x \in X$. Thus $\{T^n x\}$ is a Cauchy sequence *in a complete space* Y, so it has a limit $y \in Y$. Call T the mapping $x \to y$. It is trivial that T is linear. Since

$$\|(T^n - T^m)x\| < \epsilon \|x\|, \qquad n, m \geqslant N_\epsilon,$$

and $T^n x \to Tx$, $n \to \infty$, the continuity of the norm makes

$$\|(T - T^m)x\| < \epsilon \|x\|, \qquad m \geqslant N_\epsilon.$$

Thus $T^m \to T$ in the uniform topology, and $T - T^m \in [X \to Y]$, for all $m \geqslant N_\epsilon$. Since $\{T^n\}$ is a Cauchy sequence, we must have

$$\|T^n\| - \|T^m\| \leqslant \|T^n - T^m\| \leqslant \epsilon, \qquad n, m \geqslant N_\epsilon,$$

whence

$$\|T^n\| \leqslant \epsilon + \|T^{N_\epsilon}\|, \qquad n \geqslant N_\epsilon,$$

so $\|T^n\|$ is uniformly bounded. Thus T is a bounded linear operator, so $T \in [X \to Y]$. ■

In the special case that $X = Y$, the normed linear vector space $[X]$ is also an *algebra*: that is, if $A, B \in [X]$, then $AB \in [X]$ where AB is the operator

$$(AB)x = A(Bx).$$

It is obvious that AB is a linear operator and that

$$\|ABx\| \leqslant \|A\|\,\|Bx\| \leqslant \|A\|\,\|B\|\,\|x\|,$$

Then $\|AB\| \leqslant \|A\|\,\|B\| < \infty$, so $AB \in [X]$. If $Ix = x$ for all x (the identity operator) then $\|I\| = 1$; $\|0\| = 0$, where $0x = 0$ for all x. Also $[X]$ is complete if and only if X is complete. In this algebra we have the following very useful theorem.

THEOREM 6.2.2. *Let X be a Banach space, $A \in [X]$ with $\|A\| = R$. Then any convergent power series in z,*

$$f(z) = \sum_{m=0}^{\infty} \alpha_n z^n,$$

absolutely convergent for $z = R$, generates a power series in $[X]$,

$$f(A) = \sum_{m=0}^{\infty} \alpha_n A^n,$$

which converges in the uniform topology.

PROOF. Since $\|A\| = R$,

$$\sum_{n=0}^{\infty} |\alpha_n|\|A\|^n < \infty. \tag{1}$$

But $\|A^n\| = \|A^{n-1}A\| \leqslant \|A\|\,\|A^{n-1}\| \leqslant \ldots \leqslant \|A\|^n$ so

$$\sum_{n=0}^{\infty} \alpha_n \|A^n\| < \infty.$$

Also for $m < p$

$$\left\| \sum_{n=0}^{m} \alpha_n A^n - \sum_{n=0}^{p} \alpha_n A^n \right\| \leqslant \sum_{n=m+1}^{p} |\alpha_n|\|A\|^n. \tag{2}$$

The convergence of eq. 1 implies that the right side of eq. 2 does not exceed ϵ if $m,p \geqslant M_\epsilon$. Since X is complete, $[X]$ is complete, so the power series converges in the uniform topology to $f(A)$. ■

COROLLARY 1. *If* $\|A\|<1$, *then* $\Sigma_{n=0}^{\infty} A^n$ *converges in the uniform topology to* $(I-A)^{-1}$.

PROOF. Both the series

$$\frac{1}{1-z}=\sum_{n=0}^{\infty} z^n$$

and

$$\frac{z}{1-z}=\sum_{n=0}^{\infty} z^{n+1}$$

converge on $|z| \leqslant R < 1$. If $B=\Sigma_{n=0}^{\infty} A^n$, $B \in [X]$ and $BA=AB=\Sigma_{n=0}^{\infty} A^{n+1}$. Now $B-BA=B-AB=\Sigma_{n=0}^{\infty} A^n - \Sigma_{n=0}^{\infty} A^{n+1} = \lim_{m\to\infty} I-A^m=I$. Thus $B(I-A)=(I-A)B=I$. Thus B is a bounded linear operator inverse to $I-A$. Also

$$\|(I-A)^{-1}\| = \lim_{n\to\infty}\left\|\sum_{i=0}^{n} A^i\right\| \leqslant \lim_{n\to\infty}\frac{1-\|A\|^n}{1-\|A\|}=\frac{1}{1-\|A\|}$$

when we appraise by a geometric series. ■

THEOREM 6.2.3. *Let X be a Banach space, $R \subset [X]$ be the set of all invertible bounded linear operators. Then R is an open set and A^{-1} is a continuous function of A in the uniform topology.*

PROOF. We must show that if $A \in R$ and $\|A-B\|$ is sufficiently small, then $B \in R$, and $\|A^{-1}-B^{-1}\| \to 0$ as $\|A-B\| \to 0$. We write

$$\begin{aligned} B &= A-(A-B)=A-AA^{-1}(A-B) \\ &= A\left[I-A^{-1}(A-B)\right]. \end{aligned}$$

But $\|A^{-1}(A-B)\| \leqslant \|A^{-1}\|\,\|A-B\|$. If $\|A-B\| < \|A^{-1}\|^{-1}$, then $\|A^{-1}(A-B)\| < 1$, so $(I-A^{-1}(A-B))^{-1}$ exists. It follows that $(I-A^{-1}(A-B))^{-1}A^{-1}=B^{-1}$.

Further from our estimate in Theorem 6.2.1,

$$\|B^{-1}\| \leqslant \|(I-A^{-1}(A-B))^{-1}\| \, \|A^{-1}\|$$

$$\leqslant \frac{\|A^{-1}\|}{I-\|A^{-1}(A-B)\|}$$

$$\leqslant \frac{\|A^{-1}\|}{1-\|A^{-1}\| \, \|A-B\|}.$$

Thus $B^{-1} \in R$, and the whole ball $\|B-A\|<1/\|A^{-1}\|$ is in R. Thus R is open. Finally

$$B^{-1}-A^{-1}=A^{-1}(A-B)B^{-1}$$

so

$$\|B^{-1}-A^{-1}\| \leqslant \|A^{-1}\| \, \|A-B\| \, \|B^{-1}\|$$

or

$$\|B^{-1}-A^{-1}\| \leqslant \frac{\|A^{-1}\| \, \|A-B\| \, \|A^{-1}\|}{(1-\|A^{-1}\| \, \|A-B\|)}.$$

Then $\|B-A\| \to 0$ implies $\|B^{-1}-A^{-1}\| \to 0$. ■

EXERCISES TO SECTION 6.2

1. Let X be a Banach space, L a (closed) linear subspace of X. Write $x_1 \sim x_2$ if $x_1 - x_2 \in L$. Show that

   (a) $x_1 \sim x_2, y_1 \sim y_2$ implies $x_1 + y_1 \sim x_2 + y_2$;
   (b) $x_1 \sim x_2$ implies $\alpha x_1 \sim \alpha x_2$.

   Define $[x]=\{y \in X | x \sim y\}$. The set $\{[x]\}=X/L$ is called the *factor* space. Its elements $[x]$ are called *cosets*. Define

   $$[x]+[y]=[x+y], \qquad \alpha[x]=[\alpha x].$$

   Show that X/L is a Banach space when we define

   $$\|[x]\| = \inf_{y \in [x]} \|y\|.$$

   Show that if X is separable, so is X/L. If L is not closed, then X is only a linear vector space.

2. Let X, Y be Banach spaces. Define $X\otimes Y=\{(x,y)|x\in X, Y\in Y\}$. Define

$$(x_1,y_1)+(x_2,y_2)=(x_1+x_2,y_1+y_2)$$

$$\alpha(x_1,y_1)=(\alpha x_1,\alpha y_1)$$

$$\|(x,y)\|=\begin{cases}\|x\|+\|y\| \\ \text{or} \\ \max(\|x\|,\|y\|) \\ \text{or} \\ (\|x\|^2+\|y\|^2)^{1/2}.\end{cases}$$

Show that $X\otimes Y$ is a Banach space. $X\otimes Y$ is called the *product* space of X and Y. What if X and Y are only normed linear vector spaces?

3. Let X, Y be normed linear vector spaces. Let $T\in[X\to Y]$, $R_T=Y$. Let $N=\{h\in X|Th=0\}$. Then T induces a mapping T_0 of $X_0=X/N$ to Y. Prove that $T_0\in[X_0\to Y]$ and that T_0 possesses a bounded inverse. If $X=\mathbf{E}^n$, $Y=\mathbf{E}^m$, $n>m$, what does this mean? This result is at the basis of the so called "generalized inverse."

4. Let $y\in\mathbf{E}^n$. Let $A(t)$ be an $n\times n$ matrix of continuous functions $a_{ij}(t)$ on $t\geqslant 0$. Prove that the successive approximations

$$y^{m+1}(t)=\int_0^t A(s)y^m(s)\,ds+\xi$$

converge to a solution of the system of differential equations $y'=A(t)y$, $y(0)=\xi$, and that the solution is unique, on any finite t-interval. Use any convenient vector norm in $\mathbf{E}^n$, and the related $n\times n$ matrix norm.

5. Consider the matrix differential equation for $n\times n$ matrices,

$$M'(t)=A(t)M(t), \qquad M(\tau)=[m_{ij}(\tau)]$$

and prove the same theorem as in Exercise 6.2.4. Show that if $\det[m_{ij}(\tau)]\neq 0$, then $\det M(t)\neq 0$ for any t. Call the solution for

$$m_{ij}(\tau)=\begin{cases}1, & i=j,\\ 0, & i\neq j,\end{cases}$$

$\Phi(t,\tau)$. Show that the solution of Exercise 6.2.2 is given by

$$y(t)=\Phi(t,\tau)y(\tau).$$

We call $\Phi(t,\tau)$ a *fundamental matrix*. If $A(t)$ is constant, show that

$$\Phi(t,\tau)=e^{A(t-\tau)}$$

where

$$e^{A(t-\tau)}=\sum_{m=0}^{\infty} A^m\frac{(t-\tau)^m}{m!}.$$

6. Repeat Exercises 6.2.4 and 6.2.5 for the differential equation

$$y'=A(t)y+b(t), y(0)=\xi.$$

(Hint: Use the method of variation of parameters to determine a particular solution.)

$$y(t)=\Phi(t,\tau)y(\tau)+\int_{\tau}^{t}\Phi(t,s)b(s)\,ds$$

furnishes the *unique* solution.

7. Now repeat Exercises 6.2.4 and 6.2.5 for the case that $\|A(t)\|\in L_1(t)$, on $t\geqslant 0$. (For more along this line see Porter [1], 119–129.)
8. If $A\in[X]$ and $\|A\|<1$, then $A^n\to 0$, $n\to\infty$.
9. Prove that if $A\in R$, the bounded invertible operators on X, and $B\in[X]$ satisfies $\|B\|<1/\|A^{-1}\|$, then $A+B$ is invertible and

$$\|[A+B]^{-1}\|\leqslant\frac{\|A^{-1}\|}{(1-\|B\|\,\|A^{-1}\|)}.$$

6.3 SEQUENCES OF LINEAR OPERATORS

In addition to convergence in the uniform topology for elements of $[X\to Y]$, there are certain very useful properties of sequences of operators associated with a weaker form of convergence.

DEFINITION 6.3.1. We say that a sequence of operation $T^n\in[X\to Y]$ converges *strongly* to an operator $T\in[X\to Y]$ if for any $x\in X$, $\|T^nx-Tx\|\to 0$, $n\to\infty$.

We write $T^n\rightrightarrows T$ if $\|T^n-T\|\to 0$, $n\to\infty$ while for strong convergence we write $T^n\to T$, $n\to\infty$, to distinguish the two cases.

The following result is known variously as the *principle of uniform boundedness* or the:

BANACH-STEINHAUS THEOREM 6.3.1. *Let $\{T\}$ be a set in $[X \to Y]$, for two Banach spaces X, Y. If $\|Tx\|$ is bounded for each fixed $x \in X$, as T ranges over $\{T\}$, then there exists $M > 0$ such that $\|T\| \leqslant M$ for all of $\{T\}$.*

PROOF. If there exists an open ball S with radius ρ about x^0 such that $\|Tx\| \leqslant \overline{M}$ for all $x \in S$, all $T \in \{T\}$, then $\|x\| \leqslant \rho$ implies

$$\begin{aligned} \|Tx\| &= \|T(x+x^0) - Tx^0\| \\ &\leqslant \|T(x+x^0)\| + \|Tx^0\| \\ &\leqslant \overline{M} + \overline{M} = 2\overline{M}. \end{aligned}$$

This shifts the ball to $x^0 = 0$. For any x, $\|\rho x / \|x\|\,\| \leqslant \rho$ so

$$\|Tx\| = \left\| \frac{\|x\|}{\rho} T \frac{\rho x}{\|x\|} \right\| \leqslant \frac{2\|x\|}{\rho} \overline{M}.$$

If we set $M = 2\overline{M}/\rho$, we find $\|T\| \leqslant M$ for all T.

We need only show the existence of such a ball. Suppose the theorem false. Hence for any $a > 0$, and any open ball S about $x^0 = 0$ we can find $x^a \in S$, and $T^a \in \{T\}$ with $\|T^a x^a\| > 1/a$; in fact, by the continuity of the norm, each x^a is contained in a ball $K_a \subset S$ on which $\|T^a x\| > 1/a$. We let a run through the positive integers and find K_n, a ball of radius $\rho_n < 1/n$, a T^n and an x^n with $K_n \supset K_{n+1}$, $x^n \in K_n$ and

$$\|T^n x\| > n, \qquad x \in K_n.$$

The center of the ball K_n we call $\bar{\chi}^n$. Since $\rho_n < 1/n$, the sequence $\bar{\chi}^n$ is a Cauchy sequence. X is complete, so $\bar{\chi}^n \to \bar{\chi}$, $n \to \infty$ where $\bar{\chi} \in K_n$ for all n. Then $\|T^n \bar{\chi}\| > n, \|T^n \bar{\chi}\| \to \infty$, $n \to \infty$. This violates the boundedness of $\|Tx\|$ for each fixed x, and so proves the existence of the ball S. ∎

COROLLARY 1. *If $T^n \in [X \to Y]$, $n = 1, 2, 3, \ldots$, and $Tx = \lim_{n \to \infty} T^n x$ exists for all $x \in X$, then $\|T^n\|$ is bounded for all n, $T \in [X \to Y]$ and $T^n \to T$, $n \to \infty$.*

PROOF. The family $\{T^n\}$ satisfies the conditions of the Banach-Steinhaus theorem, so there is an M with $\|Tx\| \leqslant M\|x\|$ for all $x \in X$. ∎

The following result is most useful in separable spaces. It is a reduction theorem, for we need only investigate convergence on some dense set S, not on the whole of X.

THEOREM 6.3.2. *Let X, Y be Banach spaces. Let T^n, T be in $[X \to Y]$ for all n. Then $T^n x \to Tx$, $n \to \infty$ for all $x \in X$ if and only if*

1. $\|T^n\| \leqslant M < \infty$ *for some* $M > 0$, *and for some* S *with* $\bar{S} = X$;
2. $T^n s \to Ts$, $n \to \infty$ *for all* $s \in S$.

PROOF OF NECESSITY. Since $T^n x \to Tx, n \to \infty$, for all $x \in X$, condition 1 is trivial; condition 2 follows from the corollary to the Banach-Steinhaus theorem.

PROOF OF SUFFICIENCY. Choose $\epsilon > 0$. If $\|T^n\| \leqslant M < \infty$, for any $x \in X$ find an $s \in S$ with $\|x - s\| < \epsilon/3M$. Since $T^n s$ converges for $n \to \infty$, we must have N with

$$\|T^{N+P}s - T^N s\| < \frac{\epsilon}{3}, \qquad P = 1, 2, 3, \ldots .$$

Then

$$\begin{aligned}\|T^{N+P}x - T^N x\| &\leqslant \|T^{N+P}x - T^{N+P}s\| + \|T^{N+P}s - T^N s\| + \|T^N s - T^N x\| \\ &\leqslant \frac{M\epsilon}{3M} + \frac{\epsilon}{3} + \frac{M\epsilon}{3M} \\ &\leqslant \epsilon, \qquad P = 1, 2, 3, \ldots .\end{aligned}$$

Hence $T^n x \to Tx$, $n \to \infty$. ∎

We next study some applications of these ideas. The first to numerical integration requires the following:

DEFINITION 6.3.1. A numerical integration formula on $[0, 1]$, with $0 \leqslant t_0 < t_1 < \cdots < t_n \leqslant 1$,

$$\sum_{k=0}^{n} A_k x(t_k),$$

is said to be *exact* for x with respect to the nonnegative weight function $p(t)$ if

$$\sum_{k=0}^{n} A_k x(t_k) = \int_0^1 p(t) x(t)\, dt.$$

THEOREM 6.3.3. *Given is an infinite sequence of numerical integration formulas*

$$\int_0^1 p(t)x(t)\,dt \sim \sum_{k=0}^{n} A_{k,n}x(t_{k,n}), \qquad n=1,2,3,\ldots$$

the nth *of which is exact for all polynomials of degree* $\leqslant n$. *Then*

$$\sum_{k=0}^{n} A_{k,n}x(t_{k,n}) \to \int_0^1 p(t)x(t)\,dt, \qquad n\to\infty,$$

for any $x(t)\in C[0,1]$ *if and only if for some* $M<\infty$

$$\sum_{k=0}^{n} |A_{k,n}| \leqslant M, \qquad n=1,2,\ldots.$$

PROOF. Each of

$$\sum_{k=0}^{n} A_{k,n}x(t_{k,n})$$

is clearly a linear functional mapping $X=C[0,1]$ to $Y=\mathbf{R}$. In X we use the sup norm, in Y the absolute value norm. If $x(t)$ is a polynomial of degree m, the limit is correct; in fact for $n\geqslant m$ all the functionals take values identical to the integral. The set of polynomials is dense in $C[0,1]$. On the one hand the norm of the functionals satisfies

$$\left\| \sum_{k=0}^{n} A_{k,n}x(t_{k,n}) \right\| \leqslant \sum_{k=0}^{n} |A_{k,n}|\,\|x\|.$$

On the other hand if $x(t)$ is any continuous function with $x(t_{k,n})=$ signum $A_{k,n}$ we find

$$\left\| \sum_{k=0}^{n} A_{k,n}x(t_{k,n}) \right\| = \sum_{k=0}^{n} |A_{k,n}|.$$

Thus this is actually the norm of the given linear functional. We can apply the reduction theorem. ■

COROLLARY 1. *If further* $A_{k,n} \geqslant 0$, *then convergence occurs for any* $x \in C[0,1]$.

In this case $\Sigma_{k=0}^{n}|A_{k,n}| = \Sigma_{k=0}^{n} A_{k,n}$. Also each formula is exact for polynomials of degree 0. But if $x(t)=1$,

$$\sum_{k=0}^{n} A_{k,n} = \int_0^1 p(t)\,dt$$

so uniform boundedness is immediate.

Next we apply our results to certain classes of singular integrals.

THEOREM 6.3.4. *Let* $\{H_n(s,t)\}$ *be a sequence of functions continuous on* $0 \leqslant s,t \leqslant 1$ *such that for every* α, β *with* $0 \leqslant \alpha < s < \beta \leqslant 1$,

$$\lim_{n\to\infty} \int_\alpha^\beta H_n(s,t)\,dt = 1. \tag{1}$$

Then, for every function $x \in L_1(0,1)$,

$$\lim_{n\to\infty} \int_0^1 \left| \int_0^1 H_n(s,t)x(t)\,dt - x(s) \right| ds = 0,$$

if and only if for some $M < \infty$

$$\int_0^1 |H_n(s,t)|\,ds \leqslant M, \qquad n=1,2,3,\ldots. \tag{2}$$

PROOF. Since each $H_n(s,t)$ is continuous,

$$T^n(x) = \int_0^1 H_n(s,t)x(t)\,dt$$

is defined for every $x \in L_1(0,1)$. We wish to show that for every $x \in L_1$, $T^n x \to x$, $n \to \infty$. Here the identity operator I is the limiting operator and clearly $I \in [L_1]$. Also, if $\mathbf{\Phi}_I$ is the characteristic function of any interval $[\alpha,\beta]$, eq. 1 means that $T^n \mathbf{\Phi}_I \to \mathbf{\Phi}_I$, $n \to \infty$. Then, since any step function is a linear combination of such $\mathbf{\Phi}_I$, it follows that $T^n \phi \to \phi$, $n \to \infty$ on the dense set of all step functions in L_1. Thus Theorem 6.3.2 may be applied and we need only establish that $\|T^n\|$ is bounded in n. We calculate $\|T^n\|$

explicitly as follows:

$$\begin{aligned}\|T^n x\| &= \int_0^1 \left| \int_0^1 H_n(s,t)x(t)\,dt \right| ds \\ &\leqslant \int_0^1 \int_0^1 |H_n(s,t)||x(t)|\,dt\,ds \\ &\leqslant \int_0^1 \int_0^1 |H_n(s,t)|\,ds|x(t)|\,dt \\ &\leqslant \operatorname*{ess\,sup}_{t\in[0,1]} \int_0^1 |H_n(s,t)|\,ds\,\|x(t)\|.\end{aligned}$$

when we apply Fubini's theorem. Thus

$$\|T^n\| \leqslant \operatorname*{ess\,sup}_{t\in[0,1]} \int_0^1 |H_n(s,t)|\,ds.$$

On the other hand the fundamental theorem of the calculus implies, for any $s \in (0,1)$,

$$k\int_s^{s+1/k} H_n(s,t)\,dt(\rightarrow) H_n(s,t), \qquad k\rightarrow\infty$$

while Fatou's lemma implies

$$\int_0^1 |H_n(s,t)|\,ds(\leqslant) \liminf_{k\rightarrow\infty} \int_0^1 \left| k\int_s^{s+1/k} H_n(s,t)\,dt \right| ds.$$

Then the sequence of functions

$$x(t) = \begin{cases} k, & s \leqslant t \leqslant s+1/k, \\ 0, & 0 \leqslant t < s \text{ or } s+1/k \leqslant t < 1, \end{cases}$$

satisfies

$$\liminf_{k\rightarrow\infty} \|T^n x_k\| (\geqslant) \int_0^1 |H_n(s,t)|\,ds.$$

Hence

$$\|T^n\| = \operatorname*{ess\,sup}_{t\in[0,1]} \int_0^1 |H_n(s,t)|\,ds.$$

Then eq. 2 is exactly the needed uniform boundedness condition that implies the truth of the theorem. ■

Although proven above for $H_n(s,t)$ continuous, much weaker conditions can be used. What is needed are conditions that guarantee $T^n x$ is defined for all $x \in L_1$, for each fixed n. The remaining arguments of the proof are independent of the continuity assumption.

COROLLARY 1. *The theorem remains true if we assume about* $\{H_n(s,t)\}$ *only that for each fixed* n, $H_n(s,t) \in L_1$ $(0 \leqslant s,t \leqslant 1)$ *and that for every* $x \in L_1$

$$\int_0^1 H_n(s,t)x(t)\,dt$$

is defined and is in L_1.

The interest of this theorem lies in the applications. Thus in the exercise, it is used to show constructively the existence of a sequence of trigonometric polynomials that converges to any given $x \in L_1$; then the trigonometric polynomials are dense in L_1. It can also be used to prove that the ordinary polynomials are dense in L_1.

For similar theorems in $C[0,1]$ see Korovkin [1], Chapter 1, and the following exercises.

EXERCISES TO SECTION 6.3

1. Let

$$K_n(u) = A_n\left[\cos\tfrac{1}{2}u\right]^{2n}$$

with

$$1 = A_n\int_0^{2\pi}\left[\cos\tfrac{1}{2}u\right]^{2n}du, \qquad n = 0, 1, \ldots .$$

(Note:

$$\left[\cos\tfrac{1}{2}u\right]^{2n} = 2^{-2n}\sum_{k=0}^{2n}\binom{2n}{k}\cos(n-k)u.)$$

Let

$$f_n(x) = \int_0^{2\pi} f(\tau)K_n(\tau - x)\,d\tau.$$

Show that:

(a) $K^n f = f^n$ maps L_1 into L_1;
(b) $f^n(x)$ is a trigonometric polynomial;
(c) $A_n \int_{\alpha}^{\beta} \left[\cos \frac{1}{2}(\tau - x)\right]^{2n} d\tau \to 1, n \to \infty$ if $0 \leqslant \alpha < x < \beta \leqslant 2\pi$;
(d) $\|f^n - f\| \to 0$, $n \to \infty$ in L_1.
(e) If further $f \in C[0,1]$, $\max_{[0,1]} |f^n - f| \to 0$, $n \to \infty$. (Weirstrass trigonometric approximation theorem.)

2. Does the same approach work if

$$K_n(u) = A_n[1 - u^2]^n$$

(i.e., to prove the Weirstrass polynomial approximation theorem)?

6.4 ONE-SIDED AND TWO-SIDED INVERSES

The mapping T of the space of all infinite sequences $\alpha : (\alpha_1, \alpha_2, \alpha_3, \ldots)$ that sends α into $\beta = T\alpha$ where

$$\beta_1 = \alpha_2, \quad \beta_2 = \alpha_3, \ldots, \quad \beta_n = \alpha_{n+1}, \ldots$$

has a sort of inverse: a mapping R_λ that sends α into $\gamma = R_\lambda \alpha$ where λ is some scalar and

$$\gamma_1 = \lambda, \quad \gamma_2 = \alpha_1, \quad \gamma_3 = \alpha_2, \ldots, \quad \gamma_n = \alpha_{n-1}, \ldots.$$

That is $\alpha = TR_\lambda \alpha$ for all α. This T has an infinite number of *right inverses* R_λ; it has no *left inverses*, that is, mappings L such that $LT\alpha = \alpha$ for all α. Similarly R_λ is a left inverse of T on the domain with $\alpha_0 = \lambda$. Sometimes it is possible to establish directly only the existence of a left inverse. The following theorem then permits us to infer that the left inverse is a two-sided or ordinary inverse.

THEOREM 6.4.1. *Let $T \in [X \to Y], R_T = Y, L \in [Y \to X]$, with $LTx = x$ for all $x \in X$. Then $LT = TL = I$.*

PROOF. Since $\|Tx\| = \|x\| \leqslant \|L\| \, \|Tx\|$,

$$\|Tx\| \geqslant \frac{1}{\|L\|} \|x\|.$$

Hence T^{-1} exists and $\|T^{-1}\| \leqslant \|L\|$. Then

$$(LT)T^{-1} = L(TT^{-1}) = L = (I)T^{-1} = T^{-1}.$$

Hence $L = T^{-1}$ and L has T as a left inverse. ■

The requirement $STx = x$ for all $x \in X$ is violated in the example above.

If T maps X to Y, then the existence of a *right inverse* R implies the solvability of $Tx = y$ for a right inverse satisfies $TRy = T(Ry) = y$ for all $y \in D_R$. The existence of a *left inverse* L with domain D_L implies that the solution of $Tx = y$, if it exists, is unique.

EXERCISES TO SECTION 6.4

1. Prove that if $T \in [X \to Y]$ and $L \in [R_X \to X]$ such that L is a left inverse, then the solution of $Tx = y$, if it exists, is unique.
2. Prove that if T has both a left and a right inverse, then $L = R = T^{-1}$.
3. Let X be a Banach space, $T \in [X]$. If there exists $M \in [X]$ such that $\|I - MT\| < 1$, then $(MT)^{-1}M$ is a left inverse of T, with $\|(MT)^{-1}M\| \leqslant \|M\|/(1 - \|I - MT\|)$. If $Tx = y$, then $\|x - My\| \leqslant \{\|I - MT\| \times (1 - \|I - MT\|)\} \|My\|$. That is, all solutions of $Tx = y$ lie in a bounded sphere about My. (Hint: $(MT)^{-1}MT = I$.) Similarly, if $M \in [X]$ and $\|I - TM\| < 1$, then $M(TM)^{-1}$ is a bounded right inverse of T (see Noble [1]).
4. If X, Y are Banach spaces and $A \in [X \to Y]$ is invertible with $A^{-1} \in [Y \to X]$, then $B \in [X \to Y]$ and $\|B - A\| \leqslant 1/\|A^{-1}\|$ imply the existence of a left inverse $[A^{-1}B]^{-1}A^{-1}$ and a right inverse $A^{-1}[A^{-1}B]^{-1}$ for B.

7

A THEORY OF APPROXIMATE METHODS

7.0 INTRODUCTION

Even when the contraction mapping principle guarantees the existence of a solution to a functional equation, it is rarely possible to carry out the prescribed operations for the calculation of the solution, or of an approximate solution for an arbitrarily small error. In other cases, where we know the existence of a solution, the conditions for the convergence of successive approximations may be lacking. It is also true that usually computers permit us only to solve approximately a finite number of linear or nonlinear equations in a finite number of variables (each capable of taking on only a discrete set of values). In consequence, many approximate methods have been developed on an ad hoc basis for the approximate solution of problems that can be stated in terms of functional equations. We present here the general approach developed during the 1940s by L. V. Kantorovich [1], first announced in 1948.

In the present chapter we consider linear problems, leaving nonlinear problems until we develop more technology than is needed for this chapter.

To fix ideas, suppose that we wish to solve a functional equation $F(x)=0$ where x is an element of some normed linear vector space. It is possible to think of methods of approximate solution in at least the following two ways. From one point of view we seek an $\hat{x}$ such that $F(\hat{x})$ is near 0. Another point of view is that there is some approximating mapping $\hat{F}$ such that $\hat{F}(\hat{x})=0$; it is the latter view that L. V. Kantorovich exploited in 1948 and that leads to very powerful general approaches to the problem of approximate solution. For reasons that will become apparent in Chapter 9, these methods are often called projection methods.

7.1 PROBLEM STATEMENT

We consider a normed linear vector space X and a functional equation of the form

$$Kx = x - \lambda Tx = y, \tag{1}$$

where $x, y \in X$, and T is a linear mapping from X to X. In X we have a distinguished subspace $\hat{X} \subset D_T$ that is *complete* (so $\hat{X}$ is a Banach space) and a linear projection operator $P: X \to \hat{X}$. (Eventually both $\hat{X}$ and P will be indexed by n and we shall then consider convergence questions as $n \to \infty$.) About P we assume that

$$P^2 = P, D_P = X, R_P = \hat{X}.$$

Associated in some ad hoc manner with eq. 1 in X, we consider an equation in $\hat{X}$

$$\hat{K}\hat{x} = \hat{x} - \lambda \hat{T}\hat{x} = Py. \tag{2}$$

This, in some sense to be specified in application, is an *approximating equation* for eq. 1, which we call the *exact* equation. We assume $\hat{K} \in [\hat{X}]$, $PK \in [X]$.

At various times we use the following assumptions connecting eqs. 1 and 2:

1. $\hat{T}$ *approximates* T uniformly on $\hat{X}$. That is, $D_T \supset \hat{X}$, and there is an $\epsilon > 0$ such that

$$\|PT\hat{x} - \hat{T}\hat{x}\| \leqslant \epsilon \|\hat{x}\|, \qquad \hat{x} \in \hat{X}.$$

   Since $P\hat{x} = \hat{x}$, this is equivalent to

$$\|PK\hat{x} - \hat{K}\hat{x}\| \leqslant |\lambda| \epsilon \|\hat{x}\|, \qquad \hat{x} \in \hat{X}.$$

2. *Uniform approximation of* R_T: There is an $\epsilon_1 > 0$ such that for every $x \in D_T$ there is an $\hat{x} \in \hat{X}$

$$\|Tx - \hat{x}\| \leqslant \epsilon_1 \|x\|.$$

3. *Approximation of* y: given y, there are $\hat{y}$ and $\epsilon_2 = \epsilon_2(y)$ such that

$$\|y - \hat{y}\| \leqslant \epsilon_2 \|y\|.$$

With these assumptions we can connect the solvability of eqs. 1 and 2. We make use of the following generalization of the Banach fixed point theorem:

ITERATION LEMMA 7.1.1. *Let* X, Y *be Banach spaces,* $V \in [X \to Y]$. *For every* $y \in Y$ *let there be an* $x \in X$ *such that*

$$\|Vx - y\| \leqslant q\|y\|, \qquad \|x\| \leqslant N\|y\|$$

where q, N *are constants and* $q < 1$. *Then the equation* $Vx = y$ *is solvable for every* $y \in Y$ *and the solution* x *satisfies*

$$\|x\| \leqslant \left\{ \frac{N}{(1-q)} \right\} \|y\|.$$

PROOF. Let $y^1 = y$. There exists $x^1 \in X$ such that

$$\|Vx^1 - y^1\| \leqslant q\|y^1\|, \qquad \|x^1\| \leqslant N\|y^1\|.$$

Let $y^2 = y^1 - Vx^1$. Consider $Vx = y^2$. There exists $x^2 \in X$ such that

$$\|Vx^2 - y^2\| \leqslant q\|y^2\|, \qquad \|x^2\| \leqslant N\|y^2\|.$$

Also

$$\|Vx^2 - y^2\| \leqslant q^2\|y^1\|, \qquad \|x^2\| \leqslant qN\|y^1\|.$$

Inductively suppose we set

$$y^{k+1} = y^k - Vx^k, \qquad k = 1, 2, 3, \ldots$$

with

$$\|Vx^k - y^k\| \leqslant q^k\|y^1\|, \qquad \|x^k\| \leqslant q^{k-1}N\|y^1\|.$$

Then there exists x^{k+1} such that

$$\|Vx^{k+1} - y^{k+1}\| \leqslant q\|y^{k+1}\|, \qquad \|x^{k+1}\| \leqslant N\|y^{k+1}\|.$$

Hence

$$\|Vx^{k+1} - y^{k+1}\| \leqslant q^{k+1}\|y^1\|, \qquad \|x^{k+1}\| \leqslant q^kN\|y^1\|.$$

It is clear that

$$y^{k+1} = y^k - Vx^k = y^{k-1} - Vx^{k-1} - Vx^k$$
$$= \cdots = y^1 - V(x^k + x^{k-1} + \cdots + x^1).$$

But

$$\sum_{k=1}^{n} \|x^k\| \leqslant N\|y^1\| \sum_{k=1}^{n} q^{k-1}.$$

Hence

$$\sum_{k=1}^{\infty} \|x^k\| \leqslant N\|y^1\| \sum_{k=1}^{\infty} q^{k-1} = \left\{\frac{N}{(1-q)}\right\}\|y^1\|.$$

Also $\Sigma_{k=1}^{\infty} x^k$ converges in X to x. Since V is bounded, so continuous, and y^{k+1} satisfies $\|y^{k+1}\| \leqslant q^k\|y^1\| \to 0$, $k \to \infty$, we have $Vx = y$ and $\|x\| \leqslant \{N/(1-q)\}\|y\|$. ■

7.2 THE DIRECT THEOREMS

We assume the solvability of eq. 1 of Section 7.1 and study the solvability of eq. 2 of the same section.

THEOREM 7.2.1. Let there exist $K^{-1} \in [X]$. *Let Assumptions* 1 *and* 2 *of Section* 7.1 *hold. If*

$$q = |\lambda|\{\epsilon[1 + |\lambda|\epsilon_1] + \epsilon_1\|PK\|\}\|K^{-1}\| < 1,$$

then for every $\hat{y} \in \hat{X}$ *there is at least one* $\hat{x}^*$ *such that*

1. $\hat{K}\hat{x}^* = \hat{y}$;
2. $\|x^*\| \leqslant N\|\hat{y}\|/(1-q)$, *where* $N = (1 + |\lambda|\epsilon_1)\|K^{-1}\|$.

PROOF. The proof proceeds by showing that the conditions of the iteration lemma hold. Since $K^{-1} \in [X]$, we can always solve $Kx = \hat{y}$ by $x^0 = K^{-1}\hat{y} \in D_T$. If $z = x^0 - \hat{y} = x^0 - Kx^0 = \lambda Tx^0$ then we can find a $\hat{z} \in \hat{X}$ such that

$$\|z - \hat{z}\| \leqslant |\lambda|\epsilon_1\|x^0\|.$$

We do this using Assumption 2 and approximating Tx^0 by $\hat{z}/\lambda$ with

$$\left\| Tx^0 - \frac{\hat{z}}{\lambda} \right\| \leqslant \epsilon_1 \|x^0\|.$$

Then set $\hat{x} = \hat{z} + \hat{y}$. We have

$$\begin{aligned} \|\hat{K}\hat{x} - \hat{y}\| = \|\hat{K}\hat{x} - P\hat{y}\| &= \|\hat{K}\hat{x} - PKx^0\| \\ &\leqslant \|\hat{K}\hat{x} - PK\hat{x}\| + \|PK(\hat{x} - x^0)\| \\ &\leqslant |\lambda|\epsilon\|\hat{x}\| + \|PK\|\,\|\hat{x} - \hat{x}^0\|, \end{aligned}$$

using Assumption 1. Since $\hat{x} - x^0 = \hat{z} - z$ we have $\|\hat{x} - x^0\| = \|\hat{z} - z\| \leqslant |\lambda|\epsilon_1\|x^0\|$. But $x^0 = K^{-1}\hat{y}$ so $\|x^0\| \leqslant \|K^{-1}\|\,\|\hat{y}\|$ while

$$\begin{aligned} \|\hat{x}\| \leqslant \|x^0\| + \|\hat{x} - x^0\| &\leqslant [1 + |\lambda|\epsilon_1]\,\|x^0\| \\ &\leqslant [1 + |\lambda|\epsilon_1]\,\|K^{-1}\|\,\|\hat{y}\| \\ &\leqslant N\|\hat{y}\|. \end{aligned}$$

Thus

$$\begin{aligned} \|\hat{K}\hat{x} - \hat{y}\| &\leqslant \{|\lambda|\epsilon[1 + |\lambda|\epsilon_1] + |\lambda|\epsilon_1\|PK\|\}\,\|K^{-1}\|\,\|\hat{y}\| \\ &= q\|\hat{y}\|. \end{aligned}$$

If $q < 1$, then we may apply Lemma 7.1.1. ■

COROLLARY 1. *If, further, $\hat{K}$ and $\hat{X}$ are such that $\hat{K}\hat{x} = \hat{y}$ has at most one solution for every $\hat{y} \in \hat{X}$ then there exists $\hat{K}^{-1} \in [\hat{X}]$ and*

$$\|\hat{K}^{-1}\| \leqslant \frac{N}{(1-q)}.$$

In this case $\hat{K}$ is 1 : 1 onto and the bound is a consequence of condition 2 in Theorem 7.2.1.

COROLLARY 2. *If, instead of Assumption 1, we have $T \in [X]$, so $K \in [X]$ and $\|T\hat{x} - \hat{T}\hat{x}\| \leqslant \epsilon'\|\hat{x}\|$, for all $\hat{x} \in \hat{X}$, then we may take*

$$q = |\lambda|\{\epsilon'[1 + |\lambda|\epsilon_1] + \|K\|\epsilon_1\}\,\|K^{-1}\|.$$

independent of P.

PROOF. To see this, consider

$$\begin{aligned}\|\hat{K}\hat{x}-\hat{y}\| &= \|\hat{K}\hat{x}-Kx^0\| \\ &\leqslant \|\hat{K}\hat{x}-K\hat{x}\|+\|K(\hat{x}-x^0)\| \\ &\leqslant |\lambda|\,\|\hat{T}\hat{x}-T\hat{x}\|+\|K\|\,\|\hat{x}-x^0\| \\ &\leqslant |\lambda|\epsilon'\|\hat{x}\|+\|K\|\,|\lambda|\epsilon_1\|x^0\| \\ &\leqslant |\lambda|\{\epsilon'[1+|\lambda|\epsilon_1]+\|K\|\epsilon_1\}\,\|K^{-1}\|\,\|\hat{y}\|. \quad \blacksquare\end{aligned}$$

Notice that this gives a q independent of P. Since $P\hat{T}=\hat{T}$, we can always take $\epsilon \leqslant \|P\|\epsilon'$. If for any $\hat{x}\in\hat{X}$ there is a $\hat{z}\in\hat{X}$ with

$$\|T\hat{x}-\hat{z}\| \leqslant \epsilon_1'\|\hat{x}\|,$$

then

$$\begin{aligned}\|\hat{T}x-\hat{T}\hat{x}\| &= \|T\hat{x}-PT\hat{x}+PT\hat{x}-\hat{T}\hat{x}\| \\ &\leqslant \epsilon+\|T\hat{x}-\hat{z}+\hat{z}-PT\hat{x}\| \\ &= \epsilon+\|(I-P)(T\hat{x}-\hat{z})\| \leqslant \epsilon+\|(I-P)\|\epsilon_1'.\end{aligned}$$

Thus we have

$$\frac{\epsilon}{\|P\|} \leqslant \epsilon' \leqslant \epsilon+\|I-P\|\epsilon_1'.$$

The error committed by using the solution of eq. 2 of Section 7.1 in place of the solution of eq. 1 of the same section is estimated by the following:

THEOREM 7.2.2. *If Assumptions* 1, 2, *and* 3 *of Section* 7.1 *hold, while* K^{-1} *exists and* $Kx^*=y$, *then*

$$\|x^*-\hat{x}^*\| \leqslant p\|x^*\|$$

where $\hat{K}\hat{x}^*=Py$, $\nu(x^*)=\|Kx^*\|/\|x^*\|$, *and,*

$$p=2|\lambda|\epsilon\|\hat{K}^{-1}\|+(\epsilon_1|\lambda|+\epsilon_2\nu(x^*))(1+\|\hat{K}^{-1}PK\|).$$

PROOF. We approximate Tx^* and y in $\hat{X}$ so that

$$\|Tx^*-\hat{z}\| \leqslant \epsilon_1\|x^*\|,$$

$$\begin{aligned}\|y-\hat{y}\| \leqslant \epsilon_2\|y\| &= \epsilon_2\left(\frac{\|Kx^*\|}{\|x^*\|}\right)\|x^*\| \\ &= \epsilon_2\nu(x^*)\|x^*\|.\end{aligned}$$

Now $x^*-\lambda Tx^*=y$; set $\hat{x}=\lambda\hat{z}+\hat{y}$ and calculate

$$\begin{aligned}\|x^*-\hat{x}\| &= \|\lambda(Tx^*-\hat{z})+y-\hat{y}\|\\ &\leqslant (\epsilon_1|\lambda|+\nu(x^*)\epsilon_2)\|x^*\|.\end{aligned}$$

Since $0\in X$ and $\|x^*-0\|=1\,\|x^*\|$ we conclude that there is an $\hat{x}$ such that

$$\|x^*-\hat{x}\|\leqslant \min(1,\epsilon_1|\lambda|+\nu(x^*)\epsilon_2)\|x^*\|=\eta\|x^*\|.$$

Now

$$\begin{aligned}\|x^*-\hat{x}^*\| &\leqslant \|x^*-\hat{x}\|+\|\hat{x}-\hat{x}^*\|\\ &\leqslant \eta\|x^*\|+\|\hat{x}-\hat{x}^*\|.\end{aligned}$$

Since $Kx^*=y, \hat{K}\hat{x}^*=Py$, we have

$$\begin{aligned}\|\hat{x}-\hat{x}^*\| &= \|\hat{x}-\hat{K}^{-1}PKx^*\|\\ &\leqslant \|\hat{x}-\hat{K}^{-1}PK\hat{x}\|+\|\hat{K}^{-1}PK(\hat{x}-x^*)\|\\ &\leqslant \|\hat{x}-\hat{K}^{-1}PK\hat{x}\|+\eta\|\hat{K}^{-1}PK\|\,\|x^*\|.\end{aligned}$$

Last,

$$\begin{aligned}\|\hat{x}-\hat{K}^{-1}PK\hat{x}\| &= \|\hat{K}^{-1}\hat{K}\hat{x}-\hat{K}^{-1}PK\hat{x}\|\\ &\leqslant \|\hat{K}^{-1}\|\,\|\hat{K}\hat{x}-PK\hat{x}\|\\ &\leqslant |\lambda|\epsilon\|\hat{K}^{-1}\|\,\|\hat{x}\|\end{aligned}$$

while

$$\begin{aligned}\|\hat{x}\| &\leqslant \|x^*\|+\|\hat{x}-x^*\|\\ &\leqslant (1+\eta)\|x^*\|.\end{aligned}$$

Summing up, using the definition of η we have

$$\begin{aligned}\|x^*-\hat{x}^*\| &\leqslant (\eta+|\lambda|\epsilon\|\hat{K}^{-1}\|(1+\eta)+\eta\|\hat{K}^{-1}PK\|)\|x^*\|\\ &\leqslant p\|x^*\|. \quad \blacksquare\end{aligned}$$

Remark 1. For a fixed problem and a fixed solution x^* it may be possible to find $\eta=\eta(x^*)$ by some other procedure than that above. In this case we need not make Assumptions 2 and 3.

Remark 2. The error estimate of the theorem is unfortunate in that it assumes the knowledge of at least $\|x^*\|$, if not x^* itself. But if $p<1$, we can avoid this by observing that

$$\|x^*\| \leqslant \|\hat{x}^*\| + \|x^* - \hat{x}^*\|$$

so

$$\|x^* - \hat{x}^*\| \leqslant p\|\hat{x}^*\| + p\|x^* - \hat{x}^*\|$$

or

$$\|x^* - \hat{x}^*\| \leqslant \frac{p}{1-p}\|\hat{x}^*\|.$$

Remark 3. If $K \in [X]$ then we may replace $\nu(x^*)$ by $\|K\|$, in the definition of n and p.

Usually the procedure for determining P and $\hat{X}$ will involve a parameter n. We will have spaces $\hat{X}^n$, projections P^n, sequences of scalars ϵ^n, ϵ_1^n, ϵ_2^n, q^n, and p^n, and sequences of approximating operators $\hat{T}^n$ (and $\hat{K}^n$), $n=1,2,3,\ldots$. There may result sequences of approximate solutions $\hat{x}^{*n}$; the convergence of these to x^* is established in the following theorem.

THEOREM 7.2.3. *Let* $\|P^nK\| \geqslant \alpha > 0$ *for all n and let*

1. $K^{-1} \in [X]$;
2. $\hat{K}^n$ *be such that* $\hat{K}^n\hat{x} = \hat{y}$ *has at most one solution for every* $\hat{y} \in \hat{X}^n$;
3. *for each n, Assumptions* 1, 2, *and* 3 *be satisfied in such a way that*

$$\lim_{n\to\infty} \epsilon^n = 0,$$

$$\lim_{n\to\infty} \epsilon_1^n \|P^nK\| = 0,$$

$$\lim_{n\to\infty} \epsilon_2^n \|P^nK\| = 0.$$

Then for all sufficiently large n the equation $\hat{K}^n\hat{x} = P^ny$ *is solvable by* $\hat{x}^{*n}$ *such that* $\|x^* - \hat{x}^{*n}\| \to 0$, $n\to\infty$. *Further for suitable constants* Q_0, Q_1, Q_2 *we have*

$$\|x^* - x^{*n}\| \leqslant p^n\|x^*\| \leqslant Q_0\epsilon + Q_1\epsilon_1^n\|P^nK\| + Q_2\epsilon_2^n\|P^nK\|,$$

PROOF. Both $\epsilon_1^n, \epsilon_2^n \to 0$, $n\to\infty$, so $q^n \to 0$, $n\to\infty$. For all n sufficiently large, $q^n < 1-\sigma < 1$. Then Corollary 1 to Theorem 7.2.1 implies that $\hat{K}^n$ is invertible and that

$$\|(\hat{K}^n)^{-1}\| \leqslant \frac{N^n}{1-\sigma}.$$

Since $N^n=(1+|\lambda|\epsilon_1^n)\|K^{-1}\|$, N^n is a bounded sequence and $p^n \to 0$, $n\to\infty$. ■

Remark 1. If $K\in[X]$, then

$$\|x^*-\hat{x}^{*n}\| \leqslant (\eta^n+2|\lambda|\epsilon^n\|(K^n)^{-1}\| + \eta^n\|(\hat{K}^n)^{-1}\|\|P^n\|\,\|K\|)\,\|x^*\|;$$

we could assume $\epsilon^n\to 0$ and $\eta^n\|P^n\|\to 0$, $n\to\infty$ and obtain

$$\|x^*-x^{*n}\| \leqslant R_1\epsilon^n+R_2\eta^n\|P^n\|.$$

In this situation we may omit Assumption 3.

Remark 2. If $K\in[X]$, then in Theorem 7.2.3 we need *not* assume $\|P^nK\|\geqslant\alpha>0$ for all n. Then $P^2=P$ implies $\|P\|\geqslant 1$, so the convergence criteria, modified to

$$\lim_{n\to\infty}\epsilon^n=0,$$
$$\lim_{n\to\infty}\epsilon_1^n\|P^n\|=0,$$
$$\lim_{n\to\infty}\epsilon_2^n\|P^n\|=0,$$

may be used to imply $\epsilon_1^n\to 0$, $\epsilon_2^n\to 0$ and $q^n\to 0$, $n\to\infty$. We then conclude

$$\|x^*-x^n\| \leqslant Q_0\epsilon^n+Q_1\epsilon_1^n\|P^n\|+Q_2\epsilon_2^n\|P^n\|.$$

Remark 3. The results of this section become somewhat simpler if we form the approximate operator $\hat{K}$ as PK, which is frequently the natural thing to do. *This automatically insures* $\epsilon=0$. Then the approximation $\hat{T}$ is determined as PT, and, $\|(\hat{T}-PT)\hat{x}\|=0$, for all $\hat{x}\in\hat{X}$. In fact $\|(\hat{T}-PT)x\|=0$, for all $x\in X$. Thus we always have $\epsilon=0$. Then the various constants determined above become

$$q=|\lambda|\epsilon_1\|PK\|\,\|K^{-1}\|,$$
$$N=(1+|\lambda|\epsilon_1)\|K^{-1}\|,$$
$$p=[\epsilon_1|\lambda|+\epsilon_2\nu(x^*)](1+\|\hat{K}^{-1}PK\|).$$

The convergence conditions become

$$\epsilon_1^n\|P^nK\|\to 0,\qquad \epsilon_2^n\|P^nK\|\to 0,\qquad n\to\infty.$$

EXERCISES TO SECTION 7.2

1. Implicit in the method is a process of *regularization*. For an equation $(I-\lambda T)x=y$, we seek a solution $x=z+y$; show that z satisfies $z-\lambda Tz=\lambda Ty$. When $y\in D_T$, and T is an integral operator or other smoothing operator, the problem of approximating z may be easier than a direct attack on x, if only because z is smoother than x!

7.3 CONVERSE THEOREMS

Since $\hat{K}$ approximates K, it is also true that K approximates $\hat{K}$. It is possible to reverse the direction of the implications of the direct theorems above. That is, if an approximate method leads to an approximate solution, then it may be possible to infer the existence of a solution to the original problem. The existence of a bounded linear operator mapping R_K to D_K, which is a *left inverse* of K, is established in the following

THEOREM 7.3.1. *Let* $\hat{K}=PK$. *Let Assumptions* 1 *and* 2 *be true. Let* $\hat{K}^{-1}\in[\hat{X}]$. *Let*

$$r=|\lambda|\epsilon(1+|\lambda|\epsilon_1)\|\hat{K}^{-1}\|+|\lambda|\epsilon_1(1+\|K^{-1}PK\|)<1.$$

Then there exists a linear operator L such that

$$\|Ly\|\leqslant\frac{1+\|\hat{K}^{-1}P\|+|\lambda|\epsilon\|\hat{K}^{-1}\|+\|\hat{K}^{-1}PK\|}{1-r}\|y\|$$

for all $y\in R_K$, *while* $LKx=x$ *for all* x *in* D_K.

PROOF. For any $x^*\in D_T$ there is at least one solution of $Kx=Kx^*$, namely x^*. Using Assumption 2 we find $\hat{x}$ such that

$$\|Tx^*-\hat{x}\|\leqslant\epsilon_1\|x^*\|.$$

Then

$$\begin{aligned}\|x^*-\lambda\hat{x}\|&=\|\lambda Tx^*+Kx^*-\lambda\hat{x}\|\\&\leqslant|\lambda|\epsilon_1\|x^*\|+\|Kx^*\|.\end{aligned}$$

Apply Remark 1 following Theorem 7.2.2 to this situation with

$$\eta=|\lambda|\epsilon_1+\frac{\|Kx^*\|}{\|x^*\|}=|\lambda|\epsilon_1+\nu(x^*).$$

Now $\hat{K}\hat{x}=PKx^*$ is solvable by $\hat{x}^*=\hat{K}^{-1}PKx^*$ so

$$\begin{aligned}\|x^*-\hat{K}^{-1}PKx^*\| &\leqslant \{|\lambda|\epsilon(1+|\lambda|\epsilon_1+\nu(x^*))\|\hat{K}^{-1}\| \\ &\quad +[|\lambda|\epsilon_1+\nu(x^*)](1+\|\hat{K}^{-1}PK\|)\}\|x^*\|. \\ &\leqslant r\|x^*\|+\{|\lambda|\epsilon\|\hat{K}^{-1}\|+1+\|K^{-1}PK\|\}\|x^*\|.\end{aligned}$$

Then

$$\begin{aligned}\|x^*\| &\leqslant \|x^*-\hat{K}^{-1}PKx^*\|+\|\hat{K}^{-1}PKx^*\| \\ &\leqslant r\|x^*\|+\{1+\|\hat{K}^{-1}P\|+|\lambda|\epsilon\|\hat{K}^{-1}\| \\ &\quad +\|\hat{K}^{-1}PK\|\}\|Kx^*\|.\end{aligned}$$

Then

$$\|Kx^*\| \geqslant \frac{1-r}{1+\|\hat{K}^{-1}P\|+|\lambda|\epsilon\|\hat{K}^{-1}\|+\|\hat{K}^{-1}PK\|}\|x^*\|.$$

If $r<1$ we can apply Theorem 6.1.2 to imply the existence of the left inverse L and the estimate for $\|Ly\|$ claimed above. ■

Remark. The existence of a left inverse implies that the solution of $Kx=y$ is unique.

COROLLARY 1. *If the uniqueness of the solution x^* of $Kx=y$ implies solvability of Equation* 1 *of Section* 7.1 *for every y (so K is onto), then $L=K^{-1}\in[X]$.*

The proof of the corollary depends upon Theorem 6.4.1. The degree of the approximation of x^* by $\hat{x}$ is the subject of the following:

THEOREM 7.3.2. *Let Assumptions* 1 *and* 2 *be satisfied; let $\hat{K}^{-1}\in[\hat{X}]$ and let $Kx^*=y\in D_T$. Then*

$$\|x^*-[y+\lambda\hat{K}^{-1}PTy]\|\leqslant p'\|x^*\|$$

where $\delta(x^*)=\|Tx^*\|/\|x^*\|$ and

$$p'=|\lambda|[\epsilon_1(1+\|\hat{K}^{-1}PK\|)+|\lambda|\epsilon\|\hat{K}^{-1}\|(\delta(x^*)+\epsilon_1)].$$

PROOF. Set $x^*=y+z^*$ in $Kx^*=y$; then $Kz^*=\lambda Ty$ Since $\hat{K}^{-1}$ exists there is a $\hat{z}^*=\hat{x}^*-y$ with $\hat{K}\hat{z}^*=\lambda PTy$. Then $Kz^*=z^*-\lambda Tz^*=\lambda Ty$ so $z^*=\lambda T(z^*+y)=\lambda Tx^*$. Using Assumption 2, we approximate z^* by $\hat{z}$ with

$\|z^* - \hat{z}\| \leqslant |\lambda|\epsilon_1\|x^*\|$. Let

$$\hat{z}^0 = \hat{K}^{-1}PK\hat{z}.$$

Since $\hat{K}\hat{z}^* = PKz^*$, $\hat{z}^* = \hat{K}^{-1}PKz^*$ so

$$\begin{aligned}\|\hat{z}^* - \hat{z}\| &\leqslant \|\hat{K}^{-1}PK\|\,\|z^* - \hat{z}\| \\ &\leqslant |\lambda|\epsilon_1\|\hat{K}^{-1}PK\|\,\|x^*\|.\end{aligned}$$

Then, by Assumption 1,

$$\begin{aligned}\|\hat{z} - \hat{z}^0\| &= \|\hat{K}^{-1}\hat{K}\hat{z} - \hat{K}^{-1}PK\hat{z}\| \\ &\leqslant \|\hat{K}^{-1}\|\,\|\hat{K}\hat{z} - PK\hat{z}\| \leqslant |\lambda|\epsilon\|\hat{K}^{-1}\|\,\|\hat{z}\| \\ &\leqslant |\lambda|\epsilon\|\hat{K}^{-1}\|(\|z^*\| + \|\hat{z} - z^*\|) \\ &\leqslant |\lambda|\epsilon\|\hat{K}^{-1}\|(|\lambda|\delta(x^*) + |\lambda|\epsilon_1)\|x^*\|.\end{aligned}$$

So

$$\begin{aligned}\|x^* - (y + \hat{z}^*)\| &= \|y + \lambda Tx^* - y - \hat{z}^*\| \\ &= \|z^* - \hat{z}^*\| \\ &= \|x^* - \hat{x}^*\| \\ &\leqslant \|z^* - \hat{z}\| + \|\hat{z} - \hat{z}^0\| + \|\hat{z}^0 - \hat{z}^*\| \\ &\leqslant \{|\lambda|\epsilon_1 + |\lambda|^2\epsilon\|\hat{K}^{-1}\|[\delta(x^*) + \epsilon_1] \\ &\qquad + |\lambda|\epsilon_1\|\hat{K}^{-1}PK\|\}\,\|x^*\| \\ &= p'\|x^*\|. \quad \blacksquare\end{aligned}$$

If $T \in [X]$, then we may replace δ by $\|T\|$ above.

THEOREM 7.3.3. *If K is bounded and $R_K = X$, Assumptions* 1 *and* 2 *hold, $\hat{K}^{-1} \in [\hat{X}]$ and $p' < 1$, then*

1. *the solution of $Kx = y$ is always unique*;
2. *K^{-1} exists and*

$$\|K^{-1}\| \leqslant \frac{1 + |\lambda|\,\|\hat{K}^{-1}PT\|}{1 - p'};$$

3. *also*

$$\|K^{-1}-(I+\lambda\hat{K}^{-1}PT)\|\leqslant\frac{p'}{1-p'}\{1+|\lambda|\,\|\hat{K}^{-1}PT\|\}.$$

PROOF. If $Kx^*=0$, then

$$\begin{aligned}\|x^*-(0+\hat{z}^*)\|&=\|x^*-0+\lambda K^{-1}PT0\|\\&=\|x^*\|\leqslant p'\|x^*\|<\|x^*\|\end{aligned}$$

implies $\|x^*\|=0$. Thus the solution of $Kx=y$ exists and is unique for every y. Set $Kx^*=y$. Then

$$\begin{aligned}\|x^*-(y+\lambda\hat{K}^{-1}PTy)\|&=\|x^*-Kx^*-\lambda\hat{K}^{-1}PTKx^*\|\\&=\|[I-(I+\lambda\hat{K}^{-1}PT)K]x^*\|\\&\leqslant p'\|x^*\|.\end{aligned}$$

Then

$$\|I-(I+\lambda\hat{K}^{-1}PT)K\|\leqslant p'<1.$$

Thus we have an inverse G to $(I+\lambda\hat{K}^{-1}PT)K$ and $\|G\|\leqslant 1/(1-p')$, or

$$G(I+\lambda\hat{K}^{-1}PT)K\equiv I,$$

That is $G(I+\lambda\hat{K}^{-1}PT)$ is a bounded left inverse of K defined on $R_K=X$, so is a two-sided, bounded inverse $=K^{-1}\in[X]$, with

$$\begin{aligned}\|K^{-1}\|&\leqslant\|G\|\,\|I+\lambda\hat{K}^{-1}PT\|\\&\leqslant\frac{1+|\lambda|\,\|\hat{K}^{-1}PT\|}{1-p'}.\end{aligned}$$

Also $Kx=y$ makes $x=K^{-1}y$ and

$$\begin{aligned}\|K^{-1}y-(y+\lambda\hat{K}^{-1}PTy)\|&\leqslant p'\|K^{-1}y\|\\&\leqslant p'\|K^{-1}\|\,\|y\|\\&\leqslant\frac{p'}{1-p'}(1+|\lambda|\,\|\hat{K}^{-1}PT\|)\|y\|.\quad\blacksquare\end{aligned}$$

It is also possible to develop a theorem that parallels Theorem 7.3.2.

7.4. THE NUMERICAL SPACE

It will sometimes occur that the space in which the actual computations occur is not $\hat{X}$, but an "equivalent" space $\bar{X}$. We associate with our problem a numerical problem in a Banach space $\bar{X}$. We assume that there exists a homeomorphism, that is a mapping $\psi_0 \in [\hat{X} \to \bar{X}]$ which is invertible: $\psi_0^{-1} \in [\bar{X} \to \hat{X}]$. We assume that there is a *linear* extension of $\psi_0, \psi \in [X \to \bar{X}]$ such that $\psi \hat{x} = \psi_0 \hat{x}$ on $\hat{X}$. If we start with P, then we may always define $\psi = \psi_0 P$. If however we start with ψ, we take $P = \psi_0^{-1}\psi$. Then exactly equivalent to

$$\hat{x} - \hat{T}\hat{x} = Py$$

is the equation

$$\bar{x} - \lambda \psi_0 \bar{T} \bar{x} = \psi_0 P y$$

where $\bar{x} = \psi_0 \hat{x}, \hat{x} = \psi_0^{-1}\bar{x}$. We define another approximate equation,

$$\bar{K}\bar{x} = \bar{x} - \lambda \bar{T}\bar{x} = \psi y,$$

where $\bar{K} = \psi_0 \hat{K} \psi_0^{-1}, \bar{T} = \psi_0 \hat{T} \psi_0^{-1}$. In terms of these new entities we can recompute our parameters. Assumption 1 is

$$\|\psi_0^{-1}\bar{T}\psi_0\hat{x} - \psi_0^{-1}\psi T\hat{x}\| \leqslant \epsilon \|\hat{x}\|, \qquad x \in X.$$

This will be satisfied if (Assumption 1a):

$$\|\bar{T}\psi_0\hat{x} - \psi T\hat{x}\| \leqslant \bar{\epsilon}\|\hat{x}\|, \qquad \hat{x} \in \hat{X},$$

and we set $\epsilon = \bar{\epsilon}\|\psi_0^{-1}\|$. Assumptions 2 and 3 are not changed. We can restate our theorems as follows:

THEOREM 7.4.1. *Let $K^{-1} \in [X]$. Let Assumptions 1a and 2 hold. If*

$$\bar{q} = |\lambda| \left[\bar{\epsilon}(1 + |\lambda|\epsilon_1)\|\psi_0^{-1}\| + \epsilon_1 \|\psi_0^{-1}\psi K\| \right] \|K^{-1}\| < 1$$

then for every $\bar{y} \in \bar{X}$ there is at least one $\bar{x}^*$ such that

1. $\bar{K}\bar{x}^* = \bar{y}$;
2. $\|\bar{x}^*\| \leqslant \bar{N}\|\bar{y}\|/(1 - \bar{q})$,

where

$$\bar{N} = (1 + |\lambda|\epsilon_1)\,\|\psi_0\|\,\|\psi_0^{-1}\|\,\|K^{-1}\|.$$

This is a translation obtained by setting $\bar{x}^* = \psi_0 \hat{x}^*$ where $\hat{K}\hat{x}^* = \psi_0^{-1}\bar{y}$ above.

COROLLARY 1. *If further $\bar{K}$ and $\bar{X}$ are such that $\bar{K}\bar{x} = \bar{y}$ has at most one solution for every $\bar{y} \in \bar{X}$, then there exists $\bar{K}^{-1} \in [\bar{X}]$ and*

$$\|\bar{K}^{-1}\| \leqslant \frac{\bar{N}}{1-\bar{q}}.$$

We also have the following:

THEOREM 7.4.2. *If Assumptions* 1*a*, 2, *and* 3 *hold*, $Kx^* = y$ *and* $\bar{K}^{-1} \in [\bar{X}]$, *then*

$$\|x^* - \psi_0^{-1}\bar{x}^*\| \leqslant \bar{p}\|x^*\|$$

where $\bar{K}\bar{x}^* = Py$ *and*

$$\bar{p} = 2|\lambda|\bar{\epsilon}\|\psi_0^{-1}\bar{K}^{-1}\psi_0\|\,\|\psi_0^{-1}\| + \eta\big(1 + \|\psi_0^{-1}\bar{K}^{-1}\psi K\|\big).$$

However, in this case we may also take

$$\bar{p} = (1+\eta)|\lambda|\bar{\epsilon}\|\psi_0^{-1}\bar{K}^{-1}\| + \eta\big(1 + \|\psi_0^{-1}\bar{K}^{-1}\psi K\|\big).$$

PROOF. Reverting to the proof of Theorem 7.2.2, we note that

$$\begin{aligned}\|\hat{x} - \hat{K}^{-1}PK\hat{x}\| &= \|\psi_0^{-1}\bar{K}^{-1}\bar{K}\psi_0\hat{x} - \psi_0^{-1}\bar{K}^{-1}\psi_0\psi_0^{-1}\psi K\hat{x}\| \\ &\leqslant \|\psi_0^{-1}\bar{K}^{-1}\|\,\|\bar{K}\psi_0\hat{x} - \psi K\hat{x}\| \\ &\leqslant |\lambda|\bar{\epsilon}\|\psi_0^{-1}\bar{K}^{-1}\|\,\|\hat{x}\|\end{aligned}$$

On using this the alternate form of $\bar{p}$ is established. ■

THEOREM 7.4.3. *Assume*

1. $K^{-1} \in [X]$;
2. *from the existence of a solution of $\bar{K}\bar{x} = \bar{y}$ for every $\bar{y}$, it follows that the solution is unique*;
3. *Assumptions* 1*a*, 2, *and* 3 *hold for all n and*

$$\begin{aligned}&\bar{\epsilon}^n\|(\psi_0^n)^{-1}K\| \to 0, && n \to \infty,\\ &\epsilon_1^n\|(\psi_0^n)^{-1}\psi^n K\| \to 0, && n \to \infty,\\ &\epsilon_2^n\|(\psi_0^n)^{-1}\psi^n K\| \to 0, && n \to \infty.\end{aligned}$$

Then $\bar{K}^n\bar{x}=\psi^n y$ *is solvable for all sufficiently large n and its solution* $\bar{x}^n$ *satisfies*

$$\|x^*-\hat{x}^{n*}\|\leqslant \bar{Q}\bar{\epsilon}\|(\psi_0^n)^{-1}K\|+\bar{Q}_1\epsilon_1\|(\psi_0^n)^{-1}\psi^n K\|$$
$$+\bar{Q}_2\epsilon_2^n\|(\psi_0^n)^{-1}\psi^n K\|$$

where $\hat{x}^{n*}=(\psi_0^n)^{-1}\bar{x}^{n*}$, $\bar{Q},\bar{Q}_1,\bar{Q}_2$ *are constants.*

Note that the convergence is established for $\hat{x}^{*n}$ and no meaning is attributed to "convergence of $\bar{x}^{n*}$." We may also restate the converse theorems in the new terminology and, for instance, determine in Theorem 7.3.1 that

$$\bar{r}=|\lambda|\{(1+|\lambda|\epsilon_1)\bar{\epsilon}\|\psi_0^{-1}\bar{K}^{-1}\|+\epsilon_1\|\psi_0^{-1}\bar{K}^{-1}\psi K\|\}$$

and show that

$$\|K^{-1}\|\leqslant\frac{[1+\|\psi_0^{-1}\bar{K}^{-1}\psi\|+|\lambda|\bar{\epsilon}\|\psi_0^{-1}\bar{K}^{-1}\|+\|\psi_0^{-1}\bar{K}^{-1}\psi K\|]}{1-\bar{r}}.$$

EXERCISES TO SECTION 7.4

1. Show that if $\|\psi Tx-\bar{T}\psi x\|\leqslant\epsilon\|x\|$, for all $x\in X$, then $\bar{K}^{-1}$ exists and $\|\bar{K}^{-1}\|\leqslant\|\psi\|\,\|K^{-1}\|\,\|\psi_0\|/(1-q)$ with $q=\epsilon|\lambda|\,\|K^{-1}\|\,\|\psi_0^{-1}\|$.
2. In l_2, define $\psi^n x=\{x_1,x_2,\ldots,x_n\}$ for $x=\{x_1,x_2,x_3\ldots\}$. Then show that $\|\psi^n\|=\|\psi_0^n\|=\|(\psi_0^n)^{-1}\|=1$. Here $\hat{X}^n=\{x|x_i=0, i>n\}$. Show that if

$$\sum_{i=1}^{\infty}\sum_{k=1}^{\infty}(a_{ik})^2<\infty,$$

then

$$Ax=\left\{\alpha|\alpha_i=\sum_{j=1}^{\infty}a_{ij}x_j\right\}\in l_2$$

3. Let $X=C[0,1]$; $\hat{X}$ is the set of polygonal functions formed from (t_k,δ_k), $0=t_0<t_1<\ldots<t_n=1$ by joining (t_k,δ_k), to (t_{k+1},δ_{k+1}) by a line segment. Let $\bar{X}=\mathbf{E}^{n+1}$, $\|\bar{x}\|=\max_i|\delta_i|$ if $\bar{x}=[\delta_0,\delta_1,\ldots,\delta_n]$. Define $\psi x=[x(t_0),\ldots,x(t_n)]$. Show that

$$\|\psi\|=\|\psi_0\|=\|\psi_0^{-1}\|=1.$$

4. Let $X = L_2(0,1), \bar{X} = \mathbf{E}^n$. Let $w_i(s) \in L_2(0,1)$, and let

$$\int_0^1 w_i(s) w_j(s)\, ds = \begin{cases} 0, & i \neq j, \\ 1, & i = j. \end{cases}$$

Let $\hat{X}$ be the elements of X of the form $x = \delta_1 w_1 + \delta_2 w_2 + \cdots + \delta_n w_n$. Let $\psi x = \bar{x} = (\delta_1, \ldots, \delta_n)$ where

$$\delta_i = \int_0^1 x(s) w_i(s)\, ds.$$

Let $\|\bar{x}\| = (\delta_1^2 + \cdots + \delta_n^2)^{1/2}$. Show that $\|\psi\| = \|\psi_0\| = \|\psi_0^{-1}\| = 1$.

5. Let $w_i(t), i = 1, 2, \ldots, n$, be functions in $L_2(0,1)$. Show that

   (a) $\Sigma_{i=1}^n \alpha_i w_i(t) (=) 0$ implies $\alpha_1 = \alpha_2 = \cdots = \alpha_n = 0$ holds if and only if
   (b) the Grammian matrix $H = (h_{ij})$ is nonsingular, where

$$h_{ij} = \int_0^1 w_i(t) w_j(t)\, dt.$$

6. Restate Theorems 7.3.1, 7.3.2, and 7.3.3 in terms of the numerical space $\bar{X}$.

7.5 THE CASE $X \neq Y$

One of the least obvious but most fruitful parts of this theory is an easy generalization to the case in which $X \neq Y$. We assume, as in the figure that there are two normed linear vector spaces X, Y, each with subspaces $\hat{X}, \hat{Y}$, which are Banach spaces.

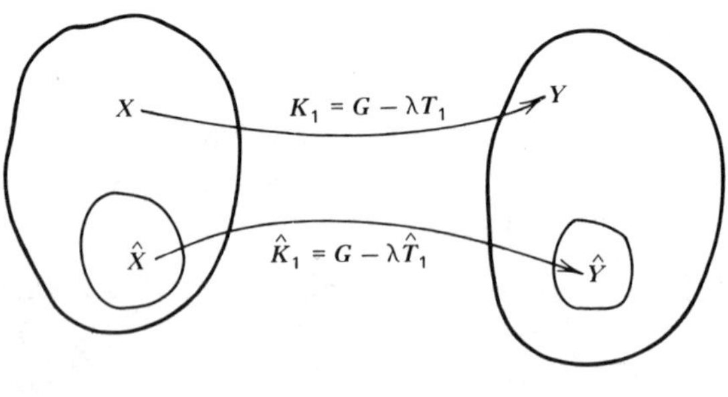

We assume that there exists a projection $\Phi \in [Y \to \hat{Y}]$ leaving each $\hat{y} \in \hat{Y}$

unchanged. We assume that there is an operator G such that

$$G \in [X \to Y],$$
$$G^{-1} \in [Y \to X],$$
$$G(x) \in \hat{Y}, \qquad \text{all } \hat{x} \in \hat{X},$$
$$G^{-1}(\hat{y}) \in \hat{X}, \qquad \text{all } \hat{y} \in \hat{Y}.$$

We assume that we associate with the equation

$$K_1 x = Gx - \lambda T_1 x = y_1$$

an approximate equation

$$\hat{K}_1 \hat{x} = G\hat{x} - \lambda \hat{T}_1 \hat{x} = \Phi y_1.$$

For simplicity we consider only the case in which $T_1 \in [X \to Y]$, $\hat{T}_1 \in [\hat{X} \to \hat{Y}]$.We introduce conditions similar to Assumptions 1, 2, and 3:

1b. There is a scalar μ such that $\|\Phi T_1 \hat{x} - \hat{T}_1 \hat{x}\| \leqslant \mu \|\hat{x}\|$ for all $\hat{x} \in \hat{X}$.
2b. There is a scalar μ_1 such that, given $x \in X$ we can find a $\hat{y} \in \hat{Y}$ with

$$\|T_1 x - \hat{y}\| \leqslant \mu_1 \|x\|.$$

3b. There is a $\hat{y}_1 \in \hat{Y}$ and a scalar μ_2 such that

$$\|y_1 - \hat{y}_1\| \leqslant \mu_2 \|y_1\|.$$

Under these assumptions the problem can be reduced to one of the type studied above. We use G^{-1} to form

$$Kx = G^{-1} K_1 x = x - \lambda G^{-1} T_1 x = G^{-1} y_1$$

and

$$\hat{K}^{-1} \hat{x} = G^{-1} \hat{K}_1 \hat{x} = \hat{x} - \lambda G^{-1} \hat{T}_1 \hat{x} = G^{-1} \Phi y_1.$$

Then set $T = G^{-1} T_1$, $\hat{T} = G^{-1} \hat{T}_1$, $P = G^{-1} \Phi G$, $y = G^{-1} y_1$. We can establish Assumptions 1, 2, and 3 as follows.

1. $\|PT\hat{x} - \hat{T}\hat{x}\| = \|G^{-1} \Phi G G^{-1} T_1 \hat{x} - G^{-1} \hat{T}_1 \hat{x}\| \leqslant \|G^{-1}\| \, \mu \|\hat{x}\|$. Then set $\epsilon = \|G^{-1}\| \, \mu$.

2. Given $x \in X$, find $\hat{y} \in \hat{Y}$ according to Assumption 2b. Then set $\hat{x} = G^{-1}\hat{y}$ and calculate

$$\|Tx - \hat{x}\| = \|G^{-1}T_1x - G^{-1}\hat{y}\|$$
$$\leqslant \|G^{-1}\|\,\mu_1\|x\|.$$

Set $\epsilon_1 = \|G^{-1}\|\,\mu$.

3. Let y_1 satisfy 3b. Then set $\hat{y} = G^{-1}\hat{y}_1$ and $y = G^{-1}y_1$; calculate

$$\|y - \hat{y}\| \leqslant \|G^{-1}y_1 - G^{-1}\hat{y}_1\|$$
$$\leqslant \|G^{-1}\|\,\mu_2\|y_1\| \leqslant \|G^{-1}\|\,\|G\|\,\mu_2\|y\|.$$

Set $\epsilon_2 = \|G^{-1}\|\,\|G\|\,\mu_2$.

Notice that we could replace Assumptions 2b and 3b by the following.

2′. There is for every $x \in X$ an $\hat{x} \in \hat{X}$ such that

$$\|G^{-1}T_1x - \hat{x}\| \leqslant \mu_1'\|x\|.$$

3′. There is a $\hat{y}$ with

$$\|G^{-1}y_1 - \hat{y}\| \leqslant \mu_2'\|G^{-1}y_1\|.$$

Then Assumptions 2 and 3 are immediately satisfied with $\epsilon_1 = \mu_1', \epsilon_2 = \mu_2'$.

The analysis in the present case is simplified if we choose norms in X and Y such that

$$\|x\|_X = \|Gx\|_Y, \|y\|_Y = \|G^{-1}y\|_X.$$

Then G is *isometric* and

$$\|P\| = \|\Phi\|, \|K\| = \|K_1\|, \|\hat{K}\| = \|\hat{K}_1\|,$$
$$\|\hat{T}\| = \|\hat{T}_1\|, \|T\| = \|T_1\|.$$

The translation of the theory to this case is now immediate.

EXERCISES TO SECTION 7.5

1. Rewrite the theorems of Section 7.2 in terms of Φ, K_1, and $\hat{K}_1$.

7.6 USE OF PROPERTIES OF THE SOLUTION OF THE EXACT EQUATION

Schindler [1] developed theorems for the case in which

1. The approximate operator $\hat{K}$ is formed by PK, and
2. Some *à priori* information is available about the ability to approximate the solution x of $Kx=y$ from $\hat{X}$.

In applications to differential and integral equations, this latter condition is often satisfied.

For these theorems we do not require a structure for K of the form $I-\lambda T$. It is convenient, however, to consider the case of Section 7.5 in which K is a mapping between distinct spaces X and Y. We assume that X and Y are normed linear vector spaces, $K\in[X\to Y]$, $\hat{X},\hat{Y}$ are Banach spaces in X, Y respectively, and P is a projection from Y to $\hat{Y}$ as in Section 7.1. Then to $Kx=y$ we associate the approximate equation $\hat{K}\hat{x}=\hat{y}$ where $\hat{K}=PK, \hat{y}=Py$. Note that $PKx=\hat{y}$.

We find, corresponding to Theorem 7.1.1, the following:

THEOREM 7.6.1 *Let there exist*

1. $K^{-1}\in[Y\to X]$;
2. *A constant q, $0<q<1$, such that, for each $y\in Y$, there can be found an $x\in\hat{X}$ with*

$$\|K^{-1}y-\hat{x}\|\leqslant\frac{q}{\|P\|\,\|K\|}\|y\|.$$

Then for any $\hat{y}\in\hat{Y}$, $\hat{K}\hat{x}=\hat{y}$ is solved by an $\hat{x}^*$ with

$$\|\hat{x}^*\|\leqslant\frac{\|K^{-1}\|+q/\|P\|\,\|K\|}{1-q}\|\hat{y}\|.$$

PROOF. Given $\hat{y}$, we solve $Kx=\hat{y}$ by $x^*=K^{-1}\hat{y}$. According to condition 2 we can approximate x^* by an $\hat{x}$ with

$$\|x^*-\hat{x}\|\leqslant\frac{q}{\|P\|\,\|K\|}\|\hat{y}\|.$$

Set $z=x^*-\hat{x}$ and observe that $PK(\hat{x}+z)=P\hat{y}=\hat{y}$ or

$$\hat{K}\hat{x}-\hat{y}=-PKZ.$$

Then

$$\begin{aligned}\|\hat{K}\hat{x}-\hat{y}\| &= \|PK(x^*-\hat{x})\| \\ &\leqslant \|PK\|\,\|x^*-\hat{x}\| \\ &\leqslant \|PK\|\frac{q}{\|P\|\,\|K\|}\|\hat{y}\| \\ &\leqslant q\|\hat{y}\|.\end{aligned}$$

Also $\|\hat{x}\| = \|x^*-z\|$, while $\|x^*\| = \|K^{-1}\hat{y}\| \leqslant \|K^{-1}\|\,\|\hat{y}\|$ and $\|z\| = \|x^*-\hat{x}\|$. Thus

$$\|\hat{x}\| \leqslant \left(\|K^{-1}\| + \frac{q}{\|P\|\,\|K\|}\right)\|\hat{y}\|.$$

With $N = \|K^{-1}\| + q/\|P\|\,\|K\|$ we have fulfilled the conditions of the iteration lemma, which implies the truth of the theorem. ■

COROLLARY 1. *Let K be such that solvability of $Kx=y$ for every $\hat{y}\in\hat{Y}$ implies the uniqueness of the solution. Then $\hat{K}^{-1}$ exists and*

$$\|\hat{K}^{-1}\| \leqslant \frac{\|K^{-1}\| + q/\|P\|\,\|K\|}{1-q}.$$

Also

$$\|K^{-1}y - \hat{K}^{-1}Py\| \leqslant \frac{q(1+\|\hat{K}^{-1}PK\|)}{\|P\|}\|x^*\|.$$

PROOF. Only the last inequality requires any comment. Now

$$\|K^{-1}y - \hat{K}^{-1}Py\| = \|x^* - \hat{x}^*\|.$$

Choose $\hat{x}$ so that $\|x^* - \hat{x}\| = \|K^{-1}y - \hat{x}\| \leqslant (q/\|P\|\,\|K\|)\ \|y\| \leqslant (q/\|P\|\,\|K\|)\ \|Kx^*\| \leqslant (q/\|P\|)\ \|x^*\|$. Then

$$\begin{aligned}\|K^{-1}y - \hat{K}^{-1}Py\| &\leqslant \|x^*-\hat{x}\| + \|\hat{x}-\hat{x}^*\| \\ &\leqslant \|x^*-\hat{x}\| + \|\hat{K}^{-1}PK\hat{x} - \hat{K}^{-1}PKx^*\| \\ &\leqslant \frac{q(1+\|\hat{K}^{-1}PK\|)}{\|P\|}\|x^*\|. \quad ■\end{aligned}$$

If now the spaces $\hat{X}$, the projection P, and the scalar q are indexed with n, we can establish the following:

THEOREM 7.6.2 *Let*

1. $K^{-1} \in [Y \to X]$;
2. *For each n, there be a constant* q^n, *such that for every* $y \in Y$ *there is an* $\hat{x}^n \in \hat{X}^n$, *with*

$$\|K^{-1}y - \hat{x}^n\| \leqslant \frac{q^n}{\|P_n\| \, \|K\|} \|y\|$$

3. *The solvability of* $P_n K \hat{x} = \hat{y}$, *for every* $y \in \hat{Y}^n$ *implies the uniqueness of the solution*;
4. $\lim_{n \to \infty} q^n = 0$.

Then for all sufficiently large n, $\hat{K}_n \hat{x} = P_n K \hat{x} = P_n \hat{Y}$ *is solved uniquely by an* $\hat{x}_n^*$ *where*

$$\|x^* - \hat{x}_n^*\| = 0(q^n) \|x^*\|.$$

PROOF. For n sufficiently large, $0 < q^n < 1$, so that all that needs to be proven is $\|x^* - \hat{x}_n^*\| = 0(q^n)$. But

$$\|x^* - \hat{x}_n^*\| \leqslant \frac{q^n \left(1 + \|K_n^{-1} P_n K\|\right)}{\|P_n\|} \|x^*\|.$$

But $\|P_n\| \geqslant 1$ (why?), while $\|\hat{K}_n^{-1}\| \leqslant \|K^{-1}\| + q^n / \|P_n\| \, \|K\|$. Thus $\|\hat{K}_n^{-1}\| = 0(1), n \to \infty$. This implies the assertion. ■

EXERCISES TO SECTION 7.6

1. Let $\bar{X}, \bar{Y}$ be Banach spaces homeomorphic to $\hat{X}, \hat{Y}$ by way of 1:1 bicontinuous linear maps ψ, ϕ, respectively. Construct an approximate equation $\bar{K}\bar{x} = \bar{y}$ in the obvious manner from

$$\phi \hat{K} \psi^{-1} \bar{x} = \phi P y = \phi \hat{y} = \Phi y = \bar{y}$$

Here $\Phi = \phi P$. Prove the following modification of the theorems of Section 7.6.

THEOREM 7.6.1′. *Let there exist*

(a) $K^{-1} \in [Y \to X]$;

(b) *A constant* $q, 0<q<1$, *such that for each* $y \in Y$ *there is an* $\bar{x} \in \bar{X}$ *with* $\|K^{-1}y - \psi^{-1}\bar{x}\| \leqslant \{q/\|\psi^{-1}\|\,\|\phi\|\,\|K\|\}\,\|y\|$. *Then for any* $\bar{y} \in \bar{Y}$, $\bar{K}\bar{x} = \bar{y}$ *has a solution* $\bar{x}^*$ *with*

$$\|\bar{x}^*\| \leqslant \frac{\|\psi\|\left[\,\|K^{-1}\phi^{-1}\| + (q/\|\Phi\|\,\|K\|)\right]}{1-q}\|\bar{y}\|.$$

If $\bar{K}$ *is such that solvability of* $Kx=y$ *for every* $y \in \bar{Y}$ *implies the uniqueness of that solution, then* $\bar{K}^{-1}$ *exists and* $\|\bar{K}^{-1}\| \leqslant \|\psi\|(\|K^{-1}\phi^{-1}\| + q/\|\Phi\|\,\|K\|)/(1-q)$, *while*

$$\|K^{-1}y - \psi^{-1}\bar{x}^*\| \leqslant \frac{q(1+\|\psi^{-1}\bar{K}^{-1}\Phi K\|)}{\|\phi^{-1}\|\,\|\Phi\|}\|K^{-1}y\|.$$

2. Now index $\bar{X}$, $\bar{Y}$, ψ, ϕ, and $\Phi = \phi P$ by n. Modify Theorem 7.62 so as to prove that $\|K^{-1}y - \psi^{-1}\bar{x}_n^*\| = 0[(1+\|\psi_n\|\,\|\psi_n^{-1}\|)q]$.

8

APPLICATIONS OF THE THEORY OF APPROXIMATE METHODS

8.0 INTRODUCTION

The value of a general theory of approximate methods lies in its applications in two different ways. Clearly the method, if applicable, leads to considerable knowledge of the exact and approximate problems and the suitability of the approximate solution obtained. Less obvious, but equally as valuable, is the structuring of the *process* of approximation, a restoration of mathematical unity to a field that is usually practiced as an art. We give several applications of the structuring to illustrate the variety of problems that can be attacked. In so doing, we will discover a need for a better understanding of the "geometry" of function spaces than we have at present; we need to understand approximation of elements of a space from a subset of the space.

8.1 INFINITE SYSTEMS OF EQUATIONS IN AN INFINITE NUMBER OF UNKNOWNS

Since early in the nineteenth century mathematicians involved with the solution of ordinary and partial differential equations not solvable in closed form developed techniques of expanding unknown solution functions as infinite or doubly infinite series of known functions with undetermined coefficients. (Compare, e.g., F. Riesz [1], and Kantorovich and

Krylov [1].) The resulting constraints, which "determine" the undetermined coefficients, often take the form $Kx=x-\lambda Ax=y$ or

$$x_i-\lambda\sum_{k=1}^{\infty}a_{i,k}x_k=y_i, \qquad i=1,2,3,\ldots.$$

A natural approximate equation is obtained by the *method of sections* in which we neglect all but the first n components of x and y and the entries $a_{i,k}, i,k=1,\ldots,n$ of the infinite matrix. It is possible to interpret the functional equation $x-\lambda Ax=y$ in various sequence spaces; first we develop a general formalism, and then to fix ideas, we illustrate it in l_2. We follow Section (7.4).

Let λ be a complex number and A an infinite matrix that is a bounded mapping from X into itself. Let $\hat{X}=\{x|x_i=0,\ i>n\}$ and $\overline{X}=\mathbf{E}^n$. Then for any $x\in X$ set

$$\begin{aligned}Px&=(x_1,x_2,\ldots,x_n,0,0,\ldots),\\ Qx&=(I-P)x\\ &=(0,\ldots,0,x_{n+1},x_{n+2},\ldots),\end{aligned}$$

and

$$\psi x=(x_1,x_2,\ldots,x_n).$$

For any $\hat{x}\in\hat{X}$, let

$$\psi_0\hat{x}=(\hat{x}_1,\hat{x}_2,\ldots,\hat{x}_n).$$

Obviously ψ_0 is invertible and $\|\psi\|=\|\psi_0\|=\|\psi_0^{-1}\|=1$. (Why?) The approximate equation $\hat{K}\hat{x}=\hat{y}$ is $\hat{x}-\lambda PA\hat{x}=Py$ while the numerical equation $\overline{K}\overline{x}=\overline{y}$ is

$$x_i-\lambda\sum_{i=1}^{n}a_{i,k}x_k=y_i, \qquad i=1,\ldots,n.$$

We evaluate the parameters of Assumptions 1, 2, and 3 from Section 7.1 as follows:

$$\begin{aligned}\left(\overline{A}\psi_0x-\psi Ax\right)_i&=\sum_{k=1}^{n}a_{i,k}x_k-\sum_{k=1}^{n}a_{i,k}x_k\\ &=0, \qquad i=1,\ldots,n.\end{aligned}$$

Thus we may take $\bar{\epsilon}=\epsilon=0$ in Assumption 1. To determine ϵ_1 for Assumption 2 we can set $\hat{x}=\psi_0^{-1}\psi Ax=PAx$ (why)? so that

$$\|Ax-\hat{x}\|=\|QAx\|\leqslant\|QA\|\,\|x\|.$$

Thus we can take $\epsilon_1=\|QA\|$. For Assumption 3 we can always take $\hat{y}=Py$ so

$$\epsilon_2=\frac{\|y-\hat{y}\|}{\|y\|}=\frac{\|Qy\|}{\|y\|}.$$

We can now assert the following general theorem about the method of sections:

THEOREM 8.1.1. *Let λ be such that $(I-\lambda A)^{-1}$ exists (for instance, if $\lambda\|A\|<1$). Then whenever*

$$\bar{q}=\lambda\|I-\lambda A\|\,\|(I-\lambda A)^{-1}\|\,\|QA\|<1$$

the numerical problem is solvable for all $\bar{y}$ and $\bar{K}^{-1}$ exists. If $Kx^*=y$ and $\bar{K}\bar{x}=Py$, then

$$\|x^*-\psi_0^{-1}\bar{x}\|\leqslant\left(|\lambda|\,\|QA\|+\frac{\|I-\lambda A\|\,\|Qy\|}{\|y\|}\right)$$
$$\times\left(1+\|I-\lambda A\|\,\|\bar{K}^{-1}\|\right)\|x^*\|,$$

where

$$\|\bar{K}^{-1}\|\leqslant\frac{(1+|\lambda|\,\|QA\|)\,\|K^{-1}\|}{1-\bar{q}}.$$

We can also state the following converse theorem:

THEOREM 8.1.2. *If $\bar{K}\bar{x}=0$ has no solution $\bar{x}\neq0$, and if*

$$\bar{r}=|\lambda|\,\|QA\|\left(1+\|\bar{K}^{-1}\|\,\|K\|\right)<1,$$

then $Kx=0$ has no solution $x\neq0$, K has a bounded inverse, and

$$\|K^{-1}\|\leqslant\frac{1+|\lambda|\,\|A\|\,\|\bar{K}^{-1}\|}{1-\bar{r}},$$

while

$$\|K^{-1}-(I+\lambda\psi_0^{-1}\bar{K}^{-1}\psi A)\| \leqslant \frac{\bar{r}}{1-\bar{r}}(1+|\lambda|\,\|A\|\,\|\bar{K}^{-1}\|).$$

It should be noted that it is not the smallness of λ that is involved in the various assumptions that are in the hypotheses of the theorems above; QA is an infinite matrix whose first n rows are identically zero while the remaining entries are identical with those of A. If, for instance, $X=l_2$, and we know that A is bounded because

$$\|A\| \leqslant \left(\sum_{i=1}^{\infty}\sum_{k=1}^{\infty}|a_{i,k}|^2\right)^{1/2} < \infty,$$

then

$$\|QA\| \leqslant \left(\sum_{i=n+1}^{\infty}\sum_{k=1}^{\infty}|a_{i,k}|^2\right)^{1/2} \to 0, \qquad n\to\infty,$$

and

$$\|Qy\| = \left(\sum_{i=n+1}^{\infty}|y_i|^2\right)^{1/2} \to 0, \qquad n\to\infty.$$

That is, the convergence theorems all apply, and we will be able to estimate the value of n beyond which $\bar{q}<1$, and so on.

EXERCISES TO SECTION 8.1

1. We can extend the analysis in Section 8.1 to the case in which $X=l_p$, $1\leqslant p\leqslant\infty$, provided that we can find conditions that guarantee that the matrix A defines an operator in $[X]$. Find conditions that suffice for this. (Further extensions may be found in Kantorovich and Krylov [1], 42–44, and the works referred to there. A different treatment of some problems in l_∞ is given there, 20–42).
2. Can we apply the analysis of Section 7.5 to the method of sections? (Hint: If $1\leqslant p<q<\infty$, then $l_q\subset l_p$, but $l_q\neq l_p$. What does this imply about G)?
3. We can apply the method of *backward error analysis* to study the method of sections as follows. If K, K^{-1}, L are all in $[X]$, and $Kx=y$ while $(K+L)(x+d)=y+r$, and if $\|L\|\leqslant 1/\|K^{-1}\|$, then one can show

(e.g., Isaacson and Keller [1], 37–39) that

$$\|d\| \leqslant \|K^{-1}\| \, \|K\| \, (1 - \|L\| \, \|K^{-1}\|)^{-1} \left(\frac{\|r\|}{\|y\|} + \frac{\|L\|}{\|K\|} \right) \|x\|.$$

Show that if $Kx^* = y$ and $\hat{K}\hat{x} = \hat{y} = Py$, then $(K + \lambda QA)[x^* + (\hat{x} - x^*)] = y - Qy$. From this find conditions that guarantee the existence of a solution of the approximate equation and an estimate of the error in $\hat{x}$. State and prove a converse theorem or else explain its nonexistence.

8.2 INTEGRAL EQUATIONS: APPROXIMATION BY DEGENERATE KERNELS

Let X be a Banach space of functions $x(t)$ where, to fix ideas, we take $t \in [0,1]$. We consider the integral equation $Kx = y$, where

$$x(s) - \lambda \int_0^1 k(s,t)x(t)\,dt = y(s),$$

with kernel $k(s,t)$ assumed to be such that the operator

$$Tx = \int_0^1 k(s,t)x(t)\,dt$$

is in $[X]$. We take $X = \hat{X}, P = I$. As approximate equation, we take the integral equation with *degenerate kernel* (see Exercise 6.1.10)

$$\hat{x}(s) - \lambda \int_0^1 \sum_{i=1}^n \phi_i(s)\psi_i(t)\hat{x}(t)\,dt = y(s).$$

We assume that the functions ϕ_i, ψ_i are chosen so that the operator

$$\hat{T}\hat{x} = \int_0^1 \sum_{i=1}^n \phi_i(s)\psi_i(t)\hat{x}(t)\,dt$$

is again in $[X]$. Then for Assumption 1 we note that

$$\|\hat{T}x - Tx\| = \left\| \int_0^1 \left[\sum_{i=1}^n \phi_i(s)\psi_i(t) - k(s,t) \right] x(t)\,dt \right\|$$
$$\leqslant \|\hat{T} - T\| \, \|x\|.$$

We may take $\epsilon = \|\hat{T} - T\|$, which value will depend upon the space X and the choice of the degenerate kernel (see below).

For Assumption 2 we note that we can always take $\hat{x}=Tx$, and thus $\|Tx-\hat{x}\|=0$; that is ϵ_1 may be taken to be zero. For the same reason, we can take $\hat{y}=y$ and $\epsilon_2=0$. Our results about this case can be summarized as follows:

If K^{-1} exists and $q=|\lambda|\,\|K^{-1}\|\epsilon<1$, then, for every y, $\hat{K}\hat{x}=y$ is solvable for $\hat{x}=\hat{x}_0$. If $\hat{K}$ is 1:1 then $\hat{K}^{-1}\in[X]$ and $\|\hat{K}^{-1}\|\leqslant\|K\|/(1-q)$. *If $Kx^*=y$ then $\|x^*-\hat{x}_0\|\leqslant 2|\lambda|\,\|\hat{K}^{-1}\|\epsilon\|x^*\|$.*

Conversely, if $\hat{K}$ is 1:1, $\hat{K}^{-1}\in[X]$ and $p=2|\lambda|\epsilon\|\hat{K}^{-1}\|<1$, then K^{-1} exists. Also $\|x^*-\hat{x}_0\|\leqslant[(p/(1-p)]\,\|\hat{x}_0\|$. Last, if $\hat{K}^{-1}$ exists, and instead $p'=\|\hat{K}^{-1}\|\,|\lambda|^2\,\|T\|\epsilon<1$, then the solution of $Kx=y$ is always unique, K^{-1} exists, and

$$\|K^{-1}\|\leqslant\frac{1+|\lambda|\,\|\hat{K}^{-1}\|}{1-p'},$$

while

$$\|K^{-1}-(I+\lambda\hat{K}^{-1}T)\|\leqslant\frac{p'}{1-p'}(1+|\lambda|\,\|K^{-1}T\|).$$

If $X=C[0,1]$, then we have

$$\epsilon=\|\hat{T}-T\|\leqslant\max_{0\leqslant s\leqslant 1}\int_0^1\left|\sum_{i=1}^n\phi_i(s)\psi_i(t)-k(s,t)\right|dt.$$

If $X=L_p(0,1)$, $1<p<\infty$ and $1/p+1/q=1$, then we have

$$\epsilon=\|\hat{T}-T\|\leqslant\left\{\int_0^1\left[\int_0^1\left|\sum_{i=1}^n\phi_i(s)\psi_i(t)-k(s,t)\right|^q dt\right]^{p/q}ds\right\}^{1/p}.$$

If $X=L_1(0,1)$, then we have

$$\epsilon=\|\hat{T}-T\|\leqslant\int_0^1\operatorname*{ess\,sup}_{0<t<1}\left|\sum_{i=1}^n\phi_i(s)\psi_i(t)-k(s,t)\right|ds.$$

The question of guaranteeing $\epsilon\to 0, n\to\infty$, and measuring the rate of convergence becomes one of approximating the kernel $k(s,t)$ in some specific form by a degenerate kernel. From Exercise 6.1.10 we can conclude that if the functions $\phi_i(s)$ are linearly independent, the functions $\psi_i(t)$ are linearly independent and a certain determinant does not vanish, then $\hat{K}$ is 1:1.

EXERCISES TO SECTION 8.2

1. If $X = L_\infty$, find a value for ϵ.
2. Find the determinant whose nonvanishing is essential in order that $\hat{K}$ be 1:1.
3. It is not *a priori* obvious that a given kernel $k(s,t)$ determines an operator $T \in [X]$. Find some reasonable conditions that guarantee this for $X = C[0,1]$, and $L_p(0,1)$, $1 \leqslant p \leqslant \infty$.
4. The various appraisals of ϵ in Section 8.2 require justification, and make certain requirements on $k(s,t)$. Determine these.
5. Consider the integral equation

$$x(s) - \lambda \int_0^1 \sin\left(\frac{st}{2}\right) x(t)\, dt = y(s)$$

and the degenerate approximate kernel $st/2$ obtained from $\sin(st/2)$ by way of Taylor's theorem:

$$x(s) - \frac{\lambda}{2} \int_0^1 stx(t)\, dt = y(s).$$

Compute the error $\|x^* - \hat{x}^0\|$ for this case, when $X = C[0,1]$. Also compute the inverse of $\hat{K}$ and both the approximate inverse of K, and the error estimate for

$$\|K^{-1} - (I + \lambda \hat{K}^{-1} T)\|.$$

6. For the integral equation of Exercise 8.2.5, compute a bound for $\|\hat{T} - T\|$ if $X = C[0,1]$ and if we approximate $\sin u$ by $u - u^3/3!$.

8.3 INTEGRAL EQUATIONS: SOLUTION BY NUMERICAL QUADRATURE

We consider again

$$x(s) - \lambda \int_0^1 k(s,t) x(t)\, dt = y(s)$$

where $x, y \in X = C[0,1]$ and k is continuous on $0 \leqslant t, s \leqslant 1$; choose a division of $[0,1]$ into n subintervals by points $t_{j,n}, 0 \leqslant t_{0,n} < t_{1,n} < \cdots < t_{n,n} \leqslant 1$, and any approximate integration formula

$$\int_0^1 f(t)\, dt \approx \sum_{j=0}^{n} A_{j,n} f(t_{j,n}).$$

Then as approximate system we choose

$$x(t_{i,n})-\lambda\sum_{j=0}^{n}A_{j,n}k(t_{i,n},t_{j,n})x(t_{j,n})=y(t_{i,n}),\qquad i=0,\ldots,n.$$

If we do this for each n, we produce an infinite sequence of approximating equations and of approximating spaces. For convenience, we drop the subscript n and assume it fixed for the following calculations. Also for convenience, to fix ideas, we adopt the integration scheme known as the *trapezoidal rule*: we drop the index n and set

$$t_j=\frac{j}{n},\qquad j=0,\ldots,n,$$

$$A_j=\begin{cases}\dfrac{1}{2n} & j=0,n,\\[2mm] \dfrac{1}{n}, & j=1,\ldots,n-1.\end{cases}$$

We let $\hat{X}\subset X$ be the *polygonal functions*; that is, the functions formed from the points (t_j,δ_j) in $\mathbf{R}^2$ by joining (t_j,δ_j) to (t_{j+1},δ_{j+1}) with a line segment. We set $\bar{X}=\mathbf{E}^{n+1}$, $\|\bar{x}\|=\max_i|\delta_i|$ if $\bar{x}=(\delta_0,\ldots,\delta_n)$. Then set

$$\Psi x=(x(t_0),\ldots,x(t_n))$$

and

$$\Psi_0^{-1}\bar{x}=\text{polygonal function with vertices }(t_k,\delta_k).$$

Clearly $\|\Psi\|=\|\Psi_0\|=\|\Psi_0^{-1}\|=1$. We let $\bar{T}$ be the (matrix) operator on $\mathbf{E}^{n+1}$ given by

$$(\bar{T})_{i,k}=A_k k(t_i,t_k).$$

Let $\omega_k(\nu;s),\omega_k(\nu;t)$ denote the oscillation (Section 4.2) or modulus of continuity of k as a function of s,t respectively, while $\omega_y(\nu)$ is the oscillation or modulus of continuity of y. Recall that

$$\omega_f(\nu)=\sup_{|s-s'|\leqslant\nu}|f(s)-f(s')|.$$

Then

$$\|\Psi T\hat{x} - \bar{T}\Psi\hat{x}\| = \max_{0 \leqslant i \leqslant n} \left| \int_0^1 k(t_i, t)\hat{x}(t)\,dt - \sum_{k=1}^{n-1} k(t_i, t_j)\hat{x}(t_j) - \frac{1}{2n} k(t_i, t_0)\hat{x}(t_0) - \frac{1}{2n} k(t_i, t_n)\hat{x}(t_n) \right| .$$

But

$$\begin{aligned}\int_0^1 k(t_i, t)\hat{x}(t)\,dt &= \sum_{j=0}^{n-1} \int_{t_j}^{t_{j+1}} k(t_i, t)\hat{x}(t)\,dt \\ &= \sum_{j=0}^{n-1} \int_{t_j}^{t_{j+1}} [k(t_i, t) - k(t_i, \xi_j)]\hat{x}(t)\,dt \\ &\quad + \sum_{j=0}^{n-1} k(t_i, \xi_j) \int_{t_j}^{t_{j+1}} \hat{x}(t)\,dt,\end{aligned}$$

where $\xi_j = (t_j + t_{j+1})/2$. Now

$$\int_{t_k}^{t_{k+1}} \hat{x}(t)\,dt = \frac{[\hat{x}(t_k) + \hat{x}(t_{k+1})]}{2n} .$$

Thus

$$\begin{aligned}\|\Psi T\hat{x} - \bar{T}\Psi\hat{x}\| = \max_{0 \leqslant i \leqslant n} | &\sum_{j=0}^{n-1} \int_{t_j}^{t_{j+1}} [k(t_i, t) - k(t_i, \xi_j)]\hat{x}(t)\,dt \\ &+ \frac{1}{2n} \sum_{j=0}^{n-1} k(t_i, \xi_j)[\hat{x}(t_j) + \hat{x}(t_{j+1})]/2n \\ &- \frac{1}{2n} k(t_i, t_0)\hat{x}(t_0) - \frac{1}{2n} k(t_i, t_n)\hat{x}(t_n) \\ &- \sum_{j=1}^{n-1} \frac{1}{n} k(t_i, t_j)\hat{x}(t_j)|.\end{aligned}$$

So

$$\|\Psi T\hat{x}-\bar{T}\Psi\hat{x}\| \leqslant \max_{0\leqslant i\leqslant n}\left|\sum_{j=0}^{n-1}\int_{t_j}^{t_{j+1}}[k(t_i,t)-k(t_i,\xi_j)]\hat{x}(t)\,dt\right|$$

$$+\max_{0\leqslant i\leqslant n}\left|\sum_{j=0}^{n-1}\frac{1}{2n}[k(t_i,\xi_j)-k(t_i,t_j)]\hat{x}(t_j)\right|$$

$$+\max_{0\leqslant i\leqslant n}\left|\sum_{j=0}^{n-1}\frac{1}{2n}[k(t_i,\xi_j)-k(t_i,t_{j+1})]\hat{x}(t_{j+1})\right|$$

$$\leqslant\left[\omega_k\left(\frac{1}{2n};t\right)+\frac{1}{2}\omega_k\left(\frac{1}{2n};t\right)+\frac{1}{2}\omega_k\left(\frac{1}{2n};t\right)\right]\|\hat{x}\|$$

$$=2\omega_k\left(\frac{1}{2n};t\right)\|\hat{x}\|.$$

Thus in Assumption 1 we may take $\epsilon=2\omega_k(1/2n;t)$. If we approximate Tx by $\hat{x}=\Psi_0^{-1}\Psi Tx$, then

$$\hat{x}(s)=(t_{i+1}-s)\int_0^1 k(t_i,t)x(t)\,dt-(s-t_i)\int_0^1 k(t_{i+1},t)x(t)\,dt$$

on $[t_i,t_{i+1}]$; so on this interval

$$|Tx-\hat{x}|=\left|\int_0^1[k(s,t)-(t_{i+1}-s)k(t_i,t)-(s-t_i)k(t_{i+1},t)]x(t)\,dt\right|$$

$$\leqslant\omega_k\left(\frac{1}{n};s\right)\int_0^1|x(t)|\,dt.$$

Then

$$\|Tx-\hat{x}\|\leqslant\omega_k\left(\frac{1}{n};s\right)\|x\|=\epsilon_1\|x\|,$$

and similarly,

$$\|y-\hat{y}\|\leqslant\left(\frac{\omega_y(1/n)}{\|y\|}\right)\|y\|=\epsilon_2\|y\|.$$

Our result may now be stated as follows:
If K^{-1} exists and if

$$q=|\lambda|\left\{\|K\|\omega_k\left(\frac{1}{n};s\right)+2\omega_k\left(\frac{1}{2n};t\right)+2|\lambda|\omega_k\left(\frac{1}{n};s\right)\omega_k\left(\frac{1}{2n};t\right)\right\}\|K^{-1}\|<1,$$

then $\bar{K}^{-1}$ *exists and*

$$\|\bar{K}^{-1}\| \leqslant \left[\frac{1+|\lambda|\omega_k(1/n;s)}{1-q}\right]\|K^{-1}\|.$$

Also if $Kx^* = y, \bar{K}\bar{x} = \bar{y} = \Psi y$, *then*

$$\|x^* - \Psi_0^{-1}\bar{x}\| \leqslant \|x^*\| \left\{ 4|\lambda|\omega_k\left(\frac{1}{2n};t\right)\|\bar{K}^{-1}\| + \left[|\lambda|\omega_k\left(\frac{1}{n};s\right) + \frac{\|K\|\omega_y(1/n)}{\|y\|}\right]\left[1+\|K\|\,\|\bar{K}^{-1}\|\right]\right\}.$$

Similar translations can be made for the other applicable theorems of Chapter 7. It is more illuminating to note that the trapezoidal rule for numerical integration is exact (gives the correct integral) when $\hat{X}$ is the space of polygonal functions. If we chose another method for numerical integration, say one that was exact for *algebraic polynomials* of degree $\leqslant n$, a parallel development is possible, with the oscillation measure replaced by measures of best approximation by polynomials of degree $\leqslant n$. We continue instead with a numerical integration scheme which is exact for *trigonometric polynomials*. In order to do this, we first restrict our problem to one exhibiting periodicity.

Let X be the subspace of $C[0,1]$ of functions x with $x(0)=x(1)$, and let $k(0,t)\equiv k(1,t), k(s,0)\equiv k(s,1)$. Choose for $\hat{X}$ the subspace spanned by $1, \cos 2\pi s, \sin 2\pi s, \ldots, \cos 2\pi ns, \sin 2\pi ns$. We take $\bar{X} = \mathbf{R}^{2n+1}$, and $\Psi x = (x(\tau_0),\ldots,x(\tau_{2n}))$. Then $\|\Psi\| = 1$. But $\Psi_0^{-1}\bar{x}$ must be the trigonometric polynomial $\hat{x}(t)$ taking the values

$$\hat{x}(\tau_j) = \bar{x}_j, \qquad j=1,\ldots,2n.$$

From a well known estimate of Lebesgue (see, e.g., Natanson [1], Vol. 3, 47–51), if $\tau_j = j/(2n+1), j=0,\ldots,2n$,

$$|\hat{x}(t)| \leqslant (A\log n + B)\max_i |\bar{x}_i|.$$

Hence $\|\Psi_0^{-1}\| \leqslant A\log n + B$.

We suppose that we can find $V(s,t), U(s,t)$, such that V is a trigonometric polynomial of degree n in s with coefficients depending continuously on t, and U is a trigonometric polynomial of degree n in t, with coefficients depending continuously on s, and bounds $M_k(n;t), M_k(n;s)$ such that

$$|k(s,t)-V(s,t)|\leqslant M_k(n;t)$$
$$|k(s,t)-U(s,t)|\leqslant M_k(n;s).$$

(For ways of estimating M_k, see below).

We note that when $A_j=1/(2n+1), t_j=j/(2n+1)+1/2(2n+1), j=0,\ldots,2n$,

$$\sum_{j=0}^{2n} A_j f(t_j)=\int_0^1 f(t)\,dt$$

for all trigonometric polynomials of degree not greater than $2n$, that is for all of $\hat{X}$. To see this, compute (why?) for $j=1,\ldots,2n$

$$\sum_{m=0}^{2n} \exp[2\pi ijm/(2n+1)]=\sum_{m=0}^{2n}\{\exp[2\pi ij/(2n+1)]\}^m$$
$$=\frac{1-\exp[2\pi ij]}{1-\exp[2\pi ij/(2n+1)]}=0.$$

Thus for $1\leqslant j\leqslant 2n$, the integral formula is exact since

$$\int_0^1 e^{2\pi ijs}\,ds=\frac{e^{2\pi ij}-1}{2\pi ij}=0.$$

For $j=0$, the result is obvious.

We now investigate the quantities $\epsilon,\epsilon_1,\epsilon_2$. Since U is a trigonometric polynomial in t of degree n, and $\hat{x}(t)$ is of degree n, $U(s,t)\hat{x}(t)$ is a polynomial of degree $2n$. Then the numerical integration formula is exact and

$$\|\Psi T\hat{x}-\bar{T}\Psi\hat{x}\|=\max_i\left|\int_0^1 k(t_i,t)\hat{x}(t)\,dt-\sum_{j=0}^{2n}A_j k(t_i,t_j)\hat{x}(t_j)\right|$$
$$\leqslant \max_i\left|\int_0^1[k(t_i,t)-U(t_i,t)]\hat{x}(t)\,dt\right|$$
$$+\max_i\left|\sum_{j=0}^{2n}A_j[k(t_i,t_j)-U(t_i,t_j)]\hat{x}(t_j)\right|$$
$$\leqslant 2M_k(n;t)\,\|\hat{x}\|.$$

Thus we may take $\epsilon = 2M_k(n;t)$. If we define

$$\hat{x}(s) = \int_0^1 V(s,t)x(t)\,dt$$

then $x(s) \subset \hat{X}$, since V is a trigonometric polynomial in s, and

$$\|Tx - \hat{x}\| = \max_s \left| \int_0^1 [k(s,t) - V(s,t)]x(t)\,dt \right|$$
$$\leqslant M_k(n;s)\,\|x\|,$$

and we may take $\epsilon_1 = M_k(n;s)$. Also, we may set $\epsilon_2 = M_y(n)/\|y\|$.

We conclude: *If $k(s,t)$ and y are sufficiently smooth, and n is so large that*

$$q = |\lambda|[\,\|K\|M_k(n;s) + 2M_k(n;t) + 2|\lambda|M_T(n;s)M_T(n;t)]$$
$$\times (A \log n + B)\,\|K^{-1}\| < 1,$$

then the approximate system $\bar{K}\bar{x} = \bar{y}$ is solvable for δ such that the interpolating trigonometric polynomial $\hat{x}(s)$ given by $\hat{x}(t_k) = \delta_k$ is an approximate solution to $Kx = y$. Also, the solution for a given n satisfies

$$\|x^* - \hat{x}^n\| \leqslant \|x^*\| \left\{ 4|\lambda|M_k(n;t)(A\log n + B)\,\|\bar{K}^{-1}\| \right.$$
$$+ \left[|\lambda|M_k(n;s) + \|K\|\frac{M_y(n)}{\|y\|} \right]$$
$$\left. \times [1 + \|K\|\,\|\bar{K}^{-1}\|(A\log n + B)] \right\}.$$

That $\bar{K}^{-1}$ is a bounded operator can be shown as in the case of the trapezoidal integration rule, and the same bound found there is valid here. That bound,

$$\|\bar{K}^{-1}\| \leqslant \frac{1 + |\lambda|\omega_k(1/n;s)}{1-q}\|K^{-1}\|$$

shows that $\bar{K}^{-1}$ is uniformly bounded for all n so large that $q^n < 1$.

Then, if $M_k(n;t), M_k(n;s)$ and $M_k(n)$ all approach 0 as $n \to \infty$, it follows that $\hat{x}^n$ converges uniformly to x^* and that

$$\|x^* - \hat{x}^n\| = 0\{(\log n)[\,M_k(n;s) + M_k(n;t) + M_y(n)]\}.$$

Conversely, if $\bar{K}^{-1}$ *exists, and if*

$$p' = \{ M_k(n;s) + (A \log n + B)\,\|\bar{K}^{-1}\|\,[\,\|K\|\,M_k(n;s)$$

$$+ 2|\lambda|[\,\|T\| + M_k(n;s)\,]\,M_k(n;t)\,]\}|\lambda| < 1,$$

then K^{-1} *exists. If* $x_1 = y + \Psi_0^{-1}\delta$, *where*

$$\delta - \lambda \bar{T}\delta = \lambda \Psi T y,$$

then

$$\|x^* - x_1\| \leqslant p'\|x^*\|.$$

Here we are utilizing regularization (see Exercise 7.2.1). In order to utilize these results, we need means of estimating $M_k(n;t)$, $M_k(n;s)$ and $M_y(n)$. For this, we use Jackson's theorem (compare, e.g. Akhiezer [1], 187–201, or Natanson [1], Vol. 1, 89.):

If $f(x)$ *is a continuous periodic function of period* 2π, *with* mth *derivative* $m > 1$ *satisfying*

$$\left\| \frac{d^m f}{dx^m} \right\| \leqslant 1,$$

then there exists an approximating trigometric polynomial $\hat{f}$ *of degree* $\leqslant n-1$ *such that*

$$\|f(x) - \hat{f}(x)\| \leqslant \frac{A_m}{n^m},$$

where A_m *depends only on* m.

This implies for instance that if $k(s,t)$ and y have bounded mth order derivatives with respect to their arguments, then

$$\|x^* - \hat{x}^n\| = 0\left\{ \log n \left[\left\| \frac{\partial^m k}{\partial t^m} \right\|^{-1} + \left\| \frac{\partial^m k}{\partial s^m} \right\|^{-1} + \left\| \frac{d^m y}{ds^m} \right\|^{-1} \right] \right\}$$

$$= 0\left\{ \frac{\log n}{n^m} \right\}$$

EXERCISE TO SECTION 8.3

1. Instead of the trapezoidal rule, we could use a more sophisticated method for numerical integration, say Gaussian quadrature. Define the

corresponding space $\hat{X}$ of functions for which a Gaussian quadrature is exact and develop the equivalent results for this case. Compare your results with Kantorovich and Krylov [1], 91–110.

8.4 INTEGRAL EQUATIONS: THE MOMENT METHOD

Let $w_1, w_2, w_3, \ldots$ be an infinite sequence of functions on $0 \leqslant s \leqslant 1$ satisfying "orthogonality conditions":

$$\int_0^1 w_i(s) w_k(s)\, ds = \begin{cases} 0, & i \neq k, \\ 1, & i \neq k. \end{cases}$$

We seek a solution of

$$x(s) - \lambda \int_0^1 k(s,t) x(t)\, dt = y(s)$$

of the form

$$\hat{x}(s) = \delta_1 w_1(s) + \delta_2 w_2(s) + \cdots + \delta_n w_n(s),$$

such that

$$\int_0^1 \left[\hat{x}(s) - \lambda \int_0^1 k(s,t) x(t)\, dt - y(s) \right] w_i(s)\, ds = 0, \qquad i = 1, \ldots, n.$$

Choose $X = L_2, \bar{X} = \mathbf{E}^n, \hat{X} =$ the set of linear combinations of the $w_i(s)$. It follows from the orthogonality conditions that

$$\delta_i - \lambda \int_0^1 \int_0^1 k(s,t) \sum_{j=1}^n \delta_j w_j(t)\, dt\, w_i(s)\, ds = \int_0^1 y(s) w_i(s)\, ds.$$

or

$$\delta - \lambda \bar{T} \delta = \eta$$

where

$$(\Psi y)_i = \eta_i = \int_0^1 y(s) w_i(s)\, ds,$$

$$\delta = \Psi_0 \hat{x},$$

$$\eta = \Psi y,$$

and

$$\bar{T}_{i,j} = \int_0^1 \int_0^1 k(s,t) w_i(s) w_j(t)\, dt\, ds.$$

Hence

$$\hat{x} - \lambda \Psi_0^{-1} \bar{T} \Psi \hat{x} = \Psi_0^{-1} \eta = \Psi_0^{-1} \Psi y = Py.$$

The operator $\hat{T}$ given by $\Psi_0^{-1} \bar{T} \Psi$ is

$$\hat{T}\hat{x} = \int_0^1 \left[\sum_{i=1}^{n} \sum_{j=1}^{n} \bar{T}_{i,j} w_i(s) w_j(t) \right] \hat{x}(t)\, dt.$$

Since $\Psi x = \bar{x} = (\delta_1, \ldots, \delta_n)$, where $\delta_i = \int_0^1 x(s) w_i(s)\, ds$ and $\hat{x} = \Psi_0^{-1} \bar{x} = \delta_1 w_1 + \cdots + \delta_n w_n$, we can take $\|\delta\| = (\delta_1^2 + \cdots + \delta_n^2)^{1/2}$, and then $\|\Psi\| = \|\Psi_0^{-1}\| = 1$.

Since $\Psi T \hat{x} - \bar{T} \Psi \hat{x} = 0$, we take $\epsilon = 0$. If we set $\hat{x} = \hat{T} x$ we find

$$\|Tx - \hat{x}\|^2 = \int_0^1 \left\{ \int_0^1 [k(s,t) - \hat{k}(s,t)] x(t)\, dt \right\}^2 ds$$

$$\leqslant \left\{ \int_0^1 \int_0^1 [k(s,t) - k(s,t)]^2\, ds\, dt \right\} \|x\|^2,$$

where $\hat{k}(s,t) = \sum_{i=1}^{n} \sum_{j=1}^{n} \bar{T}_{i,j} w_i(s) w_j(t)$. Then we may choose

$$\epsilon_1 = \left(\int_0^1 \int_0^1 [k(s,t) - \hat{k}(s,t)]^2\, ds\, dt \right)^{1/2}.$$

We may also choose $\hat{y} = \Psi_0^{-1} \Psi y$ and

$$\epsilon_2 = \frac{\|y - \hat{y}\|}{\|y\|}.$$

Then:

If K^{-1} exists and $\epsilon^n \to 0, n \to \infty$, then

$$\delta - \lambda \bar{T} \delta = \eta$$

has a unique solution δ for all sufficiently large n and $\hat{x} = \Psi_0^{-1} \delta$ satisfies

$$\|x^* - \hat{x}^n\| = 0(\epsilon_1^n + \epsilon_2^n).$$

If $\epsilon_1^n, \epsilon_2^n \to 0$, *then*

$$\int_0^1 |x^*(s) - \hat{x}^n(s)|^2 \, ds \to 0, \qquad n \to 0.$$

Since $x^* \in L_2$ and

$$\left| \int_0^s [x^*(t) - \hat{x}^n(t)] \, dt \right| \leqslant \left[\int_0^s |x^*(s) - \hat{x}^n(t)|^2 \, dt \right]^{1/2} \sqrt{s}\,,$$

by the Cauchy-Schwarz inequality, it even follows that

$$\int_0^s \hat{x}^n(t) \, dt \to \int_0^s x^*(t) \, dt, \qquad n \to \infty$$

uniformly in s on $[0,1]$. (Under slightly stronger conditions, it can be shown that $\hat{x}^n \to x^*$ uniformly on $[0,1]$. See Kantorovich and Akilov [1], 578–579.).

Conversely, if $(\bar{I} - \lambda \bar{T})^{-1}$ *exists and*

$$p' = |\lambda| \, \|T - \hat{T}\| \big(1 + \|\bar{K}^{-1}\| \, \|K\|\big) < 1$$

then $Kx = y$ *has a unique solution* x^*; *if* $x_1 = y + \Psi_0^{-1}\bar{z}, \bar{z} = \bar{K}^{-1}\lambda\Psi Ty$, *then*

$$\|x^* - x_1\| \leqslant p' \|x^*\|.$$

8.5 INTEGRAL EQUATIONS: SOLUTION BY COLLOCATION

The solution by numerical quadrature may be thought of in terms of two different processes: One looks for a solution in a certain subspace of X convenient for the numerical integration, while the other requires satisfaction of the equation $Kx = y$ only at certain distinct points, a process known as *collocation*. In Section 8.2, the numerical integration scheme determined the "collocation points." Here we separate the two processes, choosing $\hat{X}$ for one reason, the collocation points for another.

This time we approximate the solution of

$$x(s) - \lambda \int_0^1 k(s,t) x(t) \, dt = y(s)$$

for periodic y, k of period 1 in all variables by

$$\hat{x}(s) = \sum_{k=0}^{n-1} \xi_{k+1} \cos 2\pi k s = \sum_{k=1}^{n} \xi_k w_k(s),$$

but we determine $\hat{x}$ by collocation:

$$\hat{x}(t_i)-\lambda\int_0^1 k(t_i,t)\hat{x}(t)\,dt=y(t_i),\qquad i=1,\ldots,n$$

for points $0\leqslant t_1<t_2<\cdots<t_n\leqslant 1$. We choose $X=Y=\Pi[0,1]=\{x(t)\in C[0,1]\,|\,x(0)=x(1)\}$. Let $\hat{X}$ and $\hat{Y}$ be the subspace spanned by $w_1,\ldots,w_n$. Define a space $\overline{Y}=\mathbf{E}^n$ and a mapping $\Psi: Y\to\overline{Y}$ by

$$\Psi y=(y(t_1),y(t_2),\ldots,y(t_n)).$$

On $\overline{Y}$ we choose the max norm. Define Ψ_0 as Ψ on $\hat{Y}$. Then $\Psi_0^{-1}\eta,\eta\in\mathbf{E}^n$ yields a trigonometric polynomial in $\hat{Y}$ which *interpolates* η_i at $t_i, i=1,\ldots,n$.

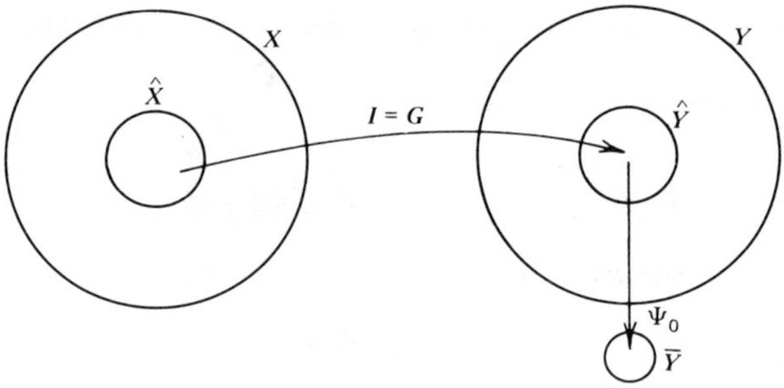

We recall that $\|\Psi_0^{-1}\|\leqslant A\log n+B$, while $\|\Psi\|=\|\Psi_0\|=1$.
Then

$$\Psi_0\hat{x}-\lambda\Psi\int_0^1 k(s,t)\hat{x}(t)\,dt=\Psi y,$$

or

$$\hat{x}-\lambda\Psi_0^{-1}\Psi T\hat{x}=\Psi_0^{-1}\Psi y.$$

Define $\hat{T}=\Psi_0^{-1}\Psi T, P=\Psi_0^{-1}\Psi$. Assumption 1 becomes

$$\|\Psi_0^{-1}\Psi T\hat{x}-\Psi_0^{-1}\Psi T\hat{x}\|=0$$

for all $\hat{x}$, so we may take $\epsilon=0$. Let

$$V(s,t)=\sum_{k=1}^{n}\delta_k(t)w_k(s)$$

where $V(s,t)$ yields a best approximation to $k(s,t)$ in supnorm for each fixed t, and $|V(s,t)-k(s,t|\leqslant M_k(n;s)$.

Set

$$\hat{x}=\int_0^1 V(s,t)x(t)\,dt.$$

Then $\|Tx-\hat{x}\|\leqslant M_k(n;s)\|x\|$, so we may take $\epsilon_1=M_k(n;s)$ in Assumption 2. For Assumption 3 we can choose y as the trigonometric polynomial of best approximation to y in supnorm. (*Not* $\hat{y}=\Psi_0^{-1}\Psi y$; then $\|y-\hat{y}\|\to\infty$ for many functions y. To understand this see, for instance, Natanson [1], Vol. 3, 23–46, in particular the theorem on p. 43). Let $\|y-\hat{y}\|\leqslant M_y(n)\|y\|, \epsilon_2=M_y(n)$. We can conclude the following:

Let K^{-1} exist, while $M_k(n;s)\log n\to 0$, $n\to\infty$ and $M_y(n)\log n\to 0$, $n\to\infty$. Then for all sufficiently large n, $K\hat{x}=Py$ is solvable; the approximate solutions converge uniformly to the solution x^* of $Kx=y$; $\hat{K}^{-1}$ exists and

$$\|\hat{K}^{-1}\|\leqslant\frac{1+|\lambda|M_k(n;s)}{1-|\lambda|M_k(n;s)(A\log n+B)\,\|K\|\,\|K^{-1}\|\,\|K^{-1}\|}.$$

Also

$$\|x^*-\hat{x}^n\|\leqslant|\lambda|\left[M_k(n;s)|\lambda|+M_y(n)\,\|K\|\right]\times\left[1+\|\hat{K}^{-1}\|(A\log n+B)\|K\|\right]\|x^*\|,$$

where $\hat{x}^n=\sum_{k=1}^n\xi_k\omega_k$ and ξ solves

$$\sum_{k=1}^n\xi_k\omega_k(t_i)-\lambda\int_0^1 k(t_i,t)\sum_{k=1}^n\xi_k\omega_k(t)\,dt=y(t_i),\qquad i=1,\dots,n.$$

If we regularize the problem and solve instead

$$\sum_{k=1}^n\xi_k\omega_k(t_i)-\lambda\int_0^1 k(t_i,t)\sum_{k=1}^n\xi_k\omega_k(t)\,dt=\lambda\int_0^1 k(t_i,t)y(t)\,dt$$

then we may approximate

$$\int_0^1 T(s,t)y(t)\,dt$$

by

$$\int_0^1 V(s,t)y(t)\,dt.$$

Then $\epsilon_2 = M_k(n;s)$. This time $x^n = y + \Sigma_{k=1}^n \xi_k \omega_k$ and we need no smoothness assumptions on y to insure $M_y(n)\log n \to 0$, $n \to \infty$.

EXERCISES TO SECTION 8.5

1. Let $X = \{x(t) | d^{2m}x/dt^{2m} \in C[0,1],\ x(0) = x'(0) = \cdots = x^{(m-1)}(0) = 0;\ x(1) = x'(1) = \cdots = x^{(m-1)}(1) = 0\}$. Define $\|x\| = \max_{0 \leqslant t \leqslant 1} |d^{2m}x/dt^{2m}|$. Show that X is a normed linear vector space. Is it a Banach space? Prove your answer!

   Consider the operator $G: X \to C[0,1]$ defined by $Gx = d^{2m}x/dt^{2m}$. Show that G is onto $C[0,1]$. Is it $1:1$? Is it bounded? Show that any solution in X of $Gx = y$ is given by

$$x(s) = \int_0^1 g(s,t) y(t)\, dt.$$

   That is, show that there exists a Green's function for the boundary conditions. Does this mean that G has a bounded inverse?

2. Consider the boundary value problem

$$\frac{d^{2m}x}{dt^{2m}} - \lambda \left[p_1 \frac{d^{2m-1}x}{dt^{2m-1}} + \cdots + p_{2m}x \right] = y$$

   subject to the boundary conditions

$$x(0) = x'(0) = \cdots = x^{(m-1)}(0) = 0,$$

$$x(1) = x'(1) = \cdots = x^{(m-1)}(1) = 0.$$

   Let $\hat{X} = \{\hat{x} | \hat{x}(t) = t^m(1-t)^m \Sigma_{k=1}^n \delta_k t^{k-1}\}$. Then on $\hat{X}$, the boundary conditions are satisfied (verify). Determine the vector $\delta = (\delta_1, \ldots, \delta_n)$ by the requirement (collocation)

$$\frac{d^{2m}\hat{x}}{dt^{2m}} - \lambda \left[p_1 \frac{d^{2m-1}\hat{x}}{dt^{2m-1}} + \cdots + p_1 \hat{x} \right] \Bigg|_{t=t_j} = y(t_j)$$

   for $j = 1, 2, \ldots, n$, where $0 \leqslant t_1 < t_2 < \cdots < t_n \leqslant 1$. Using the theory of Section 7.6, consider the solution of

$$Gx - \lambda T_1 x = y$$

   where $y \in C[0,1]$. (Shindler [1].)

8.6 GALERKIN'S METHOD (FOR DIFFERENTIAL EQUATIONS)

Let $f(t), p(t), q(t), r(t)$ be real valued continuous functions of t on $[0,1]$, with p and q continuously differentiable. Consider the differential equation

$$L(x) = (px')' + (qx)' + rx = f(t)$$

subject to the boundary conditions $x(0) = x(1) = 0$, where we assume $p(t) \geqslant \delta^2 > 0$, $0 \leqslant t \leqslant 1$. We also assume a normalization:

$$\int_0^1 \frac{dt}{p(t)} = 1.$$

Assume the existence of a solution.

Let w_k, $k = 1, 2, 3, \ldots$ be elements of $L_2(0,1)$ which are differentiable and such that

$$w_k(0) = w_k(1) = 0, \qquad k = 1, 2, 3, \ldots,$$

while

$$\int_0^1 p(t) w_k'(t) w_j'(t)\, dt = \begin{cases} 0, & k \neq j, \\ 1, & k = j. \end{cases}$$

Let $w_0'(t) = 1/p(t)$.

Galerkin's method, in this case, is that of finding an approximate solution $\hat{x}(t) = \sum_{i=1}^n \delta_i w_i(t)$ where $\delta_1, \ldots, \delta_n$ are determined by the conditions

$$\int_0^1 [L(\hat{x}(t)) - f(t)] w_i(t)\, dt = 0, \qquad i = 1, \ldots, n.$$

(This scheme applies directly to any differential equation, ordinary or partial, *without first passing through an integral equation reformulation*.) We establish solvability and convergence with the structure of Section 7.5. A considerable amount of analytic formalism is present in this example. Not until we study the geometric structure of normed linear vector spaces can this analytic formalism be grasped simply.

Let X be the space of twice continuously differentiable functions $x(t)$, $0 \leqslant t \leqslant 1$, such that $x(0) = x(1) = 0$ and define

$$\|x\|^2 = \int_0^1 p(t) [x'(t)]^2\, dt.$$

That this is a norm follows from the fact that $p(t) > 0$, so $\|x\| = 0$ implies

$x'=0$, so $x(t)\equiv x(0)=0$. The triangle inequality follows from the Cauchy-Schwarz inequality as in Section 1.1. We take as Y the range of $(px')'$; that is, $y\in Y$ means that y can be written as

$$y(t)=\frac{d}{dt}[p(t)z(t)+c]$$

where $z\in C^1[0,1]$, and the constant c is chosen so that $\int_0^1 z(t)\,dt=0$. On Y we take as norm

$$\|y\|^2=\int_0^1 p(t)[z(t)]^2\,dt.$$

With this choice of spaces $y=(px')'$ defines an isometry since we have $\|y\|=\|x\|$ for all such pairs x,y.

In X we take

$$\hat{X}=\left\{\sum_{i=1}^n \delta_i w_i(t)\,\middle|\,\delta\in\mathbf{R}^n\right\}$$

while in Y we take

$$\hat{Y}=\left\{\left[p(t)\sum_{i=1}^n \eta_i w_i'(t)\right]'\,\middle|\,\eta\in\mathbf{R}^n\right\}.$$

Notice that $\hat{Y}\subset Y$ since

$$\int_0^1 \sum_{i=1}^n \eta_i w_i'(t)\,dt=\sum_{i=1}^n \eta_i[w_i(1)-w_i(0)]=0.$$

Also

$$\|\hat{x}\|^2=\int_0^1 p(t)\left[\sum_{i=1}^n \delta_i w_i'(t)\right]\left[\sum_{j=1}^n \delta_j w_j'(t)\right]dt=\sum_{i=1}^n \delta_i^2$$

and

$$\|\hat{y}\|^2=\int_0^1 p(t)\left[\sum_{i=1}^n \eta_i w_i'(t)\right]\left[\sum_{j=1}^n \eta_j w_j'(t)\right]dt=\sum_{i=1}^n \eta_i^2.$$

We take $\bar{X}=\mathbf{R}^n$, $\bar{Y}=\mathbf{R}^n$ with the Euclidean distance as norm in both $\bar{X}$ and $\bar{Y}$.

We must utilize the ideas of Section 7.4 as well as 7.5; define two mappings $J_0: \hat{X} \to \overline{X}, H_0: \hat{Y} \to \overline{Y}$ by $J_0\hat{x} = \delta, H_0\hat{y} = \eta$. Now the numerical algorithm is determined by

$$\int_0^1 \left\{ \left[p \sum_{i=1}^n \delta_i w_i'(t) \right]' + \left[q \sum_{i=1}^n \delta_i w_i \right]' + r \sum_{i=1}^n \delta_i w_i \right\} w_j(t)\, dt = \int_0^1 f(t) w_j(t)\, dt,$$

where $j = 1, \ldots, n$. Integrating by parts we have

$$\int_0^1 \left[p \sum_{i=1}^n \delta_i w_i'(t) \right]' w_j(t)\, dt = \left[p \sum_{i=1}^n \delta_i w_i'(t) \right] w_j(t)|_0^1 - \int_0^1 p \sum_{i=1}^n \delta_i w_i'(t) w_j'(t)\, dt$$

$$= - \sum_{i=1}^n \delta_i \int_0^1 p(t) w_i'(t) w_j'(t)\, dt = -\delta_j$$

since $w_j(1) = w_j(0) = 0$. Thus the numerical problem will be of the form $\delta - \overline{S}\delta = \eta$ if we define

$$(Hy)_j = - \int_0^1 y(t) w_j(t)\, dt, \qquad j = 1, \ldots, n.$$

Then

$$(H\hat{y})_j = - \int_0^1 \frac{d}{dt} \left[p(t) \sum_{i=1}^n \eta_i w_i'(t) \right] w_j(t)\, dt$$

$$= -p(t) \sum_{i=1}^n \eta_i w_i'(t) w_j(t)|_0^1 + \int_0^1 p(t) \sum_{i=1}^n \eta_i w_i'(t) w_j'(t)\, dt$$

$$= \sum_{i=1}^n \eta_i \int_0^1 p(t) w_i'(t) w_j'(t)\, dt$$

$$= \eta_j.$$

Thus H extends H_0. Now above we defined our norm so that $\|H_0\| = \|H_0^{-1}\| = 1$ and $\|J_0\| = \|J_0^{-1}\| = 1$. Also

$$\|Hy\|^2 = \sum_{i=1}^n \left\{ \int_0^1 [p(t)z(t) + c] w_i'(t)\, dt \right\}^2$$

$$= \sum_{i=1}^n \left\{ \int_0^1 p(t) z(t) w_i'(t)\, dt \right\}^2$$

since

$$\int_0^1 cw_i'(t)\,dt = c[w_i(1) - w_i(0)] = 0.$$

Then the Cauchy-Schwarz inequality implies

$$\|Hy\|^2 \leqslant \int_0^1 p(t)[z(t)]^2\,dt \sum_{i=1}^n \int_0^1 p(t)[w_i'(t)]^2\,dt$$

$$= n\|y\|^2,$$

so $\|H\|^2 = n$. Now we define the mappings G, S as

$$Gx = \frac{d}{dt}[p(t)x'(t)]$$

$$Sx = \frac{d}{dt}[q(t)x(t)] + r(t)x(t).$$

Then G maps X onto Y and $\|G\| = 1$. The Galerkin equations

$$\int_0^1 [L(\hat{x}) - f]w_i\,dt = 0, \qquad i = 1, \ldots, n$$

can be rewritten as

$$HLJ_0^{-1}\delta = HGJ_0^{-1}\delta + HSJ_0^{-1}\delta = Hf.$$

We take

$$\bar{G} = HGJ_0^{-1},$$

and

$$\bar{S} = HSJ_0^{-1},$$

so that

$$HS - \bar{S}J_0 = HS - HSJ_0^{-1}J_0 = HS - HS = 0.$$

Thus we may take $\mu = 0$ and reduce the whole convergence question to studying μ_1 in the estimate, uniform for $x \in X$,

$$\|Sx - \hat{y}\| \leqslant \mu_1\|x\|$$

for some $\hat{y}$.

First we write $\int_0^t (Sx)\,dt$ as an integral transform, integrating by parts to get

$$\begin{aligned}\int_0^t (Sx)(\tau)\,d\tau &= \int_0^t \frac{d}{d\tau}\left[q(\tau)x(\tau)+r(\tau)x(\tau)\right]d\tau\\ &= q(t)x(t)-q(t)x(0)+\int_0^t\int_u^t r(\tau)\,d\tau\,x'(u)\,du\\ &= \int_0^t\left[q(t)x'(u)+\int_u^t r(\tau)\,d\tau\,x'(u)\right]du\\ &= \int_0^t\left[q(t)+\int_u^t r(\tau)\,d\tau\right]x'(u)\,du\\ &= \int_0^1 p(t,u)x'(u)\,du\end{aligned}$$

where

$$p(t,u)=\begin{cases} q(t)+\int_u^t r(\tau)\,d\tau, & u<t,\\ 0, & u>t.\end{cases}$$

Then, for some c,

$$Sx-\hat{y}=\frac{d}{dt}\left\{p(t)\left\{\frac{1}{p(t)}\left[\int_0^t (Sx)(\tau)\,d\tau+c\right]-\sum_{i=1}^{n}\eta_i w_i'(t)\right\}\right\},$$

so

$$\begin{aligned}\|Sx-\hat{y}\|^2 &= \int_0^1 p(t)\left\{\frac{1}{p(t)}\int_0^t (Sx)(\tau)\,d\tau+\frac{c}{p(t)}-\sum_{i=1}^{n}\eta_i w_i'(t)\right\}^2 dt\\ &= \int_0^1 p(t)\left\{\int_0^1 \frac{p(t,u)}{p(t)}x'(u)\,du+\frac{c}{p(t)}-\sum_{i=1}^{n}\eta_i w_i'(t)\right\}^2 dt.\end{aligned}$$

Notice that

$$\begin{aligned}\int_0^1 p(t)\left[\frac{1}{p(t)}\right]w_k'(t)\,dt &= \int_0^1 w_k'(t)\,dt\\ &= w_k(1)-w_k(0)=0.\end{aligned}$$

Thus if $w_0'=1/p(t)$, the set $w_0', w_1', w_2', \ldots$ is an orthogonal set. We show in the next chapter that this implies that the minimum of $\|Sx-\hat{y}\|^2$ occurs when

$$\begin{cases} c = -\int_0^1 p(t)\int_0^1 \dfrac{p(t,u)}{p(t)}x'(u)\,du\left[\dfrac{1}{p(t)}\right]dt, \\ \eta_i = \int_0^1 p(t)\int_0^1 \dfrac{p(t,u)}{p(t)}x'(u)\,du\,w_i'(t)\,dt, \qquad i=1,2,\ldots,n; \end{cases}$$

thus

$$c = -\int_0^1\int_0^1 \frac{p(t,u)}{p(t)}\,dt\,x'(u)\,du = -\int_0^1 h_0(u)x'(u)\,du,$$

and

$$\begin{aligned} \eta_i &= \int_0^1\int_0^1 p(t,u)w_i'(t)\,dt\,w'(u)\,du, \\ &= \int_0^1 h_i(u)x'(u)\,du. \end{aligned}$$

The above equations define $h_i(u)$, $i=0,1,2,\ldots,n$. Then

$$\|Sx-\hat{y}\|^2 = \int_0^1 p(t)\left\{\int_0^1\left[\frac{p(t,u)}{p(t)} - \frac{h_0(u)}{p(t)} - \sum_{i=1}^n h_i(u)w_i'(t)\right]x'(u)\,du\right\}^2 dt.$$

Now by the Cauchy-Schwarz inequality

$$\begin{aligned} &\left\{\int_0^1 \frac{\sqrt{p(u)}}{\sqrt{p(u)}}\left[\frac{p(t,u)}{p(t)} - \frac{h_0(u)}{p(t)} - \sum_{i=1}^n h_i(u)w_i'(t)\right]x'(u)\,du\right\}^2 \\ &\qquad \leqslant \int_0^1 p(u)[x'(u)]^2\,du\int_0^1\left[\frac{1}{\sqrt{p(u)}}\right]\left[\frac{p(t,u)}{p(t)}\right. \\ &\qquad\qquad \left. - \frac{h_0(u)}{p(t)} - \sum_{i=1}^n h_i(u)w_i'(t)\right]^2 du. \end{aligned}$$

Thus

$$\begin{aligned} \|Sx-\hat{y}\|^2 &\leqslant \|x\|^2\int_0^1\int_0^1 p(t)\left[\frac{p(t,u)}{p(t)} - \frac{h_0(u)}{p(t)} - \sum_{i=1}^n h_i(u)w_i'(t)\right]^2 dt\,\frac{du}{p(u)} \\ &\leqslant \mu_1^2\|x\|^2. \end{aligned}$$

If $(p(t,u)-h_0(u))/p(t)$ can be represented as a linear combination of the infinite sequence of functions $w_i'(t)$, $i=0,1,\ldots$ which converges in the norm in Y, uniformly in u, $0\leqslant u\leqslant 1$, then $\mu_1\to 0$. (The uniformity is in fact more than is needed here since, in the integration with respect to u in the definition of μ_1, the integrand is monotonely decreasing in n, as we show in the next chapter, and we can use monotone convergence theorems to imply the same result.) For specific sequences $w_i'(t)$ the order of convergence can be determined. (See, for instance, Kantorovich and Akilov [1], 586–595.) We further note that if x^* is the solution and $\hat{x}_n$ the approximate solution, then

$$\begin{aligned}|x^*-\hat{x}_n|^2 &= \left|\int_0^1 [x^{*\prime}(u)-\hat{x}_n'(u)]\,du\right|^2\\ &\leqslant \int_0^1 \frac{dt}{p(t)}\int_0^1 p(t)[x^{*\prime}(t)-\hat{x}_n'(t)]^2\,dt\\ &= \|x^*-\hat{x}\|^2;\end{aligned}$$

thus the convergence is *uniform.*

There are several points to clear up. We must show that c was chosen so that

$$\int_0^1 \frac{1}{p(t)}\left[\int_0^t (Sx)(\tau)\,d\tau + c\right]dt = 0.$$

But this becomes

$$\begin{aligned}c &= -\int_0^1 \frac{1}{p(t)}\int_0^1 p(t,u)x'(u)\,du\,dt\\ &= -\int_0^1 h_0(u)x'(u)\,du.\end{aligned}$$

Also, the normalization of $p(t)$ makes

$$\int_0^1 p(t)\frac{1}{p(t)}\,\frac{1}{p(t)}\,dt = \int_0^1 p(t)[w_0'(t)]^2\,dt = 1.$$

It is also necessary to investigate the approximation to $G^{-1}y$ for an arbitrary y. But $x=G^{-1}y$ means $(px')'=y$ or $x'=[1/p(t)]\int_0^t y(s)\,ds$, so x is smooth enough for good approximation.

9

MORE ABOUT NORMED LINEAR VECTOR SPACES

9.0. INTRODUCTION

In order to provide structure to some of the apparently ad hoc procedures followed in applying the Kantorovich theory, we will study the geometry of one special class of normed linear spaces, the inner product spaces, which generalize our customary intuitions from Euclidean space. A first approach generalizes the algorithm for finding the minimum distance from a point to a line by drawing a perpendicular line. To extend this generalized minimization problem beyond the inner product space, we introduce the ideas of convexity and the dual space of bounded linear functionals.

9.1 INNER PRODUCT SPACES AND HILBERT SPACES

Among the normed linear vector spaces, one class plays a special role, the *inner product spaces*. These are defined as follows:

DEFINITION 9.1.1. A space X is called an *inner product space* if it is a linear vector space over the (real or complex) scalars $\mathcal{C}$ and if, in addition, there exists a form (or function) (x,y), mapping $X \otimes X$ into the field of scalars $\mathcal{C}$, satisfying

1. $(x,y)=\overline{(y,x)}$;
2. $(x_1+x_2,y)=(x_1,y)+(x_2,y)$;
3. $(\lambda x,y)=\lambda(x,y)$ for every $\lambda \in \mathcal{C}$;
4. $(x,x)>0$ if $x \neq 0$.

The form (x,y) is known as an *inner product* or *scalar product*.

DEFINITION 9.1.2. If an inner product space X is complete in the metric $\rho(x,y)=(x-y,x-y)^{1/2}$ then X is called a *Hilbert space*.

Definition 9.1.2 can be justified by proving some lemmas.

LEMMA 9.1.1. (*The Schwarz inequality*) *For any x,y in an inner product space X,*

$$|(x,y)|^2 \leqslant (x,x)(y,y).$$

(This is also known as the *Cauchy-Schwarz inequality*.)

PROOF. If $y=0$, there is nothing to show. Suppose $y \neq 0$; then $(x+\alpha y, x+\alpha y)=(x,x)+\bar{\alpha}(x,y)+\alpha(y,x)+\alpha\bar{\alpha}(y,y)\geqslant 0$ for all α. Set $\alpha=-(x,y)/(y,y)$. Then

$$0 \leqslant (x,x)-\frac{(y,x)(x,y)}{(y,y)}-\frac{(x,y)(y,x)}{(y,y)}+\frac{(x,y)(y,x)}{(y,y)^2}(y,y),$$

and

$$(x,x)-\frac{(x,y)(y,x)}{(y,y)} \geqslant 0,$$

or

$$(x,y)(y,x) \leqslant (x,x)(y,y).$$

That is $|(x,y)|^2 \leqslant (x,x)(y,y)$. Equality holds if and only if $x+\alpha y=0$, so if and only if x and y are linearly dependent. ■

LEMMA 9.1.2. (*The Minkowski Inequality*) *In addition,*

$$(x+y,x+y)^{1/2} \leqslant (x,x)^{1/2}+(y,y)^{1/2}.$$

PROOF. If $x+y=0$, there is nothing to show. If $x+y \neq 0$, then $(x+y,x+y) \leqslant |(x+y,x)|+|(x+y,y)| \leqslant (x+y,x+y)^{1/2}(x,x)^{1/2}+(x+y,x+y)^{1/2}(y,y)^{1/2}$. ■

LEMMA 9.1.3. *Any inner product space (Hilbert space) is a normed linear vector space (Banach space), if we set $\|x\|=(x,x)^{1/2}$.*

PROOF. Since $\|\alpha x\|^2 = (\alpha x, \alpha x) = \alpha\bar{\alpha}(x,x) = |\alpha|^2\|x\|^2$ we have $\|\alpha x\| = |\alpha|\,\|x\|$. The Minkowski inequality is $\|x+y\| \leqslant \|x\| + \|y\|$, the triangle inequality, while $\|x\| > 0$ if $x \neq 0$. ■

LEMMA 9.1.4. *The inner product is a continuous function on $X \otimes X$ with respect to norm convergence.*

PROOF. For any x^1, x^2, y^2, y^2,

$$|(x^1,y^1)-(x^2,y^2)| \leqslant \|x^1-x^2\|\,\|y^1\| + \|y^1-y^2\|\,\|x^1\| + \|x^2-x^1\|\,\|y^2-y^1\|.$$

Thus as $x^2 \to x^1, y^2 \to y^1$, $(x^2,y^2) \to (x^1,y^1)$. ■

The following identities always hold in an inner product space:

$$\Re(x,y) = \frac{\{\|x+y\|^2 - \|x-y\|^2\}}{4};$$

$$\Im(x,y) = \frac{\{\|ix+y\|^2 - \|ix-y\|^2\}}{4}.$$

Which Banach spaces are also Hilbert spaces with the same norm? The identities above are necessary. It can be shown (Von Neumann and Jordan [1]) that

1. *If X is a normed linear vector space, and if we set*

$$2C(x,y) = \frac{(\|x+y\|^2 + \|x-y\|^2)}{(\|x\|^2 + \|y\|^2)},$$

and

$$\alpha = \inf_{x,y\neq 0} C(x,y), \qquad \beta = \sup_{x,y\neq 0} C(x,y),$$

then $1/2 \leqslant \alpha \leqslant \beta \leqslant 2$.

2. *Necessary and sufficient that there exist an inner product is $\alpha = \beta = 1$. If the inner product exists it is unique.*

The identity $\|x+y\|^2 + \|x-y\|^2 = 2\|x\|^2 + 2\|y\|^2$ is known as the parallelogram identity and is just one of many *equivalent* geometric identities that characterize the Euclidean nature of inner product spaces.

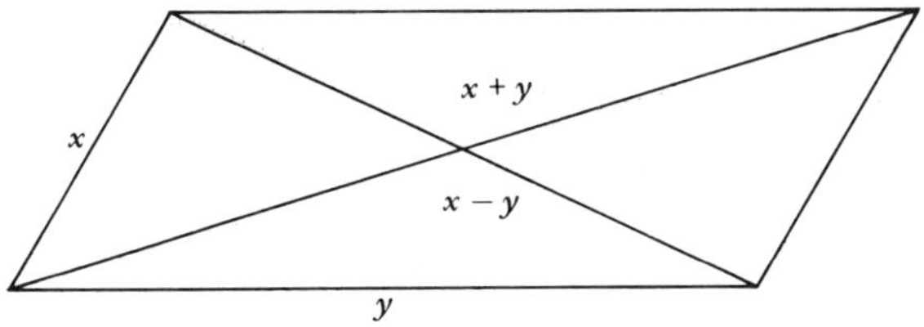

With its aid we can show that $C[0,1]$ is not an inner product space in the sup norm. For let $x=1+\sin 2\pi t, y=1+\cos 2\pi t$ so $\sup|x|=2, \sup|y|=2, \sup|x+y|=2+\sqrt{2}, \sup|x-y|=\sqrt{2}$. Then

$$\frac{\|x+y\|^2+\|x-y\|^2}{\|x\|^2+\|y\|^2}=1+\frac{\sqrt{2}}{2}.$$

This violates the parallelogram identity.

Remark. In the literature, the inner product is occasionally referred to as a *bilinear form* and as a *sesquilinear form*. The former is literally correct only if the scalar field is real, in which case (x,y) is linear in both vectors x and y. A sesquilinear form is one that satisfies conditions 1, 2, and 3 of Definition 9.1.1.

EXERCISES TO SECTION 9.1

1. In an inner product space X show that the inner product (x,y) satisfies:

   (a) (x,x) is always real as is $(x,y)(y,x)$;
   (b) $(x,y^1+y^2)=(x,y^1)+(x,y^2)$;
   (c) $(x,\alpha y)=\bar{\alpha}(x,y)$;
   (d) $(x,0)=0$;

   for all x,y,y^1,y^2 in X and all α in $\mathcal{C}$.
2. In an inner product space, show that

$$\Re(x,y)=\frac{\|x+y\|^2-\|x-y\|^2}{4},$$

$$\Im(x,y)=\frac{\|ix+y\|^2-\|ix-y\|^2}{4},$$

   and

$$\|x+y\|^2+\|x-y\|^2=2\|x\|^2+2\|y\|^2.$$

3. In a normed linear vector space show that $1/2 \leqslant \alpha \leqslant \beta \leqslant 2$. (Hint: $0 \leqslant (\|x\| - \|y\|)^2$, and $C(x+y, x-y) = 1/C(x,y)$.)
4. Let S_p be the set of pairs (z_1, z_2) of complex numbers; denote the pair (z_1, z_2) by $z = (z_1, z_2)$. Define a linear metric

$$\rho_p(z,z') = \rho_p(z - z', 0)$$

where

$$\rho_p(z,0) = \|z\|_p = \left(|z_1|^p + |z_2|^p\right)^{1/p}, \qquad p \geqslant 1$$

and

$$\rho_\infty(z,0) = \|z\|_\infty = \max(|z_1|, |z_2|).$$

Let

$$2C_p(z,z') = \frac{\|z+z'\|_p^2 + \|z-z'\|_p^2}{\|z\|_p^2 + \|z'\|_p^2}.$$

Show, for $p=1, p=\infty$, that $\alpha = \frac{1}{2}, \beta = 2$. Show, for $p=2$, that $\alpha = \beta = 1$, using as inner product $(z,z') = (z_1\bar{z}_1' + z_2\bar{z}_2')$.
5. Let X be an inner product space. Let $\langle x^n \rangle$ denote a Cauchy convergent sequence $x^n \in X$. Call two sequences $\langle x^n \rangle, \langle y^n \rangle$ equivalent if $\langle x^n - y^n \rangle$ converges to the zero element of X. If $\langle x^n \rangle$ has a limit in X, $x^n \to x$, set $\|\langle x^n \rangle\| = \|x\|$. If $\langle x^n \rangle$ has no limit in X, set $\|\langle x^n \rangle\| = \lim_{n\to\infty} \|x^n\|$.
Show that the space $\bar{X}$, whose elements are the *families of equivalent sequences* $\langle x^n \rangle$, becomes a Hilbert space under these definitions (i.e. show that

$$(\langle x^n \rangle, \langle y^n \rangle) = \lim_{n\to\infty} (x^n, y^n)$$

defines an inner product on the *complete* space $\bar{X}$).
6. Let $p(t)$ be a monotone nondecreasing function on $-\infty < t < \infty$ with $p(t) \geqslant 0$. Define an inner product by

$$(f,g) = \int_{-\infty}^{\infty} f(t)\bar{g}(t)\,dp(t)$$

where the integral is taken in the Riemann-Stieltjes sense, and f,g are continuous functions vanishing outside of some intervals (the interval

is different for different functions). Show that this is an inner product space, and determine its completion in the sense of Exercise 9.1.4. If $\lim_{t\to-\infty} p(t)=0$ and $\lim_{t\to\infty} p(t)=1$ we can interpret p to be a cumulative distribution function and (f,g) to be the expectation $E(f\bar{g})$, also known as the covariance. Interpret $\|f\|^2$.
If $p(t)$ is differentiable with derivative $q(t)$, the space we obtain is called $L_2[q(t)]$.

7. H_p: Let z be a single complex variable $D=\{z \mid |z|<1\}$. Let $f(z)$ denote a holomorphic (or analytic) function of z on D.
Let

$$N_p(f)=\sup_{0\leqslant r<1}\left[\frac{1}{2\pi}\int_{-\pi}^{\pi}|f(re^{i\theta})|^p\,d\theta\right]^{1/p} \quad \text{if } 0<p<\infty$$

and

$$N_\infty(f)=\sup_{|z|<1}|f(z)|.$$

Let H_p denote the space of functions with $N_p(f)<\infty$. Show that

(a) For $1<p<\infty$ $N_p(f-g)$ is a metric,
(b) For $1\leqslant p\leqslant\infty$, $N_p(f)$ is a norm. (Use the Minkowski inequality.)

For $p=2, H_2$ (*the Hardy space*) let $\|f\|=N_2(f)$. Expand $f(z)$ in a Taylor series about $z=0$, with coefficients $c_0, c_1, c_2, \ldots$ and show that

$$\frac{1}{2\pi}\int_{-\pi}^{\pi}|f(re^{i\theta})|^2\,d\theta=\sum_{n=0}^{\infty}|c_n|^2 r^{2n}$$

for $0\leqslant r<1$. Conclude that

$$\|f\|^2=\sum_0^\infty |c_n|^2,$$

and that if $f=\sum_0^\infty c_n z^n, g=\sum_0^\infty d_n z^n$ that

$$(f,g)=\sum_0^\infty c_n\bar{d}_n=\lim_{r\to1-0}\frac{1}{2\pi}\int_{-\pi}^{\pi}f(re^{i\theta})\,\overline{g(re^{i\theta})}\,d\theta$$

defines an inner product and that H_2 is a Hilbert space.

8. If D is a domain in the complex plane, then $HL_2(D)$ is the set of functions $f(z)$ holomorphic (or analytic) on D with

$$\iint_D |f(z)|^2 \, dx\, dy < \infty.$$

Show that

$$(f,g) = \iint_D f(z)\, \overline{g(z)}\, dx\, dy$$

is an inner product and $HL_2(D)$ is a Hilbert space. (Compare, e.g., Edwards [1], 93–94.) In $HL_2(D)$ show that *norm convergence implies uniform convergence*. (See for instance Davis [1], 207–214).
If $f \in HL_2(D)$, is $f' \in HL_2(D)$? Is $\int^z f(z)\, dz \in HL_2(D)$?

9. Let X be inner product space with $\mathscr{C}$ = real numbers. Show that $\|x+y\| = \|x-y\|$ if and only if $(x,y)=0$. What happens to this if $\mathscr{C}$ = complex numbers. If $\mathscr{C}$ is real, $\|x\| = \|y\|$ implies $(x, x-y) = (y, y-x)$.

10. In $C^1[0,1]$ define

$$(f,g)_1 = \int_0^1 f'(x) g'(x)\, dx,$$

$$(f,g)_2 = \int_a^b f'(x) g'(x)\, dx + f(0) g(0).$$

Which of these is an inner product?

11. In a complex inner product space two definitions of an angle are possible:

$$\cos \Phi = \frac{|(x,y)|}{\|x\|\,\|y\|} \tag{1}$$

or

$$\cos \Phi = \frac{\Re(x,y)}{\|x\|\,\|y\|} \tag{2}$$

For eq. 1, $\Phi = \pi/2$ if and only if $(x,y)=0$, but the law of cosines does not hold; while for eq. 2 the reverse is true. Show this.

12. Let $n>2$. "Solve" the overdetermined system

$$a_i x + b_i y = e_i, \qquad i=1,\ldots,n$$

by minimizing $\|e - xa - yb\|^2$ in an inner product space X.

9.2 ORTHOGONALITY

The geometry of inner product spaces is based on the notion of orthogonality.

DEFINITION 9.2.1. Two elements x,y of an inner product space X are *orthogonal*, $x \perp y$, if $(x,y)=0$, while $\|x\|, \|y\| \neq 0$. A collection S of nonzero elements x of an inner product space X is called *orthogonal* if $x^1, x^2 \in S$, $x^1 \neq x^2$ implies $(x^1, x^2)=0$. The collection S is called *orthonormal* if $x^1, x^2 \in S$ implies $(x^i, x^j) = \delta_i^j$. If W is any collection of x in X, then

$$[W] = \bigcap_{\alpha} M_\alpha$$

for all linear manifolds M_α in X with $W \subset M_\alpha$. By $\{W\}$ is meant the closure, in the norm topology, of $[W]$. If for two collections W^1, W^2 of elements of X we write $W^1 \perp W^2$ we mean $x \in W^1, y \in W^2$ implies $(x,y)=0$.

LEMMA 9.2.1. (*E. Schmidt*) *If $x^1,\ldots,x^n$ are linearly independent elements in an inner product space X there exists a matrix (a_{ik}) where $a_{ik}=0$ if $k>1$, while $a_{ii} \neq 0$, such that*

$$\Phi^i = \sum_{k=1}^{n} a_{ik} x^k$$

is an orthonormal set.

PROOF. Since $x^1,\ldots,x^n$ are independent, $x^1 \neq 0$. Set

$$\Phi^1 = \frac{x^1}{\|x^1\|}.$$

Given $\Phi^1,\ldots,\Phi^{j-1}$, set

$$\Phi^j = \frac{x^j - \sum_{k=1}^{j-1} (x^j,\Phi^k)\Phi^k}{\left\| x^j - \sum_{k=1}^{j-1} (x^j,\Phi^k)\Phi^k \right\|}.$$

By virtue of independence, the denominator never vanishes. A simple calculation shows that the Φ^j are an orthonormal set. ■

One interesting inner product space is the following: Consider the set of complex valued functions $f(t) \in L_2(-T,T)$ for every finite T, with norm

$$\|f\| = \limsup_{T\to\infty} \left(\frac{1}{2T} \int_{-T}^{T} |f|^2 \, dt \right)^{1/2}.$$

This is a Banach space (Zaanen [1], 103–108). If we set

$$(f,g) = \lim_{T\to\infty} \frac{1}{2T} \int_{-T}^{T} f(t)\bar{g}(t)\, dt,$$

we obtain a Hilbert space. Then $f(t) = e^{i\lambda t}, g(t) = e^{i\mu t}$ makes

$$\begin{aligned} (f,g) &= \lim_{T\to\infty} \frac{1}{2T} \int_{-T}^{T} e^{i(\lambda-\mu)t}\, dt \\ &= \lim_{T\to\infty} \frac{1}{T} \int_0^T \cos(\lambda-\mu)t\, dt \\ &= 0 \text{ if } \lambda \neq \mu. \end{aligned}$$

Here the set of functions $e^{i\lambda x}$, $-\infty < \lambda < \infty$ constitute a *nondenumerable* orthonormal set.

Let S be *any* orthonormal set in an inner product space X. Investigation of $\|x - \sum_{i=1}^{n}(x,u^i)u^i\|$ as u^i ranges over S leads to *Bessel's* inequality:

LEMMA 9.2.2. *If $u^i \in S$, $i = 1,\ldots,n$ and $x \in X$, then*

$$\sum_{i=1}^{n} |(x,u^i)|^2 \leqslant \|x\|^2.$$

PROOF

$$0 \leqslant \left\| x - \sum_{i=1}^{n} (x,u^i)u^i \right\|^2$$

$$= \|x\|^2 - \sum_{i=1}^{n} (\overline{x,u^i})(x,u^i) - \sum_{i=1}^{n} (x,u^i)(u^i,x)$$

$$+ \sum_{i=1}^{n} \sum_{j=1}^{n} ((x,u^j)u^j,(x,u^i)u^i)$$

$$= \|x\|^2 - 2\sum_{i=1}^{n} |(x,u^i)|^2 + \sum_{i=1}^{n} |(x,u^i)|^2$$

$$= \|x\|^2 - \sum_{i=1}^{n} |(x,u^i)|^2. \quad \blacksquare$$

COROLLARY 1. *For any n, at most a finite number of elements of S satisfy* $|(x,u^i)| \geqslant 1/n$.

LEMMA 9.2.3. *At most a denumerable number of* u^i *satisfy* $|(x,u^i)| > 0$.

PROOF. At most a finite number of u^i satisfy $1/n \geqslant |(x,u^i)|^2 \geqslant 1/(n+1)$. ■

THEOREM 9.2.1. (*Bessel's Inequality*) *For any* $x \in X$, *an inner product space, and any orthonormal set* $S \subset X$,

$$\sum_{u \in S} |(x,u)|^2 \leqslant \|x\|^2,$$

and

$$\sum_{u \in S} |(x,u)(y,u)| \leqslant \|x\| \, \|y\|.$$

Of course the sum is only over the nonzero terms.

PROOF. Then for any enumeration of $u^1, u^2, \ldots$ such that $(x,u^i) > 0$ we have

$$\sum_{i=1}^{n} |(x,u^i)(y,u^i)| \leqslant \left[\sum_{i=1}^{n} |(x,u^i)|^2 \right]^{1/2} \left[\sum_{i=1}^{n} |(y,u^i)|^2 \right]^{1/2}$$

$$\leqslant \|x\| \, \|y\|. \quad \blacksquare$$

An alternative statement and proof of Bessel's inequality is instructive.

THEOREM 9.2.2. *Let $u^1,\ldots,u^n$ be an orthonormal set in an inner product space X. Then the nonnegative quadratic function*

$$Q(z^1,\ldots,z^n)=\left\|x-\sum_{k=1}^{n} z_k u_k\right\|^2$$

achieves its minimum for $z_k=(x,u^k)$. Alternatively, $\mathbf{S}\{u^1,\ldots,u^n\}$ contains a unique point at a minimum distance from any x.

PROOF. Using orthogonality we can show that $Q(z^1,\ldots,z^n)=\|x\|^2-2\Re\sum_{k=1}^{n}\bar{z}^k(x,u^k)+\sum_{k=1}^{n}|z_k|^2$. Complete the square:

$$Q=\|x\|^2-\sum_{k=1}^{n}\left|(x,u^k)\right|^2+\sum_{k=1}^{n}\left|z^k-(x,u^k)\right|^2$$

Clearly Q is minimal if and only if $z^k=(x,u^k)$, $k=1,\ldots,n$ and then Bessel's inequality holds. ■

Observe that $(x-(x,u^k)u^k,u^k)=(x,u^k)-(x,u^k)(u^k,u^k)$ in order to recognize that $(x,u^k)u^k$ is the "projection" of x along u^k, and that $x-\sum_{k=1}^{n}(x,u^k)u^k$ is "perpendicular" to $\mathbf{S}\{u^1,\ldots,u^n\}$, the linear manifold spanned by $u^1,\ldots,u^n$. In many approximate methods, including several in Chapter 8, the orthonormal set is infinite, and it is only natural to seek extensions of the preceding analysis to infinite sums, with the goal of representing x as the sum of its projections. To do this we need the following:

DEFINITION 9.2.2. A set W in a linear vector space X is *convex* if $x\in W, y\in W$ implies $tx+(1-t)y\in W$ for $0\leqslant t\leqslant 1$.

The following theorem, valid for *inner product spaces*, and the subsequent theorem, which characterizes dense sets in inner product spaces, are preliminary lemmas to our present interest, although fundamental theorems for understanding generalizations beyond the inner product spaces.

THEOREM 9.2.3. (*F. Riesz*) *Let W be a convex set in an inner product space X.* Then

$$d=\inf_{x\in W}\|x\|,$$

exists, is nonnegative, and if $w^n\in W$, and $\|w^n\|\to d, n\to\infty$, then $\{w^n\}$ is a Cauchy sequence.

PROOF. In an inner product space we have the parallelogram law

$$\|w^n - w^m\|^2 + \|w^n + w^m\|^2 = 2\|w^n\|^2 + 2\|w^n\|^2.$$

Then

$$\|w^n - w^m\|^2 = 2\|w^n\|^2 + 2\|w^m\|^2 - \|w^n + w^m\|^2.$$

Now W is convex, so $(w^n + w^m)/2 \in W$. Hence $\|w^n + w^m\|^2 \geqslant 4d^2$, so

$$\|w^n - w^m\|^2 \leqslant 2\|w^n\|^2 + 2\|w^m\|^2 - 4d^2 \to 0, n, m \to \infty. \quad \blacksquare$$

THEOREM 9.2.4. *Let W be a linear manifold in a Hilbert space X. Then W is not dense in X if and only if there exists $y \neq 0, y \in X$, such that $y \perp W$.*

PROOF. Suppose first that there existed an element of X, $y \neq 0$, with $y \perp W$. Then if $w^n \in W$ and $\|w^n - y\| \to 0$, $n \to \infty$, we get $|(w^n, y) - (y,y)| = \|y\|^2 = |(w^n - y, y)| \leqslant \|w^n - y\|\,\|y\| \to 0$, $n \to \infty$. Hence $y = 0$, a contradiction. (In this we did not need X to be complete.) Now we prove Riesz's converse. Let $y^* \in X$ and let W be a linear manifold in X with $y^* \notin W$. The set $\{y^* - w \mid w \in W\}$ is a convex set since W is a linear manifold.

Let $d = \inf_{w \in W} \|y^* - w\|$. Then $d > 0$. (Why?) There is a sequence w^n in W such that $\|y^* - w^n\| \to d, n \to \infty$. By Riesz's theorem $y^* - w^n$ is a Cauchy sequence. Since X is a Hilbert space there is a limit point g, $y^* - w^n \to g, n \to \infty$, and $\|g\| = d > 0$. Suppose that for some $w \in W$, $(w,g) \neq 0$. Let z be a complex number. Then $d^2 \leqslant \|g - zw\|^2$ since $d^2 \leqslant \|y^* - w^n - zw\|^2$ for all n. Hence $d^2 \leqslant \|g\|^2 - 2\Re \bar{z}(g,w) + \|w\|^2 |z|^2$. But $\|g\|^2 = d^2$, so $0 \leqslant |z|^2 \|w\|^2 - 2\Re \bar{z}(g,w)$. Let $z = \lambda e^{i\theta}, \lambda > 0, \theta = \arg(g,w)$. Then $\lambda^2 \|w\|^2 - 2\lambda |(g,w)| \geqslant 0$, or

$$2|(g,w)| \leqslant \lambda \|w\|^2 \text{ for all } \lambda > 0.$$

Hence $|(g,w)| = 0$, a contradiction. $\blacksquare$

We identify first the separable inner product spaces.

THEOREM 9.2.5. *In a separable space inner product space X, an orthonormal set S is denumerable.*

PROOF. If x^1, x^2 are distinct elements of S, then $\|x^1 - x^2\|^2 = 2$. Surround each $x \in S$ by a circle of radius $\sqrt{2}/4$. Then each circle contains exactly one point of S. Since X is separable, there is a dense denumerable set $\{y^n\}$. Each circle contains at least one of the y^n, so there are at most a denumerable number of circles. ■

THEOREM 9.2.6. *Let X be an inner product space, $\{u^n\}$ an orthonormal sequence in X. Then $\sum_{i=1}^{\infty} a_i u^i$ converges if and only if $\sum_{i=1}^{\infty} |a_i|^2 < \infty$.*

PROOF. Direct calculation shows that

$$\left\| \sum_{i=1}^{n} a_i u^i - \sum_{i=1}^{m} a_i u^i \right\|^2 = \sum_{i=m+1}^{n} |a_i|^2.$$

Thus the partial sums of the one series are a Cauchy sequence if and only if the partial sums of the other series are a Cauchy sequence. ■

COROLLARY 1. *If $\sum_{i=1}^{\infty} a_i^2 < \infty$, and $x = \sum_{i=1}^{\infty} a_i u^i$, then $(x, u^k) = a_k$.*

THEOREM 9.2.7. *Let X be a Hilbert space, S an orthonormal set in X. For any $x \in X$, set*

$$Px = \sum_{u \in S} (x, u) u.$$

Let $M = \{S\}$. Then $x \in M$ if and only if $x = Px$.

PROOF. Px is defined for any given sequencing of those u's in S for which $(x,u) \neq 0$. Suppose now v^n is another sequencing of the u's. Then

$$\left\| \sum_{n=1}^{\infty} (x, u^n) u^n - \sum_{n=1}^{\infty} (x, v^n) v^n \right\|^2$$

$$= \sum_{n=1}^{\infty} (\overline{x, u^n}) \left(\sum_{m=1}^{\infty} (x, u^m) u^m - \sum_{m=1}^{\infty} (x, v^m) v^m, u^n \right)$$

$$+ \sum_{n=1}^{\infty} (\overline{x, v^n}) \left(\sum_{m=1}^{\infty} (x, u^m) u^m - \sum_{m=1}^{\infty} (x, v^m) v^m, v^n \right)$$

$$= \sum_{n=1}^{\infty} (\overline{x, u^n})(x, u^n) - \sum_{n=1}^{\infty} (\overline{x, u^n}) \left(\sum_{m=1}^{\infty} (x, v^m) v^m, u^n \right)$$

$$+ \sum_{n=1}^{\infty} (\overline{x, v^n}) \left(\sum_{m=1}^{\infty} (x, u^m) u^m, v^n \right) - \sum_{n=1}^{\infty} (x, v^m)(x, v^n).$$

Since the sequence of the v^m is a rearrangement of the sequence u^n,

$$\left(\sum_{m=1}^{\infty}(x,v^m)v^m,u^n\right)=(x,u^n)$$

and

$$\left(\sum_{m=1}^{\infty}(x,u^m)u^m,v^n\right)=(x,v^n)$$

so

$$\left\|\sum_{n=1}^{\infty}(x,u^n)u^n-\sum_{n=1}^{\infty}(x,v^n)v^n\right\|=0.$$

Hence Px is well defined by the collection S. Since (x,y) is norm continuous, $x-Px\perp M=\{S\}$ if and only if $x-Px\perp S$. But $(x-Px,s)=(x,s)-(\Sigma_{u\in S}(x,u)u,s)=(x,s)-(x,s)(s,s)=0$ for any $s\in S$. Hence $x-Px\perp M$. If $x\in M$, then $x-Px$ is also in M. But $x-Px\perp x-Px$ means $(x-Px,x-Px)=0$ or $x=Px$. Conversely if $x=Px$, then $Px\in M$ implies $x\in M$. ■

DEFINITION 9.2.3. An orthonormal set is *maximal* (or *complete*) if there exists no orthonormal set of which is a proper subset.

THEOREM 9.2.8. *If X is a separable inner product space then it possesses a maximal orthonormal set.*

PROOF. X separable implies the existence of a dense denumerable set $f^1,f^2,f^3,\ldots$. Form a subset $x^1,x^2,x^3,\ldots$ by discarding f^j if it is linearly dependent upon $x^1,\ldots,x^{j-1}$, otherwise keep it. Then apply the Schmidt orthogonalization to $x^1,x^2,x^3,\ldots$ to obtain an orthonormal set $\Phi^1,\Phi^2,\Phi^3,\ldots$. If $x\perp\Phi^n$ for all n, $x\perp x^n$ for all n, and hence $x\perp f^n$ for all n. Since (f^n) is a dense set, $x=0$. ■

If X is not separable, a proof would require the use of *Zermalo's axiom, well-ordering, or transfinite induction*. (See, e.g., Kuratowski [1], 100–105.)

THEOREM 9.2.9. *If the orthonormal set S is maximal in the inner product space X, then $(x,u)=(y,u)$ for all $u\in S$ implies $x=y$.*

PROOF. Suppose $x\neq y$. Then $x-y\neq 0$, while $(x-y,u)=0$ for all $u\in S$. Hence we could find a larger orthonormal set by adjoining $(x-y)/\|x-y\|$ to S. This is a contradiction. ■

Finally, we can give the full characterization of maximal orthonormal sets in the following:

THEOREM 9.2.10. *Let S be an orthonormal set in a Hilbert space X. Then the following are equivalent.*

1. *S is maximal.*
2. *For any $x \in X, x = \Sigma_{u \in S}(x,u)u$.*
3. *Polar form of Parseval's relation: For any $x,y \in X$,*

$$(x,y) = \sum_{u \in S} (x,u)(\overline{y,u}).$$

4. *Parseval's relation: For any $x \in X$*

$$\|x\|^2 = \sum_{u \in S} |(x,u)|^2.$$

PROOF. We show that assertion 1 implies assertion 2 implies 3 implies 4 implies 1. Let S be maximal. For any $x \in X$

$$\sum_{u \in S} |(x,u)|^2 \leqslant \|x\|^2.$$

Thus the sequence u^n such that $(x,u^n) \neq 0$ leads to a Cauchy convergent series

$$\sum_{n=1}^{\infty} (x,u^n)u^n.$$

Since X is complete, there is a g in X with $g = \Sigma_{n=1}^{\infty}(x,u^n)u^n$. Hence

$$(x,s) - \left(\sum_{n=1}^{\infty} (x,u^n)u^n, s \right) = (x,s) - \lim_{n \to \infty} \left(\sum_{k=1}^{n} (x,u^k)u^k, s \right)$$
$$= (x,s) - (g,s) = 0$$

for all $s \in S$. Hence $x = g$ and

$$x = \sum_{u \in S} (y,u)u.$$

Thus assertion 1 implies assertion 2. Suppose 2 holds. Then

$$y = \sum_{u \in S} (y,u)u$$

implies

$$(x,y) = \lim_{n\to\infty}\left(\sum_{k=1}^{n}(x,u^k)u^k, \sum_{k=1}^{n}(y,u^k)u^k\right)$$

where u^k is the sequence such that either $(x,u^k)\neq 0$, or $(y,u^k)\neq 0$. Then

$$(x,y) = \sum_{n=1}^{\infty}(x,u^n)(\overline{y,u^n}) = \sum_{u\in S}(x,u)(\overline{y,u}).$$

So assertion 2 implies assertion 3. If, in the polar form of Parseval's relation, we set $x=y$, then

$$\|x\|^2 = \sum_{n=1}^{\infty}|(x,u^n)|^2,$$

which is assertion 4. Now suppose the last equation holds for all $x\in X$; we will show that S is maximal. Suppose S is not maximal. Then there is an $x\in X$ with $x\neq 0, x\perp S$. But

$$\|x\|^2 = \sum_{u\in S}|(x,u)|^2 = 0,$$

a contradiction. ■

We give another form for the case where X is not complete.

COROLLARY 1. *If X is an inner product space, then assertions* 2, 3, *and* 4 *are equivalent to*

1′. $[S]$ *is dense in* X.

PROOF. If $[S]$ is dense in X, then $[S]$ is dense in $\bar{X}$. But for any $x\in X$,

$$\sum_{u\in S}(x,u)u$$

converges in $\bar{X}$ to a g with

$$(g,s) = \lim_{m\to\infty}\left(\sum_{n=1}^{m}(x,u^n)u^n, s\right) = (x,s)$$

for all $s\in S$, hence for all $s\in[S]$. Thus $x-g\perp S$, while $\overline{[S]}=\bar{X}$. Now

$x-g\in\overline{X}$ means $x-g=\lim_{n\to\infty}s^n$, $s^n\in[S]$. Then

$$(x-g,x-g)=\lim_{n\to\infty}(x-g,s^n)=0.$$

Hence $x=g\in X$. This proves assertion 2, which in turn implies assertions 3 and 4 as above.

Now suppose assertion 4 is true. Suppose $[S]$ not dense in X. Then there is some $d>0$ and some $x\in X$ with $\|x-s\|\geqslant d$ for every $s\in[S]$. Let $(x,u^n)\neq 0$, $n=1,2,3,\ldots$ and

$$Q^n=\sum_{i=1}^{n}(x,u^i)u^i.$$

Then $\|x\|^2=\|Q^n\|^2+\|x-Q^n\|^2\geqslant\sum_{i=1}^{n}|(x,u^i)|^2+d^2$. Hence $\|x\|^2\geqslant\sum_{i=1}^{\infty}|(x,u^i)|^2+d^2$, which contradicts assertion 4. ■

The geometry of subspaces of an inner product space is also connected with orthogonality.

THEOREM 9.2.11. *Let L be any subset of an inner product space X. The set M of all elements orthogonal to L is a closed linear manifold of X.*

PROOF. If $x,y\in M, x\perp L, y\perp L$, then $\alpha x+\beta y\perp L$, so $\alpha x+\beta y\in M$. If $x^n\in M$ and $x^n\to x$, $n\to\infty$, then $(x^n,x)\to(x,y)=0$ for any $y\in L$. Thus $x\in M$, and M is a closed linear manifold. ■

DEFINITION 9.2.4. The manifold orthogonal to L is called the *orthogonal complement* of L, and we denote it by $L^{\perp}$ or by $X\ominus L$.

THEOREM 9.2.12. *Let L be a Hilbert space that is a subspace of an inner product space X. Then X is the direct sum of L and $X\ominus L$: every element of X can be expressed uniquely as $x=y+w, y\in L, w\in X\ominus L, y\perp w$.*

PROOF. Let $x\in X$, and let

$$d=\lim_{y\in L}\|x-y\|.$$

Let $y^n\in L, \|x-y^n\|\to d$, $n\to\infty$. Since the set $\{x-y|y\in L\}$ is convex, $x-y^n$ is a Cauchy sequence. Since L is complete there is a $y\in L$ such that $x-y^n\to x-y$, $n\to\infty$, and $\|x-y\|=d$. We also know from the proof of Theorem 9.2.4 that $(x-y)\perp L$. Thus $x=(x-y)+y$. To show that the representation is unique we observe that $z\in L\cap X\ominus L$ makes $(z,z)=0$, so $z=0$. If $x=y^1+w^1=y^2+w^2$, then $y^1-y^2+w^1-w^2=0$. But $y^1-y^2\in L$ and $w^1-w^2\in X\ominus L$. Hence both vanish. ■

The fundamental representation theorem for separable inner product spaces is the following:

THEOREM 9.2.13. *Every real (or complex) separable inner product space X is congruent to a subspace of real (or complex) l_2.*

PROOF. By Theorem 9.2.8, X possesses a maximal orthonormal set, S: $\Phi^1, \Phi^2, \Phi^3, \ldots$. For any $x \in X$, Theorem 9.2.10 implies that

$$x = \sum_{i=1}^{\infty} (x, \Phi^i)\Phi^i,$$

and

$$\|x\|^2 = \sum_{i=1}^{\infty} |(x, \Phi^i)|^2.$$

Hence the mapping $T: x \to ((x, \Phi^1), (x, \Phi^2), \ldots)$ is a 1:1 mapping of X into a subspace of l_2, preserving norms. ■

EXERCISES TO SECTION 9.2.

1. In $L_2(0, 2\pi)$, where $f \in L_2$ means $f: (0, 2\pi) \to$ complex numbers and

$$\|f\| = \int_0^{2\pi} |f(\theta)|^2 \, d\theta < \infty,$$

let

$$(f, g) = \int_0^{2\pi} f(\theta)\bar{g}(\theta)\, d\theta.$$

Show that the set

$$\left(\frac{1}{\sqrt{2\pi}}\right)e^{in\theta}, \qquad n = 0, \pm 1, \cdots$$

is orthonormal.

2. In real $L_2(0, 2\pi)$ (the functions $f(\theta)$ are real valued) the set

$$\frac{1}{\sqrt{2\pi}}, \frac{1}{\sqrt{\pi}} \cos n\theta, \frac{1}{\sqrt{\pi}} \sin n\theta,$$

$n = 1, 2, \ldots$ is orthonormal.

3. In $L_2(-\infty, \infty)$ let

$$H_n(t) = (-1)^n e^{t^2} \frac{d^n}{dt^n}(e^{-t^2}), \qquad n = 0, 1, 2, \ldots.$$

Show that H_n is a polynomial in t. H_n is the nth Hermite polynomial. Let

$$\Psi_n(t) = (2^n n! \sqrt{\pi})^{-1/2} H_n(t) e^{-t^2/2}.$$

Then Ψ_n forms an orthonormal set (Hint: Wiener [1], 64 or Courant and Hilbert [1], Vol. 1, 91–93).

4. In $L_2(0, \infty)$, let

$$L_n(t) = e^t \frac{d^n}{dt^n}(t^n e^{-t}).$$

Show that $L_n(t)$ is a polynomial in t (the Laguerre polynomials). Let

$$\Phi_n(t) = (n!)^{-1} e^{-t/2} L_n(t), \qquad n = 0, 1, 2, \ldots.$$

Then $\Phi_n(t)$ is orthonormal in $L_2(0, \infty)$ (See Courant and Hilbert [1] Vol. 1, 81).

5. The nth Lengendre polynomial is

$$P_n(t) = \frac{1}{2^n n!} \frac{d^n}{dt^n}\left[(t^2 - 1)^n\right].$$

Then

$$\eta_n(t) = \sqrt{(n + \tfrac{1}{2})}\, P_n(t), \qquad n = 0, 1, 2, \ldots$$

are orthonormal in $L_2(-1, 1)$. (Hint: Churchill [1], 202–216; also, Courant and Hilbert [1], Vol. 1, 82–87.)

6. The orthonormal sequences in Exercises 9.2.1–5 are *maximal*. (Hint: Courant and Hilbert [1], Vol. 1, Chapter 11.)

7. (The Rademacher functions) Let

$$\Psi_n(t) = (-1)^k \text{ if } \frac{k}{2^n} < t < \frac{k+1}{2^n}$$

where $k = 0, 1, 2, \ldots, 2^{n-1}, n = 0, 1, 2, \ldots$. This system is orthogonal in

$L_2(0,1)$. Show that the system is not complete since the function $t^2+\frac{1}{6}$ is orthogonal to all $\Psi_n(t)$.

8. Let $M=\{\text{Polynomials } p(x) \text{ of degree } \leqslant n \mid p''(x)\geqslant 0 \text{ on } [0,1]\}$ (i.e., $p(x)$ is convex on $[0,1]$). Let $f\in L_2[0,1]$. Then

$$\min_{p\in M} \|f-p\|$$

possesses a unique solution. (Hint: M is a convex closed set.)

9. Prove the following theorem of Birkhoff: *Let X be a Hilbert space, x^n, y^n two orthonormal systems.* If

$$\sum_{n=1}^{\infty} \|x_n-y_n\|^2<\infty,$$

then $\{x^1,x^2,\ldots\}=\{y^1,y^2,\ldots\}$.
(Hint: see, e.g., Davis [1], 287–8.)

10. Let X be a Hilbert space, $x^1,\ldots,x^n$ elements of X and $G=\mathbf{S}\{x^1,\ldots,x^n\}$. The vectors $x^1,\ldots,x^n$ are linearly independent if and only if the Grammian matrix

$$G(x^1,\ldots,x^n)=\begin{bmatrix}(x^1,x^1) & (x^1,x^2) & \cdots & (x^1,x^n)\\ (x^2,x^1) & (x^2,x^2) & \cdots & (x^2,x^n)\\ & \cdots & & \\ (x^n,x^1) & (x^n,x^2) & \cdots & (x^n,x^n)\end{bmatrix}$$

is nonsingular. For any $x\in X$, if $y\in G$ satisfies

$$\|x-y\|=\min_{g\in G}\|x-g\|$$

then $y=\sum_{i=1}^{n}\lambda_i x^i$ where

$$\sum_{i=1}^{n}\lambda_i(x^i,x^j)=(x,x^j), \qquad j=1,\ldots,n.$$

Also

$$\|x-y\|^2=\frac{\det G(x,x^1,x^2,\ldots,x^n)}{\det G(x^1,\ldots,x^n)}.$$

(See for instance Akhieser [1], 15–20.)

11. If S is an orthonormal set in an inner product space X, then show that the mapping

$$Px = \sum_{u \in S} (x,u)u$$

is a projection and that S is maximal if and only if $P = I$.

9.3 LINEAR FUNCTIONALS ON NORMED LINEAR VECTOR SPACES

If X is a linear vector space with norm $\|\cdot\|$, over a field $\mathscr{C}$ and $\mathscr{C}$ itself is considered as a Banach space (since it is complete), then the set of all bounded linear operators from X to $\mathscr{C}$, $[X \to \mathscr{C}]$, is also a Banach space. We denote this space by X'; notice that it is a subspace of the conjugate space X^f, defined as all linear functionals, bounded or not. If $x' \in X'$, then the norm induced from X is given by

$$\|x'\| = \sup_{\|x\| \leqslant 1} |\langle x, x' \rangle|.$$

We call X' the *normed conjugate* of X, or the *dual space*. Notice that

$$\langle x, x' \rangle = \|x\| \langle \frac{x}{\|x\|}, x' \rangle,$$

so

$$|\langle x, x' \rangle| \leqslant \|x\| \|x'\|.$$

DEFINITION 9.3.1. If $p(x)$ maps a normed linear vector space X to $\mathbf{R}$, and satisfies

1. $p(x+y) \leqslant p(x) + p(y)$, all $x, y \in X$;
2. $p(tx) = tp(x)$, $0 < t < \infty$;

then $p(x)$ is called a *sublinear functional.*

The construction of bounded linear functionals is rooted in

THEOREM 9.3.1. (*Banach*) *Given a real separable linear vector space X, a linear manifold M in X and a sublinear function $p(x)$ on X. If f is a linear function on M satisfying*

$$\langle x, f \rangle \leqslant p(x),\ x \in M,$$

then there exists a denumerable set $\mathring{X}$ dense in X, and a linear functional F on $\mathring{X}$ with

1. $\langle x,F\rangle \leqslant p(x),\ x\in \mathring{X}$,
2. $\langle x,F\rangle = \langle x,f\rangle,\ x\in M$.

PROOF. If $X-M$ is empty, there is nothing to show. Suppose $x^0\in X-M$. Then if $x^1,x^2\in M$,

$$\begin{aligned}\langle x^1,f\rangle-\langle x^2,f\rangle=\langle x^1-x^2,f\rangle &\leqslant p(x^1-x^2)\\ &\leqslant p(x^1+x^0-x^0-x^2)\\ &\leqslant p(x^1+x^0)+p(-x^0-x^2).\end{aligned}$$

Thus

$$-p(-x^2-x^0)-\langle x^2,f\rangle \leqslant p(x^1+x^0)-\langle x^1,f\rangle.$$

Hence $m=\sup_{x\in M}[-p(-x-x^0)-\langle x,f\rangle]$ and $\mu=\inf_{x\in M}[p(x+x^0)-\langle x,f\rangle]$ are both finite and $m\leqslant \mu$. If $m\leqslant v_0\leqslant \mu$, then

$$-p(-x-x^0)-\langle x,f\rangle \leqslant v_0 \leqslant p(x+x^0)-\langle x,f\rangle.$$

Consider the linear manifold $[M,x^0]$. Any $y\in[M,x^0]$ has a unique representation $y=x+tx^0$, for if $y=x^1+t^1x^0=x^2+t^2x^0$, then $0=(x^1-x^2)+(t^1-t^2)x^0$. If $x^1-x^2\neq 0$, then $x^0\in M$, an impossibility. If $x^1-x^2=0$, then $t^1-t^2=0$. Define ϕ by $\langle x,\phi\rangle=\langle x,f\rangle$, $x\in M$ and $\langle x+tx^0,\phi\rangle=\langle x,f\rangle+tv_0$. Then ϕ is linear and ϕ extends f to $[M,x^0]$. Now if $y=x+tx^0$, $x\in M, t\neq 0$, then for $t>0$ $tv_0\leqslant tp(x/t+x^0)-t\langle x/t,f\rangle=p(x+tx^0)-\langle x,f\rangle$. Thus $\langle y,\phi\rangle=\langle x,f\rangle+tv_0\leqslant p(y)$. If $t\leqslant 0$, then $tv_0\leqslant tp(-x/t-x^0)-t\langle x/t,f\rangle < p(x+tx^0)-\langle x,f\rangle$. Again $\langle y,\phi\rangle\leqslant p(y)$. By induction we can extend f to each member of a dense denumerable set. ■

Since $\langle x,m'\rangle$ is a continuous function of x, inasmuch as $|\langle x,m'\rangle-\langle y,m'\rangle|\leqslant \|x-y\|\,\|m'\|$, this means that we can extend a given m' to the closure of the dense set $\mathring{X}$, hence to all of X. The restriction to separable spaces here is fundamental to the method of proof by ordinary induction. Extensions to nonseparable spaces invoke one form or another of transfinite induction. (See, e.g., Rudin [1], 57.) It is possible, without the aid of transfinite induction, to prove the preceding theorem for all Hilbert spaces (see, e.g., Zaanen [1], 146; also the discussion of well-ordering, 147.)

We shall *use* Theorem 9.3.1 in both the separable and nonseparable cases, to avoid overcomplication in hypotheses.

THEOREM 9.3.2. (*Hahn-Banach*) *If X is a separable real normed linear vector space, M a proper subspace, while M' is the space of all bounded linear functionals on M, then any $m' \in M'$ is associated with at least one $x' \in X'$ such that $\|x'\| = \|m'\|$ and $\langle x, x' \rangle = \langle x, m' \rangle$* for all $x \in M$.

PROOF. Set $p(x) = \|m'\| \, \|x\|$, $x \in X$. Then $|\langle x, m' \rangle| \leqslant \|m'\| \, \|x\| = p(x)$. By the preceding theorem we can extend the functional m' to $\mathring{X}$ dense in X as x' satisfying $|\langle x, x' \rangle| \leqslant \|m'\| \, \|x\|$. But on M, $\langle x, x' \rangle = \langle x, m' \rangle$, so $\|x'\| \geqslant \|m'\|$. Hence $\|x'\| = \|m'\|$. ■

For convenience we have given the proof for the case of a real space X and a real functional m'. For the complex case see, for instance, Zaanen [1], 145–146. We shall use the theorem in both cases.

COROLLARY 1. *Let x^0 be any element of a normed linear vector space X. Then there exists an element x' of X' satisfying*

1. $\langle x^0, x' \rangle = \|x^0\|$, *and,*
2. $\|x'\| = 1$.

PROOF. Assume $x^0 \neq 0$. On $[x^0]$, set

$$\langle x, \phi \rangle = t\|x^0\|.$$

Clearly ϕ satisfies conditions 1 and 2. We now use the extension theorem. ■

COROLLARY 2. *If $\langle x, x' \rangle = 0$ for all $x' \in X'$, then $x = 0$.*

COROLLARY 3. *In a normed linear vector space X let M be a linear manifold and $y \in M$ an element of X such that*

$$d = \inf_{x \in M} \|y - x\| > 0.$$

Then there is a linear functional $x' \in X'$ with

1. $\langle x, x' \rangle = 0$, *on* $x \in M$,
2. $\langle y, x' \rangle = 1$,
3. $\|x'\| = 1/d$.

PROOF. Consider the set $[M,y]$. If $n \in [M,y]$, $n = x + \alpha y$ for a unique α, $x \in M$. Let n' be the linear functional on $[M,y]$ given by $\langle n,n' \rangle = \alpha$. Then n' is linear on $[M,y]$, $n' = 0$ on M, and $\langle y,n' \rangle = 1$. Now $\|n\| = \|x + \alpha y\| = |\alpha|\,\|x/\alpha + y\| \geqslant |\alpha| d$. Hence $|\langle n,n' \rangle| = |\alpha| \leqslant \|n\|/d$, so $\|n'\| \leqslant 1/d$. On the other hand, choose a sequence $x^j \in M$ with $\|y - x^j\| \to d$, $j \to \infty$. Then $\langle -x^j + y, n' \rangle = \langle y,n' \rangle = 1 \leqslant \|n'\|\,\|x^j - y\| \to \|n'\| d$ as $j \to \infty$. Hence $1/d \leqslant \|n'\|$, so $\|n'\| = 1/d$. Now extend n' to X. ■

COROLLARY 4. *Let $\{x^n\}$ be a sequence in a normed linear vector space X. Then $y \in \mathbf{S}\{x^n\}$ if and only if $\langle y,x' \rangle = 0$ for all x' satisfying*

$$\langle x^n, x' \rangle = 0, \qquad n = 1,2,3,\ldots$$

PROOF. By continuity, $y \in \mathbf{S}\{x^n\}$ implies $\langle y,x' \rangle = 0$ for all x' such that $\langle x^n,x' \rangle = 0$, $n = 1,2,\ldots$. Conversely, suppose $y \notin \mathbf{S}\{x^n\}$. Then

$$d = \inf_{x \in \{x^n\}} \|y - x\| > 0.$$

Hence there exist x' such that $\langle x^n,x' \rangle = 0$, $n = 1,2,\ldots$, while $\langle y,x' \rangle = 1$. This is a contradiction. ■

Since X' is itself a normed linear vector space, it possesses a dual space $(X')'$, called the *second dual space.* Since for any $x \in X, \langle x,x' \rangle$ with x held fixed is a linear functional on X', it corresponds to some element of $(X')^f$, say Jx:

$$\langle x,x' \rangle = \langle x', Jx \rangle.$$

Thus X is imbedded in $(X')^f$. If $x'' = Jx$, then from $|\langle x,x' \rangle| \leqslant \|x\|\,\|x'\|$, we conclude $\|x''\| = \|Jx\| \leqslant \|x\|$. That is $x'' = Jx$ is actually in $(X')' = X''$ and $\|x''\| \leqslant \|x\|$. On the other hand, for any x there is x' such that $\langle x,x' \rangle = \|x\|$, $\|x'\| = 1$. Then $\langle x',x'' \rangle = \langle x',Jx \rangle = \langle x,x' \rangle = \|x\| = \|x\|\,\|x'\|$. Hence $\|x''\| \geqslant \|x\|$. Hence if $x'' = Jx$, $\|x''\| = \|x\|$. Thus J is isometric. If $J(X) = X''$, X is called *norm reflexive.*

THEOREM 9.3.3. *If X' is separable, so is X.*

PROOF. Let $\{x^{n\prime}\} = \{x' \mid \|x'\| = 1\}$. Choose $x^n \in X$ such that $\|x^n\| = 1$ but $|\langle x^n, x^{n\prime} \rangle| \geqslant \alpha > 0$. Let $M = \{\{x^n\}\}$. If $M \neq X$, then there is an $x^0 \in X - M$. Hence there exists an x' such that $\|x'\| = 1, \langle x^0,x' \rangle = \|x^0\| \neq 0$ and $\langle m,x' \rangle = 0$ for all $m \in M$. Then

$$\begin{aligned} |\langle x^n, x^{n\prime} \rangle| &= |\langle x^n, x' \rangle + \langle x^n, x^{n\prime} - x' \rangle| \\ &\leqslant |\langle x^n, x^{n\prime} - x' \rangle| \leqslant \|x^{n\prime} - x'\|. \end{aligned}$$

But this means $0<\alpha \leqslant \|x^{n'}-x'\|$. On the other hand $\{x^{n'}\}$ is dense in $\|x'\|=1$, so there is a subsequence n_j such that $\|x^{n_j'}-x'\|\to 0$. This is a contradiction. ■

In Exercise 9.3.5 it is shown that $(l_1)'=m$, the space of bounded sequences. l_1 is separable but m is not. Hence $(l_1')'=l_1''$, is not separable by virtue of the above theorem. It follows that $l_1''\neq l_1$. Similarly one can show that L_1 and $C[0,1]$ are *not* norm reflexive.

The very special nature of inner product spaces and the relation of inner products to linear functionals is explored in the following, known as the Riesz representation theorem:

THEOREM 9.3.4. *Let X be a Hilbert Space, $\langle x,x'\rangle$ a bounded linear functional on X. Then there exists a uniquely determined element $y\in X$ such that*

$$\langle x,x'\rangle=(x,y)$$

for all $x\in X$, and $\|y\|=\|x'\|$.

[If X is only an inner product space, then it can be completed by adding the ideal limits of all Cauchy sequences to it (Exercise 9.1.4). Then $\overline{X}$, the completion, is a Hilbert space, and it is here that we look for the linear functionals on X to be represented.]

PROOF. The roots in X of $\langle x,x'\rangle=0$ are a closed linear manifold M by virtue of linearity and the continuity of $\langle x,x'\rangle$. Either $M=X$, in which case $\langle x,x'\rangle=(x,0)$, or there exists a $g\in X, g\perp M, g\neq 0$. Let

$$y=\frac{g\langle g,x'\rangle}{\|g\|^2}.$$

Then

$$(x,y)=\frac{(x,g)\langle g,x'\rangle}{(g,g)}=\begin{cases}\langle g,x'\rangle, & x=g,\\ 0, & x\in M.\end{cases}$$

For any $x\in X$, set $x=x^1+\lambda g$ where

$$\lambda=-\frac{\langle x,x'\rangle}{\langle g,x'\rangle},$$

so $\langle x^1,x'\rangle=\langle x,x'\rangle-\lambda\langle g,x'\rangle=0$. Hence

$$\langle x,x'\rangle=\langle x^1,x'\rangle+\lambda\langle g,x'\rangle=(x^1,y)+\lambda(g,y)=(x,y).$$

Thus at least one y exists. If two existed, then $(x,y^1)=(x,y^2)$ for all $x \in X$ implies $y^1 - y^2 = 0$. Lastly

$$|\langle x, x' \rangle| = |(x,y)| \leqslant \|y\| \, \|x\|$$

while

$$|\langle y, x' \rangle| = |(y,y)| = \|y\| \, \|y\|.$$

so

$$\|x'\| = \|y\|. \quad \blacksquare$$

It is clear from this that a Hilbert space is isomorphic to its own conjugate space and to the second conjugate space, and is norm reflexive.

EXERCISES TO SECTION 9.3

X is a normed linear vector space in the exercises that follow.

1. Let $x^{1\prime}, \ldots, x^{n\prime} \in X'$ be linearly independent. Then there exists $x^1, \ldots, x^n \in X$ such that

$$\langle x^i, x^{j\prime} \rangle = \delta_i^j.$$

2. If $\langle x, x' \rangle = \langle y, x' \rangle$ for all $x' \in X'$, then $x = y$.
3. If $x^1, \ldots, x^n \in X$ and are linearly independent, then there exist elements $x^{1\prime}, x^{2\prime}, \ldots, x^{n\prime}$ in X' such that

$$\langle x^i, x^{j\prime} \rangle = \delta_i^j.$$

4. The dual space of $C[0,1]$: The most general linear functional is given by

$$\langle x, f \rangle = \int_0^1 x(t)\, dg(t)$$

where the integral is a Riemann-Stieltjes integral, $g(t)$ is a function of bounded variation and

$$\|f\| \leqslant T_0^1 g(t).$$

The function g can be chosen so that

$$\|f\| = T_0^1 g(t).$$

(See, e.g., Vulikh [1], 252–263 or Kantorovich and Akilov [1], 214–224.)

5. The dual space of $l_1 = m$, the space of bounded sequences. The dual spaces of $l_p = l_q, 1/p + 1/q = 1$, $1 < p, q < \infty$. (See, e.g., Taylor [1], 193–195.)
6. The dual space of $L_p = L_q, 1/p + 1/q = 1$, $1 < p, q < \infty$. The dual space of $L_1 = L_\infty$. (See, e.g., Taylor [1], 195–201, or Sobolev [1], 16–28.)
7. Reversed Hölder and Minkowski inequalities: Let $0 < p < 1$. Consider the curve $y = x^{p-1}$. Let $p' = p/(p-1)$. Show that

$$\frac{x^p}{p} + \frac{y^{p'}}{p'} \leqslant xy$$

by considering the area of a figure bounded by the x axis, the y axis, $y = y_0, x = x_0$, and the curve, and observe that

$$\int_0^x x^{p-1}\,dx - \int_y^\infty y^{p'-1}\,dy \leqslant xy.$$

From these conclude for any $P(x) > 0$

$$\int_\Omega f(x)g(x)P(x)\,dx \geqslant \left(\int_\Omega f^p(x)P(x)\,dx\right)^{1/p}\left(\int_\Omega g^{p'}(x)P(x)\,dx\right)^{1/p'}.$$

and

$$\left(\int_\Omega (f+g)(x)P(x)\,dx\right)^{1/p} \geqslant \left(\int_\Omega f^p(x)P(x)\,dx\right)^{1/p} + \left(\int_\Omega g^p(x)P(x)\,dx\right)^{1/p}$$

(see e.g., Sobelev [1], 7–8).
8. If M is a closed subspace of X, a normed linear vector space, then let $\hat{M} = \{x' \in X' | \langle x, x' \rangle = 0$, for all $x \in M\}$. M, considered as a Banach space has a dual M'. Show that M' is congruent to $X'/\hat{M}$. (Hint: See Exercise 6.2.1.)
9. If X is norm reflexive, then so is every closed subspace M in X.
10. *Definition*: Let X be a normed linear vector space, $x_\alpha \in X$, α contained in an index set I. We say that $x \in X$ can be *approximated by* $[x_\alpha]$ if for any $x \in X$ and any $\epsilon > 0$ there is a linear combination of elements of $[x_\alpha]$ with

$$\left\| x - \sum_{\alpha=1}^{n} a_\alpha x_\alpha \right\| < \epsilon.$$

Show that $x \in X$ can be approximated by $[x_\alpha]$ if and only if every $x' \in X'$ satisfying $\langle x_\alpha, x' \rangle = 0, \alpha \in I$ also satisfies $\langle x, x' \rangle = 0$. (Banach [1], 58).

11. *Definition*: We call $[x_\alpha]$ *total* if every $x \in X$ can be approximated by $[x_\alpha]$.
Show that $[x_\alpha]$ is total if and only if $\langle x_\alpha, x' \rangle = 0, \alpha \in I$ implies $x' = 0$. (Banach [1], 58).
12. *Definition*: $x_\alpha \in X, \alpha \in I$ is *norm linearly independent* if for any β, x_β cannot be approximated by $[x_\alpha]$, $\alpha \in I, \alpha \neq \beta$. Show that $[x_\alpha]$, $\alpha \in I$ is norm linearly independent if and only if there is a set $x'_\beta \in X'$, $\beta \in I$ such that $\langle x_\alpha, x'_\beta \rangle = \delta_\alpha^\beta$. The set $[x'_\beta]$ is unique if and only if $[x_\alpha]$, $\alpha \in I$, is total.
13. Show that in $C[0, 2\pi]$ the functions $1, \sin x, \cos x, \sin 2x, \cos 2x, \ldots$, which are linearly independent, are also norm linearly independent; however, the set of functions, $1, x, \sin x, \cos x, \sin 2x, \cos 2x, \ldots$ is linearly independent but not norm linearly independent. [The function x, in fact any function in $C[0, 2\pi]$, may be approximated uniformly by the trigonometric functions (Weirstrass approximation theorem).]
14. Let X be a normed linear vector space. Let $x^i \in X$, $i = 1, \ldots, n$. Show that for any $y \in X$ there is an $x = \Sigma_{i=1}^n a_i x^i$ such that

$$\min_{a_1, \ldots, a_n} \|y - (a_1 x^1 + \cdots + a_n x^n)\|$$

is achieved. (Such an x is called *a best approximate* to y.)
15. Show that if S is the set of best approximates to y, then S is convex; either S contains exactly one element, or else an infinite number of elements.
16. *Definition*: X is called *strictly convex* if

$$\|x\| \leqslant \gamma, \|y\| \leqslant \gamma \text{ imply } \|x + y\| < 2\gamma, \text{ if } x \neq y.$$

Show that: $\mathbf{E}^n$ with Euclidean distance as norm is strictly convex; l_p is strictly convex if $1 < p < \infty$; $L_p(0, 1)$ is strictly convex if $1 < p < \infty$; $C[0, 1]$ is not strictly convex. Show that $\mathbf{E}^2$ with $\|(x_1, x_2)\| = \max(|x_1|, |x_2|)$ is not strictly convex.
17. If X is strictly convex, then S, the set of best approximates to any fixed element $x \in X$, contains one element. (For more in this direction see, e.g., Davis [1], Chapter 8.)

9.4 MORE TOPOLOGIES

A given normed linear vector space may be the dual of some other normed linear vector space; it always possesses a dual space. Either of these two aspects of duality may be used to define a topology.

DEFINITION 9.4.1. Let X be a normed linear vector space, X' its dual, and $\{x^n\}$ a sequence in X. The sequence x^n is called a *weak Cauchy sequence* if $\langle x^n, x'\rangle$ is a Cauchy sequence for every x' in X'. We say $\{x^n\}$ is *weakly convergent to* x, written $x^n \rightharpoonup x$, $n \to \infty$, if $\langle x^n, x'\rangle \to \langle x, x'\rangle$, $n \to \infty$ for every $x' \in X'$.

THEOREM 9.4.1. *Every weak Cauchy sequence is bounded.*

PROOF. Let $x^n \in X, \{x^n\}$ a weak Cauchy sequence. Then $\langle x^n, x'\rangle$ is bounded in n for every fixed $x' \in x'$, so $Jx^n \in X''$ is a family of linear functionals on X', bounded individually on all of X'. The Banach-Steinhaus theorem implies $\|Jx^n\| \leqslant M$, $n = 1, 2, 3, \ldots$. But $\|Jx^n\| = \|x^n\|$, so $\|x^n\| \leqslant M$. ■

COROLLARY 1. *If every subset of a set S in X possesses a weak Cauchy subsequence, then S is bounded.*

PROOF. Suppose S not bounded; then there exist $x^n \in S, \|x^n\| \geqslant n$, $n = 1, 2, 3, \ldots$. But this subset must possess a weakly convergent subsequence, hence a bounded subsequence, which is impossible. ■

Thus, every bounded set is *weakly precompact.*

A weak limit, if it exists, is unique: $x^n \rightharpoonup f$, $n \to \infty$ and $x^n \rightharpoonup g$, $n \to \infty$ implies $\langle f, x'\rangle = \langle g, x'\rangle$ for all $x' \in X'$. By virtue of a corollary to the Hahn-Banach theorem, this implies $f = g$. In norm reflexive spaces we have the following:

THEOREM 9.4.2. *If X is norm reflexive then every weak Cauchy sequence possesses a weak limit.*

PROOF. Let x^n be weakly convergent. Then $\langle x', Jx^n\rangle$ converges for all $x' \in X'$; it is easy to see that $\lim_{n\to\infty}\langle x', Jx^n\rangle$ is a linear functional y'' on X'. But since $\|Jx^n\|$ is uniformly bounded, $\|y''\| < \infty$, so $y'' \in X''$. Then $y'' = Jy$ for some $y \in X$ and

$$\begin{aligned}\lim_{n\to\infty}\langle x', Jx^n\rangle &= \langle x', Jy\rangle = \langle y, x'\rangle \\ &= \lim_{n\to\infty}\langle x^n, x'\rangle.\end{aligned}$$

Hence $x^n \rightharpoonup y$, $n \to \infty$. ■

Partial converses to the corollary to Theorem 9.4.1 exist and are of considerable importance in the calculus of variations, other optimization theories, and differential equations. In Section 4.2 we considered minimizing sequences for metric convergence. In normed linear vector spaces this corresponds to convergence in the norm, and the compact sets found in Section 4.2 are not, as in finite dimensions, merely closed and bounded. However we next establish that in the weak topology, in certain spaces, the weakly closed, bounded sets are weakly compact.

THEOREM 9.4.3. *If X is a reflexive normed linear vector space, then every bounded infinite subset of X contains a weak Cauchy sequence.*

PROOF. We may take as our model a sequence $\{x^n\}$ in X with $\|x^n\| \leqslant B < \infty$. Form the smallest closed linear manifold M containing $\{x^n\}$. Then M is separable (why?) and by Exercise 9.3.9 is norm reflexive, so $M = (M')'$. By Theorem 9.3.3, M' is separable and contains a dense denumerable sequence $\{m^{n\prime}\}$. Since $|\langle x^n, m^{1\prime}\rangle| \leqslant B\|m^1\|$, we can choose a subsequence x^{n_1} such that $\langle x^{n_1}, m^{1\prime}\rangle$ converges. Then for each successive j we choose a subsequence of $x^{n_{j-1}}$ such that $\langle x^{n_j}, m^{j\prime}\rangle$ converges, $j = 2,3,4,\ldots$. The sequence x^{n_n} is such that $\langle x^{n_n\prime}, m^{j\prime}\rangle$ converges for all j.

Now the sequence $\{m^{n\prime}\}$ is dense in M', while $\|x^{n_n}\| \leqslant B$, $n = 1,2,3,\ldots$. From this it follows that $\langle x^{n_n}, m'\rangle$ converges for all $m' \in M'$. But every element x' of X' coincides with some $m' \in M'$ on M (why?), so that $\langle x^{n_n}, x'\rangle$ is convergent for all $x' \in X'$. ■

A sort of dual of this theorem requires another definition of convergence.

DEFINITION 9.4.2. Let X be a normed linear vector space with dual X'. If $\langle x, x^{n\prime}\rangle$ converges for each x in X, the sequence $\{x^{n\prime}\}$ is called a *weak* Cauchy sequence*. If there is an $x' \in X'$ such that $\langle x, x^{n\prime}\rangle \to \langle x, x'\rangle, n \to \infty$, for all $x \in X$, then we say that $x^{n\prime}$ *converges to x' in the weak* topology*, and write $x^{n\prime} \xrightarrow{*} x'$, $n \to \infty$.

Then the next theorem concerns *weak* precompactness*:

THEOREM 9.4.4. *If X is a separable normed linear vector space, then every bounded infinite set S' in X' contains a weak* Cauchy sequence with a weak* limit in X'.*

PROOF. A construction dual to that of the proof of Theorem 9.4.3 provides the weak* convergent sequence $\{x^{n\prime}\}$. Since every element s' of S' satisfies $\|s'\| \leqslant B < \infty$, $|\langle x, x^{n\prime}\rangle| \leqslant B\|x\|$ for all n. Then $\lim_{n\to\infty}\langle x, x^{n\prime}\rangle$ defines an element of X', say x', $\|x'\| \leqslant B$, and $x^{n\prime} \xrightarrow{*} x'$, $n \to \infty$. ■

In the Exercises to Section 9.4 it is established that a (norm) convergent sequence is weakly convergent (and is weak* convergent if applicable), and that the topologies all coincide in finite dimensional spaces. The best converse is the following:

THEOREM 9.4.5. *Let X be an inner product space, $x^n \rightharpoonup x$, $n\to\infty$ and $\|x^n\|\to\|x\|$, $n\to\infty$. Then $x^n\to x$, $n\to\infty$.*

PROOF. Since $(x^n,x)\to(x,x)$, $n\to\infty$ and $(x,x^n)\to(x,x)$, $n\to\infty$,

$$\|x-x^n\|^2=(x,x)-(x,x^n)-(x^n,x)+(x^n,x^n)$$

$$=\|x\|^2+\|x^n\|^2-(x,x^n)-(x^n,x)\to 0,\qquad n\to\infty. \quad \blacksquare$$

The following example shows that the convergence of norms is essential: In $L_2(0,1)$, let $x^n=\sqrt{2}\sin n\pi t$. Then $\|x^n\|=1$, $\|x^n\|\to 1, n\to\infty$. But every linear functional here can be written as

$$\langle x^n,x'\rangle=\sqrt{2}\int_0^1 \sin n\pi t\, f(t)\,dt=C_n$$

for some function $f(t)\in L_2(0,1)$. Here C_n is the Fourier coefficient of $f(t)$. In $L_2(0,1)$, $\Sigma_{m=1}^{\infty}|C_n|^2=\|f\|^2<\infty$, so $\langle x^n,x'\rangle\to 0, n\to\infty$ for every $x'\in X'$. Hence $x^n\rightharpoonup 0$, $n\to\infty$, and $\|0\|=0$. But x^n does not converge strongly to zero.

If we look back to the discussion of minimizing a functional $\mathcal{F}(x)$ in Section 4.2, it is clear that in the weak topology on a reflexive space X, boundedness of a sequence $\{x^n\}$ such that $\mathcal{F}(x^n)\geqslant\mathcal{F}(x^{n+1})$ is sufficient for the existence of a subsequence x^{n_j} that converges (weakly) to a weak limit. That is, we need no longer restrict candidates for a minimizing sequence to some compact set S. However, some analysis of weak continuity is needed to imply that the weak limit is a local minimum.

EXERCISES TO SECTION 9.4

1. If $x^n\to x$, $n\to\infty$, then $x^n\rightharpoonup x$, $n\to\infty$.
2. If $x^{n\prime}\to x'$, $n\to\infty$, then $x^{n\prime}\overset{*}{\rightharpoonup}x'$, $n\to\infty$.
3. If X (X') is finite dimensional, then $x^n\rightharpoonup x$, $n\to\infty$ ($x^{n\prime}\overset{*}{\rightharpoonup}x'$, $n\to\infty$) implies $x^n\to x$, $n\to\infty$ ($x^{n\prime}\to x'$, $n\to\infty$).
4. If X is a normed linear vector space, then X' has two convergences, one defined as the weak* convergence from X, the other as the weak convergence from $(X')'$. Show that if X is norm reflexive, the two definitions lead to the same convergent sequences with the same limits.

5. If M is a linear subspace of a normed linear vector space X, and if M is closed under weak convergence (i.e., $x^n \in M, x^n \rightharpoonup x$, $n \to \infty$ implies $x \in M$) then $M = \overline{M}$, its closure under norm convergence. (Hint: If $\hat{M}$ is the closure under weak convergence, then $\overline{M} \subseteq \hat{M}$. If $M \neq \hat{M}$ there is $\hat{m} \in \hat{M}$, $\hat{m} \notin \overline{M}$. Use a corollary to the Hahn-Banach theorem to produce a contradiction.)
6. If M is a linear subspace of a normed linear vector space X, we can define its *annihilator* M^0, generalizing Exercise 5.3.5 as

$$M^0 = \{x' \in X' | \langle m, x' \rangle = 0 \text{ for all } m \in M\}.$$

Similarly, if $N' \subset X'$, its annihilator ${}^0N'$ is

$${}^0N' = \{x \in X | \langle x, n' \rangle = 0 \text{ for all } n' \in N'\}.$$

Show that:

(a) M^0 is a weak* closed linear subspace of X';
(b) ${}^0N'$ is a norm closed linear subspace of X;
(c) ${}^0(M^0) = \overline{M}$;
(d) $({}^0N')^0$ is the closure of N' with respect to weak* convergence.

7. Show that $\|x\| = \sup_{\|x'\|=1} \langle x, x' \rangle$ and $\|x'\| = \sup_{\|x\|=1} \langle x, x' \rangle$, where x, x' are elements of a normed linear vector space X and its dual X'.
8. If $X = l_p$, $1 < p < \infty$, then $x^n \rightharpoonup x$, $n \rightharpoonup \infty$ with $x, x^n \in X$ if and only if

(a) $\sup_n \|x^n\| < \infty$;
(b) if $x^n = \{\xi_1^n, \xi_2^n, \xi_3^n, \ldots\}, x = \{\xi_1, \xi_2, \xi_3, \ldots\}$ then $\xi_i^n \to \xi_i$, $i = 1, 2, 3, \ldots$, $n \to \infty$.

(Hint: Every linear functional on X is of the form $\Sigma_{i=1}^{\infty} \xi_i \eta_i$ where $\eta = (\eta_1, \eta_2, \eta_3, \ldots) \in l_q, 1/p + 1/q = 1$.)
9. If $X = l_1$, then $x^n \rightharpoonup x \in l_1$, $n \to \infty$ if and only if $\|x^n - x\| \to 0$, $n \to \infty$ (i.e., $x^n \to x$, $n \to \infty$). (Banach [1], 137–138).
10. If $X = C[0, 1]$, then $x^n \rightharpoonup x$, $n \to \infty$, $x, x^n \in X$, if and only if

(a) $\sup_n \|x^n\| < \infty$; and
(b) $x^n(t) \to x(t)$, $n \to \infty$, $0 \leqslant t \leqslant 1$.

11. If $X=L_p[0,1]$, then $x^n \rightharpoonup x$, $n\to\infty$, $x, x^n \in X$, $1<p<\infty$ if and only if

(a) $\sup\|x^n\|<\infty$; and

(b) $\int_E x^n(t)\,dt \to \int_E x(t)\,dt$, $n\to\infty$,

for all sets E such that

$$\Phi_E(t)=\begin{cases} 1, & t\in E,\\ 0, & t\notin E,\end{cases}$$

is Lebesgue integrable and

$$\int_{-\infty}^{\infty} \Phi_E(t)\,dt<\infty.$$

For $p=1$ this is not true. (Banach [1], 135–136).

10

MORE ABOUT LINEAR OPERATORS

10.0 INTRODUCTION

We continue the study of linear operators begun in Chapter 6. Our interests here concern unbounded but still tractable operators, the connection to duality, and special types of operators in inner product spaces.

Summaries of the theory of the spectrum, and of the spectral theory of compact operators are provided.

10.1 CLOSED OPERATORS

An inspection of common Lagrangian functions in applications of the calculus of variations shows that they involve differentiation, which can be made a bounded operation only by such contortions and gyrations as we invoked in discussing the Galerkin method for differential equations. In other cases the structures needed for these manipulations may not be available, but we still wish to apply limiting processes. To do this some aspects of continuity must be preserved.

DEFINITION 10.1.1. Let T be a linear mapping from a subspace D_T of a normed linear vector space X to a normed linear vector space Y. T is called a *closed operator* or a *closed transformation* if for every sequence $x^n \in D_T$, $n = 1, 2, 3, \ldots$ such that both $x^n \to x$, $n \to \infty$ and $Tx^n \to y$, $n \to \infty$, it follows that $x \in D_T$ and $Tx = y$.

A more geometric view of this definition can be obtained in the product space $X \otimes Y$. We introduce a norm on $X \otimes Y$ as in Exercise 6.2.2, and we make the following:

DEFINITION 10.1.2. To any operator $T: X \to Y$ we associate a *graph* in $X \otimes Y$: the set of points $\{x, T(x)\}$ as x ranges over D_T.

Then the following theorems are obvious.

THEOREM 10.1.1. *A linear transformation $T: X \to Y$ is closed if and only if its graph is a closed set in $X \otimes Y$.*

THEOREM 10.1.2. *If T is a closed linear transformation and T^{-1} exists, then T^{-1} is closed.*

In the special case that X, Y are Banach spaces, there is the following powerful result, of which we need one consequence:

THEOREM 10.1.3. (*Closed Graph*) *If X, Y are Banach spaces, and $T: X \to Y$ is a closed linear transformation whose domain $D_T = X$, then T is bounded.*

(Proofs of this rather deep theorem may be found, for instance, in Reed and Simon [1], 79–84; Schechter [1], 62–69; or Taylor [1], 175–181.) Arguments based on the closed graph theorem are common; one important consequence which we shall use is the following:

THEOREM 10.1.4. (*Hellinger-Toeplitz*) *If X is a Hilbert space, $T: X \to X$ is a linear operator, $D_T = X$, and*

$$(Tx, y) = (x, Ty)$$

for all $x, y \in X$, then T is bounded.

PROOF. We will show that the graph of T is closed; then $D_T = X$ implies T bounded by the closed graph theorem. Let $x^n \to x, Tx^n \to y, n \to \infty$; then for every $z \in X$, $(z, Tx^n) = (Tz, x^n)$. If we let $n \to \infty$, the continuity of the inner product implies $(z, y) = (Tz, x) = (z, Tx)$. Thus $y - Tx$ is perpendicular to every element of X, and must vanish. Then the graph of T is closed. ■

(Direct proofs of this theorem may be found, for instance, in Riesz-Nagy [1], 296–297.)

Most frequently encountered differential equations lead to closed operators on a suitably defined domain. If, by virtue of existence and uniqueness theorems, the range of the differential operator is the whole space, it will possess a bounded inverse. (For a discussion of this, see Schechter [1], 336–368.)

EXERCISES TO SECTION 10.1

1. If $D_T = X$ and $T \in [X \to Y]$, show that T is a closed linear transformation.
2. If T is a linear mapping from X to Y, where Y is a Banach space, and $R_T = Y$, then show that

   (a) T^{-1} exists if and only if $Tx = 0$ implies $x = 0$;
   (b) T^{-1} is a linear mapping from Y to X.
   (c) T^{-1} is bounded if and only if there exists an $m > 0$ such that $\|Tx\| \geqslant m\|x\|$ for all $x \in D_T$, in which case $m \leqslant 1/\|T^{-1}\|$.

3. Let $X = L_2(0,1)$. Let T be differentiation and

$$D_T = \left\{ x(t) \middle| \frac{dx}{dt} \in L_2(0,1),\ x(0) = x(1) \right\}.$$

   Show that T is a closed linear transformation from X to X, but T is not bounded. Show that T^{-1} exists and is bounded.
4. Let $X = C[0,1]$, let T be differentiation and

$$D_T = \left\{ x(t) \middle| \frac{dx}{dt} \in C[0,1] \right\}.$$

   Show that T is a closed linear transformation from X to X, but T is not bounded. Show that T^{-1} exists and is bounded.
5. Consider the differential equation

$$p(t)x''(t) + q(t)x'(t) + r(t)x(t) = y(t), \qquad 0 \leqslant t \leqslant 1,$$

   on some linear space characterized by boundary conditions $x(0) = x(1) = 0$. On $C[0,1]$ this is a linear operator. If p, q, r are all in $C[0,1]$, then $Tx = px'' + qx' + rx$ is defined on the set $D_T = \{x(t) | x, x', x'' \in C[0,1], x(0) = x(1) = 0\}$ with range in $C[0,1]$. Show that T is closed. If $R_T = C[0,1]$ (an existence theorem), then show that T^{-1} is bounded. If instead we norm D_T by

$$\|x\| = \max_{0 \leqslant t \leqslant 1} (|x(t)|, |x'(t)|, |x''(t)|)$$

   show that T is a bounded operator from the Banach space D_T to $C[0,1]$, and find a bound. Again if $R_T = C[0,1]$ show that T^{-1} is bounded.
6. If P is a projection of X which is closed, then $R_P = \bar{R}_P$.
7. If P is a projection of a Banach space X, such that both $R_P, N(P)$ are closed, then $P \in [X]$. (Hint: $D_P = X$; it is sufficient to prove P is closed.)

10.2 CONJUGATE OF A BOUNDED LINEAR OPERATOR

Let X, Y be normed linear spaces and $T \in [X \to Y]$. To each of X, Y we associate the conjugate spaces X', Y':

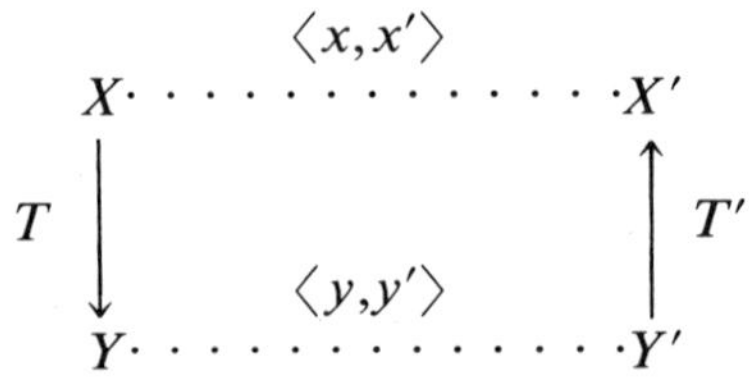

Then for each $y' \in Y'$, the functional $\langle Tx,y' \rangle$ maps x into $\mathscr{C}$. Since $\langle y,y' \rangle$ is linear in y, this is a linear functional in X^f. Further $|\langle Tx,y' \rangle| \leqslant \|T\|\,\|x\|\,\|y'\|$ so this linear functional in X^f is actually an element x' of X': $\langle Tx,y' \rangle = \langle x,x' \rangle$ for all $x \in X$. Hence we induce a mapping $y' \to x'$, which we call the *conjugate* of T, and which we denote by $x' = T'y'$, satisfying

$$\langle Tx,y' \rangle = \langle x, T'y' \rangle \text{ for all } x \in X, \text{ all } y \in Y'.$$

Since $\langle y,y' \rangle$ is linear and distributive in y', this is a linear mapping of Y' into X' that we show is bounded as follows.

LEMMA 10.2.1. *If* $T \in [X \to Y]$, *then*

$$\|T\| = \sup_{\substack{\|x\|=1 \\ \|y'\|=1}} \langle Tx,y' \rangle.$$

PROOF. From the definitions and Exercise 9.4.7

$$\|T\| = \sup_{\|x\|=1} \|Tx\| = \sup_{\|x\|=1} \sup_{\|y'\|=1} \langle Tx,y' \rangle. \quad \blacksquare$$

LEMMA 10.2.2. *If* $T' \in [Y' \to X']$, *then*

$$\|T'\| = \sup_{\|x\|=1, \|y'\|=1} \langle x, T'y' \rangle.$$

PROOF. As in Lemma 10.2.1,

$$\|T'\| = \sup_{\|y'\|=1} \|T'y'\| = \sup_{\|y'\|=1, \|x\|=1} \langle x, T'y' \rangle. \quad \blacksquare$$

THEOREM 10.2.1. *If $T \in [X \to Y]$, then $T' \in [Y' \to X']$ and $\|T\| = \|T'\|$.*

PROOF. For all x, y',

$$|\langle x, T'y' \rangle| = |\langle Tx, y' \rangle| \leqslant \|T\| \, \|x\| \, \|y'\|.$$

Then

$$\|T'y'\| = \sup_{\|x\|=1} \langle x, T'y' \rangle \leqslant \|T\| \, \|y'\|.$$

That is, T' is bounded and $\|T'\| \leqslant \|T\|$. But

$$\begin{aligned} \|Tx\| &= \sup_{\|y'\|=1} \langle Tx, y' \rangle = \sup_{\|y'\|=1} \langle x, T'y' \rangle \\ &\leqslant \sup_{\|y'\|=1} \|x\| \, \|T'y'\| = \|x\| \, \|T'\|, \end{aligned}$$

so $\|T\| \leqslant \|T'\|$. ■

This definition of conjugate operator is the natural translation to normed linear vector spaces of the algebraic concept of transposed operator. The names "dual operator" (rarely) and "adjoint operator" (very common) are also used. However, it is more convenient to reserve the name of adjoint for the related concept in Hilbert spaces.

EXERCISES TO SECTION 10.2

1. If 0 is the zero operator, then $0'$ is also the zero operator.
2. If $T_1, T_2 \in [X \to Y]$, then $(T_1 + T_2)' = T_1' + T_2'$.
3. If $T \in [X \to Y]$, then $(\alpha T)' = \alpha(T')$ for any scalar α.
4. If $T \in [X \to Y]$ and $T^{-1} \in [Y \to X]$, then $(T')^{-1} = (T^{-1})'$.
5. If $X = Y$, then $I' = I$ where I is the identity operator. Further, if $T_1, T_2 \in [X]$, then $(T_1 T_2)' = T_2' T_1'$.
6. As in Exercises to Section 5.3, let $N(T) = \{x \in X \mid Tx = 0\}$. Then show that

   (a) $N(T') = \overline{R}_T^0$;
   (b) $N(T) = {}^0\overline{R}_{T'}$; and
   (c) $(T')^{-1} \in [X' \to Y']$ if and only if $\overline{R}_T = Y$.

7. If both X and Y are finite dimensional, then $X' = X^f$, $Y' = Y^f$ and $T' = T^T$.

8. *Definition*: A Banach space X has a *countable basis* u^n, $n=1,2,\ldots$ if each $x\in X$ has a unique representation $x=\sum_{k=1}^{\infty} w_k u^k, w_k\in\mathcal{C}$. Show that no $u^k=0$, so that $u^k/\|u^k\|$ is also a countable basis.
9. Let X, Y be Banach spaces with countable bases u^n and v^n, respectively. Let $u'^n\in X', v'^n\in Y'$ be the linear functionals such that

$$x\in X, x=\sum_{k=1}^{\infty} w_k u^k \text{ makes } \langle x,u'^n\rangle=w_n; \tag{1}$$

$$y\in Y, y=\sum_{k=1}^{n} z_k v^k \text{ makes } \langle y,v'^n\rangle=z_n. \tag{2}$$

   (a) Show that u'^n, v'^n can be taken as countable bases for X', Y' (Hint: Taylor [1], 207, Exercise 2).
   (b) Let $A\in[X\to Y]$, and $\alpha_{i,j}=\langle Au^j, v'^i\rangle$. Then $y=Ax$ is equivalent to

$$z_i=\sum_{j=1}^{\infty}\alpha_{i,j}w_j, \qquad i=1,2,3,\ldots.$$

   Show that $x'=A'y'$ is given by

$$\lambda_j=\sum_{i=1}^{\infty}\alpha_{i,j}\mu_i, \qquad j=1,2,3,\ldots$$

   where

$$\langle x,x'\rangle=\sum_{j=1}^{\infty}\lambda_j w_j$$

   and

$$\langle y,y'\rangle=\sum_{i=1}^{\infty}z_i\mu_i.$$

10. Let $k(s,t)\in C([0,1]\otimes[0,1])$, and let $y=Kx$ be the linear mapping from $X=C[0,1]$ to $Y=C[0,1]$ given by

$$y(s)=\int_0^1 k(s,t)x(t)\,dt.$$

Every $x' \in X'$ and $y' \in Y'$ has the form

$$\langle x, x' \rangle = \int_0^1 x(t)\, d\alpha(t)$$

and

$$\langle y, y' \rangle = \int_0^1 y(t)\, d\beta(t).$$

Show that

$$\alpha(s) = \int_0^s \left[\int_0^1 k(s,t)\, d\beta(s) \right] dt$$

is the representation of $x' = K'y'$. (Banach [1], 101–103.)

11. Repeat Exercise 10.2.10 for the case $X = Y = L_2(0,1)$.

10.3 ADJOINT OPERATORS

In a Hilbert space X there is an isometry between X and X': to each $y \in X$ we can associate a linear functional $y' \in X'$ such that $(x,y) = \langle x, y' \rangle$. Call this isometry $\mathcal{J}$; that is $y' = \mathcal{J}y$. Now if α, β are scalars and y^1, y^2 are elements of X, then we must have

$$\begin{aligned}(x, \alpha y^1 + \beta y^2) &= \langle x, (\alpha y^1 + \beta y^2)' \rangle \\ &= \bar{\alpha}(x, y^1) + \bar{\beta}(x, y^2) \\ &= \bar{\alpha}\langle x, y^{1\prime} \rangle + \bar{\beta}\langle x, y^{2\prime} \rangle.\end{aligned}$$

Thus $\mathcal{J}(\alpha y^1 + \beta y^2) = \bar{\alpha}\mathcal{J}y^1 + \bar{\beta}\mathcal{J}y^2$, so $\mathcal{J}$ is not linear, and we cannot quite identify X with X' by means of $\mathcal{J}$. However, $\mathcal{J}^{-1}$ exists and for any $T \in [X]$ and any $y \in X$, the linear functional $T'\mathcal{J}y$ is associated by $\mathcal{J}^{-1}$ with a unique element $h \in X$, where $h = \mathcal{J}^{-1} T' \mathcal{J} y$.

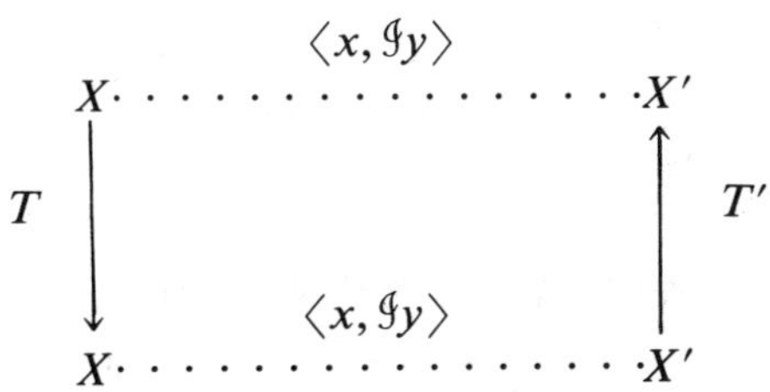

For every $x \in X$,

$$\begin{aligned}(Tx,f) &= \langle Tx, \mathcal{G}f\rangle \\ &= \langle x, T'\mathcal{G}f\rangle \\ &= (x, \mathcal{G}^{-1}T'\mathcal{G}f) \\ &= (x,h).\end{aligned}$$

DEFINITION 10.3.1. For any $T \in [X]$, X a Hilbert space, the mapping $y \to \mathcal{G}^{-1}T'\mathcal{G}y$ is a linear mapping of X into X which we call the *Hilbert adjoint*, more commonly the *adjoint* of T, and which we denote by T^*.

The defining identity for T^* is

$$(Tx,y) = (x, T^*y) \text{ for all } x,y \in X.$$

This definition can be extended to unbounded closed operators T where D_T is dense in X by virtue of the following lemmas.

LEMMA 10.3.1. *If $\bar{D}_T = X$, then the equation*

$$(Tx,y) = (x,y^*) \textit{ for all } x \textit{ in } D_T$$

possesses, for each $y \in X$ at most one solution y^*.

PROOF. If there were a second solution $\hat{y}$, then $(x,y^* - \hat{y}) = 0$ for a dense set of x's implies $y^* = \hat{y}$. ∎

DEFINITION 10.3.2. If $\bar{D}_T = X$, the solution y^* of $(Tx,y) = (x,y^*)$ for all $x \in D_T$ defines a mapping $y \to y^*$ called the *adjoint* of T, written T^*.

LEMMA 10.3.2. *If T is a closed densely defined linear operator, then T^* is a closed linear operator.*

PROOF. We need only show that T^* is closed. Suppose that $y^n \to y$, $T^*y^n \to g$, $n \to \infty$. Then

$$\lim_{n\to\infty} (Tx,y^n) = (Tx,y)$$

and

$$\lim_{n\to\infty} (x, T^*y^n) = (x,g)$$

for all $x \in D_T$, which is dense in X. That is, $y \in D_{T^*}$ and $T^*y = g$. ∎

When T is a closed, densely defined linear operator, it follows that T^* is also a closed *densely defined* linear operator (see, e.g., Riesz-Nagy [1], 305). In the cases that will interest us, this indirect guarantee that D_{T^*} includes more than the null element will not be needed.

EXERCISES TO SECTION 10.3

1. If X is a Hilbert space and $T \in [X]$, then $T^* \in [X]$ and $\|T^*\| = \|T\|$. (Hint: this can be proved directly or from $T^* = \mathcal{G}^{-1}T'\mathcal{G}$.)
2. Further, $(T^*)^*$ exists and equals T.
3. Show that $\|T^*T\| = \|T\|^2$.
4. Show that

   (a) $0^* = 0$;
   (b) $I^* = I$;
   (c) $(\alpha T_1 + \beta T_2)^* = \bar{\alpha} T_1^* + \bar{\beta} T_2^*$;
   (d) $(T_1 T_2)^* = T_2^* T_1^*$.

5. If T is a densely defined closed linear operator in a Hilbert space X, then the existence of T^*, and of $T^{-1} \in [X]$, implies $(T^*)^{-1} = (T^{-1})^*$.
6. If X is finite dimensional, show that T^* is the conjugate transpose of T.
7. If X is a real Hilbert space, show that $T' = T^*$.
8. The operator T defined by

$$Tx = p(t)x''(t) + q(t)x'(t) + r(t)x(t)$$

   on the set of functions $x(t)$ in $L_2(0, 1)$ that are continuous together with their derivatives on $[0, 1]$, possess second derivatives $x''(t) \in L_2(0, 1)$, and satisfy $x(0) = x(1) = x'(0) = x'(1) = 0$, possesses an adjoint in $L_2(0, 1)$. Find it, under the assumption that p, q, r are in $C[0, 1]$, and specify the domain of the adjoint.

10.4 SOME OPERATORS PECULIAR TO HILBERT SPACE

The inner product in Hilbert space has produced the adjoint operator. Various relationships between an operator and its adjoint are worth distinguishing by name.

DEFINITION 10.4.1. Let X be a complex inner product space, $U \in [X]$. If $UU^* = U^*U = I$, then U is called *unitary*.

It is easy to see that a unitary operator U has a bounded inverse $U^{-1}=U^*$, and that U is an isometry of X onto itself.

DEFINITION 10.4.2. Let X be an inner product space, T a linear operator with dense domain. If $(Tx,y)=(x,Ty)$ for all x,y in D_T, T is called a *symmetric* operator. If X is a Hilbert space and $D_T=X$, then T is called *selfadjoint*.

The Hellinger-Toeplitz theorem asserts that a selfadjoint operator is bounded. If T is a densely defined symmetric operator on an inner product space which is bounded, then we may as well extend it to be selfadjoint as follows.

If X is completed in the norm to a Hilbert space $\overline{X}$, and T is extended to an operator $\overline{T}$ on $\overline{X}$ (which is possible since T is bounded), we find $(\overline{T}x,y)=(x,\overline{T}y)$ for all x,y in $\overline{X}$, so that $\overline{T}$ is selfadjoint. On $\overline{X}$ we can define $\overline{T}^*$, and $\overline{T}=\overline{T}^*$; for bounded operators, we may as well assume this completion. The next three theorems are valid for both symmetric and selfadjoint operators, though stated for the selfadjoint case. (Compare this with Schechter [1], 336–368.)

THEOREM 10.4.1. *If X is a Hilbert space and $T\in[X]$, then T is selfadjoint if and only if (Tx,x) is real for all $x\in X$.*

PROOF. If T is selfadjoint, then $(Tx,x)=(x,Tx)=\overline{(Tx,x)}$ for all x. Then $(T(x+\lambda y),x+\lambda y)=(x+\lambda y,T(x+\lambda y))$ for all complex λ, all $x,y\in X$, so $\lambda(Ty,x)+\bar{\lambda}(Tx,y)=\lambda(y,Tx)+\bar{\lambda}(x,Ty)$. If $\lambda=1$, then $(Tx,y)+(Tx,y)=(y,T,x)+(x,Ty)$. If $\lambda=i$, then $i(Ty,x)-i(Tx,y)=i(y,Tx)-i(x,Ty)$. Multiply the last equation by i and add to the preceding equation to get $(Tx,y)=(x,Ty)$ for all x and y in X. ■

THEOREM 10.4.2. *If T is a bounded selfadjoint operator on a Hilbert space X, then*

$$\begin{aligned}\|T\| &= \sup_{\|x\|=1} |(Tx,x)| \\ &= \sup_{\substack{\|x\|=1\\ \|y\|=1}} |(Tx,y)|.\end{aligned}$$

PROOF. The second formulation follows from Lemma 10.2.1. The first is a consequence of the following definition and theorem. ■

DEFINITION 10.4.3. If A, B are selfadjoint operators on a Hilbert space X such that $(Ax,x) \geqslant (Bx,x)$ for all $x \in X$, we write $A \geqslant B$. If $(Ax,x) > (Bx,x)$ for all $x \neq 0$, we write $A > B$. If A is selfadjoint and bounded, we can always find two real numbers m_A, M_A such that $m_A I \leqslant A \leqslant M_A I$, but no such inequality holds for real numbers s, S with $m_A < s \leqslant S < M_A$.

(This definition is also valid for bounded symmetric operators on an inner product space.)

THEOREM 10.4.3. *If A is selfadjoint, then*

$$\|A\| = \max(|m_A|, |M_A|).$$

PROOF. Clearly $m = \sup_{\|x\|=1} |(Ax,x)| \leqslant \|A\|$. On the other hand, for any $x,y \in X$,

$$(A(x \pm y), x \pm y) = (Ax,x) + (Ay,y) \pm (Ay,x) \pm (Ax,y).$$

Thus

$$\begin{aligned}(A(x+y),x+y)-(A(x-y),x-y) &= 2(Ay,x)+2(Ax,y)\\ &\leqslant m\|x+y\|^2 + m\|x-y\|^2\\ &\leqslant 2m(\|x\|^2+\|y\|^2).\end{aligned}$$

Now we set $x = \lambda z, y = Az/\lambda$ for $z \in X$, $\|z\| = 1, \lambda$ real, $\lambda \neq 0$. Then

$$(Ay,x) = (y,Ax) = (Az,Az)$$

and

$$(Ax,y) = (x,Ay) = (Az,Az).$$

That is

$$4\|Az\|^2 \leqslant 2m(\lambda^{-2}\|Az\|^2 + \lambda^2\|z\|^2)$$

Set $\lambda^2 = \|Az\|$ and we read

$$4\|Az\|^2 \leqslant 2m(2) = 4m.$$

Hence, $\|A\| \leqslant m$, so $\|A\| = m = \max(|m_A|, |M_A|)$. ■

DEFINITION 10.4.4. A is a bounded selfadjoint operator on a Hilbert space. If $m_A \geqslant 0$, A is called *positive*. If $M_A \leqslant 0$, A is called *negative*. If $(Ax,x)>0$ when $x \neq 0$, A is called *positive definite*. If $m_A>0$, A is called *strictly positive*.

This terminology has no standard form in the literature. The case of finite dimensions is standardized, but the language cannot be carried over without some ambiguity to infinite dimensions since the set $\{x \mid \|x\| \leqslant 1\}$ is not compact there.

DEFINITION 10.4.5. A bounded operator T on a Hilbert space X is called *normal* if $T^*T = TT^*$.

This generalizes the terminology from finite matrix theory. In many applications, particularly those arising from classical mathematical physics, the operators exhibiting symmetry and normality are unbounded, and the theory of bounded operators is developed for their bounded inverses. However, it is possible, natural, and even desirable to define selfadjoint and normal unbounded operators.

The central difficulty in the definitions and the subsequent analysis is the matter of domains. Thus, an adjoint T^* is uniquely defined only for a densely defined operator. Further we cannot freely add or multiply unbounded operators, since $D_{S+T} = D_S \cap D_T$ while $D_{ST} = \{x \in D_T \mid Tx \in D_S\}$. To see the implications of this, consider the problem of defining $(ST)^*$; we require that ST be densely defined, of course. But is $(ST)^* = T^*S^*$? If $x \in D_{ST}$, then $x \in D_T$. Suppose S, T are densely defined and $y \in D_{T^*S^*}$. Then $y \in D_{S^*}$ and

$$(Tx, S^*y) = (x, T^*S^*y) = (STx, y).$$

That is, $D_{T^*S^*} \subset D_{(ST)^*}$ and $T^*S^*y = (ST)^*y$ for all $y \in D_{T^*S^*}$. We make the following:

DEFINITION 10.4.6. If two operators S, T satisfy $D_S \subset D_T$ and $Sx = Tx$ for all $x \in D_S$, we say that T *extends* (*is an extension of*) S and write $S \subset T$.

In terms of this definition, we have just proved the first part of the following:

THEOREM 10.4.4. *If S, T, and ST are closed and densely defined on a Hilbert space X, then $T^*S^* \subset (ST)^*$. If $S \in [X]$, then $T^*S^* = (ST)^*$.*

PROOF. If $S \in [X]$ then $S^* \in [X]$. Let $y \in D_{(ST)^*}$; then $y \in D_{S^*}$, and for all $x \in D_{ST}$

$$(STx,y)=(Tx,S^*y)=(x,(ST)^*y).$$

Hence, $S^*y \in D_{T^*}$ and $y \in D_{T^*S^*}$. That is, $D_{(ST)^*}=D_{T^*S^*}$. ∎

DEFINITION 10.4.7. If T is a densely defined operator such that $(Tx,y)=(x,Ty)$ for all $x,y \in D_T$, T is called *symmetric*. If T is a densely defined symmetric operator with $D_T=D_{T^*}$, so that $T=T^*$, we call T *selfadjoint*.

An immediate consequence of this definition is the observation that a densely defined symmetric operator satisfies $T \subset T^*$. In Exercise 10.4.2, it is shown that some symmetric operators possess selfadjoint extensions, as well as extensions that are not even symmetric; others possess many selfadjoint extensions, still others none. However, most of the differential operators that arise in classical mathematical physics are densely defined and are extendable through appropriate choice of boundary conditions to selfadjoint operators. (E.g., see, Stone [1], 448–462, or Schechter [1], 336–368.) We will assume that those differential operators with which we deal in Chapter 11, for instance, are capable of such selfadjoint extensions. A selfadjoint operator has no proper symmetric extensions. This is seen as follows:

THEOREM 10.4.5. *If A is a selfadjoint operator and B is a symmetric operator that extends A, then $B=A$.*

PROOF. Since D_A is dense in $H, x \in D_B$ implies that there is a sequence $x^n \in D_A$ and $x=\lim_{n\to\infty} x^n$. Then

$$(Ax^n,y)=(x^n,Ay)=(Bx^n,y)=(x^n,By),$$

for any $y \in D_A$; hence

$$(x,Ay)=(x,By)=(Bx,y)$$

is a bounded linear functional on y, by the Cauchy-Schwarz inequality. Then $x \in D_{A^*}$ and $(A^*x,y) = (x,Ay)$. But $A=A^*$, so $x \in D_A$ and $D_B=D_A$. ∎

Even when A is an unbounded symmetric operator, it may still be *half bounded* in the sense that for some real δ, we have either

$$(Ax,x) \geqslant \delta(x,x) \text{ for all } x \in D_A$$

or

$$(Ax,x) \leqslant \delta(x,x) \text{ for all } x \in D_A;$$

when $\delta = 0$ we can extend the nomenclature of Definition 10.4.4 as follows.

DEFINITION 10.4.8. If A is a closed, densely defined, symmetric operator such that

$$(Ax,x) \geqslant 0, \text{ all } x \in D_A,$$

then A is called *positive*. If $x \neq 0$ implies $(Ax,x) > 0$, A is called *positive definite*. If there is a $\delta > 0$ such that

$$(Ax,x) \geqslant \delta(x,x), \text{ all } x \in D_A,$$

then A is called *strictly positive.*

Commutation of bounded and unbounded operators is specified by the following:

DEFINITION 10.4.9. On a Hilbert space X, let T be a linear operator and $B \in [X]$. If $BT \subset TB$, then we say that B *commutes with* T. That is, if $x \in D_T$, then BTx is defined, hence TBx is defined and $BTx = TBx$.

EXERCISES TO SECTION 10.4

1. Show that a unitary operator on a Hilbert space X is an isometry. (Hint: $(U^*Ux,x) = (Ux, Ux$.)
2. Show that if X is a finite dimensional Hilbert space and T is an isometry, then T is unitary.
3. In l_2, the mapping

$$T\{x_1, x_2, x_3, \ldots\} = \{0, x_1, x_2, x_3, \ldots\}$$

   is an isometry but is not unitary. Why?
4. Show that for any Hilbert space X, T is an isometry if and only if $T^*T = I$, but that T is unitary if and only if $T^*T = TT^* = I$.
5. If X is a finite dimensional Hilbert space, and A is a symmetric operator, then the matrix representation of $A = (a_{i,j})$ with respect to any orthonormal basis $\phi^1, \ldots, \phi^n$ satisfies $a_{i,j} = \bar{a}_{j,i}$, $i,j = 1, \ldots, n$.
6. If X is $L_2(0,1)$ and $k(s,t)$ is the kernel of an integral operator

$$Tx = \int_0^1 k(s,t)x(t)\,dt$$

such that

$$\int_0^1 \int_0^1 |k(s,t)|^2 \, ds \, dt < \infty,$$

then T is selfadjoint if and only if $k(x,t) = \overline{k(t,s)}$, $0 \leqslant s,t \leqslant 1$.

7. If X is a Hilbert space, and $S \subset [X]$ is the set of bounded selfadjoint operators, then S is closed in the uniform topology.
8. If A, B are bounded selfadjoint operators on a Hilbert space X, then $A + B$ is selfadjoint, αA is selfadjoint if and only if α is real, while AB is selfadjoint if and only if $AB = BA$.
9. If $A \leqslant B$ and $B \leqslant A$, then $A = B$ for any selfadjoint A, B.
10. Let X be a Hilbert space. Let $A, T \in [X]$. Then

   (a) If A is selfadjoint, T^*AT is selfadjoint. If $A > 0$, then $T^*AT \geqslant 0$;
   (b) If R_T is dense in X and T^*AT is selfadjoint then A is selfadjoint. If T^*AT is also positive, then $A \geqslant 0$.
   (c) T^*T and TT^* are always positive.
   (d) T is normal if and only if $\|Tx\| = \|T^*x\|$, all $x \in x$.
   (e) If T is normal, then $\|T^2\| = \|T\|^2$. (Hint: $\|TT^*\| = \|T^2\|$ by (d).
   (f) T is normal if and only if $T = A + iB$ where $AB = BA$ and A, B are selfadjoint.

11. Using Definition 10.4.9 show that:

   (a) If B commutes with T_1, T_2, then B commutes with $T_1 + T_2$.
   (b) If T^{-1} exists and B commutes with T, then B commutes with T^{-1}.
   (c) If B and a sequence $B^n \in [X]$, each B^n commutes with T, and $\|B^n - B\| \to 0$, $n \to \infty$, then B commutes with T.

12. Let $X = L_2(0,1)$, and A, B, C, be the unbounded operation of differentiation. Then complete the specification by choosing appropriate domains as follows:

$$D_A = \{x(t) \in L_2(0,1) | x'(t) \in L_2(0,1)\}$$

$$D_B = \{x(t) \in L_2(0,1) | x'(t) \in L_2(0,1), x(0) = x(1)\}$$

$$D_C = \{x(t) \in L_2(0,1) | x'(t) \in L_2(0,1), x(0) = x(1) = 0\}.$$

Then $D_C \subset D_B \subset D_A$, so $C \subset B \subset A$. Also A, B, C are densely defined.

Show that $A^*=C$, $B^*=B$, and $C^*=A$. (Hint: Integration by parts yields $\int_0^1 [ix'(t)]\overline{y(t)}\,dt=\int_0^1 x(t)\overline{[iy'(t)]}\,dt+x(i)\overline{y(i)}-x(0)\overline{y(0)}$.) (See also Rudin [1], 331–332, or Riesz-Nagy [1], 308–313.) Notice that B is a selfadjoint extension of C, while A, which extends both C and B, is not even symmetric.

13. If $P\in[H]$, where H is a Hilbert space, show that the following four conditions are equivalent:

 (a) $(Px,x)=\|Px\|^2$ for all $x\in H$;
 (b) $P^*=P$ (i.e., P is selfadjoint);
 (c) P is normal;
 (d) $R_P=N(P)^\perp$.

 If any of these conditions hold, P is called an *orthogonal projection.* Show that the mapping $Px=x^*$ such that

$$\min_{m\in M}\|x-m\|=\|x-x^*\|,$$

 where M is a closed linear manifold in H, is an orthogonal projection, as is $I-P$.

14. Show that: if T is a densely defined operator on a Hilbert space H, then T^* is closed; if T is symmetric, then $T\subset T^{**}\subset T^*$; if T is closed, then $T=T^{**}\subset T^*$; if T is selfadjoint, then $T=T^{**}=T^*$. (Hint: from $A\subset B$ show that $B^*\subset A^*$.)

10.5 SQUARE ROOTS OF POSITIVE OPERATORS

As a technical aid in the study of minimizing quadratic functionals on a Hilbert space, we will establish the existence of a square root operation for positive selfadjoint operators. First we have the following:

LEMMA 10.5.1. *On $|z|\leqslant 1$, the power series for $\sqrt{(1-z)}$ converges absolutely.*

PROOF. We expand $\sqrt{(1-z)}=\sum_{n=0}^{\infty}a_n z^n$, $a_0=+1$. Then $a_0=1$, $a_n<0$, $n\geqslant 1$. Thus

$$-\sum_{n=1}^{m}a_n z^n=\sum_{n=1}^{m}|a_n|z^n\leqslant 1-\sqrt{(1-z)},\qquad 0\leqslant z<1.$$

Let $z \to +1$; we find $\Sigma_{n=1}^{m}|a_n| \leqslant 1$ for all m, so the series representation of $\sqrt{(1-z)}$ converges absolutely at $z=1$, and then for $|z| \leqslant 1$. ■

With this we prove the following:

THEOREM 10.5.1. *Let X be a Hilbert space, A a positive selfadjoint operator. Then there exists a unique positive selfadjoint operator B such that*

1. $B^2 = A$, *and,*
2. *B commutes with any operator that commutes with A.*

If $A > 0$, then $B > 0$. If $A \geqslant \delta^2 I$ for some real $\delta > 0$, then $B \geqslant \delta I$.

PROOF. PART 1. At first, we assume $A \in [X]$; B also will be in $[X]$. Choose $k > 0$ so that $k^2\|A\| < 1$. Let $T = I - k^2 A$. Then $((I - k^2 A)x, x) = \|x\|^2 - k^2(Ax, x) \geqslant 0$, while $((I - T)x, x) = k^2(Ax, x) \leqslant \|x\|$. Thus $0 \leqslant T \leqslant I$. Now $\|T\| = \sup_{\|x\|=1}((I - k^2 A)x, x) = 1 - \inf_{\|x\|=1} k^2(Ax, x) \leqslant 1$. If $m_A = \delta^2 > 0$, $((I - k^2 A)x, x) = (x, x) - k^2(Ax, x) \leqslant (1 - k^2\delta^2)(x, x)$, so $\|T\| \leqslant (1 - k^2\delta^2) < 1$.

Then define an operator B by

$$kB = I + \sum_{n=1}^{\infty} a_n T^n.$$

The theorem on convergence of operator series applies; $kB \in [X]$, and as a uniform limit of selfadjoint operators kB is selfadjoint. The same is true for B. Since

$$(T^n x, x) \leqslant \|T^n\| \, \|x\|^2 \leqslant \|T\|^n \, \|x\|^2,$$

$$(Bx, x) = k^{-1}\left[(x, x) + \sum_{n=1}^{\infty} a_n (T^n x, x)\right]$$

$$\geqslant k^{-1}\left[(x, x) + \sum_{n=1}^{\infty} a_n \|T\|^n \, \|x\|^2\right]$$

$$= k^{-1}\|x\|^2 \sqrt{(1 - \|T\|)} \geqslant 0.$$

Then $B \geqslant 0$. If $m_A = \delta^2 > 0$, then $B \geqslant k^{-1}\sqrt{(1 - k^2\delta^2)} I \geqslant \delta I$.

If an operator commutes with A, it commutes with $I + \Sigma_{n=1}^{m} a_n T^n$ for all n, hence with kB, and so with B.

That we have a solution of $B^2=A$ goes as follows:

$$\left\|k^2B^2-\left(I+\sum_{n=1}^{m}a_nT^n\right)^2\right\| \leqslant \left\|kB-\left(I+\sum_{n=1}^{m}a_nT^m\right)\right\|\left\|kB+I+\sum_{n=1}^{m}a_nT^n\right\|$$

$$\to 0, \qquad n\to\infty.$$

But, with $0<k, j<n$ in what follows,

$$\left(\sum_{n=0}^{m}a_nT^n\right)^2=\sum_{n=0}^{2m}\left(\sum_{k+j=n}a_ka_j\right)T^n.$$

Also

$$1-z=\sum_{n=0}^{\infty}\left(\sum_{k+j=n}a_ka_j\right)z^n,$$

so

$$I-T=\sum_{n=0}^{\infty}\left(\sum_{k+j=n}a_ja_k\right)T^n.$$

By the identity theorem for power series

$$\sum_{k+j=n}a_ka_j=\begin{cases}-1, & n=1,\\ 0, & n>1,\\ 1, & n=0.\end{cases}$$

If $n>1$, since $a_k<0, k>1$, this gives

$$0\leqslant -2a_n=\sum_{k+j=n}a_ka_j.$$

Then, if $n>m$, this implies

$$\left|\sum_{k+j=n}a_ka_j\right|\leqslant 2|a_n|.$$

Thus

$$\left\|\sum_{n=0}^{m}\left(\sum_{k+j=n}a_ja_k\right)T^n-\left(\sum_{n=0}^{m}a_nT^n\right)^2\right\|\leqslant 2\sum_{n=m+1}^{2m}|a_n|\|T\|^n\to 0, \qquad m\to\infty.$$

Hence $I-T=I-(I-k^2A)=k^2B^2$, so $A=B^2$.

If $C \geqslant 0$ is another selfadjoint square root, then $CA = CC^2 = C^2C = AC$. Then $BC = CB$. Let y be arbitrary, $x = [B - C]y$. Then

$$\begin{aligned}([B+C]x,x) &= ([B+C][B-C]y,x)\\ &= ([B^2+CB-BC-C^2]y,x)\\ &= ([A-A]y,x) = 0.\end{aligned}$$

Since $B \geqslant 0, C \geqslant 0$, there exist $B^{1/2}, C^{1/2}$, both positive, and

$$\begin{aligned}(B^{1/2}x,B^{1/2}x)+(C^{1/2}x,C^{1/2}x) &= (Bx,x)+(Cx,x)\\ &= ([B+C]x,x).\end{aligned}$$

Hence $B^{1/2}x = 0$, $C^{1/2}x = 0$, so $Bx = Cx = 0$. Now $\|x\|^2 = (x,x) = (x,[B-C]y) = ([B-C]x,y) = (0,y) = 0$. Hence $B = C$.

Now suppose $(Ax,x) = 0$. Then $(Ax,x) = (B^2x,x) = (Bx,Bx)$, so $\|Bx\|^2 = 0$. Then $Ax = B(Bx) = 0$. Since $B \geqslant 0$, B also has a unique nonnegative square root. Thus, if $(Bx,x) = 0$, then $Bx = 0$; and $Ax = 0$, so $(Ax,x) = 0$. From this we conclude that if A is positive definite, so is B. Above we showed that if $A \geqslant \delta^2 I$ then $B \geqslant \delta I$. ■

PROOF. PART 2. We now remove the requirement that $A \in [X]$, replacing it at first with the assumption $(Ax,x) \geqslant \gamma^2(x,x)$ for $x \in D_T$, $\gamma > 0$; that is, A is strictly positive. We will show the existence of a square root $A^{1/2} \geqslant \gamma I$. From the Cauchy-Schwarz inequality we have

$$0 \leqslant \gamma^2\|x\|^2 \leqslant (Ax,x) \leqslant \|Ax\|\,\|x\|.$$

Hence A is one-to-one from D_A to R_A. Suppose R_A is not dense in X. Then there is a $y \in X$ such that $(y,Ax) = 0$ for all $x \in D_A$; also $(Ax,y) = 0$, so $y \in D_{A^*}$ and $(x,A^*y) = 0$ for all $x \in D_A$ which is dense in X, hence in D_{A^*}. Then $(y,A^*y) = 0$. But it is clear that $(Ax,x) \geqslant \gamma^2(x,x)$ on D_A implies first $(x,A^*x) \geqslant \gamma^2(x,x)$ on D_A, then, since A^* is closed, on D_{A^*}. This guarantees $y = 0$.

Now R_A is dense in X. If f is any element of X there is a sequence f^n in R_A with $f^n \to f$, $n \to 0$. Let $Ax^n = f^n$; then

$$\|f^n - f^m\| = \|Ax^n - Ax^m\| \to 0, \qquad n,m \to \infty,$$

and

$$\gamma^2\|x^n - x^m\|^2 \leqslant (A[x^n - x^m], x^n - x^m) \leqslant \|Ax^n - Ax^m\|\,\|x^n - x^m\|.$$

Thus x^n is a convergent sequence with limit x; A is closed, so $x \in D_A$ and $Ax=f$. We have shown that $R_A = X$, so that $A^{-1} \in [X]$ (why?) and $\|A^{-1}\| \leqslant 1/\gamma^2$. Clearly A^{-1} is a positive selfadjoint operator and $0 \leqslant A^{-1} \leqslant (1/\gamma^2)I$.

Actually A^{-1} is positive definite, since for any $x \in X$, $x = Af$ and $(A^{-1}x, x) = (f, Af) \geqslant \gamma^2 \|f\|^2 \geqslant 0$. Then $(A^{-1}x, x) = 0$ implies $f=0$, and so $x=0$. However A^{-1} is not strictly positive unless A is bounded.

At any rate we conclude that $B = A^{-1}$ is a positive definite bounded selfadjoint operator and hence possesses a unique positive definite selfadjoint square root $B^{1/2}$ (with $\|B^{1/2}\| \leqslant 1/\gamma$), which commutes with any bounded operator that commutes with B. Since B commutes with any bounded operator that commutes with A, so does $B^{1/2}$. Since $B^{1/2}B^{1/2} = B$, $R_{B^{1/2}} \supseteq R_B$, so that $B^{1/2}$ is one-to-one from X to a set D_C which contains D_A and possesses an (unbounded selfadjoint) inverse C with domain D_C and range X. It is easy to see that if $x \in D_A$ then $Ax=f$ implies $x = B^{1/2}B^{1/2}f$, whence $Cx = B^{1/2}f$ and $C^2x = f$, so that $D_{C^2} \supseteq D_A$. If $x \in D_{C^2}$, then $Cx = y$, and $C^2x = Cy = z$. Then $B^{1/2}y = x, B^{1/2}z = y$ so $x = Bz$. Then $x \in D_A$ and $Ax = z$. Thus $D_{C^2} \subseteq D_A$, so $C^2 = A$. Last, $x \in D_A$ implies $x = B^{1/2}g$. Then $(Cx, x) = (CB^{1/2}g, B^{1/2}g) = (B^{1/2}g, g) = (x, Cx) \geqslant 0$; that is, C is a positive selfadjoint square root of A. Since C commutes with $B^{1/2}$, $CB^{1/2}$ is positive. Also $CB^{1/2}$ is selfadjoint. For let E, F be the square roots of $C, B^{1/2}$ respectively. Then E and F commute on D_C, and

$$(CB^{1/2}x, x) = (EFx, EFx) \geqslant 0$$

for all $x \in D_{B^{1/2}} = X$, while

$$(B^{1/2}Cx, x) = (FEx, FEx)$$

on all $x \in D_C$. Then

$$0 \leqslant (B^{1/2}Cx, x)^2 \leqslant (B^{1/2}Cx, Cx)(B^{1/2}x, x);$$

since $B^{1/2}Cx = x$, this is

$$\|x\|^4 \leqslant (x, Cx)\|B^{1/2}\| \cdot \|x\|^2 \leqslant \frac{1}{\gamma}(x, Cx)\|x\|^2$$

or

$$\gamma(x, x) \leqslant (Cx, x)$$

for all $x \in D_C$. Thus C is strictly positive. If L is any positive selfadjoint square root of A, then for all $x \in D_A$,

$$(Ax,x)=(L^2x,x)=(Lx,Lx) \geqslant \gamma^2(x,x).$$

Hence $\|Lx\| \geqslant \gamma\|x\|$, so L possesses a bounded selfadjoint inverse L^{-1}. Then $(L^{-1})^2Ax=(L^{-1})^2L^2x=x$ for all $x \in D_A$, while $A(L^{-1})^2x=L^2(L^{-1})^2x=x$ for all $x \in X$. Thus $(L^{-1})^2=B$. The uniqueness of the square root for B implies $L^{-1}=B^{1/2}$, $L=C$.

Last, if T commutes with A, T commutes with $B^{1/2}$ and hence with $C=(B^{1/2})^{-1}$.

PROOF. PART 3. We now remove the requirement that A be strictly positive. It is now only a positive operator. We associate with A a bounded positive selfadjoint operator B via

$$B=A(A+I)^{-1}=I-(A+I)^{-1}.$$

Since $A \geqslant 0$, $A+I \geqslant I$, so that $(A+I)^{-1}$ exists and, by Part 2 of the proof is a bounded positive definite selfadjoint operator possessing a unique positive definite square root C, which commutes with any bounded operator that commutes with $A+I$, or therefore with A. Since $\|(A+I)^{-1}\| \leqslant 1$, B is also a positive operator, which by the second representation is a bounded operator. Hence, B has a unique square root. We find now the equation

$$B^{1/2}B^{1/2}=ACC.$$

Since $B(A+I)=A$ on D_A, while $(A+I)B=A$ on X, B commutes with $(A+I)$, hence with $(A+I)^{-1}$, hence with C. Since B and C are both bounded, $B^{1/2}$ commutes with C. But if we inspect Part 2 of the proof it shows that C possesses an (unbounded strictly positive selfadjoint) inverse C^{-1} that must also commute with $B^{1/2}$. Observe that $D_{C^{-1}C^{-1}}=R_{C^2}=D_{A+I}=D_A$. Thus, on D_A we obtain the representation

$$(B^{1/2}C^{-1})^2=A.$$

Hence, A has a positive selfadjoint square root that clearly commutes with any bounded operator that commutes with A, since both $B^{1/2}$ and C have this property. Notice that the square root is positive definite if A is positive definite.

If A has another positive selfadjoint square root L, which commutes with every bounded operator commuting with A, then L commutes with B, $B^{1/2}$ and C. Since $B=AC^2$, $B=L^2C^2=LCLC$, and LC is a positive and selfadjoint square root of B, so $B^{1/2}=LC$ and $B^{1/2}C^{-1}=L$. Thus the square root is unique. ■

EXERCISES TO SECTION 10.5

1. If A is an invertible positive selfadjoint operator, then its unique positive square root is invertible. (Hint: A^{-1} commutes with A.)
2. Any bounded invertible operator T on a Hilbert space can be written as $T=UP=PV$ where P is a positive selfadjoint operator and U and V are unitary operators. (Hint: Since T is invertible, TT^* is invertible as is $P=(TT^*)^{1/2}$. Consider TP^{-1} and $P^{-1}T$.) Last, show that T is normal if and only if U and P commute with each other and T, while T is selfadjoint if and only if $U=V^*$. (This representation is known as the *polar decomposition*.)
3. If X is a Hilbert space and A is a positive definite bounded operator, then $[x,y]_A=(Ax,y)$ defines a new inner product on X. From this fact establish the inequality

$$(Ax,y)^2 \leqslant (Ax,x)(Ay,y)$$

used in Part 2 of the proof of Theorem 10.5.1.
4. Prove the following along the lines of Part 2 of the proof of Theorem 10.5.1:

THEOREM. *Let X be a Hilbert space, A a densely defined symmetric operator that is strictly positive: $(Ax,x)\geqslant\gamma^2(x,x), x\in D_A$. Then*

(a) *R_A is dense in X;*
(b) $\|A^{-1}y\|\leqslant\|y\|/\gamma, \quad y\in R_A$;
(c) *If we extend A^{-1} to $\hat{A}^{-1}$ by defining*

$$\hat{A}^{-1}y=\lim_{n\to\infty}A^{-1}y^n$$

for any sequence $y^n\to y, n\to\infty$ then $\hat{A}^{-1}$ is a bounded selfadjoint operator and $A^{-1}\subset\hat{A}^{-1}$.
(d) *$(\hat{A}^{-1})^{-1}$ exists and is a selfadjoint extension of A.*

5. Extend the inequality of Exercise 10.5.3 to the case in which A is only positive. (Hint: (Ax,y) is a continuous function of A in the uniform topology, and for any $\epsilon>0, A+\epsilon I>0$.)

10.6 SPECTRUM AND RESOLVENT

Although we will make little use of more than the terminology to be introduced below, it is so widespread in physical theories that a knowledge of it is essential in the reading of source materials in functional analysis. We recall from linear algebra that a linear operator on $\mathbf{E}^n$ appears in the form of a finite square matrix, say M. The set of (at most n points) λ in the complex plane at which $M-\lambda I$ does *not* possess an inverse, together with the linear manifolds invariant under the mapping $M-\lambda I$, determine the structure of the linear operator. The Jordan normal form of M exhibits this structure. The infinite dimensional equivalents of these objects are what we summarize here.

DEFINITION 10.6.1. Let $T:X\to X$ be a linear operator on a complex Banach space X. Let $T_\lambda = T-\lambda I$, where λ is a complex scalar. The set of points λ such that T_λ possesses an inverse $R_\lambda \in [X]$ is called *the resolvent set* of T, written $\rho(T)$, and R_λ is called the *resolvent* of T. The remaining set of values of λ is called *the spectrum* of $T, \sigma(T)$. If $\lambda \in \sigma(T)$ because T is not one-to-one, λ is called an *eigenvalue* (*characteristic value*, *proper value*) of T, and the solutions $x \neq 0$ of $T_\lambda x = 0$ are called *eigenvectors* (*characteristic vectors*, *proper vectors*). The set of eigenvalues is called the *point spectrum*, $P\sigma(t)$. Important properties of the resolvent of closed linear operators are in the following:

THEOREM 10.6.1. *Let $T:X\to X$ be a closed linear operator on a complex Banach space X with resolvent R_λ.*

1. *Analyticity of the resolvent*: If $\lambda_0 \in \rho(T)$, *then* $|\lambda-\lambda_0| < 1/\|T_{\lambda_0}^{-1}\|$ *implies* $\lambda \in \rho(T)$ *and*

$$R_\lambda = \sum_{n=0}^{\infty} (\lambda-\lambda_0)^n [R_{\lambda_0}]^{n+1}.$$

2. *Functional equation for the resolvent*: *If* $\lambda, \mu \in \rho(T)$, *then*

$$R_\lambda - R_\mu = (\lambda-\mu) R_\lambda R_\mu = (\lambda-\mu) R_\mu R_\lambda.$$

3. *If S_λ is a family of bounded linear operators on X, for every $\lambda \in \Sigma$, a subset of the complex plane, such that*

$$S_\lambda - S_\mu = (\lambda-\mu) S_\lambda S_\mu, \textit{all } \lambda, \mu \in \Sigma,$$

and for some $\lambda_0 \in \Sigma, S_{\lambda_0}\ x=0$ implies $x=0$, then there exists a unique *transformation T of X which is closed, whose resolvent exists and is identical with S_λ for all $\lambda \in \Sigma$.*

4. *Neumann series: If $T \in [X]$ then $\lambda > \|T\|$ implies $\lambda \in \rho(T)$ and*

$$R_\lambda = -\sum_{n=1}^{\infty} \lambda^{-n} T^{n-1}.$$

About the spectrum we note some typical results in the following.

THEOREM 10.6.2. *Under the same hypotheses as in Theorem* 10.6.1,

1. *If $T \in [X]$ then $\rho(T) = \rho(T')$ and $\sigma(T) = \sigma(T')$.*
2. *If X is a Hilbert space, then $\lambda \in \rho(T)$ implies $\bar{\lambda} \in \rho(T^*)$;*
3. *If X is a Hilbert space, and T is selfadjoint then $\sigma(T)$ contains only real numbers.*
4. *If X is a Hilbert space and T is unitary, then $\lambda \in \sigma(T)$ implies $|\lambda| = 1$.*

Although we will make no use of the spectral theory that can be based on our definitions, the reader is encouraged to attempt to establish these theorems, using a favorite comprehensive source such as Zaanen [1] or Taylor [1] as a reference.

10.7 THE RIESZ-SCHAUDER THEORY OF COMPACT OPERATORS

DEFINITION 10.7.1 Let $T: X \to Y$, where X, Y are normed linear vector spaces and $D_T = X$. T is called *compact* (*completely continuous*) if it maps every infinite bounded set in X into a *precompact* set in Y. It will be seen in the exercises that the compact operators are well represented among the integral operators, and as such, play an important role in classical mathematical physics. We collect some facts about compact operators in the following:

THEOREM 10.7.1. *Let $T: X \to Y$ and $D_T = X$, where X, Y are normed linear vector spaces. Then*

1. *If T is compact, it is bounded.*
2. *If T is compact, then every weak Cauchy sequence x^n in X is mapped into a strong Cauchy sequence Tx^n in Y.*
3. *If T is compact, then R_T is separable.*

4. *If T is compact, then T' is compact. If Y is a Banach space and T' is compact, then T is compact.*
5. *If T^n is a sequence of compact operators $T^n: X \to Y$, where Y is a Banach space, and $T \in [X \to Y]$, then $\|T^n - T\| \to 0, n \to \infty$ implies T is compact.*

For compact operators mapping a space X into itself, there is a theory of the spectrum which we summarize in the following:

THEOREM 10.7.2. *Let $T: X \to X$ be a compact operator on a normed linear vector space X. Then*

1. If $\lambda \neq 0$, *then T_λ maps every bounded closed set $B \subset X$ into a closed set in X. Also, T_λ maps every closed linear manifold $M \subset X$ into a closed linear manifold in X.*
2. *If $\lambda \neq 0$ set*

$$L_j = R_{(T_\lambda)^j}, \qquad j = 0, 1, 2, \ldots .$$

Then all L_j are closed linear manifolds and $L_0 \supset L_1 \supset L_2 \supset \ldots$. Let M_j be the set of $x \in X$ such that $(T_\lambda)^j x = 0$. Then all M_j are finite dimensional and $M_0 \subset M_1 \subset M_2 \subset \ldots$.

3. *The number of distinct λ's in $P\sigma(T)$ is either finite or denumerably infinite; in the latter case $\lambda_n \to 0, n \to \infty$.*
4. *If $\lambda \neq 0$, then for each λ there is an index $v(\lambda)$ such that $M_j = M_v, j \geqslant v$, while M_j is a proper subset of M_{j+1} if $j < v$.*
5. *If $\lambda \neq 0$, then*

$$L_j = L_v, \qquad j \geqslant v(\lambda).$$

while L_j is a proper subset of L_{j+1} for $j < v(\lambda)$. We call $v(\lambda)$ the index of λ for T.

6. *Every $x \in Y$ has a unique decomposition $x = m + z$ where $m \in M_{v(\lambda)}$, $z \in L_{v(\lambda)}$.*
7. If $\lambda \neq 0$, and $v(\lambda) = 0$, then $\lambda \in \rho(T)$.
8. *Let $v'(\lambda)$ be the index of λ with respect to T' the adjoint of T, M_j' and L_j' the corresponding manifolds. Then*

$$\dim M_j = \dim M_j'$$

and

$$v(\lambda) = v'(\lambda).$$

9. *If $y \in L_{v(\lambda)}$, then $(T_\lambda)^k x = y, k \geqslant 1$ is uniquely solvable for $x \in L_{v(\lambda)}$.*

Again, the reader is invited to prove these theorems with the aid of his favorite reference.

EXERCISES TO SECTION 10.7

1. Let $k(s,t) \in C[0 \leqslant s,t \leqslant 1]$, and

$$y(s) = \int_0^1 k(s,t)x(t)\,dt = Tx$$

map $C[0,1]$ to $C[0,1]$. Then T is *compact.*
2. Now let $\int_0^1 \int_0^1 |k(s,t)|^2\,ds\,dt < \infty$. Then $y = Tx$ defined in Exercise 10.7.1 maps $L_2(0,1)$ to $L_2(0,1)$ and T is compact.
3. Let $y = Ax$ be a mapping of a separable Hilbert space X to itself Y. If A is represented by the matrix $(\alpha_{i,j})$ with respect to a complete orthonormal sequence and

$$\sum_{i=1}^{\infty} \sum_{i=1}^{\infty} |\alpha_{i,j}|^2 < \infty,$$

then A is compact.
4. Let X be a Hilbert space; $\{\phi^n\}$ a maximal orthonormal set in X. Let T be a compact operator $X \to X, D_T = X$. Let

$$K_n f = \sum_{i=1}^{n} \sum_{j=1}^{n} (f,\phi^i)(T\phi^i,\phi^j)\phi^j.$$

Then K_n is also a compact operator and $\|K_n - T\| \to 0, n \to \infty$.
5. Let T be compact and selfadjoint and X an inner product space. Show that there exists $f \in X$ such that $\|f\| = 1$ and

$$|(Tf,f)| = \sup_{\|x\|=1} |(Tx,x)|.$$

Also $Tf = \mu_1 f$ and $|\mu_1| = \|T\|$. Let $f = \phi^1$. Show that there exists $X_2 \subset X_1 = X, X_2 \perp \phi^1$ with T compact and selfadjoint on X_2. Hence, there exists $\mu_2, |\mu_2| \leqslant |\mu_1|$, and a $\phi^2, (\phi^1,\phi^2) = 0$, with $T\phi^2 = \mu_2\phi^2$ and

$$|\mu_2| = \sup_{\substack{\|x\|=1 \\ (x,\phi^1)=0}} |(Tx,x)|.$$

In this way, construct an infinite orthonormal sequence $\{\phi^n\}$. Show that

(a) $\mu_n \to 0, n \to \infty$.
(b) $Tf = \sum_{i=1}^{\infty} (Tf, \phi^i)\phi^i$ for all $f \in X$.
(c) All μ_i are of finite multiplicity and

$$Af = \sum_{i=1}^{\infty} \mu_i (f, \phi^i)\phi^i.$$

(d) If X is complete (a Hilbert space), the set $\{\phi^n\}$ is maximal if and only if 0 is not an eigenvalue! The equation $f - \lambda Af = g$ is solved by

$$f = g + \sum_{i=1}^{\infty} \frac{\lambda\mu_i}{1 - \lambda\mu_i} (g, \phi^i)\phi^i,$$

if $\lambda\mu_i \neq 1, i = 1, 2, \ldots$. If $\lambda\mu_i = 1$, and μ_i corresponds to $\phi^{1,i}\phi^{2,i}\phi^{3,i} \ldots \phi^{k,i}$, then when $(g, \phi^{j,i}) = 0, j = 1, \ldots, k$, the same series furnishes the desired solution.

6. Let $p(t) > 0, 0 \leqslant t \leqslant 1, p(t), q(t) \in C[0, 1]$ and let $\mathfrak{L}(x) = -(px')' + qx$. Find a domain and a class of homogeneous boundary conditions for which $\mathfrak{L}$ is a symmetric operator on the complex space $L_2(0, 1)$. Show that if for some λ and some such boundary condition $-(px') + qx = \lambda x$ has no solution in $L_2(0, 1)$ that the inverse operator $\mathfrak{L}^{-1}$ exists, is given by a Green's function, and is a selfadjoint compact operator. From this obtain the eigenfunction expansion theory for the problem $-(px') + qx = y, y \in L_2$.
7. If X is finite dimensional and A is a matrix, what does Theorem 10.7.2 really say? (Hint: Consider the Jordan normal form.)

11 MINIMIZATION OF QUADRATIC FUNCTIONALS

11.0 INTRODUCTION

Classical mathematical physics starts with the assertion that certain physical systems assume states of least energy. On this is built a formalism that leads us to the calculus of variations and the relationship between the minimization problem and the solutions of the Euler-Lagrange equations. This connection permits a fruitful alternation of techniques using variational methods to approximate the solutions of the Euler-Lagrange differential equations or else using the solution theory of the Euler-Lagrange equations to study the minimization problems. The classical problems are quadratic, the related Euler-Lagrange equations, linear, and what is called nowadays formally selfadjoint. Functional analysis developed on this foundation.

In this chapter we study quadratic functionals associated with positive symmetric operators to develop techniques for the approximate solution of their related functional equations. The study takes place in Hilbert space, which is the natural place to study energy. Our treatment follows Mikhlin [1], [2] and is based on Friedrichs' extension theory for half bounded operators.

11.1 QUADRATIC FUNCTIONALS AND EQUIVALENT LINEAR PROBLEMS

Throughout this chapter H will denote a Hilbert space, and A will denote an operator on H whose domain is dense in H. A will at times be symmetric, at other times selfadjoint. The fundamental theorem of minimization is:

THEOREM 11.1.1. *Let A be a densely defined symmetric positive operator on a Hilbert space H. The linear equation*

$$Ax=f$$

has a solution x^* for a given $f \in H$ if and only if the quadratic functional

$$Q(x)=(Ax,x)-(x,f)-(f,x)$$

achieves a minimum on D_A at x^*.

PROOF. If $Ax^*=f$ and $x \in D_A$, then

$$\begin{aligned}Q(x)&=(Ax,x)-(x,Ax^*)-(Ax^*,x)\\&=(A(x-x^*),x)-(x,Ax^*)\\&=(A(x-x^*),x-x^*)+(A(x-x^*),x^*)-(Ax,x^*)\\&=(A(x-x^*),x-x^*)-(Ax^*,x^*).\end{aligned}$$

Since A is positive, $Q(x) \geqslant Q(x^*), x \neq x^*$, so that x^* is a location at which Q achieves its minimum, and

$$\min_{x \in D_A} Q(x)=-(x^*,f)=-(f,x^*)=-(Ax^*,x^*).$$

On the other hand, if

$$\min_{x \in D_A} Q(x)=Q(x^*)$$

then $Q(x^*+t\eta) \geqslant Q(x^*)$ for any complex t and any $\eta \in D_A$. Now

$$\begin{aligned}Q(x^*+t\eta)&=(A(x^*+t\eta),x^*+t\eta)-(x^*+t\eta,f)-(f,x^*+t\eta)\\&=|t|^2(A\eta,\eta)+t(A\eta,x^*)+\bar{t}(Ax^*,\eta)\\&\qquad -t(\eta,f)-\bar{t}(f,\eta)+Q(x^*).\end{aligned}$$

We note that $(A\eta,\eta) \geqslant 0$, so that the minimum can be achieved at x^* if and only if

$$t(A\eta,x^*)+\bar{t}(Ax^*,\eta)-t(\eta,f)-\bar{t}(f,\eta)=0,$$

or

$$t(\eta, Ax^* - f) + \bar{t}(Ax^* - f, \eta) = 0.$$

If t is real, we conclude that

$$\Re(\eta, Ax^* - f) = 0$$

for all $\eta \in D_A$. If $t = i|t|$, then we conclude that

$$\Im(\eta, Ax^* - f) = 0$$

for all $\eta \in D_A$. Thus $(Ax^* - f) \perp D_A$ which is dense in H. Hence $Ax^* = f$.
Notice that if A is positive definite, $Q(x) > Q(x^*)$ if $x \neq x^*$, so the minimizing x^* is unique. In this case, the solution of $Ax = f$ must also be unique. ■

As an illustration, we consider Poisson's equation:

$$-\triangle u = f(x,y)$$

where $\triangle u = u_{xx} + u_{yy}$. The problem is posed on a bounded region Ω in the x,y plane whose boundary is a Jordan curve S, and on S we assume a boundary condition $u(S) = 0$. We assume $f \in L_2(\Omega)$. As H we take the functions $u(x,y) \in L_2(\Omega)$ with

$$(u, v) = \iint_\Omega u(x,y)\, \overline{v(x,y)}\, dx\, dy$$

and

$$\|u\|^2 = \iint_\Omega |u(x,y)|^2\, dx\, dy.$$

As a domain for $-\triangle$ we take

$$D_{-\triangle} = M: \{u \in C^2(\Omega + S) | u(S) = 0\}.$$

Then M is dense in H (see Exercise 11.1.1), and

$$(-\triangle u, u) = -\iint_\Omega \bar{u} \triangle u\, dx\, dy$$

$$= \iint_\Omega (\bar{u}_x u_x + \bar{u}_y u_y)\, dx\, dy - \int_S \bar{u} \frac{\partial u}{\partial n}\, dS$$

$$= \iint_\Omega (|u_x|^2 + |u_y|^2)\, dx\, dy.$$

Here n is the exterior normal, and we use $u(S)=0$. Obviously $(-\Delta u,u) \geqslant 0$. If $(-\triangle u,u)=0$, then both $u_x=0$ and $u_y=0$ on $\Omega+S$, so u is constant there. But on $S, u=0$, so $u\equiv 0$. Thus, $(-\triangle)$ is a positive definite operator, and the solution of $-\triangle u=f$ is equivalent to minimizing, on M, the functional

$$(-\triangle u,u)-(u,f)-(f,u)=\iint_{\Omega}\{|u_x|^2+|u_y|^2-2\Re(u\bar{f})\}\,dx\,dy.$$

Actually, $-\triangle$ is *strictly positive*. To see this, enclose Ω in a square K centered at the origin, with half side a. Extend u to K by letting $u=0$ outside S. Then

$$u=\int_0^x u_x(s,y)\,ds,$$

so that

$$|u|^2\leqslant x\int_0^x |u_x(s,y)|^2\,ds\leqslant a\int_0^a |u_x(s,y)|^2\,ds,$$

and

$$\begin{aligned}\iint_K |u|^2\,dx\,dy &\leqslant a\int_0^a\int_0^a\int_0^a |u_x(s,y)|^2\,ds\,dy\,dx\\ &\leqslant a^2\int_0^a\int_0^a |u_x(s,y)|^2\,ds\,dy\\ &\leqslant a^2\iint_K |u_x(x,y)|^2\,dx\,dy.\end{aligned}$$

Since everything vanishes outside Ω we find

$$\|u\|^2\leqslant a^2\iint_{\Omega}(|u_x|^2+|u_y|^2)\,dx\,dy$$

or

$$\frac{1}{a^2}\|u\|^2\leqslant(-\triangle u,u).$$

In view of Exercise 10.5.4 this implies that there is a selfadjoint operator A that extends $-\triangle$ and possesses a bounded selfadjoint inverse A^{-1}. If $f\in C(\Omega+S)$, then it can be shown that $A^{-1}f\in D_{-\triangle}$, but this is an *analytic* fact beyond our present purposes.

EXERCISES TO SECTION 11.1

1. *Density of C^∞ functions in certain L_2 spaces*:
Let R be an open set in $\mathbf{R}^2$ with boundary Γ, a closed Jordan curve of finite length. In $L_2(R)$, define the inner product

$$(u,v)=\iint_R u\bar{v}\,dx\,dy.$$

Let Γ_ϵ be the set of points $P\in R$ such that for some point $Q\in\Gamma, |P-Q| \leqslant \epsilon$, and R_ϵ the points $P\in R, P\not\in\Gamma_\epsilon$.

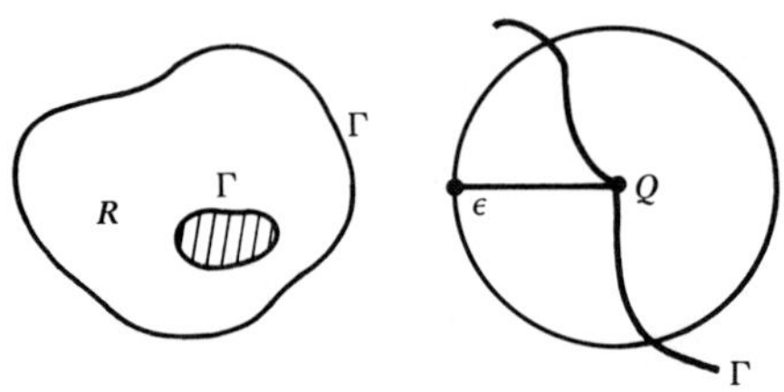

(a) Show that the set of functions $M_\epsilon: u\in L_2(R)$ with $u(P)=0, P\in\Gamma_\epsilon$ for some $\epsilon>0$, is dense in $L_2(R)$.

(b) Let

$$\rho_r(t)=\begin{cases} 0, & t\geqslant r,\\ k_r e^{-r^2/(r^2-t^2)}, & t\leqslant r.\end{cases}$$

where $k_r=k/r^2$ and

$$k^{-1}=\iint_{x^2+y^2\leqslant 1} e^{-1/[1-(x^2+y^2)]}\,dx\,dy.$$

Show that

$$\rho_r\left(\left[x^2+y^2\right]^{1/2}\right)\in C^\infty(x,y).$$

(c) Let $u(x,y)\in M_\epsilon$; form

$$\hat{u}_r(x,y)=\int_{-\infty}^{\infty}\int_{-\infty}^{\infty} u(\bar{x},\bar{y})\rho_r\left(\left[(x-\bar{x})^2+(y-\bar{y})^2\right]^{1/2}\right)d\bar{x}\,d\bar{y}$$

Show that $\hat{u}_r(x,y)\in C^\infty(x,y), \hat{u}(P)=0, P\in\Gamma.$

(d) Show that the continuous functions in M_ϵ are dense in M_ϵ.

(e) Show that if $u(x,y) \in C(R+\Gamma)$ then $\hat{u}_r(x,y) \to u(x,y), r \to \infty$ uniformly on any compact set in R.

(f) Conclude that the functions $u \in C^2(R), u \in C(R+\Gamma), u(P)=0, P \in \Gamma$ are dense in $L_2(R)$.

(see, e.g., Bremerman [1], 20–21.)

2. *Variational Methods*: Let $u \in C^2(R), u \in C^0(R+\Gamma)$ and

$$J(u) = \iint_R \left\{ \tfrac{1}{2} a u_x^2 + b u_x u_y + \tfrac{1}{2} c u_y^2 - \tfrac{1}{2} f u^2 + g u \right\} dx\, dy$$

$$+ \int_\Gamma \left\{ \tfrac{1}{2} \alpha u^2 + \gamma u \right\} ds,$$

where R is an open domain in $\mathbf{E}^2$ bounded by a rectifiable Jordan curve Γ with arc length parameter s. Here a,b,c,f,g are all C^1 functions of (x,y), while α, γ are C^1 functions of s. Using the divergence (or Green-Gauss) theorem,

$$\iint_R [P_x + Q_y]\, dx\, dy = \int_\Gamma P\, dy - Q\, dx,$$

show that

$$J(u+w) - J(u) = \iint_R [-\mathfrak{L}(u) + g] w\, dx\, dy$$

$$+ \int_\Gamma \left[\alpha u + (a u_x + b u_y) \frac{dy}{ds} - (b u_x + c u_y) \frac{dx}{ds} + \gamma \right] w\, ds$$

$$+ \tfrac{1}{2} \iint_R \left[a w_x^2 + 2 b w_x w_y + c w_y^2 - f w^2 \right] dx\, dy$$

$$+ \tfrac{1}{2} \int_\Gamma \alpha w^2\, ds,$$

where

$$\mathfrak{L}(u) = (a u_x + b u_y)_x + (b u_x + c u_y)_y + f u.$$

Now introduce the inner normal and tangential derivatives of u along Γ

by way of

$$\frac{\partial u}{\partial n} = -u_x y' + u_y x',$$

$$\frac{\partial u}{\partial s} = u_x x' + u_y y',$$

and set

$$\beta = -ay'^2 + 2bx'y' - cx'^2,$$

$$\delta = (a-c)x'y + b(y'^2 - x'^2).$$

Show that the second integral becomes

$$\int_\Gamma \left\{ \alpha u + \beta \frac{\partial u}{\partial n} + \gamma + \delta \frac{\partial u}{\partial s} \right\} w \, ds.$$

Show that $J(u+w) \geqslant J(u)$ for all $w \in C^1(R+\Gamma)$ if and only if

$$-\mathfrak{L}(u) + g = 0 \text{ on } R;$$

$$a\xi_1^2 + 2b\xi_1\xi_2 + c\xi_2^2 \geqslant 0, \text{ all } \xi_1, \xi_2;$$

$$f \leqslant 0 \text{ on } R;$$

$$\alpha \geqslant 0 \text{ on } \Gamma;$$

$$\alpha u + \gamma + \beta \frac{\partial u}{\partial n} + \delta \frac{\partial u}{\partial s} = 0 \text{ on } \Gamma.$$

If Γ is decomposed into two parts, Γ_1 and Γ_2 and we are assigned as boundary data,

$$u(x,y) = \psi(x,y), \qquad (x,y) \in \Gamma_1$$

show that the fourth condition becomes

$$\alpha u + \gamma + \beta \frac{\partial u}{\partial n} + \delta \frac{\partial u}{\partial s} = 0 \text{ on } \Gamma_2.$$

Now show that if $w \in C^2(R), C^0(R+\Gamma)$ that

$$\iint_R \{ w\mathfrak{L}(u) - u\mathfrak{L}(w) \} \, dx \, dy$$

$$= \int_\Gamma \left\{ \beta \frac{\partial u}{\partial n} + \delta \frac{\partial u}{\partial s} \right\} w - \left\{ \beta \frac{\partial w}{\partial n} + \delta \frac{\partial w}{\partial s} \right\} u \, ds$$

for all u, w satisfying

$$T(v)=\alpha_1 v+\beta_1 \frac{\partial v}{\partial n}+\delta_1 \frac{\partial v}{\partial s}=0 \text{ on } \Gamma.$$

for assigned $\alpha_1, \beta_1, \delta_1$. Show that: vanishing of this integral for all u, w satisfying $T(u)=0$, $T(w)=0$ on Γ implies that we can take $\beta_1=\beta, \delta_1=\delta$. (Thus the homogeneous problem with $\gamma=0$ is formally symmetric. We also call the inhomogeneous problem symmetric.)

Establish a Hilbert space structure for these problems which parallels the text treatment of the Poisson equation for $L_2(R)$. (See, e.g., Kantorovich and Krylov [1], 246–249, 253–255, Forsythe and Wasow [1], 159–165, 167–170, or Mikhlin [2], 160–167.)

3. Since

$$(-\triangle u, u)=\iint_{\Omega} |u_x|^2+|u_y|^2 \, dx\, dy$$

$$=\iint_{\Omega} \nabla u \cdot \overline{\nabla u} \, dx\, dy,$$

describe a candidate for the square root of any selfadjoint extension of $-\triangle$. Why is this a concrete illustration of the fact that for unbounded operators $D_{A^{1/2}} \neq D_A$?

11.2 THE METHOD OF STEEPEST DESCENT FOR BOUNDED OPERATORS

Since we have associated a minimization problem with the functional equation $Ax=f$, we might expect to apply some of the ideas discussed in Section 4.2. We will do this later on, but first we go to another attack, one that uses the quadratic nature of the minimization problem more directly. There is a sense in which the mapping $Ax-f$ is a derivative of $Q(x)$, as follows.

Let A be bounded, selfadjoint and positive definite. For any $x, z \in H$ and any real parameter t,

$$\begin{aligned} Q(x+tz) &= (A(x+tz), x+tz)-(x+tz, f)-(f, x+tz) \\ &= (Ax, x)+t(Ax, z)+t(Az, x)-t^2(Az, z) \\ &\quad -(x, f)-t(z, f)-(f, x)-t(f, z) \\ &= Q(x)+2t\Re(Ax-f, z)+t^2(Az, z). \end{aligned}$$

It is a very old idea due to Cauchy to choose a z that maximizes the rate of change of q at $t=0$. Since on any sphere $\|z\|=$constant,

$$\frac{d}{dt}Q(x+tz)|_{t=0}=2\Re(Ax-f,z),$$

this will be a maximum for $z=Ax-f$, the residual at x. Then we minimize $Q(x+tz)$ if we set

$$t=-\frac{\|z\|^2}{(Az,z)}.$$

That is, we move with the greatest speed possible to the minimum of the parabola $Q(x+tz)$. To see this notice that if $z\neq 0, (Az,z)\neq 0$, so

$$\begin{aligned} Q(x+tz) &= Q(x)+2t(z,z)+t^2(Az,z) \\ &= (Az,z)\left[\frac{Q(x)}{(Az,z)}+2t\frac{(z,z)}{(Az,z)}+t^2\right], \end{aligned}$$

and this is minimized by $t=-\|z\|^2/(Az,z)$. The minimum value is $Q(x)-\|z\|^4/(Az,z)$. The algorithm to construct the sequence of approximate solutions by the *method of steepest descent is*

$$\begin{cases} x^{n+1}=x^n+\epsilon_n z^n, \\ \epsilon_n=-\dfrac{\|z^n\|^2}{(Az^n,z^n)}, \\ z^n=Ax^n-f. \end{cases}$$

We will first prove the following:

LEMMA 11.2.1. *Let A be a bounded, strictly positive selfadjoint operator on a Hilbert space H. If $0<mI\leqslant A\leqslant MI, \alpha=(M-m)/(M+m)$, and $\beta=-2/(M+m)$, then $I+\beta A$ is a contraction, the iteration $x^{n+1}=x^n+\beta z^n$, where $z^n=Ax^n-f$, converges to the unique solution of $Ax=f$ and, for any x^0, both $\|z^{n+1}\|\leqslant\alpha\|z^n\|$ and $\|x^{n+1}-x^*\|\leqslant\alpha\|z^n\|, n=0,1,2,\ldots,$ where $Ax^*=f$.*

PROOF. For any real $\beta, I+\beta A$ is a bounded selfadjoint operator and $\|I+\beta A\|=\max(|1+\beta m|,|1+\beta M|)$. This bound will be a minimum if $1+\beta m=-(1-\beta M)$, or if $\beta=-2/(M+m)$, in which case $\|I+\beta A\|=(M-m)/M+m)<1$. The rest is obvious. ■

With the aid of this lemma, convergence of the method of steepest descent is established in the following:

THEOREM 11.2.1 (*Kantorovich* [1]) *If* $0<mI\leqslant A\leqslant MI$, *then the sequence* $\{x^n\}$ *determined by the method of steepest descent for any* x^0, *converges to the unique solution* x^* *of* $Ax=f$, *and*

$$\|x^n-x^*\|\leqslant\left[\frac{M-m}{M+m}\right]^n\frac{\|Ax^0-f\|}{m}.$$

PROOF. At any place x^n,

$$Q(x^{n+1})=\min_\beta Q(x^n+\beta(Ax^n-f)).$$

Since $Q(x)=(A(x-x^*),x-x^*)-(Ax^*,x^*)$, and A has a unique positive definite square root $A^{1/2}$,

$$\begin{aligned}Q(x^{n+1})-Q(x^*)&=\|A^{1/2}(x^{n+1}-x^*)\|^2\\&\leqslant(A(x^n+\beta(Ax^n-f)-x^*),x^n+\beta(Ax^n-f)-x^*),\end{aligned}$$

while $x^n+\beta(Ax^n-f)-x^*=x^n+\beta(Ax^n-f)-x^*-\beta(Ax^*-f)=(I+\beta A)(x^n-x^*)$. Thus, since $A^{1/2}$ commutes with $I+\beta A$, if $\beta=-2/(M+m)$, then using Lemma 11.2.1,

$$\begin{aligned}\|A^{1/2}(x^{n+1}-x^*)\|&\leqslant\|A^{1/2}(I+A)(x^n-x^*)\|\\&\leqslant\left[\frac{M-m}{M+m}\right]\|A^{1/2}(x^n-x^*)\|\\&\leqslant\left[\frac{M-m}{M+m}\right]^{n+1}\|A^{1/2}(x^0-x^*)\|.\end{aligned}$$

Since $m^{1/2}I\leqslant A^{1/2}\leqslant M^{1/2}I$, $\|x^n-x^*\|\leqslant\|A^{1/2}(x^n-x^*)\|/m^{1/2}$ and

$$\|x^n-x^*\|\leqslant\left[\frac{M-m}{M+m}\right]^n\frac{\|A^{1/2}(x^0-x^*)\|}{m^{1/2}}.$$

But $\|A^{1/2}(x^0-x^*)\|=\|A^{-1/2}A(x^0-x^*)\|\leqslant\|Ax^0-f\|/m^{1/2}$. This proves the theorem; in addition we note that

$$\|Ax^n-f\|=\|A(x^n-x^*)\|\leqslant M\|x^n-x^*\|,$$

and thus the residuals also decrease to zero. ■

An inspection of the treatment of Poisson's equation in Section 11.1 shows that for an unbounded domain, the resulting operator need not be strictly positive. Singular Sturm-Liouville problems also need not be strictly positive. On the other hand, it may still be possible to show, by analytic means, that the only function that makes the associated quadratic form vanish is the identically zero function; then the operator is still positive definite. In that case, and even if the operator is only positive, we can still obtain the following result.

THEOREM 11.2.2. (*Kantorovich* [1]) *If A is a bounded positive selfadjoint operator and $Ax^*=f$, then the algorithm for the method of steepest descent determines a minimizing sequence $\{x^n\}$, $\|x^{n+1}-x^*\| \leqslant \|x^n-x^*\|$, $n=1,2,3,\ldots$, and $Q(x^n)-Q(x^*)=(A(x^n-x^*),x^n-x^*)=0(n^{-1})$.*

PROOF. Since $Q(x)-Q(x^*)=(A(x-x^*),x-x^*)$ when $Ax^*=f$, it follows that, if $z^n=Ax^n-f$,

$$\begin{aligned} Q(x^n)-Q(x^*) &= (A(x^n-x^*),x^n-x^*) \\ &= (z^n,x^n-x^*) \\ &\leqslant \|z^n\|\,\|x^n-x^*\|. \end{aligned}$$

In the exercises we show that $\|x^n-x^*\|$ is a nonincreasing sequence, so that

$$[Q(x^n)-Q(x^*)]^2 \leqslant \|z^n\|^2\|x^0-x^*\|^2.$$

On the other hand, since $(Ax,x)\leqslant M\|x\|^2$,

$$\begin{aligned} Q(x^n)-Q(x^{n+1}) &= \frac{\|z^n\|^4}{(Az^n,z^n)} \\ &\geqslant \frac{\|z^n\|^4}{M\|z^n\|^2} = \frac{\|z^n\|^2}{M} \\ &\geqslant \frac{[Q(x^n)-Q(x^*)]^2}{M\|x^0-x^*\|^2}. \end{aligned}$$

If we set $\alpha^2=1/M\|x^0-x^*\|^2$, the excess value of the quadratic form, $Q(x^n)-Q(x^*)=E_n$ satisfies the inequality

$$\alpha^2E_n^2 \leqslant E_n-E_{n+1}, \qquad n=1,2,3,\ldots.$$

Then $0 \leqslant E_{n+1} \leqslant E_n - \alpha^2 E_n^2$, so E_n is a strictly decreasing sequence that has a limit E. But $0 \leqslant E \leqslant E - \alpha^2 E^2$, or $0 \leqslant -\alpha^2 E^2$, which is true only if $E=0$. Thus $Q(x^n) \to Q(x^*), m \to \infty$, and x^n is a minimizing sequence.

To establish the estimate $E_n = 0(n^{-1})$, observe that $t - \alpha^2 t^2$ is a monotone increasing function on $[0, \frac{1}{4}\alpha^2]$, decreasing on $[\frac{1}{4}\alpha^2, \frac{1}{2}\alpha^2]$. If $E_n \leqslant K/\alpha^2 n \leqslant \frac{1}{4}\alpha^2$, which E_2 automatically satisfies with $K \geqslant 1$, then $E_{n+1} \leqslant (K/\alpha^2 n)(1 - K/n)$. It is an easy calculation to show that $(n-K)/n^2 \leqslant 1/(n+1)$ if $K \geqslant 1$. ■

Thus we have a minimizing sequence x^n; does it converge? One can show that, if R_A is dense in H, then $x^n - x^* \rightharpoonup 0, n \to \infty$.

EXERCISES TO SECTION 11.2

1. Show that if A is selfadjoint and positive,

$$\|A(u-v)\|^4 \leqslant (AA(u-v), A(u-v))(u-v, A(u-v)).$$

(Hint: Use Exercises 10.5.3 and 10.5.4.)

2. Show that in the method of steepest descent

$$|\epsilon_n| \leqslant \frac{2(x^n - x^*, A(x^n - x^*))}{\|A(x^n - x^*)\|^2}.$$

(Hint: Exercise 11.2.1.)

3. Show that the algorithm for steepest descent generates a sequence $\{x^n\}$ satisfying

$$\|x^{n+1} - x^*\| \leqslant \|x^n - x^*\|, \qquad n = 1, 2, 3, \ldots,$$

if $Ax^* = f$. (Hint: $z^n = A(x^n - x^*)$.)

4. Show that if $K \geqslant 1$, then $(n-K)/n^2 \leqslant 1/(n+1)$. (Hint: $(n+1)/n \geqslant 1/K$).

5. Using Exercises 10.1.3 and 10.5.4 show that the minimizing sequence converges weakly to the solution if R_A is dense in H. (Hint:

$$(x^n - x^*, Ay) = (A^{1/2}(x^n - x^*), A^{1/2}y).$$

6. If A is a bounded *positive* selfadjoint operator on a Hilbert space H, then the definitions

$$[u, v] = (Au, v),$$

$$\square u \square^2 = [u, u],$$

define a pseudonorm. Show that on the factor space $H/N(A)$, a norm is defined and steepest descent is convergent. (Hint: Recall Exercise 6.2.1.)

11.3 MINIMIZATION FOR STRICTLY POSITIVE OPERATORS

The proofs of Theorems 11.2.1 and 11.2.2 cannot be extended to the case of positive unbounded operators because of immediately obvious difficulties about domains. There is no way to guarantee, for instance, that $x^{n+1}=x^n+\epsilon z^n \in D_A$, without very restrictive assumptions about A and f. An entirely different approach is needed, one with its mathematical foundation in K.O. Friedrichs' work on the energy method and halfbounded operators (Friedrichs [1]).

The starting point is the observation, made in Exercise 10.5.3, that, if A is a strictly positive densely defined operator, say $A \geqslant \gamma^2 I$ on a domain D_A dense in a Hilbert space H, then we can construct, on D_A, a new inner product:

$$[u,v]=[u,v]_A=(Au,v), \qquad u,v \in D_A.$$

This makes D_A an inner product space, with norm

$$\square u \square = \|u\|_A=(Au,u)^{1/2}, \qquad u \in D_A.$$

We often find (Au,u) called the *energy* of u. Since A is strictly positive, $\square u \square = \|u\|_A \geqslant \gamma\|u\|$. In this new norm, D_A may still not be a complete space. We can always complete D_A in the $\square \cdot \square$ norm to a Hilbert space H_A, whose elements are the limits, and ideal limits, of sequences convergent in norm (see Exercise 9.1.4). D_A is dense in H in the given norm $\|\cdot\|$, and also in H_A in the norm $\square \cdot \square$. H_A is often called the *energy space* of A. The fundamental fact here is that $H_A \subset H$.

THEOREM 11.3.1. *If A is a densely defined symmetric strictly positive operator on a Hilbert space H, then $H_A \subset H$.*

PROOF. The domain of A is common to H and H_A: $D_A \subset H \cap H_A$. Consider $u \in H_A$, but $u \notin D_A$. There must exist a sequence in D_A, $\{u^n\}$, with $\square u^n - u \square \to 0, n \to \infty$. Then u^n is a Cauchy sequence in H_A, and $\square u^n - u^m \square \to 0, n,m \to \infty$. But $\gamma\|u^n - u^m\| \leqslant \square u^r - u^m \square \to \infty$, so the sequence u^n also has a limit $\hat{u}$ in H. This establishes a mapping from $u \in H_A$ to $\hat{u} \in H$, which is readily seen to be linear, with domain H_A. We show that it is one-to-one as follows. If $u \in H_A$ and $\hat{u}=0$, then for any sequence $u^n \in D_A$ with $\square u^n - u \square \to 0, n \to \infty$, we have $\|u^n - \hat{u}\| = \|u^n\| \to 0, n \to \infty$. That is

$$\gamma\|u^n\| = \gamma\|u^n - \hat{u}\| \leqslant \square u^n - u \square \to 0, n \to \infty.$$

Now for any $v \in D_A$, $(u^n, Av) \to 0, n \to \infty$, so

$$[u^n, v] = (Au^n, v) = (u^n, Av) \to 0, n \to \infty.$$

Continuity of the inner product implies that $[u, v] = 0$ for all $v \in D_A$. Since D_A is dense in $H_A, u = 0$. Thus the mapping is one-to-one from H_A to H, and on H_A we have

$$\gamma \|u\| \leqslant \Box u \Box. \quad \blacksquare$$

Consider the minimization problem for $Q(u)$, restricted to $H_A \subset H$. For any $f \in H, |(u,f)| \leqslant \|f\| \, \|u\| \leqslant \|f\| \Box u \Box / \gamma$. Thus (u,f) is a bounded linear functional on H_A; then $(u,f) = [u, v]$ for some $v \in H_A$. Hence the general minimization problem for $u \in D_A$

$$Q(x) = (Ax, x) - (x, f) - (f, x)$$

becomes

$$\hat{Q}(u) = [u, u] - [u, v] - [v, u].$$

But in this form, $\hat{Q}(u)$ is defined not merely on D_A, but also on H_A, and on H_A

$$\hat{Q}(u) = [u - v, u - v] - [v, v].$$

Hence $\hat{Q}(u)$ always achieves its minimum on H_A at $u = v$:

$$\min_{u \in H_A} \hat{Q}(u) = \hat{Q}(v).$$

We have thus solved the problem $Ax = f$ in the sense of a generalized solution v, which may not be in D_A.

It should be noted that this analysis also shows that what we need in H_A is not $f \in H$, but the inequality $|(u,f)| \leqslant \|f\| \Box u \Box / \gamma$, which establishes the existence of a bounded linear functional on H_A. Thus we could replace $(u,f) + (f,u)$ by a linear functional $l(u)$ satisfying $|l(u)| \leqslant M(Au, u), u \in D_A$, to achieve this equivalent, solvable minimization problem for $\hat{Q}(u)$. We also note the following:

THEOREM 11.3.2. *Let $\{v^n\} \in H_A$ be a minimizing sequence for $\hat{Q}(u)$ on H_A. Then $\Box v^n - v \Box \to 0, n \to \infty$ and $\|v^n - v\| \to 0, n \to \infty$.*

PROOF. The sequence v^n satisfies $\hat{Q}(v^n) = \square v^n - v \square^2 - \square v \square^2$ while

$$\hat{Q}(v) = -\square v \square^2.$$

Then from $\hat{Q}(v^n) \to \hat{Q}(v), n \to \infty$ we conclude

$$\gamma \| v^n - v \| \leqslant \square v^n - v \square \to 0, \qquad n \to \infty. \quad \blacksquare$$

That is, a minimizing sequence converges in both H_A and H to the generalized solution v of $Ax = f$. Strangely enough, we cannot conclude that $Au^n \to f, n \to \infty$ in H, even if A is selfadjoint. For suppose $Av = f$, and that $\|Au^n - f\| = \|Au^n - Av\| \to 0$, whenever $\square u^n - v \square \to 0$; u^n is clearly minimizing. Every strictly positive selfadjoint A has a strictly positive selfadjoint square root $A^{1/2}$ whose range is the whole space H, so

$$\begin{aligned} \square u^n - v \square &= (A(u^n - v), (u^n - v)) \\ &= (A^{1/2}(u^n - v), A^{1/2}(u^n - v)) \\ &= \|A^{1/2}u^n - A^{1/2}v\|^2. \end{aligned}$$

Let $\psi^n = A^{1/2}u^n, \psi = A^{1/2}v$. Then $\square u^n - v \square \to 0$ if and only if $\|\psi^n - \psi\| \to 0$. Whenever $\|\psi^n - \psi\| \to 0$, it follows that $\square u^n - v \square \to 0$, so $\|Au^n - Av\| \to 0$, whence $\|A^{1/2}\psi^n - A^{1/2}\psi\| \to 0$. This shows that $A^{1/2}$ is continuous at ψ. A linear operator continuous at a point is continuous everywhere, so that $A^{1/2}$ is bounded; then A must be bounded! But we can prove the following weaker:

THEOREM 11.3.3. *If $\{u^n\}$ is a minimizing sequence for a strictly positive densely defined symmetric operator A and element f of a Hilbert space H, then $Au^n \rightharpoonup f, m \to \infty$.*

PROOF. For any $g \in H_A$, Theorem 10.3.2 implies that

$$\begin{aligned} |(Au^n - f, g)| &= |[u^n - u, g]| \\ &\leqslant \square u^n - u \square \square g \square \to 0, \qquad n \to \infty, \end{aligned}$$

if $\{u^n\}$ is minimizing. Thus $Au^n - f \rightharpoonup 0$ in H_A and $\|Au^n\|$ is bounded by some constant M. Since H_A is dense in H, given $\epsilon > 0$ and $h \in H$, there is a $g \in H_A$ with $\|h - g\| < \epsilon$. Then

$$(Au^n - f, h) = (Au^n - f, g) + (Au^n - f, h - g)$$

where

$$|(Au^n - f, h-g)| \leqslant \|Au^n\| \, \|h-g\| + \|f\| \, \|h-g\| \leqslant M\epsilon.$$

Thus $(Au^n - f, h) \to 0$, and $Au^n \rightharpoonup f, n \to \infty$. ■

11.4 THE RITZ METHOD

One alternative to the method of steepest descent for the computation of a minimum is the *method of Ritz*. H_A is assumed to be separable, and we select a sequence $\phi^1, \phi^2, \phi^3, \ldots$ of vectors of H_A such that

1. The vectors $\phi^1, \phi^2, \ldots, \phi^n$ are linearly independent, $n = 1, 2, 3, \ldots$.
2. The sequence $\{\phi^n\}$ is total in H_A.

(Recall Exercise 9.3.11.) The vectors ϕ^n are often called *coordinate functions*. For reasons which will become apparent, they are not chosen to be a complete orthonormal system. The nth *Ritz approximate solution* is determined by solving the problem

$$\min_{a_1, \ldots, a_n} \left[\hat{Q}(a_1\phi^1 + a_2\phi^2 + \cdots + a_n\phi^n) \right]$$

Since $\hat{Q}(w) = \square w - v \square^2 - \square v \square^2$, this problem is solved, according to Exercise 9.2.10, by solving the equivalent linear system of n equations in n unknowns

$$G(\phi^1, \phi^2, \ldots, \phi^n) a^{(n)} = v^{(n)}$$

where G is the Grammian matrix with $g_{i,j} = [\phi^j, \phi^i], i, j = 1, \ldots, n, a^{(n)}$ is the column vector $(a_1, \ldots, a_n)^T$ and $v^{(n)}$ is the column vector $([v,\phi^1],[v,\phi^2],\ldots,[v,\phi^n])^T = ((f,\phi^1),(f,\phi^n),\ldots,(f,\phi^n))^T$. Since the coordinate functions are linearly independent, the linear system is always uniquely solvable. The solution $u^n = a_1\phi^1 + a_2\phi^2 + \cdots + a_n\phi^n$ satisfies

$$\square u^n - v \square = \min\{\square w - v \square \mid w \in [\phi^1, \ldots, \phi^n]\}.$$

Since $\{\phi^n\}$ is total in H_A, for any $\epsilon > 0$ there is an n with $\square u^n - v \square < \epsilon$. Thus the Ritz approximations are a minimizing sequence.

If we choose $\{\phi^n\}$ in D_A, then we can rewrite the entries in the Grammian matrix as

$$g_{i,j} = \left[\phi^j, \phi^i\right] = (A\phi^j, \phi^i), \; i, j = 1, \ldots, n.$$

Thus our arithmetic takes place in H. If $Ax=f$ is solvable in D_A, then the Ritz solutions converge to the true solution in H, even though we cannot guarantee $Au^n \to f, n \to \infty$. How do we recognize a sequence that is total in H_A? We have the following:

THEOREM 11.4.1 *Let $\phi^n \in D_A, n=1,2,3,\ldots$, where A is a strictly positive symmetric operator on a Hilbert space H. If $\{A\phi^n\}$ is total in H, then $\{\phi^n\}$ is total in H_A.*

PROOF. If $v \in D_A$, set $u=Av$. Since $\{A\phi^n\}$ is total in H, for any $\epsilon>0$ there is a linear combination of $A\phi^n$'s with

$$\|u-\sum_{k=1}^{n} a_k A\phi^k\| = \|A\left(v-\sum_{k=1}^{n} a_k\phi^k\right)\| \leqslant \epsilon\gamma.$$

Then

$$\begin{aligned}\square v-\sum_{k=1}^{n} a_k\phi^k \square^2 &= \square v - v^n \square^2 \\ &= (A(v-v^n),(v-v^n)) \\ &\leqslant \|A(v-v^n)\| \, \|v-v^n\| \\ &\leqslant \epsilon\gamma \|v-v^n\| \\ &\leqslant \epsilon \square v-v^n \square .\end{aligned}$$

That is, $\{\phi^n\}$ is total in H_A. ■

COROLLARY 1. *If H is separable, so is H_A.*

COROLLARY 2. *If $\{\phi^n\}$ is a maximal orthonormal sequence in H_A then the minimum v of*

$$\hat{Q}(u)=[u,u]-[u,v]-[v,u]$$

may be written as

$$v=\sum_{n=1}^{\infty} [v,\phi^n]\phi^n,$$

the series converging in H_A. Also

$$v=\sum_{n=1}^{\infty} (f,\phi^n)\phi^n$$

in both H and H_A.

PROOF. Observe that $Q(u)=[u,u]-[v,u]$ and that $[v,\phi^n]=(f,\phi^n)$. ■

(For computation, then, one might desire the ϕ^n to be orthonormal in H_A. The computational cost of achieving this may exceed its worth. There is more on this in Section 11.8.)

Even if the operator A is not symmetric, we may still utilize the Ritz equations formally to define an approximate solution. In this case, our analysis is not applicable. Methods of this type are called *Galerkin methods*, or *projection methods*.

An apparently different procedure is known as the *method of least squares*. In this method, we utilize a sequence $\{\phi^n\}$ in D_A, but we change the quadratic form. In place of $Q(u)$ we choose

$$R(u)=\|Au-f\|^2,$$

which is a quadratic function of the residual $z=Au-f$. Here we must choose $\{\phi^n\}\in D_A$ in order for anything to make sense: if we find $a_1,\ldots,a_n$ to achieve

$$\min_{a_1,\ldots,a_n} R\big(a_1\phi^1+a_2\phi^2+\cdots+a_n\phi^n\big),$$

then we conclude from Exercise 9.2.10 that $a^{(n)}$, the solution, satisfies

$$\sum_{i=1}^{n}(A\phi^i,A\phi^j)a_i^{(n)}=(f,A\phi^j).$$

We assume:

1. The vectors $A\phi^1,\ldots,A\phi^n$ are linearly independent, $n=1,2,3,\ldots$.
2. The sequence $\{A\phi^n\}$ is total in H.

Then assumption 1 guarantees the existence of $a^{(n)}$ for all n. Let $u^n=a_1^{(n)}\phi^1+a_2^{(n)}\phi^2+\cdots+a_n^{(n)}\phi^n$. Then for any $\epsilon>0$ we can find an n and $\alpha_1,\ldots,\alpha_n$ such that

$$\|\alpha_1A\phi^1+\alpha_2A\phi^2+\cdots+\alpha_nA\phi^n-f\|\leqslant\frac{\epsilon}{2\gamma},$$

while

$$\|Au^n-f\|\leqslant\|\alpha_1A\phi^1+\alpha_2A\phi^2+\cdots+\alpha_nA\phi^n-f\|$$

Since $\|Au^n-f\|$ is nonincreasing with n, this can be read as

$\|Au^n - f\| \to 0, n \to \infty$, or $Au^n \to f, n \to \infty$. Then Au^n is a Cauchy sequence, and

$$\gamma\|u^n - u^m\| \leqslant \|Au^n - Au^m\| \to 0, \qquad n,m \to \infty.$$

In this form of minimization, the approximate solutions converge in norm to a vector $\hat{u}$ which is a true solution if $\hat{u} \in D_A$, a generalized solution otherwise. To see this, note that

$$(Au^n - Au^m, u^n - u^m) \to 0, \qquad n,m \to \infty,$$

so

$$[u^n - u^m, u^n - u^m] = \square u^n - u^m \square^2 \to 0, \qquad n,m \to \infty$$

and thus $\hat{Q}(\hat{u})$ is defined and minimized.

In fact, this method of least squares is not independent of the Ritz method. From

$$[A\phi^i, A\phi^j] = (A^*A\phi^i, \phi^j)$$

and

$$(f, A\phi^i) = (A^*f, \phi^i),$$

we see that it coincides with the Ritz method, applied to the equation

$$A^*Ax = A^*f,$$

if $f \in D_{A^*}$ and D_{A^*A} is dense in H. In this form we see that

$$\begin{aligned}\|Au - f\|^2 - \|f\|^2 &= (Au, Au) - (Au, f) - (f, Au) + (f, f) - \|f\|^2 \\ &= (A^*Au, u) - (u, A^*f) - (A^*f, u),\end{aligned}$$

which shows the equivalence. It should also be observed that in the method of least squares, A need not be symmetric: it is sufficient if A^*A is densely defined and strictly positive, for then it is symmetric, and $\|Au\| \geqslant \gamma\|u\|$ holds if $A^*A \geqslant \gamma^2 I$. This also suggests that, if A is a selfadjoint, strictly positive operator, we can improve on Theorem 11.3.3 as follows:

THEOREM 11.4.2 *If A is a strictly positive selfadjoint operator on a Hilbert space H and $\{u^n\}$ is a Ritz minimizing sequence for $Ax = f$, then $\|A^{1/2}u^n - A^{1/2}f\| \to 0, n \to \infty$.*

The proof is left as an exercise. It is based on the relation between the minimization problem for $Ax=f$ and the least squares problem for $A^{1/2}x = A^{-1/2}f$.

We note that for the method of least squares, the arithmetic will be simplest if

$$(A\phi^i, A\phi^j) = [\phi^i, A\phi^j] = \delta_i^j.$$

However, the cost of achieving this orthogonality in practice may exceed its worth.

EXERCISES TO SECTION 11.4

1. Prove Corollary 1 of Theorem 11.4.1.
2. Prove Theorem 11.4.2. (Hint: $A = (A^{1/2})^* A^{1/2}$.)
3. In Galerkin's method, with A not necessarily self-adjoint, show that $Au^n - f \rightharpoonup 0, n \to \infty$ if $\{\phi^n\}$ is total in H.

11.5 SELFADJOINT EXTENSIONS OF STRICTLY POSITIVE SYMMETRIC OPERATORS

A is a densely defined strictly positive symmetric operator on a Hilbert space H. The relation $(u,f) = [u,v]$ associates with each $f \in H$ an element v_f minimizing $\hat{Q}(u)$ on H_A. Since $H_A \subset H$ this establishes a mapping from H to H_A, if $f \to v_f = Gf$, where $D_G = H$ and

$$(u,f) = [u, Gf], u \in H_A.$$

The correspondence $f \to v_f$ is clearly linear. Further if we set $u = Gf$ we find

$$(Gf,f) = [Gf, Gf] = \square Gf \square^2 > 0.$$

Thus (Gf,f) assumes only nonnegative real values on all of H, so G will be selfadjoint, if we know that G is bounded. But $\square Gf \square^2 \leqslant \|f\| \, \|Gf\|$ while $\gamma^2 \|Gf\|^2 \leqslant \square Gf \square$ or $\|Gf\| \leqslant \|f\|/\gamma^2$. Thus G is a bounded selfadjoint positive operator on H. It is the same on H_A for $\|Gf\| \leqslant \square Gf \square / \gamma$ makes $\square Gf \square \leqslant \|f\|/\gamma^2, f \in H$. If $f \in H_A$, then $\|f\| \leqslant \square f \square / \gamma$, so $\square Gf \square \leqslant \square f \square / \gamma^2$ on H_A. G is actually positive definite, since $(Gf,f) = 0$ makes $\square Gf \square = 0$, so $(u,f) = 0$ for all $u \in H_A$, a dense set. Then $f = 0$. Hence G is one-to-one to its range, so G^{-1} exists. Let $\tilde{A} = G^{-1}$.

LEMMA 11.5.1. *$\tilde{A}$ is self-adjoint.*

PROOF. R_G is dense in H. To see this, suppose R_G were not dense in H. Then there is a vector $y \in H$ with $y \neq 0, (y, Gf) = 0$ for all $f \in D_G = H$. Then $(Gy, f) = 0$ for all $f \in H$; in particular, $(Gy, y) = 0$. This contradicts $(Gy, y) > 0$, $y \neq 0$. Put another way, this means $D_{\tilde{A}}$ is dense in H, and $\tilde{A}^*$ exists. If $x \in D_{\tilde{A}^*}$ and $f \in D_{\tilde{A}}$, then $(\tilde{A}f, x) = (f, x^*)$, where $x^* = \tilde{A}^* x$. But from $\tilde{A}f = y$ we have $f = Gy$ so that $(y, x) = (Gy, x^*) = (y, Gx^*)$ since G is selfadjoint. Read backwards, this says that $x = Gx^*$, since $R_{\tilde{A}} = x^*$. Thus $\tilde{A}$ extends $\tilde{A}^*$; that is, $\tilde{A} = \tilde{A}^*$ on D_{A^*}.

Conversely if we set $y = \tilde{A}f$, $z = \tilde{A}h$ into

$$(Gy, z) = (y, Gz)$$

we have

$$(f, \tilde{A}h) = (\tilde{A}f, h).$$

Thus on $D_{\tilde{A}}$, $\tilde{A}^* = \tilde{A}$. Hence $\tilde{A}^*$ extends $\tilde{A}$, so $\tilde{A}^* = \tilde{A}$. ■

We now give a second, more instructive proof of a theorem first noted in Exercise 10.5.5.

THEOREM 11.5.1. (*Friedrichs*) *Every strictly positive operator defined on a dense set possesses a selfadjoint extension.*

PROOF. We will show that $\tilde{A}$ extends A. Let $Ax = f$. Then on D_A, x minimizes the functional

$$Q(u) = (Au, u) - (u, f) - (f, u).$$

By the definition of $G, x = Gf$, so $f = \tilde{A}x$. Hence $\tilde{A}$ extends A. ■

COROLLARY 1. *If A is a densely defined symmetric operator that satisfies either $A \geqslant \delta I$ or $A < \delta I$ for some real number δ, then A possesses a selfadjoint extension.*

PROOF. We consider either $A + cI, c + \delta > 0$ or $-A + cI, c - \delta > 0$. ■

The extension need not be unique. For instance, for differential operators, each set of boundary conditions may lead to a self-adjoint operator extension, different boundary conditions to different extensions.

We can identify H_A by the following:

THEOREM 11.5.2. *If A is self-adjoint and $A \geqslant \gamma^2 I$, then $H_A = D_{A^{1/2}}$*

PROOF. We show $H_A \subset D_{A^{1/2}}$. If $x \in D_A, x \in H_A$, while $D_A \subset D_{A^{1/2}}$. If $u \in H_A, u \not\in D_A$, there exists $\{u^n\} \in D_A$ with $\square u^n - u \square \to 0, n \to \infty$. Then $\square u^n - u^m \square \to 0, n, m \to \infty$. Also $\|u^n - u^m\| \to 0, n, m \to \infty$, and

$$\begin{aligned}\square u^n - u^m \square^2 &= (A(u^n - u^m), u^n - u^m)\\ &= (A^{1/2}(u^n - u^m), A^{1/2}(u^n - u^m))\\ &= \|A^{1/2}(u^n - u^m)\|^2 \to 0.\end{aligned}$$

Now in $H, u^n \to u$, and $A^{1/2}u^n$ converges, while $A^{1/2}$ is closed. Thus $u \in D_{A^{1/2}}$ and $H_A \subset D_{A^{1/2}}$.

Next suppose $D_{A^{1/2}} \subset H_A$. Let $u \in D_{A^{1/2}}$, dense in H. Then $A^{1/2}u \in H$, so there exists $v^n \in D_{A^{1/2}}, v^n \to A^{1/2}u$, $n \to \infty$ in H. Now $A^{1/2} \geqslant \gamma I$, so $(A^{1/2})^{-1}$ exists and is bounded; set $u^n = (A^{1/2})^{-1} v^n$. Since $v^n \in D_{A^{1/2}}, Au^n$ is defined, so $A^{1/2}u^n$ is defined. Then

$$\lim_{n \to \infty} \|A^{1/2}u^n - A^{1/2}u\| \to 0.$$

The sequence $\{A^{1/2}u^n\}$ is a Cauchy sequence and

$$\square u^n - u^m \square = \|A^{1/2}u^n - A^{1/2}u^m\| \to 0, n, m \to \infty.$$

Thus u^n converges in H_A, to $\bar{u}$. On the other hand,

$$u^n - u = (A^{1/2})^{-1} A^{1/2}(u^n - u).$$

Since $(A^{1/2})^{-1} \in [H]$, $\|u^n - u\| \to 0$. But $\|u^n - \bar{u}\| \to 0, n \to \infty$, so $\bar{u} = u$. ■

Then $D_{\tilde{A}} \subset D_{\tilde{A}^{1/2}}$; but $D_{\tilde{A}} \neq D_{\tilde{A}^{1/2}}$; the later would imply $\tilde{A}$ bounded!

EXERCISES TO SECTION 11.5

1. Show that on an inner product space H, if (Ax, x) is real for all $x \in D_A$, then $(Ax, y) = (x, Ay)$ for all $x, y \in D_A$. From this conclude that every densely defined positive operator A satisfies $A \subset A^*$.

11.6 SELFADJOINT EXTENSIONS OF POSITIVE OPERATORS

The corollary to Theorem 11.5.1 insures the existence of selfadjoint extensions for positive operators. It fails to describe them as directly as we have described the extension for strictly positive operators. We investigate here the energy space of positive definite and positive operators.

First suppose that A is a positive definite operator on a Hilbert space H, and that D_A is dense in H. There is no loss in generality in assuming that A is selfadjoint. If we introduce in D_A the inner product

$$[u,v]=(Au,v), \qquad u,v\in D_A,$$

and the related norm $\square u\square^2=[u,u]$, we obtain a new inner product space X_A. When we complete X_A to H_A, by adding as ideal elements the strong limits of all Cauchy sequences in X_A, we no longer can prove Theorem 11.3.1, that is, $H_A\subset H$. *Which elements are common to H and H_A?* Clearly $D_A\subset Y\cap H_A$. Suppose $x^n\in D_A$, $n=1,2,3,\ldots,$ and for $\bar{x}\in H_A$, $y\in H$, $\square x^n-\bar{x}\square\to 0$ in H_A,.and $\|x^n-y\|\to 0$ in H. Then we can identify $\bar{x}$ and y, but the uniqueness of limits does not guarantee only one y to each x. However, for positive definite operators, we can establish uniqueness as follows:

Suppose $u^n, v^n\in D_A$, $n=1,2,3,\ldots,$ $\square u^n-\bar{x}\square\to 0$, $n\to\infty$, and $\square v^n-\bar{x}\square\to 0$, $n\to\infty$; suppose $u,v\in H$ and $\|u^n-u\|\to 0$, $\|v^n-v\|\to 0$, $n\to\infty$, with $u\neq v$. Since $D_{A^{1/2}}\subset D_A$, $\square u^n-u^m\square=\|A^{1/2}u^n-A^{1/2}u^m\|\to 0$, $n,m\to\infty$, and $\square v^n-v^m\square=\|A^{1/2}v^n-A^{1/2}v^m\|\to 0$, n, $m\to\infty$. Also $\|A^{1/2}u^n-A^{1/2}v^n\|=\square u^n-v^n\square\leqslant\square u^n-\bar{x}\square+\square v^n-\bar{x}\square\to 0, n\to\infty$. Hence $A^{1/2}u^n$, $A^{1/2}v^m$ have the same limit in H. But $A^{1/2}$ is closed, so $u^n\to u, v^n\to v$ in H implies $u,v\in D_A$ and $A^{1/2}(u-v)=0$. Since $A^{1/2}$ is positive definite, $u-v=0$.

Thus any sequence $\{x^n\}$ in D_A which is a Cauchy sequence in both the $\square\cdot\square$ and $\|\cdot\|$ norms corresponds to an element in $H_A\cap H$. However, it can occur that there are elements of H_A, not in H. *It is no longer true that (u,f) is automatically a bounded linear functional on H_A.* If it happens that on D_A, $|(u,f)|\leqslant K\square u\square$ for some constant K, then the Hahn-Banach theorem guarantees the existence of a bounded linear functional $f(u)$ that extends (u,f) to H_A; the Riesz representation theorem assures the existence of $v_f\in H_A$ with

$$f(u)=[u,v_f], \qquad u\in H_A$$

and

$$(u,f)=f(u)=[u,v_f], \qquad u\in D_A.$$

Then

$$\hat{Q}(u) = \square u - v_f \square^2 - \square v_f \square^2,$$

so v_f minimizes $\hat{Q}(u)$ on H_A. The mapping $Gf = v_f$ has domain

$$D_G = \{ f \in H \| (u,f) | \leqslant K \square u \square, u \in D_A, \text{ some K} \},$$

and range $R_G \subset H_A$. If we form an operator $\hat{G}$ from G by the additional domain constraint $\hat{G}f \in H$, (discarding all other f in H), then we have $D_{\hat{G}} = R_A$ and $\hat{G} = A^{-1}$. This can be seen as follows: If $f \in R_A$, then there is a vector $u_f \in D_A$ with $Au_f = f$, or $u_f = A^{-1}f$. Then u_f minimizes $Q(u)$ and $\hat{Q}(u)$. Hence $f \in D_G$ and $u_f = \hat{G}f$, with $u_f \in H$. Hence $\hat{G}$ extends A^{-1}. Since

$$(u,v) = [u, u_v] = [u, \hat{G}v]$$

for any $v \in D_{\hat{G}}$, and any $u \in H \cap H_A$, let $u = \hat{G}w$ for any $w \in D_{\hat{G}}$. This gives $(\hat{G}w, v) = [\hat{G}w, \hat{G}v] = \overline{[\hat{G}v, \hat{G}w]} = \overline{(\hat{G}v, w)} = (w, \hat{G}v)$. Thus $\hat{G}$ is symmetric. Now A^{-1} is a selfadjoint operator extended by a symmetric operator $\hat{G}$. But a selfadjoint operator cannot have a proper symmetric extension so $\hat{G} = A^{-1}$.

It is possible to exhibit a positive definite differential operator that is *not* strictly positive, and for which any situation occurs. That is, one can find functions f for which any one of the following is true:

1. $f \in H$, and there is a solution u to $Au = f$, but $u \notin H$ and $u \notin H_A$.
2. $f \in H$, and there is a solution to $Au = f$ with $u \in H_A, u \notin H$.
3. $f \in H$, and there is a solution to $Au = f$ with $u \in H_A \cap H$.

(See, e.g., Mikhlin [1], 15–17.)

11.7 STEEPEST DESCENT FOR UNBOUNDED OPERATORS

Let A and B be closed symmetric operators with dense domains on a Hilbert space H such that $D_A \supset D_B$. Further suppose that for some real γ, and positive m, and M,

1. $(Bx,x) \geqslant \gamma^2 (x,x)$, $x \in D_B$;
2. $m(Bx,x) \leqslant (Ax,x) \leqslant M(Bx,x)$, $x \in D_B$.

Then also we have $(Ax,x) \geqslant m\gamma^2(x,x)$, $x \in D_B$. We introduce the auxiliary Hilbert space H_B, $D_B \subset H_B \subset H$ as before with $[u,v] = (Bu,v)$ $u,v \in D_B$ and

$\square u \square^2 = [u,u]$. Then there exists a selfadjoint positive definite operator $G \in [H]$, with $\|G\| \leqslant 1/\gamma^2$, $\square G \square \leqslant 1/\gamma^2$ and a selfadjoint inverse G^{-1} which is a selfadjoint extension of B. If $\hat{B}$ denotes the selfadjoint extension of B, then $\hat{B}G = I$ on H. (Why?) Also,

$$m[x,x] \leqslant m(\hat{B}x,x) \leqslant (Ax,x) \leqslant M(\hat{B}x,x) = M[x,x], \qquad x \in D_B,$$

and

$$(Ax,y) = (\hat{B}G\,Ax,y), \qquad x \in D_B.$$

If we consider the space $H_{\hat{B}}$ constructed from the inner product $(\hat{B}x,x)$ we have that $D_B \subset H_B \subset H$, $D_B \subset D_{\hat{B}} \subset H_{\hat{B}}$, while $(\hat{B}x,x) = \|x\|_{\hat{B}}^2 = \|x\|_B^2 = (Bx,x)$, for all $x \in D_B$, which is dense in H_B, $H_{\hat{B}}$, and H. Then convergence in the $\|\cdot\|_B$ and $\|\cdot\|_{\hat{B}}$ norms are equivalent, so $H_B = H_{\hat{B}}$. On D_B, $(Ax,x) = [GAx,x]$ and $m[x,x] \leqslant [GAx,x] \leqslant M[x,x]$. Further

$$\begin{aligned} \square GAx \square^2 &= \sup_{\square y \square = 1} |[GAx,y]|^2 = \sup_{\square y \square = 1} |(Ax,y)|^2 \\ &= \sup_{y \neq 0} \frac{|(Ax,y)|^2}{(By,y)}. \end{aligned}$$

Since $(Ax,x) \geqslant m\gamma^2(x,x)$,

$$|(Ax,y)|^2 \leqslant (Ax,x)(Ay,y).$$

Then, for $x \in D_B$,

$$\begin{aligned} \square GAx \square^2 &\leqslant \sup_{y \neq 0} \frac{(Ax,x)(Ay,y)}{(By,y)} \\ &\leqslant \sup_{y \neq 0} (Ax,x)M = M(Ax,x) \\ &\leqslant M^2 \square x \square^2. \end{aligned}$$

Hence GA is a bounded and uniformly continuous operator on D_B; it has a unique bounded extension with the same bound on H_B. Since

$$\begin{aligned} [GAx,x] &= (Ax,x) = (x,Ax) \\ &= (GBx,Ax) = (Bx,GAx) \\ &= [x,GAx] \end{aligned}$$

for all $x \in D_A$, this extension has the property that the quadratic form is always real. Hence this is a bounded selfadjoint operator on H_B, which we will write as $\widehat{GA}$. To $\widehat{GA}$ we can apply the method of steepest descent:

$$\begin{cases} x^{n+1} & = x^n + \epsilon_n z^n, \\ z^n & = \widehat{GA} x^n - Gf, \\ \epsilon_n & = -\dfrac{[z^n, z^n]}{[\widehat{GA} z^n, z^n]}. \end{cases}$$

This converges to the solution of

$$\widehat{GA} x = Gf$$

and the rate of convergence we already know from Theorem 11.2.1.

Suppose now that the following hold:

1. On some set $K \subset D_A$, $x \in K$ implies $Ax \in R_B$ and $GAx \in K$;
2. $Gf \in K$.

Then, if $x \in K$, $GAx = GAx \in K$, so $z^n \in K$, and we can rewrite the method of steepest descent in the form

$$\begin{cases} x^{n+1} & = x^n + \epsilon_n z^n \\ B z^n & = A x^n - f, \\ \epsilon_n & = -\dfrac{(Ax^n - f, z^n)}{(Az^n, z^n)}. \end{cases}$$

The iteration can be accomplished within K and will converge to x^*, the solution or generalized solution of $Ax = f$, with

$$\Box x^n - x^* \Box \leqslant \frac{(M-m)^n}{(M+m)^n m} \Box z^0 \Box,$$

or

$$\|x^n - x^*\| \leqslant \frac{(M-m)^n}{(M+m)^n m\gamma} \|z^0\|.$$

EXERCISES TO SECTION 11.7

1. Let D be the subset of $L_2(0,1)$ consisting of twice continuously differentiable functions. Let B be the operator on D given by

$$Bx = -\frac{d^2x}{dt^2}.$$

Find conditions on the real numbers a, b, c, d such that, on the subset of D satisfying

$$ax(0)+bx'(0)=0, \qquad cx(1)+dx'(1)=0,$$

B is a symmetric operator. Are there any other linear boundary conditions that make B symmetric? (Hint: in (Bx, x), integrate by parts.)

2. Using D_0, the subset of D of Exercise 11.7.1 satisfying $x(0)=x(1)=0$, show that B is positive definite.
3. On D_0 as in Exercise 11.7.2, show that B satisfies

$$(Bx, x) \geqslant 2\|x\|^2.$$

(Hint: Cauchy's inequality and

$$x(t)=\int_0^t x'(s)\,ds$$

imply

$$\int_0^1 [x(t)]^2\,dt \leqslant \tfrac{1}{2}\int_0^1 [x'(t)]^2\,dt.)$$

4. On D_0, show that, if $Bx=f$, then

$$x(t)=\int_0^1 k(t,s) f(s)\,ds$$

where

$$k(t,s)=\begin{cases} t(1-s), & t \leqslant s, \\ s(1-t), & t \geqslant s. \end{cases}$$

5. Let $q(t) \in C[0,1]$, $p(t) \in C[0,1]$, and $p(t)>0$, $q(t) \geqslant 0$ on $[0,1]$. On D_0 let

$$Ax=q(t)x(t)-[p(t)x'(t)]'.$$

Use integration by parts to show that for suitable m, M,

$$m(Bx, x) \leqslant (Ax, x) \leqslant M(Bx, x).$$

Find m, M. Write out the steepest descent algorithm and estimate the rate of convergency to x. From

$$|x^m(t)-x^*(t)| \leqslant \left\{ \int_0^t |[x^m(t)-x^*(t)]'|dt \right\}^{1/2} t^{1/2},$$

show that the convergence is uniform. Have you shown the existence of a solution?

11.8 SOME AFTERTHOUGHTS ABOUT STEEPEST DESCENT: MINIMUM RESIDUALS AND CONJUGATE GRADIENTS

In the method of steepest descent, we maximize the change in $Q(x)$ by moving in the direction of the residual $Ax-f$; in the method of least squares, we do not proceed sequentially, but we minimize the residual. We may combine these two ideas to generate a method related to least squares that is sequential: *the method of minimal residuals* (Krasnosel'skii and Krein, [1]). A is a strictly positive, bounded selfadjoint operator.

We proceed in the direction $z^n = Ax^n - f$ from x^n to $x^{n+1} = x^n + \epsilon_{n+1} z^n$ where ϵ_{n+1} will be determined by minimizing $\|Ax^{n+1}-f\|^2$. Since $z^{n+1} = Ax^{n+1} - f = Ax^n + \epsilon_{n+1} A z^n - f = z^n + \epsilon_{n+1} A z^n$,

$$\begin{aligned}\|z^{n+1}\|^2 &= (z^{n+1}, z^{n+1}) \\ &= (z^n, z^n) + 2\epsilon_{n+1}(Az^n, z^n) + \epsilon_{n+1}^2 (Az^n, Az^n),\end{aligned}$$

which is minimized by choosing

$$\epsilon_{n+1} = -\frac{(Az^n, z^n)}{(Az^n, Az^n)}.$$

This establishes *the method of minimal residuals* as the algorithm

$$\begin{cases} x^{n+1} = x^n + \epsilon_{n+1} z^n, \\ \epsilon_{n+1} = -\dfrac{(Az^n, z^n)}{(Az^n, Az^n)}, \\ z^{n+1} = Ax^n - f. \end{cases}$$

THEOREM 11.8.1. *If $0 < mI \leqslant A \leqslant MI$, then the sequence $\{x^n\}$ determined by the method of minimal residuals converges to the solution x^* of $Ax = f$.*

Further, if $x^0 = Ax^0 - f$, *then*

$$\|Ax^n - f\| \leqslant \left[\frac{M-m}{M+m}\right]^n \|z^0\|$$

while

$$\|x^n - x^*\| \leqslant \left[\frac{M-m}{M+m}\right]^n \frac{\|z^0\|}{m}.$$

PROOF. Since $\|x^{n+1}\|$ is minimal, it is less than $\|z^n + \alpha A z^n\|$ for any α, in particular for $\alpha = -(M-m)/(M+m)$. Thus $\|z^{n+1}\| \leqslant |\alpha| \|z^n\| \leqslant |\alpha|^n \|z^0\|$. From $\|z^n\| = \|Ax^n - f\| = \|Ax^n - Ax^*\| \geqslant m\|x^n - x^*\|$ we conclude the second inequality. ■

Observe that in H_A,

$$(Ax-f, Ax-f) = [Ax, x] - [f, x] - [x, f] + (f, f).$$

Thus minimum residuals is steepest descent in H_A. On the other hand, we have

$$(Ax-f, Ax-f) = (A^*Ax, x) - (x, A^*f) - (A^*f, x) + (f, f),$$

so that minimal residuals replaces the equation $Ax = f$ with $A^*Ax = A^*f$. However, it still requires $A^* = A$ for the convergence proof. The minimum residuals algorithm is *not* a steepest descent algorithm for $A^*Ax = A^*f$ (Why)?

Another observation about the method of steepest descent is that it consists of two parts: first a choice of a direction in which to move; followed by a minimization over a line in the selected direction. It is conceivable, when $X = \mathbf{E}^n$, that an infinite number of steps is needed to achieve the minimum, even though only a finite number of steps are needed to solve $Ax = f$. That this is in fact true was established by Akaike [1], and Forsythe [1], and is attributable to an asymptotic restriction of the steepest descent directions to a two-dimensional subspace; thus the residuals fail to search the whole space H. A scheme that attempts to obviate this, so that the search directions are well distributed, is the *method of conjugate gradients* (Hestenes and Stieffel [1]). It is based on the following idea: if we choose an arbitrary fixed direction p, then along any line $x + \lambda p$, $Q(x + \lambda p)$ achieves its minimum at a point $x + \pi = x_\pi$, such that

$$Q(x_\pi) = Q(x) - \frac{\{\Re(Ax - f, \pi)\}^2}{(A\pi, \pi)}$$

$$= Q(x) - \frac{\{\Re(Ax - f, \pi)\}^2}{(A\pi, \pi)}.$$

Since $Q(x_\pi)=\min Q(x_\pi+\lambda\pi)$, the subset $H_\pi\subset H$ of all such points x_π is characterized by $\Re(Ax_\pi-f,\pi)=0$, a proper linear subspace of H. If we choose another direction g such that $(Ag,\pi)=(A\pi,g)=0$ (orthogonality in the energy space linear product: $[\pi,g]=[g,\pi]=0$) then $\Re(A(x_\pi+\lambda g)-f,\pi)$ $=\Re(Ax_\pi-f,\pi)+\lambda\Re(Ag,\pi)=0$ for all λ, so that further minimization occurs by mapping H_π into a proper linear subspace of itself. In a separable Hilbert space H, we can exhaust all directions by choosing, as a sequence of directions, an orthogonal basis in H_A. Any process for minimization that generates a sequence orthogonal in H_A is called a *method of conjugate direction*. A procedure that constructs the conjugate directions by orthogonalizing the residuals in H_A is known as a *method of conjugate gradients*. Since an orthogonal set in H_A is linearly independent, when $H=\mathbf{E}^n$ these methods must terminate in at most n iterations. (One generator for conjugate gradients is studied in Exercise 11.8.3.)

If we recall the Ritz method for minimization, we quickly realize that a method of conjugate directions or conjugate gradients is a special Ritz method, one in which the coordinate functions are orthogonal in H_A, since at any stage in the process we have minimized $Q(u)$ over the first n directions or coordinate functions.

If we recall the method of least squares, which is equivalent to the Ritz method with the quadratic functional $\|Ax-f\|^2$ and linear system $A^*Ax=A^*f$, we obtain another *method of conjugate directions* by applying the same procedure as above, using an orthogonal set of directions in H_{A^*A}, and also a *method of conjugate gradients* (see Exercise 11.8.4). Of course we require $A^*A\geqslant\gamma I$ for some positive γ, which will be true if $\|Ax\|\geqslant\gamma^{1/2}\|x\|$ for all x; that is, if A is invertible.

The most general algorithm in this direction replaces $\|Ax-f\|^2$ by a *weighted* residual $\|Ax-f\|_M^2$ where M is a bounded strictly positive selfadjoint operator on H. This corresponds to considering the linear system $A^*MAx=A^*Mf$, and choosing the conjugate directions in H_{A^*MA}. (See, e.g., Hestenes and Stieffel [1], or Daniel [2]. That this is the most general procedure follows from Exercise 11.8.5.)

Since the method is a special case of the Ritz method, convergence is guaranteed (see Exercise 11.8.3) but without error estimates. However, conjugate gradients is also close to steepest descent, and it is easy to prove the following:

THEOREM 11.8.2. *Let A be a selfadjoint operator on a Hilbert space H, $0<mI\leqslant A\leqslant MI$. Then $[Q(x^{n+1})-Q(x^*)]\leqslant[(M-m)/(M+m)]^2[Q(x^n)-Q(x^*)]$ and $\|x^{n+1}-x^*\|\leqslant(M/m)[(M-m)/(M+m)]^n\|x^0-x^*\|$, where $Ax^*=f$ and $\{x^n\}$ is the sequence generated by the conjugate gradients algorithm.*

PROOF. If $p^0,\ldots,p^{n-1}$ are conjugate directions: $(Ap^i,p^j)=0$, $i\neq j$, let $\mathfrak{M}_n=[Ap^0,Ap^1,\ldots,Ap^{n-1}]$, and $\mathcal{P}^n$ be the (orthogonal) projection in X onto $\mathfrak{M}_n$. Then $I-\mathcal{P}^n$ maps X into a closed linear Hilbert space $X_n\subset X$. On X_n consider the "restricted" steepest descent problem: minimize $Q(x+t(I-\mathcal{P}^n)z)$, $z\in X$. A repetition of the argument of Section 11.2 shows that $Q(x+t(I-\mathcal{P}^n)z)=Q(x)+2t\Re(Ax-f,(I-\mathcal{P}^n)z)+t^2(A(I-\mathcal{P}^n)z,(I-\mathcal{P}^n)z)$. Now $\mathcal{P}^n=\mathcal{P}^{n*}$ (why?) so $(Ax-f,(I-\mathcal{P}^n)z)=((I-\mathcal{P}^n)(Ax-f),z)$. The restricted minimization problem is solved by $z=(I-\mathcal{P}^n)(Ax-f)$ and

$$t(x)=\|(I-\mathcal{P}^n)(Ax-f)\|^2/(A(I-\mathcal{P}^n)(Ax-f),(I-\mathcal{P}^n)(Ax-f)).$$

We define $p^n=(I-\mathcal{P}^n)(Ax^n-f)$ and $x^{n+1}=x^n+a_np^n$ where $a_n=t(x^n)$. This specifies the method of conjugate gradients sufficiently for the proof. We note that on X_n:

$$mI\leqslant(I-\mathcal{P}^n)A(I-\mathcal{P}^n)\leqslant MI.$$

(Why?) Thus, at the nth stage of the method of conjugate gradients we perform, in X_n, exactly one steepest descent step. Since

$$\begin{aligned}Q(x^{n+1})&=\min_\beta Q(x^n+\beta(I-\mathcal{P}^n)(Ax^n-f))\\&\leqslant Q(x^n+\beta(I-\mathcal{P}^n)(Ax^n-f))\end{aligned}$$

for any β, it holds in particular for $\beta=-2/(M+m)$. It is also easy to see that

$$\|x^n-x^*\|=\min_{x\in\mathfrak{M}_n}\|x-x^*\|,$$

so that $(I-\mathcal{P}^n)(x^n-x^*)=x^n-x^*$. Now

$$\begin{aligned}Q(x^{n+1})-Q(x^*)&=\left[A(x^{n+1}-x^*),x^{n+1}-x^*\right]\\&\leqslant(A(x^n+\beta(I-\mathcal{P}^n)(Ax^n-f)-x^*),x^n+\beta(I-\mathcal{P}^n)(Ax^n-f)-x^*)\\&\leqslant(A(I+\beta(I-\mathcal{P}^n)A)(x^n-x^*),(I+\beta(I-\mathcal{P}^n)A)(x^n-x^*)).\end{aligned}$$

Since $x^n-x^*=(I-\mathcal{P}^n)(x^n-x^*)$ while $(I-\mathcal{P}^n)^2=(I-\mathcal{P}^n)$, we set $B^n=(I-\mathcal{P}^n)A(I-\mathcal{P}^n)$; then B^n is a bounded strictly positive self-adjoint operator on X_n, and it has a bounded strictly positive self-adjoint square root C there; further, as in Theorem 11.2.1,

$$\begin{aligned} Q(x^{n+1})-Q(x^*) &\leqslant (B^n(I+\beta B^n)(x^n-x^*),(I+\beta B^n)(x^n-x^*)),\\ &\leqslant \|(I+\beta B^n)C(x^n-x^*)\|^2,\\ &\leqslant \left[\frac{M-m}{M+m}\right]^2 (B(x^n-x^*),x^n-x^*), \end{aligned}$$

while $((I-\mathcal{P}^n)A(I-\mathcal{P}^n)(x^n-x^*),x^n-x^*)=(A(x^n-x^*),x^n-x^*)=Q(x^n)-Q(x^*)$. This proves the first inequality. The second inequality is an immediate consequence of the first. ■

A strengthened form of the theorem with M, m replaced by $\sqrt{M}$, $\sqrt{m}$ can also be proven (see, e.g., Daniel [2]). It is also possible to construct conjugate gradient methhods for certain unbounded operators in the manner of Section 11.7.

EXERCISES TO SECTION 11.8

1. If A is a selfadjoint operator on a Hilbert space H such that $0<mI\leqslant A \leqslant MI$, then on H_A show that $m[u,u]\leqslant[u,Au]\leqslant M[u,u]$. From this develop an alternate proof for Theorem 11.8.1.
2. The original conjugate gradient algorithm in a real space H is

$$\begin{cases} p^0=z^0=f-Ax^0,\\ a_i=\dfrac{\|z^i\|^2}{(Ap^i,p^i)},\\ x^{i+1}=x^i+a_ip^i,\\ z^{i+1}=z^i-a_iAp^i,\\ b^i=\dfrac{\|z^{i+1}\|^2}{\|z^i\|^2},\\ p^{i+1}=z^{i+1}+b_ip^i. \end{cases}$$

Show that: $p^i=\|z^i\|^2\Sigma_{j=0}^i z^j/\|z^j\|^2$, $i=0,1,2,\ldots$; $(z^i,z^j)=0$, $i\neq j$; $(Ap^i,p^j)=0$, $i\neq j$; $(p^i,z^j)=0$, $i<j$; $(p^i,z^j)=(z^i,z^i)$, $i\geqslant j$; $(z^i,Ap^i)=(p^i,Ap^i)$, $(z^i,Ap^j)=0$, $i\neq j, j+1$. Show that x^i is linear combination of the vectors $f, Af,\ldots,A^if, x^0, Ax^0,\ldots,A^ix^0$, also of the vectors $p^0,p^1,p^2,\ldots,p^i$, and also of the vectors $z^0,Az^0,A^2z^0,\ldots,A^iz^0$.

3. If $0 < mI \leqslant A \leqslant MI$, show that the solution of $Ax = f$ must lie in the closure of the linear subspace spanned by $f, Af, A^2 f, \ldots$. (Hint: Use $[I-(I-\alpha A)]x = \alpha f$ and Corollary 6.2.1.)
4. If A is any bounded invertible linear operator on a Hilbert space H and B is a bounded linear operator such that BA is a bounded strictly positive selfadjoint operator, then the following algorithm generates a method of conjugate gradients:

$$\begin{cases} z^0 = f - Ax^0;\ p^0 = Bz^0; \\ a_i = \dfrac{\|Bz^i\|^2}{(BAp^i, p^i)}, \\ x^{i+1} = x^i + a_i p^i, \\ z^{i+1} = z^i - a_i A p^i, \\ b_i = \dfrac{\|Bz^{i+1}\|^2}{\|Bz^i\|^2}, \\ p^{i+1} = Bz^{i+1} + b_i p^i. \end{cases}$$

Show all the formulas established in Exercise 11.8.2 hold if A is replaced by BA. This corresponds to using the equation $BAx = Bf$.
5. Show that given any bounded invertible operator A on a Hilbert space H, then there exists a bounded linear operator B such that BA is a bounded, strictly positive selfadjoint operator and that any such $B = A^*C$ where C is an arbitrary bounded strictly positive selfadjoint operator. (Hint: Exercise 10.5.2.)
6. Show that the sequence $\{x^n\}$ generated by the conjugate gradient algorithm of Exercise 11.8.3 is the Ritz sequence for the coordinate functions $p^0, p^1, p^2, \ldots$.
7. When $H = \mathbf{E}^n$, show that the method of conjugate gradients must terminate in at most n iterations. When will it terminate in less than n iterations?
8. Construct a conjugate gradients algorithm for solution of $-[p(t)x'(t)]' + g(t)x(t) = f(t)$ as in Exercises 11.2.1–5.

12
A CALCULUS FOR NONLINEAR OPERATORS

12.0 INTRODUCTION

In the preceding chapter, we utilized a very simple notion: it is possible to find, for a (bounded) quadratic functional, a direction in which its rate of change is maximal, and, on the curve to which the functional reduces over the resulting one-dimensional linear manifold, a place of maximum change. In a sense, we found a meaning for the directional derivative of a functional, and in maximizing this, for the gradient of the functional. However, because our operations occurred in a Hilbert space, the structure of the space hid some of the structure of the differentiation operation. In the present chapter we establish similar notions for general nonlinear functionals. Developing analogs for the differential and integral calculus in Banach space, we obtain a Taylor's formula, an implicit function theorem, an analysis of Newton's method for solving nonlinear functional equations, and an analytic theory of constrained minimization.

12.1 DIFFERENTIATION OF ABSTRACT FUNCTIONS

We consider two Banach spaces X and Y and a possibly nonlinear mapping F from X to Y. In order to generalize the notion of derivative, we start with a geometric definition: the tangent is the line of best approximation to the curve. We now can ask, at a fixed $x \in D_F \subset X$, if there is some linear approximation to the variation of F near X. This can be done in two ways. In the first we consider rays $x + th \in D_F$, $|t| < t_0$ as in the figure and ask if there is a best linear approximation in t. By this we mean that there

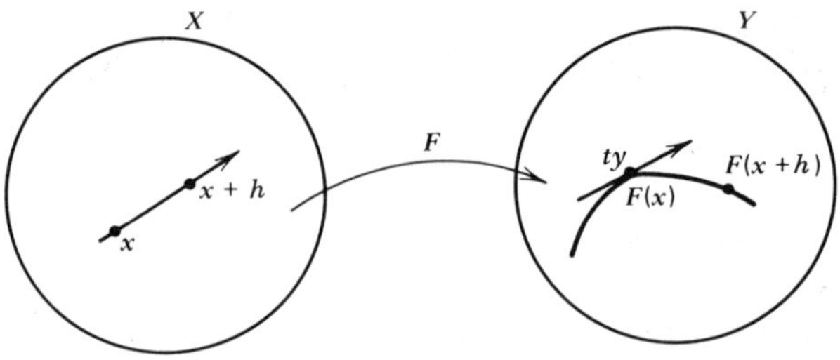

is some element y of Y such that

$$\|F(x+th)-F(x)-ty\|=o(t).$$

The particular element y depends upon both x and h, and is a "derivative in the direction h." In order to get very far here, we will assume that some neighborhood of x is in D_F, and make the following:

DEFINITION 12.1.1. If $F: X \to Y$ is such that

1. Some ball centered at $x \in X$ is in D_F;
2. For every $h \in X$ there is a $y \in Y$ satisfying

$$\|F(x+th)-F(x)-ty\|=o(t);$$

then we call the mapping $x, h \to y$ the Gâteaux differential of F at x, and write it variously as

$$GDF(x;h) = \lim_{t\to 0} \frac{F(x+th)-F(x)}{t}$$

$$= \frac{d}{dt} F(x+th)\bigg|_{t=0}.$$

The limit here is taken in the sense of the norm. It is immediate that $GDF(x,\alpha h) = \alpha GDF(x,h)$ so that the Gâteaux differential is homogeneous in h. It is not obvious that it is linear; if it should be linear as a function of h we will write $DF(x;h)$ without the G. We will also write $DF(x)h$, and call the linear operator $DF(x)$ a *gradient* or a *Gâteaux derivative*.

As an illustration of a Gâteaux differential, we consider, in the elementary calculus of variations the functional

$$\int_{t_0}^{t_1} f(t,x,x')\,dt$$

for a fixed end point problem. Proceeding formally, if $x \in C^2, h \in C^2, h(t_0) = h(t_1) = 0$, we have

$$J(x+\epsilon h) = \int_{t_0}^{t_1} f(t, x(t) + \epsilon h(t), x' + \epsilon h'(t))\, dt,$$

so

$$\left.\frac{dJ}{d\epsilon}\right|_{\epsilon=0} = \int_{t_0}^{t_1} [\, f_x(t,x,x')h(t) + f_{x'}(t,x,x')h'(t)\,]\, dt.$$

This is the Lagrange form of the first variation. It can be rewritten, by integration by parts, as

$$\left.\frac{dJ}{d\epsilon}\right|_{\epsilon=0} = \int_{t_0}^{t_1} \left[f_x(t,x,x') + \frac{d}{dt} f_{x'}(t,x,x') \right] h(t)\, dt,$$

so that the derivative is represented as a linear functional on h. If a suitable choice is made for X, Y is taken as the real line, and f is smooth enough, this formal manipulation can be justified.

As in the calculus, the secant and *some* tangent are related by the following:

MEAN VALUE THEOREM 12.1.1. *Let $f: X \to \mathbf{E}^1$. If $GDf(x;h)$ exists for all $x \in \Omega$, a convex set in X, then for all $x, x+h \in \Omega$ there is a τ, $0 < \tau < 1$, with $f(x+h) - f(x) = GDf(x + \tau h; h)$.*

PROOF. Let $\phi(t) = f(x + th)$. Then

$$\begin{aligned}\phi(t) &= \lim_{s \to 0} \frac{f(x+th+sh) - f(x+th)}{s} \\ &= GDf(x+th;h).\end{aligned}$$

Since $f: x \to \mathbf{E}^1$, f is a functional. But for fixed x, h, ϕ is an ordinary function and $f(x+h) - f(x) = \phi(1) - \phi(0) = \phi'(\tau) = GDf(x + \tau h; h)$, for some τ, $0 < \tau < 1$. ■

COROLLARY 1. (*Mean Value Theorem for Operators*) *If the operator F has a Gâteaux differential at each point of a convex set $\Omega \subset X$, then for each $x, x+h \in \Omega$,*

$$\langle F(x+h) - F(x), y' \rangle = \langle GDF(x+\tau h; h), y' \rangle$$

for any $y' \in Y'$ and some τ depending upon y', $0 < \tau < 1$.

PROOF. For each fixed $y' \in Y'$, $f(x,y') = \langle F(x), y' \rangle$ satisfies the conditions of the theorem. ■

COROLLARY 2. *If the operator F has a Gâteaux differential $GDF(x;h)$ at each point x of some open convex set $\Omega \subset X$, then for each $x, x+h \in \Omega$ there is a τ, $0<\tau<1$,* such that

$$\|F(x+h) - F(x)\| \leqslant \|GDF(x+\tau h; h)\|.$$

COROLLARY 3. *If $GDF(x;h)$ is a bounded linear operator on h at each point x of some open convex set $\Omega \subset X$, then for each $x, x+h \in \Omega$, there is a τ, $0<\tau<1$ such that*

$$\|F(x+h) - F(x)\| \leqslant \|DF(x+\tau h)\| \, \|h\|.$$

PROOF. Choose that y' for which

$$\langle F(x+h) - F(x), y' \rangle = \|F(x+h) - F(x)\|. \quad ■$$

COROLLARY 4. *On some convex closed set Ω in a Banach space X, let $f: X \to X$ have a linear Gâteaux differential $DF(x)h$. If $x \in \Omega$ implies $f(x) \in \Omega$, and*

$$\sup_{x \in \Omega} \|DF(x)\| = \alpha < 1,$$

then the functional equation $x = F(x)$ has a unique solution $x \in \Omega$.

PROOF. From Corollary 3,

$$\|F(x') - F(x)\| \leqslant \sup_{0 \leqslant t \leqslant 1} \|DF(x + t(x'-x))\| \, \|x' - x\|$$

$$\leqslant \alpha \|x' - x\|. \quad ■$$

A second approach to differentiation, due to Fréchet, is based on the best approximation of $F(x+h) - F(x)$ by a bounded linear operator $L \in [X \to Y]$ operating on h to Y.

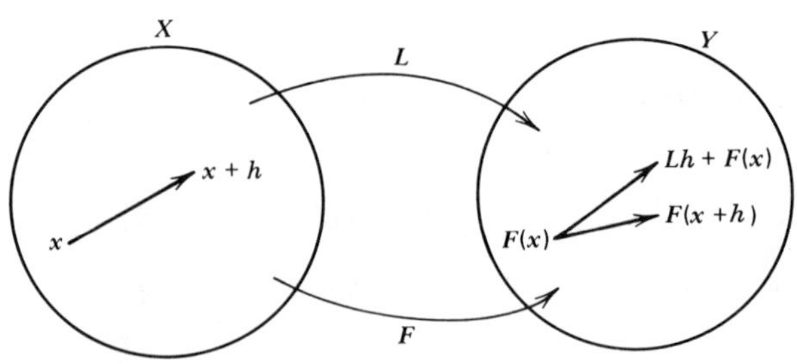

DEFINITION 12.1.2. Let X, Y be Banach spaces and $F: X \to Y$. If there exists $L \in [X \to Y]$ such that at some point $x \in D_F$

$$\|F(x+h) - F(x) - Lh\| = o(\|h\|), \qquad h \in X,$$

then Lh is called the *Fréchet differential* of $F(x)$ at x, written $dF(x;h)$. The operator L is called the *Fréchet derivative* of $F(x)$ at x, and we write variously $L = F'(x)$ and $dF(x;h) = F'(x)h$.

If $dF(x;h)$ exists so does $GDF(x;h)$, and they are equal. Then $F'(x)$ is the gradient. It is easy to see that for ordinary functions on $\mathbf{E}^n$, $F'(x)$ coincides with the ordinary gradient. In the special case that $X = \mathbf{R}$, we deal with functions on the real line with values in a Banach space Y. It is possible to generalize the ordinary differential and integral calculus of vector valued functions to this situation. First note that here the Gâteaux and Fréchet definitions coincide. (Why?) We shall use the results of the following theorem whose proof is left as an easy exercise:

THEOREM 12.1.2. *Let $y(t)$ be a function of a real variable $t \in [a,b]$ with $-\infty \leqslant a < b \leqslant \infty$, with values in a Banach space Y. Then*

1. *If $y'(t)$ exists at $t = t_0$, then $y(t)$ is continuous at t_0.*
2. *For any constant $\alpha \in \mathcal{C}$, $(\alpha y)' = \alpha(y')$.*
3. *For any $y^1(t), y^2(t)$, $(y^1 + y^2)' = (y^1)' + (y^2)'$.*
4. *If $A \in [Y \to Z]$ for some Banach space Z, then $[Ay(t)]' = A[y'(t)]$.*
5. *If $A(t): \mathbf{R} \to [Y \to Z]$, then $[A(t)y]' = A'(t)y$.*
6. *$\langle y(t), u \rangle' = \langle y'(t), u \rangle$ for any $u \in Y'$.*

A similar development for the case $X = \mathbf{R}^n$ is possible and details may be found, for instance, in Lusternik and Sobolev [1], 275–278. An integral calculus is based on the following:

DEFINITION 12.1.3. A *partition* of an interval $-\infty < a \leqslant t \leqslant b < \infty$ is any set of points $A_\tau = \{t_0, t_1, \ldots, t_n\}$ where $a = t_0 < t_1 < t_2 < \cdots < t_n = b$, for some finite n, with $\max_i (t_{i+1} - t_i) \leqslant \tau$. For any partition A_τ and function $y(t)$ with values in a Banach space Y, we define a sum

$$S[A_\tau, y(t)] = \sum_{i=0}^{n-1} y(t_i)(t_{i+1} - t_i).$$

If there exists an element $z \in Y$ such that, for any $\epsilon > 0$, there is a $\delta = \delta(\epsilon)$, with

$$\|z - S[A_\tau, y(t)]\| < \epsilon,$$

whenever $\tau < \delta$, we call z the *integral* of $y(t)$ and write

$$z(b) = \int_a^b y(t)\,dt.$$

With this definition, it is easy to translate the usual proofs from the calculus to the following:

THEOREM 12.1.3. *If $y(t)$ maps $[a,b]$ into a Banach space Y, then*

1. *If $y(t)$ is continuous on $[a,b]$, then $\int_a^b y(t)\,dt$ exists.*
2. $\int_a^b [y^1(t) + y^2(t)]\,dt = \int_a^b y^1(t)\,dt + \int_a^b y^2(t)\,dt.$
3. $\int_a^b \alpha y(t)\,dt = \alpha \int_a^b y(t)\,dt$, *for any* $\alpha \in \mathcal{R}$.
4. $\left\| \int_a^b y(t)\,dt \right\| \leqslant \int_a^b \|y(t)\|\,dt.$
5. *If $A \in [Y \to Z]$ for some Banach space Z, then*

$$\int_a^b Ay(t)\,dt = A \int_a^b y(t)\,dt.$$

6. *If $A(t) \in [Y \to Z]$, $y \in Y$ and $A(t)$ is continuous, then*

$$\int_a^b A(t)y\,dt = \left[\int_a^b A(t)\,dt \right] y.$$

7. *If $y(t)$ is differentiable on $[a,b]$ then*

$$y(b) - y(a) = \int_a^b y'(t)\,dt.$$

With the above machinery it is possible to prove the following theorems about the relationship between Gâteaux differentials and the Fréchet derivative.

THEOREM 12.1.4. *Let $F: X \to Y$, where X, Y are Banach spaces. If $GDF(x;h)$ exists on a ball $\|x^0 - x\| < \rho$ and is uniformly continuous in x and continuous in h there, then $GDF(x;h)$ is a bounded linear operator on h, the Fréchet derivative $F'(x)$ exists, and $dF(x;h) = F'(x)h = GDF(x;h)$.*

PROOF. Since

$$GDF(x+th;h) = \lim_{s\to 0} \frac{[F(x+th+sh)-F(x+th)]}{s}$$
$$= \frac{d}{dt}[F(x+th)],$$

while

$$F(x+th)-F(x) = \int_0^t \frac{d}{ds} F(x+sh)\,ds,$$

and

$$\frac{d}{ds} F(x+sh) = GDF(x+sh;h) = GDF(x;h) + \mathrm{o}(1),$$

uniformly in h, $\|h\| < \rho$, we find

$$F(x+th)-F(x) = tGDF(x;h) + \mathrm{o}(1).$$

Also, if $\|h+g\| \leqslant \rho$, then

$$F(x+t(h+g))-F(x) = tGDF(x;h+g) + \mathrm{o}(t),$$

and

$$F(x+t(h+g))-F(x+th) = tGDF(x;g) + \mathrm{o}(t),$$

by virtue of the uniform continuity in x. Therefore

$$0 = [F(x+th)-F(x)] + [F(x+t(h+g))-F(x+th)] - [F(x+t(h+g))-F(x)]$$
$$= t(GDF(x;h) + GDF(x;g) - GDF(x;h+g)) + \mathrm{o}(t).$$

Hence

$$GDF(x;h) + GDF(x;g) = GDF(x;h+g) + \mathrm{o}(1).$$

Since everything but the o(1) term is independent of t, we must have

$$GDF(x;h) + GDF(x;g) = GDF(x;h+g).$$

Thus $GDF(x;h)$ is a continuous *linear* operator on $\|h\| \leqslant \rho$, hence on h of all norms. Then $GDF(x;h)=DF(x;h)=DF(x)h$. We wish to show that $DF(x)h=dF(x;h)$. Now

$$F(x+h)-F(x)-DF(x)h=\int_0^1 \frac{d}{dt}F(x+th)\,dt-DF(x)h$$
$$=\int_0^1 DF(x+th)h\,dt-DF(x)h.$$

Since $DF(x+th)$ is uniformly continuous in x, $DF(x+th)=DF(x)+\mathrm{o}(1)$, $\|h\|\to 0$, uniformly on $0\leqslant t\leqslant 1$. Then

$$F(x+h)-F(x)-DF(x)h=\int_0^1 DF(x)h\,dt-DF(x)h+\int_0^1 \mathrm{o}(1)h\,dt$$
$$=\int_0^1 \mathrm{o}(1)h\,dt.$$

Now

$$\left\|\int_0^1 \mathrm{o}(1)h\,dt\right\|\leqslant\int_0^1 dt\,\mathrm{o}(1)\|h\|.$$

Hence $DF(x)h=dF(x;h)$, and the Fréchet derivative $F'(x)$ exists. ■

Given a linear Gâteaux differential we can show that it is a Fréchet differential under the conditions of the following:

THEOREM 12.1.5. (*Vainberg* [1], 41) *If $F(x):X\to Y$ has a Gâteaux differential $GDF(x;h)$ of the form $L(x)h$, where $L(x)\in[X\to Y]$ for $\|x-x^0\|<\delta$, and $L(x)$ is continuous in x at x^0, then $dF(x^0;h)=F'(x^0)h$ exists and $F'(x^0)=L(x^0)$.*

PROOF. Corollary 1 to the mean value Theorem 12.1.1 implies that $\langle F(x^0+h)-F(x^0)-L(x^0)h,y'\rangle=\langle L(x^0+\tau h)h-L(x^0)h,y'\rangle$ for any $y'\in Y'$, any $h\in X$, and some τ, $0<\tau<1$. If we choose y' so that $\|y'\|=1$ and

$$\langle F(x^0+h)-F(x^0)-L(x^0)h,y'\rangle=\|F(x^0+h)-F(x^0)-L(x^0)h\|,$$

then

$$\|F(x^0+h)-F(x^0)-L(x^0)h\|\leqslant\|L(x^0+\tau h)-L(x^0)\|\,\|h\|.$$

But $\|L(x^0+\tau h)-L(x^0)\|\to 0$, $\|h\|\to 0$, so the right side is $\mathrm{o}(\|h\|)$. Then $F'(x^0)$ exists and is $L(x^0)$. ■

We also will need a functional integral, which is different from the integral of an abstract functional of Definition 12.11.3.

DEFINITION 12.1.4. Let $G(x)$ be a mapping from $x = x^0 + t(x^1 - x^0) \in X$ to $[X \to Y]$. Form the approximating sums

$$S[A_\tau, G] = \sum_{i=0}^{n-1} G(x^0 + t_i(x^1 - x^0))(x^1 - x^0)(t_{i+1} - t_i)$$

for partitions A_τ of the interval $[0, 1]$. If there exists an element $y \in Y$ such that, for any $\epsilon < 0$ we have

$$\|y - S[A_\tau, G]\| < \epsilon$$

for all $\tau \leqslant \tau_\epsilon$, we call y the *integral* of G,

$$y = \int_{x^0}^{x^1} G(x)\, dx.$$

In the case that G is continuous in x it is easy to show the existence of the integral. We will use two properties of this integral.

THEOREM 12.1.5. *If* $G(x) = F'(x)$, *then*

$$\int_{x^0}^{x^1} F'(x)\, dx = F(x^1) - F(x^0).$$

PROOF. This follows from Theorem 12.1.3 and

$$\int_{x^0}^{x^1} F'(x)\, dx = \int_0^1 F'(x^0 + \tau(x^1 - x^0))(x^1 - x^0)\, d\tau$$

$$= \int_0^1 \frac{d}{d\tau} F(x^0 + \tau(x^1 - x^0))\, d\tau$$

$$= F(x^1) - F(x^0). \quad \blacksquare$$

Further we need an estimate of this type of integral:

THEOREM 12.1.6. *If* $\|G(x^0 + \tau\,(x^1 - x^0))\| \leqslant \phi(t_0 + \tau(t_1 - t_0))$ *where* $\|x^1 - x^0\| \leqslant t_1 - t_0$, *then*

$$\left\| \int_{x^0}^{x^1} G(x)\, dx \right\| \leqslant \int_{t_0}^{t_1} \phi(\tau)\, d\tau.$$

PROOF. Again, from Theorem 12.1.3,

$$\left\|\int_{x^0}^{x^1} G(x)\,dx\right\| = \left\|\int_0^1 G(x^0+\tau(x^1-x^0))(x^1-x^0)\,d\tau\right\|$$

$$\leqslant \|x^1-x^0\| \int_0^1 \phi(t_0+\tau(t_1-t_0))\,d\tau$$

$$\leqslant (t_1-t_0)\int_0^1 \phi(t_0+\tau(t_1-t_0)d\tau$$

$$= \int_{t_0}^{t_1} \phi(\tau)\,d\tau. \quad \blacksquare$$

Remark: We have defined derivatives and integrals for functions on the real line with values in a Banach space X in terms of limit processes, as is to be expected, with convergence defined by the norm in X. It is also possible to define weak derivatives and integrals by using the weak topology on X induced by X'. One also finds occasional use of the terms "weak derivative" for a Gâteaux (linear) derivative and "strong derivative" for the Fréchet derivative (e.g., Kantorovich and Akilov [1], 657–658).

EXERCISES TO SECTION 12.1

1. Assume that $f(t,u,v)$ is C^2 on $t_0 \leqslant t \leqslant t_1$, $|u| \leqslant U$, $|v| \leqslant V$ for some nonzero constants U, V. For any fixed function $x(t) \in C^2$ define $\|x(t)\| = \max_{t_0 \leqslant t \leqslant t_1}\{\max[|x(t)|, |x'(t), |x''(t)|]\}$. This makes C^2 a Banach space X. Show that any x in the Banach space X satisfying

$$|x(t)| < U, |x'(t)| < V, \qquad t_0 \leqslant t \leqslant t_1,$$

is a point at which the Gâteaux differential of $F(x)$ exists as a linear functional of h, where

$$F(x) = \int_{t_0}^{t_1} f(t, x(t), x'(t))\,dt.$$

What if $|x(t)| = U$ at a point t^*, $t_0 \leqslant t^* \leqslant t_1$? What if $|x'(t)| = V$ on an interval $[t^*, t^{**}]$, $t_0 \leqslant t^* < t^{**} \leqslant t$? Is there a Fréchet differential?

2. Let X be a Banach space and $[X]$ the space of bounded linear operators on X. Is the mapping $A \to A^n$, $n = 1, 2, \ldots$ of $[X]$ into $[X]$ differentiable? What is its Gâteaux differential? (Hint: Start with $n=1$ and work your way up.) Is the Gâteaux differential linear? When is $GDF(x;h) = nA^{n-1}h$?

3. For any $A \in [X]$, where X is a Banach space, compute the Gâteaux differential of e^A, using the results of Exercise 12.1.2.
4. Prove the following partial converse to Corollary 4: *If $DF(x)$ is continuous on Ω then $\sup_{x \in \Omega} \|DF(x)\| = \alpha < 1$ is also necessary in order that F be a contraction.* (Kantorovich and Akilov [1], 661–662.)
5. Show that if $dF(x;h)$ exists, then $DF(x)h$ exists and equals $dF(x;h)$. (The converse is not true.)
6. Prove Theorems 12.1.2 and 12.1.3.
7. Show that the Gâteaux differential in Exercise 12.1.1 is a Fréchet differential.
8. If $X = \mathbf{E}^n$, $Y = \mathbf{E}^m$ and $F(x)$ is an m-vector, show that

$$GDF(x;h) = Jh$$

where J is the Jacobian matrix:

$$\frac{\partial(f_1, \ldots, f_m)}{\partial(x_1, \ldots, x_n)}.$$

9. If $F: X \to Y$, where X, Y are Banach spaces, and $F'(x^0)$ exists, then F is continuous at x^0.
10. Consider a real Hilbert space H and the two functionals (x,x) and $\|x\|$. Compute their Gâteaux differentials and determine if they are Fréchet differentials.
11. If $L \in [X \to Y]$ where X, Y are Banach spaces, show that L', the Fréchet derivative, exists and is L.
12. If $F: X \to Y$ and $L \in [Y \to Z]$, where X, Y, Z are Banach spaces, and $F(x)$ possesses a Fréchet derivative (Gâteaux differential) at $x \in X$, then $LF(x)$ also possesses a Fréchet derivative (Gâteaux differential) at x and $[LF(x)]' = LF'(x)$ $(GDLF(x;h) = LGDF(x;H))$.
13. Let $G: X \to Y, F: Y \to Z$ where X, Y, Z are Banach spaces. Let $\Omega_\rho = \{x \mid \|x - x^0\| < \rho\} \subset D_G, G(\Omega_\rho) = \{y = G(x) \mid x \in \Omega_\rho\}$ and assume that $D_F \supset G(\Omega_\rho)$. If $G(x), F(y)$ possess Fréchet derivatives (gradients) at x^0 then $FG(x) = F(G(x))$ possesses a Fréchet derivative (gradient) at x^0 and $(FG)'(x)|_{x=x^0} = F'(G(x^0))G'(x^0)$.
14. If $Q(x) = (Ax,x) - (f,x) - (x,f)$, where A is a bounded selfadjoint operator on a Hilbert space H, then compute the Gâteaux differential of Q, the gradient, and the Fréchet derivative. Show that $Q(x)$ achieves its minimum at a root of the gradient.

15. Let X be a Banach space, f a real valued nonlinear functional on X. If f achieves its minimum on X at x^*:

$$f(x^*) = \min_{x \in X} f(x),$$

and $GDf(x^*; h)$ exists, then show that $GDf(x^*; h) = 0$ for all h, the gradient of f exists and is zero, and the Fréchet derivative, if it exists, is zero.

16. Let $F: X \otimes Y \to Z$ where X, Y, Z are Banach spaces. Show that

$$dF(x,y; h,g) = d_x F(x,y; h) + d_y F(x; y; g)$$

where $d_x F$ ($d_y F$) is the (partial) differential of F with y (x) held fixed. Show a similar result for the Gâteaux differential.

17. Let X be a Banach space, X' its dual. If $x(t): [0,1] \to X, y'(t): [0,1] \to X'$ are continuously differentiable on $[0,1]$, then

$$\int_0^1 \left[\left\langle \frac{dx(t)}{dt}, y'(t) \right\rangle + \left\langle x(t), \frac{dy'(t)}{dt} \right\rangle \right] dt = \langle x(1), y'(1) \rangle - \langle x(0), y'(0) \rangle,$$

which is the formula for integration by parts.

12.2 HIGHER DERIVATIVES AND TAYLOR'S FORMULA

Since $F(x)$, mapping X into Y, may have a Fréchet derivative $F'(x)$ at many points $x \in X$, this sets up a natural mapping $x \to F'(x) \in [X \to Y] = Z$.

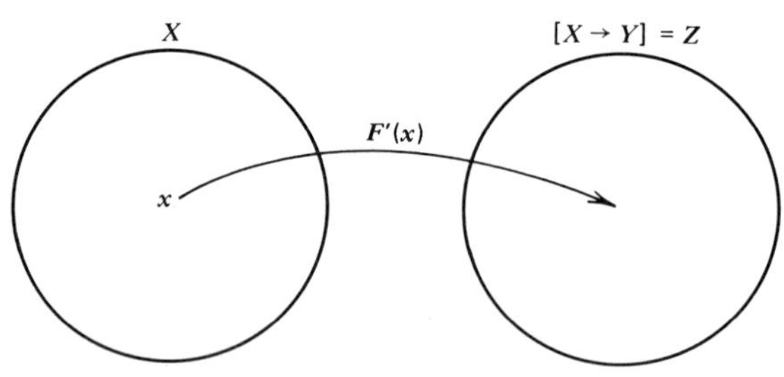

For each x, $F'(x)$ is a bounded linear operator; however $F'(x)$ need not be linear in x.

It may happen that $F'(x)$ is itself differentiable in the Fréchet sense. Then $F''(x)$ maps $X \to [X \to Z] = [X \to [X \to Y]]$. One can proceed in this way, but it is awkward. It becomes convenient to consider a different approach.

DEFINITION 12.2.1. Let $X^2 = X \otimes X$, the space of ordered pairs $\{x^1, x^2\}$ of elements $x^1, x^2 \in X$. Let B be any mapping of $X^2 \to Y$ such that

1. $B\{\alpha x' + \beta x'', x^2\} = \alpha B\{x', x^2\} + \beta B\{x'', x^2\}$, $\alpha, \beta \in \mathscr{C}$;
2. $B\{x^1, \alpha x' + \beta x''\} = \alpha B\{x^1, x'\} + \beta B\{x^1, x''\}$, $\alpha, \beta \in \mathscr{C}$;
3. $\|B\{x^1, x^2\}\| \leq M \|x^1\| \, \|x^2\|$ for some $M > 0$, all $x^1, x^2 \in X$.

We call $B\{x^1, x^2\}$ a *bilinear form*. The sum of two such bilinear forms is defined by

$$(B + C)\{x^1, x^2\} = B\{x^1, x^2\} + C\{x^1, x^2\},$$

and we define

$$(\alpha B)\{x^1, x^2\} = \alpha(B\{x^1, x^2\}).$$

The norm of a bilinear form is defined by

$$\|B\| = \sup_{\substack{\|x^1\| \leq 1 \\ \|x^2\| \leq 1}} \|B\{x^1, x^2\}\|.$$

The space of all such *bounded* bilinear forms is denoted by $[X^2 \to Y]$. It is easy to show that $[X^2 \to Y]$ is a Banach space. (It is also easy to generalize this definition to multilinear n-forms $B\{X^1, \ldots, X^n\}$ from

$$X^n = \overbrace{X \otimes X \otimes \cdots \otimes X}^{n}$$

to Y and to show that the related space $[X^n \to Y]$ is a Banach space.)

Now we can introduce a natural isometry between $[X^2 \to Y]$ and $[X \to [X \to Y]]$. Any element of $[X \to [X \to Y]]$ is a bounded linear operator from X to Y, parametrized by an element $x \in X$, say $T_x = T(x) \in [X \to Y]$. If $\hat{x} \in X$, then $y = T_x \hat{x} \in Y$, and $T_x \hat{x}$ is of course linear in $\hat{x}$. Since $T_x \in [X \to \cdot]$, T_x is also linear in x. Then $T_x \hat{x}$ is a bilinear form in $\{x, \hat{x}\}$. Also

$$\|y\| \leq \|T_x\| \, \|\hat{x}\| \leq \|T\| \, \|\hat{x}\| \, \|x\|.$$

Thus $T_x \hat{x} \in [X^2 \to Y]$, say $T_x \hat{x} = B\{x, \hat{x}\}$, where $\|B\| \leq \|T\|$. On the other hand, for a fixed x^2, any bilinear form $B\{x^1,x^2\}$ is a bounded linear operator on $x^1 \in X$ to Y, say $y = S_{x^1}x^2 \equiv B\{x^1,x^2\}$ and

$$\|S_{x^1}\| = \sup_{\|x^1\|=1} \|S_{x^1}x^2\| \leqslant \|B\|\|x^2\|.$$

Then S_{x^1} is a bounded linear operator on $x^1 \in X$ to $[X \to Y]$, $\|S_{x^1}\| \leqslant \|B\|$, and $S_{x^1} \equiv T_{x^1}$ for all x^1; that is $S = T$, and $\|T\| = \|S\| = \|B\|$. The mapping $T \to B$ is, then, one-to-one, invertible, and norm preserving. (A similar natural isometry exists between the following:

$$\big[\overbrace{X \to [X \to \cdots [X}^{n} \to Y]\cdots\big] \quad \text{and} \quad \big[\overbrace{X \otimes X \otimes \cdots \otimes X}^{n} \to Y\big]\,.)$$

We wish to describe $F''(x)$ in these terms. To do that, consider a functional $F: X \to Y$, which possesses a *Taylor's formula*

$$F(x+h) - F(x) = A(h) + B\{h,h\} + o(\|h\|^2).$$

Suppose $A(h) = Ah$, where $A \in [X \to Y]$, and $B \in [X^2 \to Y]$. Since $\|B\{h,h\}\| \leqslant \|B\|\,\|h\|^2$, $\|F(x+h) - F(x) - A(h)\| = o(\|h\|)$. That is, $F(x)$ is Fréchet differentiable, and $dF(x;h) = F'(x)h = Ah$, $F'(x) = A$. Conversely, then $F'(x)h = GDF(x;h)$. On the other hand

$$\|F(x+th) - F(x) - A(th)\| = t^2 B\{h,h\} + o(\|th\|^2).$$

Then $t^{-1}A(th) \to GDF(x;h)$. If A is homogeneous, then $A(h) = GDF(x;h) = F'(x)h$. Thus the only homogeneous term in h is linear, and the only linear term in this Taylor's formula is given by $dF(x;h) = F'(x)h$.

Having identified the linear term, we turn to the second term. Suppose now that in some ball of radius ρ about x, $F'(x)$ is itself Fréchet differentiable. Then

$$F'(x+h) - F'(x) = F''(x)h + o(\|h\|).$$

Hence

$$\begin{aligned} F(x+h) - F(x) &= \int_0^1 F'(x+th)h\,dt \\ &= \int_0^1 \big[F'(x)h + \{F''(x)th\}h + \{o(t\|h\|)\}h\big]\,dt \\ &= F'(x)h + \tfrac{1}{2}\{F''(x)h\}h + o(\|h\|^2). \end{aligned}$$

Given any other Taylor's formula, as above, then we see immediately that

$$B\{h,h\}=\tfrac{1}{2}\{F''(x)h\}h+o(\|h\|^2).$$

Then, for any $\epsilon>0$,

$$\sup_{\|h\|=1}\left\|B\{h,h\}-\tfrac{1}{2}\{F''(x)h\}h\right\|<\epsilon.$$

Thus $\{F''(x)h\}h\equiv 2B\{h,h\}$. The problem of identifying $F''(x)$ with $2B$ is complicated by one problem, which is evident from the proof of the following:

LEMMA 12.2.1. *If B,C are in $[X^2\to Y]$, and both $B\{x',x''\}\equiv B\{x'',x'\}$ and $C\{x',x''\}\equiv C\{x'',x'\}$, for all $x',x''\in X$, then*

$$B\{x,x\}=C\{x,x\},\qquad x\in X,$$

implies $B=C$.

PROOF. Consider $B\{x+h,x+h\}=C\{x+h,x+h\}$. ■

Such a bilinear form is *symmetric*. If both $\{F''(x)h\}g$ and $2B\{h,g\}$ are symmetric, then $F''(x)=2B$. For B, we make this an assumption, but the following result shows it for $F''(x)$, and the method of proof is instructive.

THEOREM 12.2.1. *Let $F:X\to Y$, where X,Y are Banach spaces. If $F'(x),F''(x)$ exist and are continuous on $\|x-x^0\|<\rho$, then*

$$\{F''(x)g\}h=\{F''(x)h\}g,\qquad g,h\in X.$$

PROOF. Let $f(x)=\langle F(x),y'\rangle$, $y'\in Y'$. Let

$$\phi=f(x+tg)-f(x),\qquad g\in X,\qquad t \text{ real},$$

and

$$\Delta=\phi(x+sh)-\phi(x),\qquad h\in X,\qquad s \text{ real}.$$

Now Δ is a function of a single real variable s, so the Mean Value Theorem 12.1.1 asserts that for some s' between 0 and s,

$$\begin{aligned}\Delta&=s\phi'(x+s'h)h\\&=s\{f'(x+tg+s'h)-f'(x+s'h)\}h.\end{aligned}$$

A second use of the mean value theorem yields

$$\Delta = st\{\{f''(s+t'g+s'h)\}g\}h,$$

for t' between 0 and t. Now

$$\Delta = f(x+tg+sh)+f(x)-f(x+tg)-f(x+sh).$$

Another grouping of the terms shows

$$\Delta = st\{\{f''(x+t''g+s''h)'\}h\}g,$$

for suitable s'', t'' as before. Dividing by st, then letting $s,t\to 0$, yields

$$\langle\{F''(x)h\}g,y'\rangle = \langle\{F''(x)g\}h,y'\rangle,$$

for all y'. ■

In the case of a Gâteaux differential, we can also define a second order differential

$$GD^2F(x;g,h) = \lim_{t\to 0}\frac{[GDF(x+th;h)-GDF(x;h)]}{t},$$

if the limit exists. Again $GD^2F(x;g,h)$ is homogeneous in g and h; if it is given by a linear operator in $[X^2\to Y]$, the operator may be called a second Gâteaux derivative. It is easy to extend the mean value theorem for operators in the form,

$$\langle F(x+h)-F(x),y'\rangle = \langle GDF(x;h),y'\rangle + \frac{\langle GD^2F(x+\tau h;h,h),y'\rangle}{2},$$

for some τ, $0<\tau<1$, by the calculus. Also we have the following:

THEOREM 12.2.2. *Let $F:X\to Y$, where X,Y are Banach spaces. If $GDF(x;h)$ and $GD^2F(x;g,h)$ exist and are continuous in x for $\|x-x^0\|\leqslant\rho$, then*

$$GD^2F(x;g,h)\equiv GD^2F(x;h,g),\qquad \textit{for all } h,g\in X.$$

The proof is an easy modification of that for Theorem 12.2.4. We illustrate the calculation of derivatives of nonlinear integral operators in the following:

THEOREM 12.2.3. *Let* $K(x,t,u), K_u(s,t,u), K_{uu}(s,t,u) \in C^0$ $(0 \leqslant s,t \leqslant 1, -\infty < u < \infty)$, *and let* $|K_{uu}(s,t,u)| \leqslant M_2$ *there. Then*

$$F(x) = y(s) = \int_0^1 K(s,t,x(t))\,dt$$

is a Fréchet differentiable mapping of $L_2(0,1)$ *into itself at every* $x(t) \in L_2(0,1)$, *and*

$$dF(x;h) = F'(x)h = \int_0^1 K_u(s,t,x(t))h(t)\,dt.$$

Also the Gâteaux differential of $F'(x)h$ *with respect to* x *exists and is given by*

$$GD^2F(x;g,h) = \int_0^1 K_{uu}(s,t,x(t))g(t)h(t)\,dt.$$

PROOF. By the ordinary Taylor's formula

$$K(s,t,x(t))(=)K(s,t,0) + K_u(s,t,0)x(t) + \tfrac{1}{2}K_{uu}(s,t,\theta x(t))x^2(t)$$

where $\theta = \theta(x(t))$, $0 < \theta < 1$. Then

$$\left|\int_0^1 K(s,t,0)\,dt\right| \leqslant \max|K(s,t,0)| \leqslant M$$

and

$$\left|\int_0^1 K_u(s,t,0)x(t)\,dt\right|^2 \leqslant \int_0^1 K_u^2(s,t,0)\,dt \int_0^1 x^2(t)\,dt$$

$$\leqslant M_1^2\|x\|^2,$$

and even

$$\left|\int_0^1 K_{uu}(s,t,\theta(t)x(t))x^2(t)\,dt\right| \leqslant M_2\|x\|^2.$$

Then

$$|y(s)| \leqslant M + M_1\|x\| + \tfrac{1}{2}M_2\|x\|^2$$

Thus $y(s)$ is bounded. Define

$$x^n(t)=\begin{cases} -n, & x(t)<-n, \\ x(t), & -n\leqslant x(t)\leqslant n, \\ n, & x(t)>n. \end{cases}$$

Then $x\in L_2(0,1)$ makes $x^n\in L_2(0,1)$ and $\|x^n-x\|\to 0$, $n\to\infty$. Let

$$y^n(x)=\int_0^1 K(s,t,x^n(t))\,dt.$$

On $\{0\leqslant s,t\leqslant 1\}\otimes\{|u|\leqslant n\}$, $K(s,t,u)$ is uniformly continuous, so $|K(s_1,t,u)-K(s_2,t,u)|=\Delta(s_1-s_2)$ satisfies:

$$\text{for any } \epsilon>0 \text{ there is a } \delta=\delta_\epsilon \text{ with } \Delta(s_1-s_2)\leqslant\epsilon$$
$$\text{for } |s_1-s_2|<\delta_\epsilon, \text{ uniformly in } t,u.$$

Then $|y^n(s_1)-y^n(s_2)|<\epsilon$, $|s_1-s_2|<\delta_\epsilon$. Thus each $y^n\in C[0,1]$, so each $y^n\in L_2(0,1)$. Also

$$y^n(s)-y^m(s)=\int_0^1 K_u(s,t,x^m(t))(x^n-x^m)\,dt$$
$$+\tfrac{1}{2}\int_0^1 K_{uu}(s,t,x^m+\theta(x^n-x^m))(x^n-x^m)^2\,dt.$$

But

$$\int_0^1 |K_u(s,t,x^m(t))|\,|x^n-x^m|\,dt\leqslant\int_0^1 |K_u(s,t,0)|\,|x^n-x^m|\,dt$$
$$+\tfrac{1}{2}\int_0^1 M_2|x^m|\,|x^n-x^m|\,dt$$
$$\leqslant M_1\|x^n-x^m\|+\tfrac{1}{2}M_2\|x^m\|\,\|x^n-x^m\|,$$

by the Schwarz inequality. Since $\{x^m\}$ is a convergent sequence, $\|x^m\|$ is bounded. Thus the term is o(1), uniformly in s. The second integral is bounded by $M_2\|x^n-x^m\|^2/2$, uniformly in s. Then

$$\int_0^1 [y^n(s)-y^m(s)]^2\,ds=o(1), \qquad n,m\to\infty.$$

Hence $\|y^n-y^m\|\to 0$, so y^n converges to an $L_2(0,1)$ function. Repeating the same calculation with $y^m(s)$ replaced by $y(s)$ shows that

$$|y^n(s)-y(s)| \leqslant M_1\|x^n-x\|+\tfrac{1}{2}M_2\|x\|\,\|x^n-x\|+\tfrac{1}{2}M_2\|x^n-x\|^2,$$

so $y^n \to y, n\to\infty, |y(s)| \leqslant M+M_1\|x\|+\frac{1}{2}M_2\|x\|^2$ and $y \in L_2(0,1)$. Thus $F: L_2(0,1)\to L_2(0,1)$.

To calculate $F'(x)$, consider

$$\begin{aligned}\int_0^1 \{K(x,t,x(t)+h(t))-K(x,t,x(t))\}\,dt \\ &= \int_0^1 K_u(s,t,x(t))h(t)\,dt \\ &\quad + \tfrac{1}{2}\int_0^1 K_{uu}(s,t,x(t)+\theta(t)h(t))h^2(t)\,dt,\end{aligned}$$

for some $\theta=\theta(t), 0\leqslant\theta\leqslant 1$, by Taylor's theorem for functions of one variable. Since

$$\left(\int_0^1 \left|\int_0^1 K_{uu}(s,t,x(t)+\theta(t)h(t))h^2(t)\,dt\right|^2 ds\right)^{1/2} \leqslant M_2\|h\|^2,$$

this shows that

$$(F'(x)h)(s)=\int_0^1 K_u(s,t,x(t))h(t)\,dt.$$

That is, $F'(x)$ is the bounded linear integral operator

$$F'(x)=\int_0^1 K_u(s,t,x(t))(\cdot)\,dt,$$

with $\|F'(x)\| \leqslant M_1+M_2$ (why?). We calculate $GD^2F(x;g,h)$, the Gâteaux differential, as follows: by Taylor's theorem, for some $\theta'=\theta'(t), 0<\theta'<1$,

$$\begin{aligned}\Big|\int_0^1 K_u(s,t,x(t)+\tau g(t))h(t)\,dt-\int_0^1 K_u(s,t,x(t))h(t)\,dt \\ &-\tau\int_0^1 K_{uu}(s,t,x(t))g(t)h(t)\,dt\Big| \\ &=\Big|\tau\int_0^1 [K_{uu}(s,t,x(t)+\theta'(t)\tau y(t))-K_{uu}(s,t,x(t))]\,g(t)h(t)\,dt\Big|.\end{aligned}$$

Since $K_{uu}(s,t,u)$ is continuous and bounded, we find

$$K_{uu}(s,t,x(t)+\theta'(t)\tau g(t))-K_{uu}(s,t,x(t))(\to)0, \qquad \tau\to 0,$$

while

$$|[K_{uu}(s,t,x(t)+\theta'(t)\tau g(t))-K_{uu}(s,t,x(t))]\,g(t)h(t)|\leqslant 2M_2|\,g(t)h(t)|.$$

We can apply Theorem 2.5.3 to assert that

$$\tau\int_0^1[K_{uu}(s,t,x(t)+\theta'(t)\tau g(t))-K_{uu}(s,t,x(t))g(t)]h(t)\,dt=\mathrm{o}(\tau),\qquad \tau\to 0.$$

This shows that the Gâteaux differential of $F'(x)h$ exists and is given by

$$\int_0^1 K_{uu}(s,r,x(t))g(t)h(t)\,dt.$$

It is a linear Gâteaux differential; that it is an element in $[X^2\to Y]$, is easily seen from the Schwarz inequality:

$$\left[\int_0^1\left|\int_0^1 K_{uu}(s,t,x(t))g(t)h(t)\,dt\right|^2 dt\right]^{1/2}\leqslant M_2\|\,g\|\,\|h\|. \quad \blacksquare$$

If we work in $C[0,1]$, it is possible to show that $GD^2F(x;g,h)$ actually is a Fréchet differential, so that $F''(x)$ exists. It is instructive to attempt to prove this in $L_2(0,1)$, where it need not be true.

EXERCISES TO SECTION 12.2

1. Show that $[X^2\to Y]$ is a Banach space when X, Y are Banach spaces.
2. Generalize Definition 12.2.1 to *multilinear n-forms* and show that $[X^n\to Y]$ is a Banach space. (See for instance, Lusternik and Sobelev [1], 303–315.)
3. If $X=\mathbf{R}^n$, $Y=\mathbf{R}$ (so $F(x)$ is one function of n variables) show that

$$F''(x)\{h,h\}=\sum_{i=1}^{n}\sum_{i=1}^{n}\left\{\frac{\partial^2F(x)}{\partial x_i\partial x_j}\right\}h_ih_j,$$

   where $h=(h_1,\ldots,h_n)$.
4. If $X=\mathbf{E}^n$, $Y=\mathbf{E}^m$, compute $F''(x)\{h,h\}$.
5. Let $X=Y=\mathbf{R}^n$, A be an $n\times n$ real symmetric matrix ($A=A^T$), and $f\in X$. Compute $F'(x), F''(x)$, for $F(x)=(Ax,x)-2(x,f)$.
6. Let $X=Y=C[0,1]$, let Ω_ρ be the ball $\sup_{0\leqslant t\leqslant 1}|x^0(t)-x(t)|\leqslant\rho$, and let G be the set $(0\leqslant s,t\leqslant 1)\otimes(|x^0-x|\leqslant\rho)$. Let $K(s,t,u), K_u(s,t,u)$,

$K_{uu}(s,t,u)$ be continuous on G. Then the integral transform

$$y(s)=\int_0^1 K(s,t,x(t))\,dt,$$

maps Ω_ρ into $C[0,1]$, is Fréchet differentiable two times at every point x such that $\|x^0-x\|<\rho$, with the derivatives as in Theorem 12.2.5.

7. Prove the following extension of the *mean value theorem*: *If $F(x): X \to Y$ possesses Gâteaux differentials $GDF(x;h)$ and $GD^2F(x;g,h)$ on a ball $\Omega_\rho=\{x|\,\|x-x^0\|<\rho\}$, then for any $y'\in Y'$, and h with $\|h\|<\rho$, there is a τ, $0<\tau<1$, with*

$$\langle F(x+h)-F(x),y'\rangle=\langle GDF(x;h),y'\rangle+\tfrac{1}{2}\langle GD^2F(x+\tau h;h,h),y'\rangle$$

and

$$\|F(x+h)-F(x)-GDF(x;h)\|\leqslant\tfrac{1}{2}\|GD^2F(x+\tau h;h,h)\|.$$

8. Reformulate and prove the theorem of Exercise 12.2.7 for the case in which $F(x)$ possesses continuous first and second Fréchet derivatives on Ω_ρ.
9. Under the conditions of Exercise 12.2.8 prove Taylor's formula

$$F(x+h)=F(x)+F'(x)h+\int_0^1\{F''(x+th)h\}h(1-t)\,dt.$$

10. Show that if $L\in[X\to Y]$, where X,Y are Banach spaces, then L possesses a second Fréchet derivative $L''=0$.
11. Let $X=C^1[0,1]$, $Y=C[0,1]$, and $F(x)=\dfrac{dx(t)}{dt}-f(t,x(t))$. Suppose that $f\in C^1([0,1]\otimes\mathbf{R})$. Show that $F(x)$ is Fréchet differentiable at all x, and compute $F'(x)$. If $f\in C^2([0,1]\otimes\mathbf{R})$, then show that F possesses a second Fréchet derivative, and compute it.
12. In Exercise 12.2.11, if we rewrite the differential operator in the form

$$F(x)=x(t)-x(0)-\int_0^t f(s,x(s))\,ds,$$

compute the same Fréchet derivatives of $F(x)$. Specify the spaces X,Y as seems appropriate.

13. If f is a real valued nonlinear functional on a Banach space X that achieves its minimum at x^*:

$$f(x^*)=\min_{x\in X} f(x),$$

and $GD^2f(x^*;h,h)$ exists, then show that $GD^2f(x^*;h,h)\geqslant 0$ for all h.

12.3 SOLVING FUNCTIONAL EQUATIONS BY NEWTON'S METHOD

We can put Taylor's formula to use in solving functional equations in a way that parallels exactly Newton's method for one continuously differentiable function $f:\mathbf{R}\to\mathbf{R}$. In that case, if we have already constructed a sequence $x^1,\ldots,x^n$ of approximations to a *root* of $f(x)=0$, then we determine x^{n+1} as the unique intersection of the tangent line $y=f(x^n)+f'(x^n)(x-x^n)$ with the x axis. That is, $x^{n+1}=x^n-[f'(x^n)]^{-1}f(x^n)$. So long as $f'(x^n)\neq 0$, this process can continue; if the sequence x^n actually converges, then it is obvious that its limit is a root. Now the equation of the tangent line is really given by the first two terms of Taylor's formula. Kantorovich proposed to solve a functional equation $F(x)=0$ where F is defined and Fréchet differentiable on some open convex set of a Banach space X, with range in a Banach space Y, in a parallel manner. The first two terms of Taylor's theorem, set equal to zero, permit the determination of x^{n+1} from the linear functional equation

$$F(x^n)+F'(x^n)(x^{n+1}-x^n)=0,$$

provided that we can solve the equation

$$F'(x^n)(x-x^n)=-F(x^n).$$

If $[F'(x^n)]^{-1}$ exists, we can rewrite this as the method of successive approximations for the equation,

$$x=x-\Gamma(x)F(x)$$

where $\Gamma(x)=[F'(x)]^{-1}$; that is

$$x^{n+1}=x^n-[F'(x^n)]^{-1}F(x^n).$$

It is in this form that our analysis proceeds. If $F'(x)$ is itself continuous, and the sequence x^n is defined for all n as a Cauchy sequence, then its limit is easily shown to be a solution of $F(x)=0$. This solution will in fact turn out to be unique in the small because $F'(x)$ is invertible.

Even though we shall assume enough to guarantee $F'(x)$ invertible, it may not be desirable to have to pay the price of computing $F'(x^n)$ at each stage of the iteration and then solving for x^{n+1}. At a sacrifice in convergence rates, one may fix $\Gamma(x)=\Gamma_0$, taken to be an approximate inverse (possibly exact at x^0) of $F'(x)$. Successive approximations in these cases are known as *modified Newton's methods.* Again, what one takes in practice is an approximation Φ to $F'(x)$ used in the linear functional equation $\Phi(x^{n+1}-x^n)=F(x^n)$. We commence with Kantorovich's main theorem:

THEOREM 12.3.1. (*Newton's Method*) *Let $F:X\to Y$, X,Y Banach spaces, be a Fréchet differentiable function for $x\in\Omega$, an open convex set in X. Let $[F'(x^0)]^{-1}\in[Y\to X]$ at some $x^0\in\Omega$, and*

1. $\|[F'(x^0)]^{-1}\|\leqslant B$,
2. $\|[F'(x^0)]^{-1}F(x^0)\|\leqslant\eta$,
3. $\|F'(x)-F'(y)\|\leqslant K\|x-y\|, \qquad x,y\in\Omega,$

for constants B,k,η satisfying $h=Bk\eta\leqslant\frac{1}{2}$. If $t^*=[1-\sqrt{(1-2h)}]\eta/h$ and $\Omega_{t^*}=\{x\mid \|x-x^0\|\leqslant t^*\}\subset\Omega$, then the successive approximations (Newton's method iterates)

$$x^{n+1}=x^n-[F'(x^n)]^{-1}F(x^n)$$

are defined for all n, $x^n\in\Omega_{t^*}$, $n=1,2,3,\ldots$ and converge to $x^*\in\Omega_{t^*}$ which satisfies $F(x^*)=0$. Further

$$\|x^*-x^n\|\leqslant\frac{\eta}{h}\frac{[1-\sqrt{(1-2h)}]^{2^n}}{2^n}, \qquad n=0,1,2,\ldots.$$

PROOF. (Ortega [1], Tapia [1]) We consider the associated problem of finding a root of

$$p(t)=\frac{h}{2\eta}t^2-t+\eta,$$

using Newton's method with

$$s(t)=t-\frac{p(t)}{p'(t)},$$

where $t_{n+1}=s(t_n)$, $t_0=0$. Since $p(t)+p'(t)[s(t)-t]=0$,

$$\begin{aligned}t_{n+1}-t_n&=-[p'(t_n)]^{-1}p(t_n)\\&=-[p'(t_n)]^{-1}[p(t_n)-p(t_{n-1})-p'(t_{n-1})(t_n-t_{n-1})]\\&=-\frac{[(h/\eta)t_n-1]^{-1}(h/\eta)(t_n-t_{n-1})^2}{2},\end{aligned}$$

using Taylor's theorem. Thus

$$t_{n+1}-t_n=\frac{(h/2)(t_n-t_{n-1})^2}{(\eta-ht_n)}.$$

We use this problem as a *majorant* for Newton's method. (For discussions of majorant techniques see Kantorovich and Akilov [1], 697–702 or Dennis [1], 428–437 and further references there. See also Exercise 12.3.11.) From

$$\begin{aligned}F(y)-F(x)&=\int_0^1 F'(x+ty)(y-x)\,dt\\&=\int_0^1(F'(x+ty)-F'(x))(y-x)\,dt+\int_0^1 F'(x)(y-x)\,dt,\end{aligned}$$

we see that, on Ω, using condition 3,

$$\begin{aligned}\|F(y)-F(x)-F'(x)(y-x)\|&\leqslant\int_0^1\|F'(x+t(y-x))-F'(x)\|\,\|y-x\|\,dt\\&\leqslant\left(\frac{K}{2}\right)\|y-x\|^2.\end{aligned}$$

Since $t^*\leqslant 2\eta\leqslant 1/BK$, if $x\in\operatorname{int}\Omega_{t^*}$,

$$\|F'(x)-F'(x^0)\|\leqslant K\|x-x^0\|\leqslant Kt^*\leqslant\frac{1}{B}.$$

Then the proof of Theorem 6.2.3 shows that $F'(x)$ is invertible on Ω_{t^*} and

$$\|F'(x)\|\leqslant\frac{B}{[1-BK\|x-x^0\|]}.$$

We now consider the Newton function $N(x)=x-[F'(x)]^{-1}F(x)$; if $N(x)\in\operatorname{int}\Omega_{t^*}$, we have

$$\begin{aligned}\|N(N(x))-N(x)\|&=\|[F'(N(x))]^{-1}F(N(x))\|\\&\leqslant\|[F'(N(x))]^{-1}\|\,\|F(N(x))\|.\end{aligned}$$

But $F(x)+F'(x)(N(x)-x)=0$, so

$$\begin{aligned}\|F(N(x))\|&=\|F(N(x))-F(x)-F'(x)(N(x)-x)\|\\&\leqslant\left(\frac{K}{2}\right)\|N(x)-x\|^2.\end{aligned}$$

Also

$$\|[F'(N(x))]^{-1}\| \leqslant \frac{B}{[1-BK\|N(x)-x^0\|]}.$$

That is, so long as $N(x) \in \operatorname{int}\Omega_{t^*}$,

$$\|N(N(x))-N(x)\| \leqslant \frac{h\|x-N(x)\|^2}{2[\eta-h\|x^0-N(x)\|]}.$$

Now t^* is the first positive root of $p(t)$, which has no negative roots, so $p(t)>0, p'(t)<0$ on $0<t<t^*$. Therefore $t_n<t_{n+1}\to t^*$, $n\to\infty$. Since $\|x^1-x^0\|=\|[F'(x^0)]^{-1}F(x^0)\|\leqslant\eta=t_1=t_1-t_0$, if we assume inductively that for $k\leqslant n$, $\|x^k-x^{k-1}\|\leqslant t_k-t_{k-1}$, it follows that $\|x^k-x^0\|\leqslant t_k-t_0=t_k<t^*$, so $x^k\in\operatorname{int}\Omega_{t^*}$, $k=0,1,2,\ldots,n$. Then $x^n=N(x^{n-1})$ makes

$$\|x^{n+1}-x^n\| \leqslant \frac{h\|x^n-x^{n-1}\|^2}{2[\eta-h\|x^0-x^n\|]}$$

$$\leqslant \frac{h(t_n-t_{n-1})^2}{2[\eta-ht_n]} = t_{n+1}-t_n.$$

That is, for all n,m, $\|x^{n+1}-x^n\|\leqslant t_{n+1}-t_n$, $\|x^n-x^0\|<t^*$, and $\|x^{n+m}-x^n\|\leqslant t_{n+m}-t_n\to 0$, $n,m\to\infty$. Thus the Newton iterations are all defined and converge to $x^*\in\Omega_{t^*}$, which is a solution, since $F(x)$ and $F'(x)$ are continuous on Ω_{t^*}. Also $\|x^*-x^n\|\leqslant t^*-t_n$.

The sequence t_n depends upon h. Both $ht_1(h)$ and $h[t_1(h)-t_0(h)]$ are increasing functions of h. (Why?) An easy inductive argument shows that $h[t_n(h)-t_{n-1}(h)]$ and $ht_n(h)$ are increasing functions of h for all n. When $h=\frac{1}{2}$, $t^*=2\eta$ and $s(t)=(t+2\eta)/2$, so $t_n=2\eta(1-2^{-n})$. Then $\eta-ht_n\geqslant\eta/2^n$ for all $h\leqslant\frac{1}{2}$, all n, with equality when $h=\frac{1}{2}$. On the other hand, since $p(t^*)=0$, $t^{*2}=(2\eta/h)(t^*-\eta)$; then a straightforward calculation shows that

$$(t^*-t_n)^2=\left[\frac{2(\eta-ht_n)}{h}\right][t^*-t_{n+1}].$$

That is,

$$t^*-t_{n+1}=\frac{h(t^*-t_n)^2}{2(\eta-ht_n)}$$

$$\leqslant \frac{h}{2\eta}(t^*-t_n)^2.$$

It is known from other proofs of this theorem that there is a bound of the form

$$t^* - t_n \leqslant \frac{(\eta/h)\alpha^{2^n}}{2^n}$$

where $\alpha = \alpha(h)$. If we assume this here for t_j, $j=0,1,\ldots,n$, then

$$t^* - t_{n+1} \leqslant \frac{(h/\eta)2^{n-1}(\eta^2/h^2)\alpha^{2^{n+1}}}{2^{2n}}$$

$$\leqslant \frac{(\eta/h)\alpha^{2^{n+1}}}{2^{n+1}}.$$

Thus it is true for all n, since $t^* - t_0 = (\eta/h)\,[1-\sqrt{(1-2h)}]$. This same identity shows that $\alpha(h) = 1-\sqrt{(1-2h)}$ is the best value we can achieve. ■

Under the same conditions as for Theorem 12.3.1, we can show that if $\Gamma(x) \equiv \Gamma_0 = [F'(x^0)]^{-1}$ and $h < \frac{1}{2}$, successive approximations, known as the modified Newton's method, converge. The proof utilizes simpler techniques and rests directly on Theorem 1.4.1.

THEOREM 12.3.2. (*Modified Newton's Method*) *Under the same conditions as in Theorem* 12.3.1, *if* $h < \frac{1}{2}$, *then the iterations*

$$\hat{x}^{n+1} = \hat{x}^n - \Gamma_0 F(\hat{x}^n)$$

are defined for all n, *for any* $\hat{x}^0 \in \Omega_{t^*}$, *and converge to a root* $x^* \in \Omega_{t^*}$. *Further,* $F(x)=0$ *has a unique root in* Ω_{t^*}. *Also,*

$$\|\hat{x}^n - x^*\| \leqslant 2\left(\frac{\eta}{h}\right)[1-\sqrt{(1-2h)}]^{n+1}; \qquad n=1,2,3,\ldots.$$

PROOF. Let $M(x) = x - \Gamma_0 F(x)$. Then $M'(x) = I - \Gamma_0 F'(x)$, so

$$\|M(x) - M(x^0)\| = \left\| \int_0^1 [I - \Gamma_0 F'(x^0 + t(x-x^0))](x-x^0)\,dt \right\|$$

$$= \left\| \int_0^1 \Gamma_0 [F'(x^0) - F'(x^0 + t(x-x^0))](x-x^0)\,dt \right\|$$

$$\leqslant BK \int_0^1 t\|x-x^0\|^2\,dt = \tfrac{1}{2} BK \|x-x^0\|^2.$$

If $\|x-x^0\| \leqslant t^*$, then $\|M(x)-M(x^0)\| \leqslant (t^*)^2h/2\eta$, while $(t^*)^2h/2\eta - t^* + \eta = 0$. That is,

$$\|M(x)-M(x^0)\| \leqslant t^* - \eta, \qquad \|x-x^0\| \leqslant t^*.$$

But

$$\|M(x^0)-x^0\| = \|\Gamma_0 F(x^0)\| \leqslant \eta.$$

Then

$$\|M(x)-x^0\| \leqslant t^*, \qquad \|x-x^0\| \leqslant t^*.$$

Thus M maps Ω_{t^*} into itself. Further, if $x,y \in \Omega_{t^*}$, then, proceeding as above we have

$$\begin{aligned}\|M(x)-M(y)\| &= \left\| \int_0^1 M'(x+t(y-x))(y-x)\,dt \right\|,\\ &\leqslant BK \int_0^1 \max(\|x-x^0\|, \|y-x^0\|)\|y-x\|\,dt,\\ &\leqslant BKt^*\|y-x\|,\\ &\leqslant \left(\frac{t^*h}{\eta}\right)\|y-x\|.\end{aligned}$$

But $(h/\eta)t^* = (h/\eta)(\eta/h)[1-\surd(1-2h)] = 1-\surd(1-2h) < 1$. We can apply Theorem 1.4.1 starting anywhere in Ω_{t^*}; the successive approximations $\hat{x}^n$ converge to a solution x^* of $F(x)=0$ unique in Ω_{t^*}, and

$$\|x^*-\hat{x}^n\| \leqslant 2\left(\frac{\eta}{h}\right)[1-\surd(1-2h)]^{n+1}, \qquad n=1,2,3,\ldots,$$

since $\|x^*-\hat{x}^0\| \leqslant 2t^*$. ■

In fact, the uniqueness statement is not the strongest possible. Using majorant methods, we can prove the following:

THEOREM 12.3.3. *Under the same hypotheses as in Theorem* 12.3.1, *let*

$$t^{**} = \left(\frac{\eta}{h}\right)[1+\surd(1-2h)]$$

and

$$\Omega_{t^{**}} = \{x \mid \|x-x^0\| \leqslant t^*\} \subset \Omega.$$

If $h<\frac{1}{2}$, $\left(h=\frac{1}{2}\right)$, *the solution of* $F(x)=0$ *is unique in* $\operatorname{int}\Omega_{t^{**}}(\Omega_{t^{**}})$, *and the modified Newton iterates converge to it for any* $x^0\in\operatorname{int}\Omega_{t^{**}}(\Omega_{t^{**}})$.

PROOF. Let $\phi(t)=p(t)+t$; then $\phi(t)=t$ is equivalent to $p(t)=0$. But $\phi(t)=(h/2\eta)t^2+\eta>t$ on $[0,t^*)$, and $\phi(t)<t$ on (t^*,t^{**}), $\phi(t^*)=t^*$, $\phi(t^{**})=t^{**}$, and $\phi(t)$ is strictly increasing on $[0,t^{**}]$. (Why?) If $0<\tilde{t}_0<t^*$, then $\tilde{t}_{n+1}=\phi(\tilde{t}^n)\uparrow t^*$, $n\to\infty$, while if $t^*\leqslant\tilde{t}_0<t^{**}$, $\tilde{t}_{n+1}=\phi(\tilde{t}^n)\downarrow t^*$, $n\to\infty$. If $h<\frac{1}{2}$, it is sufficient to consider the case $t^*\leqslant\tilde{t}_0<t^{**}$. (If $h=\frac{1}{2}$, then $t^*=t^{**}$, and this case does not occur.) Then Theorems 12.1.5 and 12.1.6 show that if $\|\tilde{x}^0-x^0\|\leqslant\tilde{t}_0-t_0=\tilde{t}_0$, then

$$M(\tilde{x}^0)-M(x^0)=\int_{x^0}^{\tilde{x}^0}M'(x)\,dx.$$

But

$$\begin{aligned}\|M'(x^0+\tau(\tilde{x}^0-x^0))\|&=\|\Gamma_0[F'(x^0)-F'(x^0+\tau(\tilde{x}^0-x^0))]\|\\&\leqslant BK\tau\|\tilde{x}^0-x^0\|\leqslant\left(\frac{h}{\eta}\right)\tau\tilde{t}_0.\end{aligned}$$

So

$$\begin{aligned}\|M(\tilde{x}^0)-M(x^0)\|&\leqslant\left[\left(\frac{h}{2\eta}\right)\tau^2+\eta\right]_0^{\tilde{t}_0}\\&\leqslant\phi(\tilde{t}_0)-\phi(t_0).\end{aligned}$$

That is $\|\tilde{x}^1-x^1\|\leqslant\tilde{t}_1-t_1$ and $\|\tilde{x}^1-x^0\|\leqslant\tilde{t}_1-t_1+t_1-t_0=\tilde{t}_1\leqslant\tilde{t}_0$, so $\tilde{x}^1\in\operatorname{int}\Omega_{t^{**}}$. Inductively we assume $\tilde{x}^0,\tilde{x}^1,\ldots,\tilde{x}^n\in\operatorname{int}\Omega_{t^{**}}$ with $\|\tilde{x}^k-x^k\|\leqslant\tilde{t}_k-t_k$. Then using Theorems 12.1.5 and 12.1.6 again we write

$$\begin{aligned}\|M(\tilde{x}^n)-M(x^n)\|&=\left\|\int_{x^n}^{\tilde{x}^n}M'(x^n+\tau(\tilde{x}^n-x^n))\,d\tau\right\|\\&\leqslant(\tilde{t}_n-t_n)\int_0^1\|M'(x^n+\tau(\tilde{x}^n-x^n))\|\,dt\\&\leqslant\int_{t_n}^{\tilde{t}_n}\phi'(\tau)\,d\tau,\end{aligned}$$

since

$$\begin{aligned}\|x^n+\tau(\tilde{x}^n-x^n)-x^0\|&\leqslant\tau(\tilde{t}_n-t_n)+(t_n-t_{n-1})+\cdots+(t_1-t_0)\\&\leqslant t_n+\tau(\tilde{t}_n-t_n).\end{aligned}$$

That is, $\|\tilde{x}^{n+1}-x^{n+1}\|=\|M(x^n)-M(x^n)\|\leqslant\phi(\tilde{t}_n)-\phi(t_n)=\tilde{t}_{n+1}-t_{n+1}$ while $\|\tilde{x}^{n+1}-x^0\|\leqslant\tilde{t}_{n+1}-t_{n+1}+(t_{n+1}-t_n)+\cdots+(t_1-t_0)\leqslant\tilde{t}_0$. Since $\tilde{t}_n\downarrow t^*$ and $t_n\uparrow t^*$, $n\to\infty$, it follows that $\{\tilde{x}^n\}$ is a Cauchy sequence converging to x^*, which is the unique solution of $F(x)=0$ on $\|x-x_0\|\leqslant\tilde{t}_0$, for any $\tilde{t}_0<t^{**}$.

When $h=\frac{1}{2}$, $t^*=t^{**}$, and $\phi(t^*)=t^*$. In this case we repeat the preceeding argument with $\tilde{t}_0=t^*$, so $\tilde{t}_n=t^*$, $n=0,1,2,\ldots$. ■

The modified Newton's method has proven to be robust under perturbation of the initial point $\hat{x}^0$. In the next theorem we establish robustness for Newton's method.

THEOREM 12.3.4. *Let $F:X\to Y$ be a Fréchet differentiable function for $x\in\Omega$, an open convex set. Let $\Gamma\in[Y\to X]$, $\Gamma^{-1}\in[X\to Y]$, and $x^0\in\Omega$ satisfy*

1. $\|\Gamma\|\leqslant B^1$,
2. $\|\Gamma F(x^0)\|\leqslant\eta^1$,
3. $\|\Gamma F'(x^0)-I\|\leqslant\delta<1$,
4. $\|F'(x)-F'(y)\|\leqslant K^1\|x-y\|, \qquad x,y\in\Omega.$

If $h^1=B^1K^1\eta^1/(1-\delta)^2\leqslant\frac{1}{2}$, $t^*=[1-\sqrt{(1-2h^1)}][\eta^1/h^1(1-\delta)]$ and $\Omega_{t^*}=\{x|\,\|x-x^0\|\leqslant t^*\}\subset\Omega$, then Newton's method starting at x^0 produces a sequence x^n converging to a root x^* of $F(x)=0$ with

$$\|x^*-x^n\|\leqslant\frac{1}{2^n}[\eta^1/h^1(1-\delta)][1-\sqrt{(1-2h^1)}]^{2^n}.$$

The solution x^* is unique in the open sphere $\|x-x^0\|<[\eta^1/h^1(1-\delta)][1+\sqrt{(1-2h^1)}]$, if $h^1<\frac{1}{2}$, and is unique in Ω_{t^*} if $h^1=\frac{1}{2}$.

PROOF. In fact what we show is that from the existence of an approximate inverse Γ to $F'(x^0)$ we conclude the existence of $[F'(x^0)]^{-1}$. Condition 2 and Theorem 6.2.2 imply that $\Gamma F'(x^0)$ is invertible with

$$\left\|[\Gamma F'(x^0)]^{-1}\right\|\leqslant\frac{1}{1-\delta}.$$

Then $\Gamma^{-1}[\Gamma F'(x^0)]=F'(x^0)$ shows that $[F'(x^0)]^{-1}\in[Y\to X]$ and is given by $[\Gamma F'(x^0)]^{-1}\Gamma$. Then

$$\left\|[F'(x^0)]^{-1}F(x^0)\right\|\leqslant\frac{\eta^1}{1-\delta}$$

and

$$\|[F'(x^0)]^{-1}\| \leqslant \frac{B^1}{1-\delta}.$$

If we set $\eta=\eta^1/(1-\delta), B=B^1/(1-\delta), K=K^1$, then the conditions of Theorems 12.3.1 and 12.3.3 hold. ■

COROLLARY 1. *Under the conditions of the theorem, the modified Newton's iteration*

$$\hat{x}^{n+1}=\hat{x}^n-\Gamma F(\hat{x}^n), \qquad n=0,1,2,\ldots,$$

is defined for all n and $\hat{x}^n \to x^*$.

The proof of the corollary is a similar modification of the proof of Theorem 12.3.3.

COROLLARY 2. *Under the conditions of Theorem* 12.3.1, *let* $\tilde{x}^0 \in \operatorname{int}\Omega_{t^*}$ *and*

$$\|\tilde{x}^0-x^0\|=\epsilon \leqslant \frac{(1-2h)}{4BK}.$$

Then Newton's method starting at $\tilde{x}^0$ converges to x^*.

PROOF. Let $\Gamma=[F'(x^0)]^{-1}$. Then

$$F(\tilde{x}^0)=F(x^0)+\int_{x^0}^{\tilde{x}^0} F'(x^0+\tau[\tilde{x}^0-x^0])\,dx,$$

$$=F(x^0)+F'(x^0)[\tilde{x}^0-x^0]+\int_{x^0}^{\tilde{x}^0}[F'(x^0+\tau[\tilde{x}^0-x^0])-F'(x^0)]\,dx.$$

So

$$\|\Gamma F(\tilde{x}^0)\| \leqslant \eta+\epsilon+\tfrac{1}{2}BK\epsilon^2=\eta^1.$$

Also $\|\Gamma F'(\tilde{x}^0)-I\|=\|\Gamma[F'(\tilde{x}^0)-F'(x^0)]\| \leqslant BK\epsilon=\delta$. Set $B^1=B, K^1=K$; a short calculation shows that $h^1=\frac{1}{2}$. Thus we can apply Corollary 1. ■

We have established that, either with exact calculation of $\Gamma(x)=[F'(x^n)]^{-1}$ at each step, or with a fixed approximate calculation of $\Gamma(x)=[F'(x^0)]^{-1}$, the Newton's or modified Newton's method are robust with respect to starting values. One last situation, that in which $\Gamma(x^n) = A_n^{-1}$ where A_n is an approximation to $F'(x^n)$, the so called *Newton-like* or *quasi-Newton* methods, has been treated extensively by Dennis [1] using majorant methods. That this is a natural formulation follows from the earlier observation that the linear system in these methods is solved, usually, without actual inversion of the linear operator involved. Results similar in structure to those above are obtained.

EXERCISES TO SECTION 12.3

1. In Banach spaces X, Y, let $F: X \to Y$ be Fréchet differentiable on a convex set $\Omega \subset X$, and let $GD^2F(x; g, h)$ exist on Ω and satisfy $\|GD^2F(x; g, h)\| \leqslant K\|g\|\,\|h\|$, $x \in \Omega$, $g, h \in X$. Show that $\|F'(x) - F'(y)\| \leqslant K\|x - y\|$ for all $x, y \in \Omega$.
2. In Exercise 12.3.1, show that if $F''(x)$ exists on Ω and $\|F''(x)\| \leqslant K$, then $\|F'(x) - F'(y)\| \leqslant K\|x - y\|$ for all $x, y \in \Omega$.
3. Show that if $b \neq 0$ and $b^2 - 4|ac| \geqslant 0$, then for $F(x) = ax^2 + bx + c$, Newton's method always converges with the starting value $x^0 = 0$.
4. In Exercise 12.3.3, how far from $x^0 = 0$ can one start with guaranteed convergence?
5. Generalize Exercise 12.3.3 to the case of a single complex valued function of a single complex variable.
6. Prove Corollary 1 to Theorem 12.3.4.
7. For the differential equation of Exercise 12.2.10, we can solve the initial value problem $x(0) = c$ by Newton's method. Set up the iteration and compute $[F'(x)]^{-1}$ explicitly. If $x(0) = 0$ and $f(t, x(t)) = 1 + \sin x(t)$, start at $x^0(t) = t$ and compute $x^1(t)$ explicitly. (Hint: Rall [1], 169–172.) Estimate $\|x^1 - x^*\|$.
8. Redo Exercise 12.3.7, for the integral equation formulation of Exercise 12.2.11.
9. By analogy with roots of functions, an x^* such that $F(x^*) = 0$ is called a *simple* root if $[F'(x^*)]^{-1} \in [Y \to X]$. In Theorem 12.3.1 show that if $h < \frac{1}{2}$, then x^* is a simple root. By considering the scalar equation $t^2 = 0$, $t_0 \neq 0$, show that when $h = \frac{1}{2}$ this is no longer automatically true.
10. Show that if x^* is a simple root and $F'(x)$ is continuous at x^*, then $\|N(x) - N(x^*)\| / \|x - x^*\| \to 0$, $x \to x^*$, hence that $N'(x^*)$ exists and is zero. (Hint: $N(x) - N(x^*) = \Gamma(x)[F'(x^*)(x - x^*) - F(x) + F(x^*)] + \Gamma(x)[F'(x) - F'(x^*)](x - x^*)$. See Krasnosel'skii, Vainikko, *et al.* [1], 139–140.)

11. The proofs of Theorems 12.3.1 and 12.3.3 contain the idea of the proofs of the following theorem which is central to the *majorant method*: *Let* $S(x)$ *be Fréchet differentiable on a ball* $\Omega_\rho = \{x \mid \|x - x^0\| \leqslant \rho\}$. *Let there exist a function* $\psi(x)$ *on* $[t_0, t_0 + \tau]$, $\tau < \rho$ *such that* $\psi'(t)$ *is continuous and*

    1. $\|S(x^0) - x^0\| \leqslant \psi(t_0) - t_0$;
    2. $\|S'(x)\| \leqslant \psi'(t)$, *if* $\|x - x^0\| \leqslant t - t_0 \leqslant \tau$.

    If $t = \psi(t)$ *has a root on* $[t_0, t_0 + \tau]$, *then the successive approximations* $x^{n+1} = S(x^n)$ *converge to an* $x^* = S(x^*)$ *with* $\|x^* - x^0\| \leqslant t^* - t_0$, *where* t^* *is the least root of* $\psi(t) = t$ *on* $[t_0, t_0 + \tau]$. *If* $t = \psi(t)$ *has only one root on* $[t_0, t_0 + \tau]$, *then the solution of* $x = S(x)$ *is unique and is achieved from any starting value in* $\|x - x^0\| \leqslant \tau$. Prove this.

12.4 IMPLICIT FUNCTION THEOREMS

Closely related to the ideas used in studying Newton's method are those used in studying implicit functions, which are solutions $y = F(x)$ of functional equations of the form $G(x,y) = 0$. As in Euclidean space, these involve the concept of partial differentiation.

DEFINITION 12.4.1. Let X, Y, Z be Banach spaces, $G(x,y)$ a functional on $X \otimes Y$ with range in Z. If, for a fixed value of $x \in X$, $G(x,y)$ is Fréchet differentiable with respect to y, we call this derivative *the partial derivative of G with respect to y* and write either $G_y(x,y)$ or $D_y G(x,y)$.

A similar definition holds for $G_x(x,y) = D_x G(x,y)$. We can prove the following implicit function theorem:

THEOREM 12.4.1. *Let* $G : X \otimes Y \to Z$ *be defined on a ball* $\Omega_r = \{x \mid \|x - x^0\| \leqslant r, \|y - y^0\| \leqslant r\}$ *and satisfy*

1. $G(x^0, y^0) = 0$;
2. $G_y(x,y)$ *exists at each* $(x,y) \in \Omega_r$ *and is continuous in* (x,y);
3. $G(x,y)$ *is continuous on* Ω_r;
4. $\left[G_y(x^0, y^0) \right]^{-1}$ *exists in* $[Z \to Y]$.

Then there exist positive numbers δ, γ *and a continuous mapping* $y = F(x) : X \to Y$ *on* $\|x - x^0\| \leqslant \delta \leqslant r$ *satisfying*

(a) $\|F(x) - y^0\| \leqslant \gamma \leqslant r$;
(b) $F(x^0) = y^0$; and
(c) $G(x, F(x)) \equiv 0$.

PROOF. We first fix x, and consider the mapping

$$\Phi(y) = y - \Gamma G(x,y)$$

where

$$\Gamma = \left[G_y(x^0,y^0) \right]^{-1}.$$

Then $\Phi(y)$ has a Fréchet derivative for $\|y-y^0\| \leqslant r$ and

$$\begin{aligned}\|\Phi'(y)\| &= \|I - \Gamma G_y(x,y)\| \\ &= \|\Gamma[G_y(x^0,y^0) - G_y(x,y)]\| \\ &\leqslant \|\Gamma\| \|G_y(x^0,y^0) - G_y(x,y)\|.\end{aligned}$$

If we choose any $\alpha > 0$, there is a $\delta = \delta(\alpha) \leqslant r$ such that $\|G_y(x^0,y^0) - G_y(x,y)\| \leqslant \alpha / \|\Gamma\|$ if $(x,y) \in \Omega_\delta$. Then on Ω_δ, $\|\Phi(y^1) - \Phi(y^2)\| \leqslant \alpha \|y^1 - y^2\|$. If we require $\alpha < 1$, Φ is a contraction. Further

$$\begin{aligned}\|\Phi(y) - y^0\| &\leqslant \|\Phi(y) - \Phi(y^0)\| + \|\Phi(y^0) - y^0\| \\ &\leqslant \alpha \|y - y^0\| + \|\Gamma\| \, \|G(x,y^0)\|.\end{aligned}$$

If we further decrease δ, then for any $\gamma > 0$, $\|G(x,y^0)\| = \|G(x,y^0) - G(x^0,y^0)\| \leqslant (1-\alpha)\gamma / \|\Gamma\|$ provided that $\|x - x^0\| \leqslant \delta$. Then $\|\Phi(y) - y^0\| \leqslant \alpha\gamma + (1-\alpha)\gamma = \gamma$ on $\|y - y^0\| \leqslant \gamma$. That is, Φ maps $\|y - y^0\| \leqslant \gamma$ into itself, provided that $\|x - x^0\| \leqslant \delta$. The contraction mapping Theorem 1.4.1 implies that in $\|y - y^0\| \leqslant \gamma$ there is a unique solution $y = F(x)$ to $y = \Phi(y)$ for each x in $\|x - x^0\| \leqslant \delta$.

Now from the contraction mapping Theorem 1.4.1 we know that the successive approximations $\hat{y}^{n+1}(x) = \hat{y}^n(x) - \Gamma G(x, \hat{y}^n(x))$, $n = 0, 1, 2, \ldots$, converge to $F(x)$ for any starting value $\hat{y}^0$ in $\|y - y^0\| \leqslant \delta$, and that

$$\|F(x) - \hat{y}^0\| \leqslant (1-\alpha)^{-1} \|\hat{y}^1(x) - \hat{y}^0\|.$$

If we set $\hat{y}^0 = F(x^*)$ for any x^* in $\|x - x^0\| \leqslant \delta$, then

$$\|F(x) - F(x^*)\| \leqslant (1-\alpha)^{-1} \|\Gamma\| \, \|G(x, F(x^*))\|.$$

But if $\|x - x^*\| \to 0$, then $\|G(x, F(x^*))\| \to \|G(x^*, F(x^*))\| = 0$, so $\|F(x) - F(x^*)\| \to 0$. Thus F is continuous. ■

COROLLARY 1. *If in addition $G_x(x,y)$ exists on Ω_r and is continuous at (x^0,y^0), then $y=F(x)$ has a Fréchet derivative at x^0 given by*

$$F'(x^0)=-\left[G_y(x^0,y^0)\right]^{-1}G_x(x^0,y^0).$$

PROOF. We set $x=x^0+h$ and $g(h)=F(x)-y^0$. Then $g(0)=0$ and

$$\|g(h)+\Gamma G_x(x^0,y^0)h\|\leqslant\|\Gamma\|\,\|G_y(x^0,y^0)g(h)+G_x(x^0,y^0)h\|,$$

while

$$\begin{aligned}&G_y(x^0,y^0)g(h)-G_x(x^0,y^0)h\\&\qquad=-G(x^0+h,y^0+g(h))+G(x^0,y^0)+G_y(x^0,y^0)g(h)+G_x(x^0,y^0)h,\end{aligned}$$

since the added terms both vanish. Then if θ_1,θ_2 are numbers in (0, 1),

$$\begin{aligned}&\|G_y(x^0,y^0)g(h)+G_x(x^0,y^0)h\|\\&\qquad\leqslant\|G(x^0+h,y^0+g(h))-G(x^0,y^0+g(h))-G_x(x^0,y^0)h\|\\&\qquad\quad+\|G(x^0,y^0+g(h))-G(x^0,y^0)-G_y(x^0,y^0)g(h)\|\\&\qquad\leqslant\sup_{\theta_1,\theta_2}\|G_x(x^0+\theta_1h,y^0+\theta_2g(h))-G_x(x^0,y^0)\|\,\|h\|\\&\qquad\quad+\sup_{\theta_1,\theta_2}\|G_y(x^0+\theta_1h,y^0+\theta_2g(h))-G_y(x^0,y^0)\|\,\|g(h)\|.\end{aligned}$$

Thus, using the continuity of G_x,G_y, for any $\epsilon>0$ there is a $\delta=\delta(\epsilon)$ such that, on $\|x-x^0\|\leqslant\delta$,

$$\begin{aligned}\|g(h)+\Gamma G_x(x^0,y^0)h\|&\leqslant\|\Gamma\|\epsilon(\|h\|+\|g(h)\|)\\&\leqslant\|\Gamma\|\epsilon(\|h\|+\|-\Gamma G_x(x^0,y^0)h\|\\&\qquad+\|g(h)+\Gamma G_x(x^0,y^0)h\|).\end{aligned}$$

Hence

$$\|g(h)+\Gamma G_x(x^0,y^0)h\|\leqslant\Gamma\epsilon\frac{1+\|\Gamma G_x(x^0,y^0)\|}{1-\|\Gamma\|\epsilon}\|h\|$$

and the coefficient of $\|h\|$ on the right can be made as small as we please as $\|h\| \to 0$. Thus

$$\|F(x)-[F(x^0)-\Gamma G_x(x^0,y^0)(x-x^0)]\| = o(\|x-x^0\|). \quad \blacksquare$$

It is possible to extend this theorem to the computation of higher derivatives of F, provided that G is sufficiently smooth. (See, e.g., Lusternik and Sobolev [1], 317–319 or Krasnosel'skii, Vainikko *et al.* [1], 329–332.)

We will apply this result to the initial value problem for ordinary differential equations. Let $C_n^1[0,1]$ denote the space of continuously differentiable functions $y(t)$ from $[0,1]$ to $\mathbf{E}^n$, normed by

$$\|y\| = \max_{1 \leqslant i \leqslant n} \max_{0 \leqslant t \leqslant 1} \max(|y_i(t)|, |y_i'(t)|).$$

Let $f(t,y)$ map $[0,1]\otimes\mathbf{E}^n$ to $\mathbf{E}^n$ and be continuous, with continuous Fréchet derivative $f_y(t,y)$ there. Then the initial value problem $dy/dt = f(t,y)$, $y(0)=x$, is equivalent to setting $z(t)=0$ in

$$G(x,y) = y(t) - x - \int_0^t f(x,y(s))\,ds = z(t).$$

G maps $\mathbf{E}^n \otimes C_n^1[0,1]$ to $C_n^1[0,1]$. Suppose now that at $x=x^0$, there is a function $y^0(t)$ such that $dy^0(t)/dt = f(t,y^0(t))$, $y^0(t)=x^0$ on $[0,1]$, so that $G(x^0,y^0)=0$. We recognize that

$$G_y(x^0,y^0)h = h(t) - \int_0^t f_y(s,y^0(s))h(s)\,ds.$$

Invertibility of $G_y(x^0,y^0)$ follows from Exercises 6.2.4–6. From $G_y(x^0,y^0)h = z$ we must have the *variational equation* $h'(t) - f_y(t,y^0(t))h(t) = z'(t)$, and $h(0)=z(0)$, which is uniquely solvable for all z in $C_n^1[0,1]$, with

$$h(t) = \Phi(t,0)z(0) + \int_0^t \Phi(t,s)z'(s)\,ds,$$

and $\|h\| \leqslant \max_{t,s}\|\Phi(t,s)\|\,\|z\|$. Thus we have all the conditions needed to imply the existence of $y=y(x)$ solving $G(x,y(x))=0$ on some region $\|x-x^0\| \leqslant \delta$, with the existence of y_x guaranteed by the corollary. That is, from the unique solvability of the *variational equation* for all initial conditions and forcing functions, there follows the unique solvability of the initial value problem for all nearby initial values; further, as a function of the initial values, $y(t,x)$ is differentiable, and satisfies the matrix differential equation

$$\frac{dy_x(t,x)}{dt} = f_y(t,y^0(t))y_x(t,x), \qquad y_x(0,x) = I.$$

(This follows by an easy calculation from Exercise 12.4.1.)

It should be clear from this example that there is a marked difference between the implicit function theorem, which assumes the existence of a solution at a point, and extends it to some ill-defined neighborhood of that point, and Newton's method which does not assume a solution, and which, under suitable conditions, gives a precise region for unique solvability. In the next theorem we find an intermediate situation, one which occurs naturally in problems dependent in a linear fashion on a parameter.

THEOREM 12.4.2. *Let X, Y, Z be Banach spaces, $P(x): X \to Z, R(x): X \to Y$, $\mu \in [Y \to Z]$, where P, Q are Fréchet differentiable for $x \in \Omega$, an open convex set in X, and*

1. $P(x^0)=0;\ x^0 \in \Omega;$
2. $\Gamma = [P'(x^0)]^{-1}$ *exists in* $[Z \to X]$;
3. $\|\Gamma\| \leqslant B;$
4. $\|R(x^0)\| \leqslant \eta;$
5. $R'(x^0)$ *exists and* $\|R'(x^0)\| \leqslant \alpha;$
6. $\|P'(x) - P'(\bar{x})\| \leqslant K\|x - \bar{x}\|,\ x, \bar{x} \in \Omega;$
7. $\|R'(x) - R'(\bar{x})\| \leqslant L\|x - \bar{x}\|,\ x, \bar{x} \in \Omega.$

If $h_\mu = B^2\eta(K + L\|\mu\|)\|\mu\|/(1-\alpha B\|\mu\|)^2 \leqslant \frac{1}{2}$, then $F(x,\mu) = P(x) + \mu R(x) = 0$ has a unique solution $x = x^*(\mu)$ on $\|\mu\| \leqslant 1/\alpha B$, provided that $\{x \mid \|x - x^0\| \leqslant [1-(1-2h_\mu)^{1/2}]B\|\mu\|\eta/h_\mu(1-\alpha B\|\mu\|)\} \subset \Omega$.

PROOF. We appeal to Theorem 12.3.1 with

$$\Gamma = [P'(x^0)]^{-1}, \text{ so } B^1 = B,$$

$$\eta^1 = \|\Gamma \mu R(x^0)\| \leqslant B\|\mu\|\eta,$$

$$\delta_1 = \|I + \Gamma\mu R'(x^0) - I\| \leqslant B\|\mu\|\alpha,$$

$$K^1 = K,$$

and

$$h_\mu = \frac{B(K + L\|\mu\|)B\|\mu\|\eta}{[(1-\alpha B)\|\mu\|]^2}. \quad \blacksquare$$

We will apply this result to a problem in the theory of nonlinear oscillations. It is known from experiment that Duffing's equation has subharmonic solutions: this is equivalent to the mathematical assertion that for any $n=2,3,4,\ldots$, the normalized equation

$$x''+x+\mu(x^3-f\cos nwt)=0,$$

has a periodic solution of period $2\pi/w$. For convenience set $T=2\pi/w$, and pose the problem of the existence of such solutions in the space X of twice continuously differentiable functions $x(t)$ that satisfy $x(t)\equiv x(t+T)$, and take as norm in X:

$$\|x\|=\max_t|x(t)|+\max_t|x'(t)|+\lambda\max_t|x''(t)|.$$

Here $\lambda\in(0,1)$ is to be specified below. We take Z to be the space of periodic continuous functions of period T, with $\|z\|=\max|z(t)|$. $P(x)$ is now the differential operator $x''+x$, and $P(x)=0$ has the unique solution $x=0$ in X, provided that $T\neq 2j\pi$, $j=1,2,3,\ldots$. With that restriction we take $x^0=0$. Since P is linear, $P'=P$, so the Lipschitz constant for P', $K=0$. (That $P'\in[X\to Z]$ follows from $|x''+x|\leqslant(1/\lambda)\|x\|$.)

To find Γ, we solve $P(x)=z$ subject to $x(0)=x(T)=a, x'(0)=x'(T)=b$. It is a simple calculation to show that

$$x(t)=a\cos t+b\sin t+\int_0^t\sin(t-s)z(s)\,ds,$$

$$x'(t)=-a\sin t+b\cos t+\int_0^t\cos(t-s)z(s)\,ds.$$

Another straightforward calculation using $a=x(T), b=x'(T)$ determines a and b in terms of z, and shows that if $\|z\|\leqslant 1$, then both $|a|$ and $|b|$ are not greater than $(|1-\cos T|+|\sin T|)/2(1-\cos T)$, while

$$\|x\|\leqslant 2(|a|+|b|+T)\lambda\max_t|x''(t)|.$$

Since $|x''|\leqslant|z|+|x|\leqslant 1+\|x\|$, we have $\|x\|\leqslant(1-\lambda)^{-1}\times[2(|a|+|b|+T)+\lambda]$. Also $2(|a|+|b|+T)\leqslant(3+\sqrt{10})T/(1-\cos T)$, so that

$$\|\Gamma\|=\left\|[P'(x^0)]^{-1}\right\|\leqslant(1-\lambda)^{-1}\left[\frac{(3+\sqrt{10})T}{(1-\cos T)}+\lambda\right].$$

We call the right side B_λ, to be used for B in applying the theorem. Now $R(x)=x^3-f\cos nwt$, so $\|R(0)\|=|f|$ which we use for η. Since $R'(x)=3x^2$,

$\|R''(x^0)\|=0$, so $\alpha=0$. On any ball $\Omega_\beta=\{x|\,\|x\|\leqslant\beta\}$, $\|R'(x)-R'(\bar{x})\|\leqslant 6\beta\|x-\bar{x}\|$. Take $L=6\beta$ and set

$$h_{\mu,\lambda}=B_\lambda^2|f|6\beta|\mu|^2.$$

If $h_{\mu,0}<\frac{1}{2}$, then for any sufficiently small $\lambda>0$, $h_{\mu,\lambda}\leqslant\frac{1}{2}$, and we can apply the theorem. But $h_{\mu,0}<\frac{1}{2}$ will hold for

$$|\mu|<\frac{2}{\sqrt{(6\beta|f|)}\,B_0}$$
$$<\frac{2(1-\cos T)}{\sqrt{(6\beta|f|)}(3+\sqrt{10}\,)T}.$$

Left as an exercise is the determination of a value of β such that for all μ in this range, the solution $x(\mu)$ is in Ω_β.

We have established the existence of subharmonic solutions for all values of $n=2,3,\ldots$, and for all $T\neq 2j\pi, j=1,2,3,\ldots$; by comparison, most existence proofs work for a specific value of n (see, e.g., the case $n\equiv 3$ in Holzman [1]). We used only the fact that the $\cos nwt$ is of period T for all n.

EXERCISES TO SECTION 12.4

1. Extend the chain rule of differentiation to the case of $G(x,y)$: $X\otimes Y\to Z, x=\phi(w): W\to X, y=\psi(w): W\to Y$ where G_x, G_y, ϕ_w, and ψ_w exist at suitable places, and show that $G_x\phi_w+G_y\psi_w=D_wG(\phi(w),\psi(w))$.
2. Let $k(s,t,u)$ be defined on $\{0\leqslant s,t\leqslant 1,\ -\infty<u<\infty\}$, continuous, and differentiable with respect to u there. Let

$$k_u(s,t,u)\leqslant k_0(s,t)[a(s)+b|u|]$$

where $a(s)\in L_2(0,1)$ and both

$$\int_0^1|k_0(s,t)|\,dt\in L_2(0,1)$$

and

$$\int_0^1 k(s,t,0)\,dt\in L_2(0,1).$$

Show that, for all sufficiently small λ, the *Uryson integral operator* equation

$$x(s)=\lambda\int_0^1 k(s,t,x(t))\,dt,$$

has a unique *small* solution $x(s;\lambda)\in L_2(0,1)$. (Compare with Exercise 1.5.3.)

3. When $G(x,y)=F(y)-x$, the implicit function theorem and its corollary become an inverse function theorem. State and prove it.
4. For the treatment of Duffing's equation, show that the system of ordinary differential equations $x_1'=-x_2, x_2'=x_1$ has a fundamental matrix,

$$\begin{bmatrix} \cos(t-s) & \sin(t-s) \\ -\sin(t-s) & \cos(t-s) \end{bmatrix}.$$

Use this to derive the formulas for $x(t), x'(t)$, determine a and b so that the solutions are of period T, and bound $|a|, |b|$. Determine a value of β so that $x(\mu)\in\Omega_\beta$.

5. Extend the argument for Duffing's equation to show the existence of solutions of period T to $x''+x+\mu\phi(x,x',t)=0$ where ϕ is of period T in t.
6. Treat van der Pol's equation,

$$x''+x=\mu\left(x'+\gamma(x')^3+f\cos wt\right),$$

by the same methods.

12.5 FURTHER APPLICATION OF THE NEWTON-KANTOROVICH METHODS

We focus on two particular applications, the first of which is fundamental to most numerical realizations of these ideas; the second permits numerical realizations of a special sort. The first is to the problem of solving n nonlinear equations in n unknowns, the second to the problem of solving nonlinear integral equations.

Suppose that we have given n functions $f_1, f_2, \ldots, f_n$ defined on a domain Ω in $\mathbf{E}^n$, where $x\in\Omega$ is $x=(z_1,z_2,\ldots,z_n)$. We seek roots x of $F(x)=0$ or of,

$$f_1(z_1,z_2,\ldots,z_n)=0,$$

$$f_2(z_1,z_2,\ldots,z_n)=0,$$

$$f_n(z_1,z_2,\ldots,z_n)=0.$$

To be specific we will take the norm in $\mathbf{E}^n$ to be

$$\|x\|=\left(\sum_{i=1}^{n}|z_i|^2\right)^{1/2}.$$

The Jacobian matrix J of the system has as its i,kth entry

$$J_{ik}(x)=\frac{\partial f_i(z_1,\ldots,z_n)}{\partial z_k}.$$

Thus Newton's method is the iteration of the following linear system of equations:

$$J(x^n)(x^{n+1}-x^n)=F(x^n).$$

If $\Gamma_0=[J(x^0)]^{-1}$, then $\|\Gamma_0\|=B$ is determined as follows

$$\begin{aligned}\|\Gamma\| &= \sup_{\|x\|=1}(\Gamma_0 x,\Gamma_0 x)^{1/2}\\ &= \sup_{\|x\|=1}(\Gamma_0^*\Gamma_0 x,x)^{1/2}\\ &=\|\Gamma_0^*\Gamma_0\|^{1/2}.\end{aligned}$$

But $\Gamma_0^*\Gamma_0$ is a compact selfadjoint operator (why?) so, by Exercise 10.7.29, $\|\Gamma_0^*\Gamma_0\|=\Lambda$ where Λ is the *largest* eigenvalue of $\Gamma_0^*\Gamma_0$. (It is also easy to see that $1/\Lambda$ is the *smallest* eigenvalue of $J(x^0)J(x^0)^*$.) Thus in applications we must estimate Λ. The quantity η is determined as $\eta=\|x^1-x^0\|$ by solving

$$J(x^0)(x^1-x^0)=F(x^0).$$

Last, if each entry in $J(x)$ satisfies the Lipschitz condition

$$\left|\frac{\partial f_i(x)}{\partial z_k}-\frac{\partial f_i(\bar{x})}{\partial z_k}\right|\leqslant L\|x-\bar{x}\|,$$

on Ω, then

$$\|J(x)-J(\bar{x})\|\leqslant L\eta\|x-\bar{x}\|;$$

we can take $K=L\eta$. If for some x^0, $h=BK\eta\leqslant\frac{1}{2}$ then Newton's method will converge, as will the modified Newton's method.

The computation of J^{-1} is not required; however, in practice, even the computation of $J(x^n)$ for each n may be too expensive so that some approximation is substituted for $J(x^n)$. Methods of this type are called variously *Newton-like methods* (Dennis [1]), *quasi-Newton methods* (Broyden [1]), and *update methods* (Rheinboldt and Vandergraft [1]). (These methods often are considered in the context of optimization, but see Exercise 12.1.15.)

Next we consider the general nonlinear integral equation $F(x)=y$:

$$x(s)-\int_0^1 K(s,t,s(t))\,dt=y(s),$$

where $K(s,t,u)\in C\{(0\leqslant s,t\leqslant 1)\otimes(|u|<\infty)\}$ and $x,y\in X=C[0,1]$. We have shown in Exercise 12.2.6 that $F'(x)$ exists, that

$$F'(x)h(s)=h(s)-\int_0^1 K_u(s,t,x(t))h(t)\,dt,$$

and that

$$GD^2F(x;g,h)=\int_0^1 K_{uu}(s,t,x(t))g(t)h(t)\,dt,$$

for any $x\in X$, under appropriate restrictions on K. Then Newton's method takes the form

$$F'(x^n)(x^{n+1}-x^n)+F(x^n)=0$$

or

$$x^{n+1}(s)-x^n(s)-\int_0^1 K_u(s,t,x^n(t))\left[x^{n+1}(t)-x^n(t)\right]dt$$
$$+x^n(s)-\int_0^1 K(s,t,x^n(t))\,dt=0.$$

The existence of the resolvent kernel (see Exercise 12.5.1) for the kernel $K_u(s,t,x^n(t))$ corresponds to the invertibility of $F'(x^n)$. We immediately obtain the following:

THEOREM 12.5.1. *Let $G(s,t)$ be the resolvent kernel for the kernel $K_u(s,t,x^0(t))$ and let*

$$\int_0^1 |G(s,t)|\,dt\leqslant B.$$

Let

$$\left\| x^0(s) - \int_0^1 K(s,t,x^0(t))\,dt \right\| \leqslant \eta'$$

and

$$\int_0^1 |K_{uu}(s,t,u)|\,dt \leqslant K'$$

if $0 \leqslant s \leqslant 1$ *and* $|u - x^0(s)| \leqslant 2(1+B)\eta'$. *Let* $h = (1+B)^2 K'\eta' \leqslant \frac{1}{2}$. *Then Newton's method and the modified Newton's method both converge to the solution of* $F(x) = 0$. *The solution* $x^*(s)$ *is in the ball*

$$\|x^* - x^0\| \leqslant \frac{[1 - \sqrt{(1-2h)}](1+B)\eta'}{h}$$

and is unique on

$$\|x - x^0\| \leqslant \frac{[1 + \sqrt{(1-2h)}](1+B)\eta'}{h}.$$

PROOF. We note that $[F']^{-1}$ is given by

$$\Gamma y = y(s) + \int_0^1 G(s,t) y(t)\,dt,$$

so that $\|\Gamma\| \leqslant 1 + B$, while $\|F'(x) - F'(\bar{x})\| \leqslant K'\|x - \bar{x}\|$, in Theorem 12.3.4 and its corollary. ■

EXERCISES TO SECTION 12.5

1. If $Z = (I - \lambda K)$ is the integral operator

$$z(s) = w(s) - \lambda \int_0^1 k(s,t) w(t)\,dt,$$

then the resolvent kernel $r(s,t,\lambda)$ such that

$$(I - \lambda K)^{-1} w = w + \lambda \int_0^1 r(s,t,\lambda) w(t)\,dt$$

must satisfy

$$\begin{aligned} r(s,t,\lambda) &= k(s,t) + \lambda \int_0^1 k(s,u) r(u,t,\lambda)\,du \\ &= k(s,t) + \lambda \int_0^1 r(s,u,\lambda) k(u,t)\,du. \end{aligned}$$

Show this if $k(s,t) \in C[0 \leqslant s,t \leqslant 1]$ and z,w are both in $C[0,1]$ or else both in $L_2(0,1)$. (Hint: Consider the Neumann series.)

2. Let $k(s,t,u) = st$ arctan $u, y(s) = 1 - 0.485s + s^2$. In $C[0,1]$ consider the nonlinear equation $F(x) = y$:

$$x(s) = \int_0^1 st \arctan x(t)\, dt + y(s).$$

Set up the Newton's iteration with $x^0 = \frac{3}{2}$ and show that the resolvent kernel for the first iteration is $r(s,t) = (12/35)st$, and that from Newton's method we conclude the existence of a solution with $\|x^* - \frac{3}{2}\| \leqslant \frac{82}{35}$. Show that the result of the first iteration satisfies $\|x^1 - x^*\| \leqslant 0.0067$, where $x^*(s) = 1 + s^2$ is the true solution. (Hint: C.f. Kantorovich and Akilov [1], 734.)

3. Let $f(t,x) \in C^2\{[0,1] \otimes (-\infty, \infty)\}$ and treat the boundary value problem $x'' = f(t,x), x(0) = A, x(1) = B$ in $L_2(0,1)$ by Newton's method after applying Exercise 11.7.4. (Hint: If $\hat{x} = x + a + bt$, then there is a set (a,b) for which $\hat{x} \in D_0$.) Compare with ordinary successive approximations for the same problem.

4. In Exercise 12.5.3, how would you treat the following boundary conditions:

   (a) $x'(0) = A, x'(1) = B$;
   (b) $\alpha x(0) + \beta x'(1) = A; \gamma x(0) + \delta x'(1) = B; \alpha + \gamma > 0, \alpha \geqslant 0, \gamma \geqslant 0$.

5. If $T \in [H]$ where H is a Hilbert space, then the eigenvalue problem $Tx = \lambda x, x \neq 0$, is equivalent to $(Tx,x) = \lambda(x,x), x \neq 0$, and also to $(x,x)Tx - (Tx,x)x = 0, x \neq 0$. Show that the latter equation holds if and only if

$$\|x\|^2 \|Tx\|^2 - (Tx,x)^2 = 0, \qquad x \neq 0.$$

Apply Newton's method to find conditions that guarantee the existence of a solution (eigenvector). What guarantees that the solution obtained is not $x = 0$? Another way to guarantee this is to consider

$$(\|x\|^2 - 1)^2 + \|x\|^2 \|Tx\|^2 - (Tx,x) = 0.$$

Using Newton's method, find conditions that guarantee the existence of a solution. (C.f. Rall [1], 201–203.)

12.6 CONTINUOUS ANALOGS OF ITERATIVE METHODS

Any iteration: $x^{n+1}=x^n+\Psi(n,x^n)$ for some functional Ψ, can also be thought of as a difference equation $x(t_{n+1})-x(t_n)=(t_{n+1}-t_n)\Psi(t_n,x(t_n))$, $t_n=n$, for a function x of the real variable t, and by analogy as a discretization of a formal differential equation $x'(t)=\Psi(t,x(t))$. The analog of convergence: $x^n\to x^*, n\to\infty$ is now the existence of the limit of $x(t)$, as t approaches its upper bound, either finite or infinite, in such a way that $x'(t)$ approaches zero. For instance, an obvious continuous analog of Newton's method would be $x'(t)=-[F'(x(t))]^{-1}F(x)$.

Another way of generating such differential equations to solve functional equations is to imbed $F(x)=0$ in some way into a one-parameter family of problems $H(t,x)=0$ on $0\leqslant t\leqslant T$, where we permit $T=\infty$, in such a way that $H(T,x)=F(x)$, we can solve $H(0,x)=0$, and $H(t,x)$ is so smooth that a solution $x=x(t)$ exists and is differentiable on $0\leqslant t<T$. For instance, this will be true if the conditions of the implicit function Theorem 12.4.1 and its corollary hold. Then also $H_t(t,x(t))+H_x(t,x(t))x'(t)=0$, so that;

$$x'(t)=-\left[H_x(t,x(t))\right]^{-1}H_t(t,x(t)).$$

Then $x(t)\to x^*, x'(t)\to 0, t\to T$ for a solution of this differential equation, will imply $H(T,x^*)=F(x^*)=0$. Of course, we do not expect to be able to solve the differential equation in closed form. This implies that some form of numerical integration of the differential equation is used. This will generate new iteration methods. (If $T<\infty$, then the accuracy of the approximate solution at T is of interest; if $T=\infty$, then we need only guarantee that the integration scheme and differential equation are both sufficiently stable to have the same limit.) The idea goes back at least to Davidenko [1], the following development to Gavurin [1]. (Al'ber [1] provides a survey of recent developments and more detailed and unifying results.) Let $\mathbf{R}^+$ be the nonnegative real numbers and X be a Banach space. By an abstract differential equation is meant $x'=\Psi(t,x)$ where $\Omega\subset X$ and $\Psi(t,x)$ maps $\mathbf{R}^+\otimes\Omega$ to Ω. The initial value problem is that of finding a differentiable function $x(t)$ such that $x(0)=x^0\in\Omega$ and $x'(t)=\Psi(t,x(t))$ on some interval $[0,T]$. Existence and uniqueness theorems familiar from $\mathbf{R}^n$ do not easily go over to the present situation. However, we can use the following easy theorem:

THEOREM 12.6.1. *Let $\Psi(t,x)$ be continuous on $[0,T]\otimes(\|x-x^0\|\leqslant p)$ and for some $L>0$ satisfy $\|\Psi(t,x)-\Psi(t,\bar{x})\|\leqslant L\|x-\bar{x}\|$ there. Then there* *exists a unique path $x(t)$ in $C^1[0,h]$ for some $h, 0<h\leqslant T$, such that*

$$\begin{cases} x'(t)\equiv\Psi(t,x(t)), \\ x(0)=x^0. \end{cases}$$

PROOF. Consider the integral transformation

$$A(x)=\int_0^t \Psi(\tau,x(\tau))d\tau+x^0.$$

If $x(t)\in C^0, x(0)=x^0$, and $\|x(t)-x^0\|\leqslant p, 0\leqslant t\leqslant h$, then $A(x)$ is defined and is continuous. Let $Y=\{x(t)|x\in C^0, x(0)=x^0\}$ and introduce the metric

$$\rho(x(t),\bar{x}(t))=\max_{0\leqslant t\leqslant h}\|x(t)-\bar{x}(t)\|.$$

Then by a trivial estimate

$$\rho(A(x),A(\bar{x}))\leqslant Lh\rho(x,\bar{x}).$$

If $Lh<1$, then A is a contraction, while in the metric ρ, Y is complete. The contraction mapping theorem implies our theorem. ■

DEFINITION 12.6.1. If $x(t)$ maps $\mathbf{R}^+\to X$, we can define an *arc length* along $x(t)$ as follows. Partition $0\leqslant a\leqslant t\leqslant b<\infty$ by $a=t_0<t_1<\cdots<t_n=b$ and form

$$\sum_{i=0}^{n-1}\|x(t_{i+1})-x(t_i)\|.$$

Clearly as we add new points to the mesh, the sum is nondecreasing. If there exists a nonnegative number $S=S[a,b]$ such that, given $\epsilon>0$, there is a $\delta=\delta(\epsilon)$ with

$$\left|S-\sum_{i=0}^{n-1}\|x(t_{i+1})-x(t_i)\|\right|<\epsilon,$$

whenever $\max|t_{i+1}-t_i|\leqslant\delta$, then we call *$S$ the arc length of $x(t)$ on $[a,b]$.*

$S[a,b]$ is a decreasing function of a or increasing function of b, clearly. Also if $x'(t)$ exists on $[a,b]$

$$\sum_{i=1}^{n-1}\|x(t_{i+1})-x(t_i)\|=\sum_{i=1}^{n-1}\left[\|x'(t_i)\|+o(1)\right](t_{i+1}-t_i).$$

If $x'(t)$ is continuous on $[a,b]$, it can be shown by way of the Heine-Borel theorem that it is uniformly continuous and that the o(1) term can be made uniformly small, hence that

$$S[a,b]=\int_a^b \|x'(t)\|\,dt.$$

Arcs of finite length on $\mathbf{R}^+$ have the following property.

THEOREM 12.6.2. *If*

$$\int_0^\infty \|x'(t)\|\,dt<\infty,$$

then

$$\lim_{t\to\infty} x(t)=x(\infty)$$

exists.

PROOF. We observe that

$$\|x(t)-x(\tau)\|\leqslant\int_\tau^t \|x'(\sigma)\|\,d\sigma\to 0, \qquad t,\tau\to\infty. \quad \blacksquare$$

For solutions of a differential equation the result can be made stronger.

THEOREM 12.6.3. *If $x(t)$ is a solution of $x'=\Psi(t,x)$ and $\Psi(t,x)$ is continuous on $(0\leqslant t\leqslant\infty)\otimes X$, then*

$$\int_0^\infty \|x'(t)\|\,dt<\infty$$

implies

$$\lim_{t\to\infty}\Psi(t,x(t))=\Psi(\infty,x(\infty))=0.$$

PROOF. Since $x(t)$ is of finite arc length, $x(t)\to x(\infty), t\to\infty$. Also $x'(t)=\Psi(t,x(t))\to x'(\infty), t\to\infty$. If $x'(\infty)\neq 0$, then for $t\geqslant T_\eta$, $\|x'(t)\|\geqslant\eta>0$ and $S[T_\eta,T]\geqslant\eta(T-T_\eta)$. This is a contradiction. Hence $x'(\infty)=0$. $\blacksquare$

In the obvious analog to Newton's method, $x'=-[F'(x)]^{-1}F(x)$, any solution $x(t)$ also satisfies

$$\frac{dF(x(t))}{dt}=F'(x(t))x'(t)=-F(x(t)),$$

a differential equation which, using Theorem 12.6.1, already has the *unique* solution $e^{-t}F(x^0)$. That is, $F(x(t))-e^{-t}F(x(0))=0$ along any solution of the Newton differential equation. $F(x)-e^{-t}F(x^0)$ is one $H(t,x)$ for the solution of $F(x)=0$ that leads to the Newton differential equations with $T=\infty, x(0)=x^0$. It is also obvious that a C^1 function $x(t)$ that satisfies $F(x(t))-e^{-t}F(x(0))=0$, and along which $F'(x(t))$ is invertible, is also a solution of the Newton differential equation. To guarantee convergence we have the following:

THEOREM 12.6.4. *Let Ω_0 be the ball $\|x-x^0\| \leqslant B\|F(x^0)\|$. On Ω_0 let $F'(x), F''(x)$ exist while*

$$\|[F'(x)]^{-1}\| \leqslant B.$$

Let $F''(x)$ be bounded in x in any open ball in Ω_0. Then $x'=\Psi(x)=-[F'(x)]^{-1}F(x), x(0)=x^0$ is solvable on $0 \leqslant t<\infty$; the solution $x(t) \in \Omega_0$ for $0 \leqslant t<\infty$; $\lim_{t \to \infty} x(t)=x^*$ exists, and $F(x^*)=0$.

PROOF. The Fréchet derivative of $[F'(x)]^{-1}F(x)$ is given by $I+\{[F'(x)]^{-1}\}'F(x)$, if the differentiation is permissible. If this is bounded on some sphere about x^0, there is a Lipschitz condition on Ψ that guarantees the existence of a solution. Since F'' exists, $F'(x)$ is continuous and invertible on some sphere $\|x-x^0\| \leqslant \eta$. Since

$$F'(x+h)-F'(x)=[F''(x)h]+o(h),$$

we have

$$[F'(x)]^{-1}-[F'(x+h)]^{-1}=[F'(x+h)]^{-1}\{[F''(x)h]+o(h)\}[F'(x)]^{-1}$$

while

$$[F'(x+h)]^{-1}=[F'(x)]^{-1}+o(1).$$

Then

$$[F'(x)]^{-1}-[F'(x+h)]^{-1}=[F'(x)]^{-1}\{F''(x)h\}[F'(x)]^{-1}+o(h).$$

Hence $\{[f'(x)]^{-1}\}'$exists and is bounded on some sphere $\|x-x^0\| \leqslant \eta_1 \leqslant \eta$. Then the differential equation has a solution $x(t)$ on some interval $0 \leqslant t \leqslant \delta$ and

$$\|x(t)-x^0\|=\|\int_0^t x'(\tau)\,d\tau\| \leqslant B\|F(x^0)\|(1-e^{-t}).$$

Thus $x(t)$ may be continued beyond any finite t and $x(t)\in\Omega_0$ for all t. Then $F(x(t))=e^{-t}F(x^0)\to 0, t\to\infty$. Also

$$\int_0^t \|x'(\tau)\|\,d\tau \leqslant B\|F(x^0)\|(1-e^{-t})$$

implies that $x(t)$ is of finite arc length, so $\lim_{t\to\infty}x(t)=x^*$ exists and $F(x^*)=0$. ∎

As another application of this idea let X be a Hilbert space and $\phi(x)=(F(x),F(x))$, where $F:X\to X$. This is, of course, only one of many functionals that are minimized at $F(x)=0$. Along a path $x(t)$

$$\frac{d}{dt}\phi(x(t))=2\Re(F'(x(t))x'(t),F(x(t))$$

$$=2\Re(x'(t),F'(x(t))^*F(x(t)))\,.$$

A differential equation for nonlinear continuous steepest descent is obtained if we set

$$x'(t)=-F'(x(t))^*F(x(t)).$$

About this method we can prove:

THEOREM 12.6.5 *In the ball* $\Omega_0:\|x-x^0\|\leqslant MB^2\|F(x^0)\|$, *let* $F'(x),F''(x)$ *exist and let*

$$\|F'(x)\|\leqslant M,$$

and

$$\|[F'(x)]^{-1}\|\leqslant B,$$

there, while $F''(x)$ *is locally bounded near any point in* Ω_0. *Then* $x'=-F'(x)^*F(x), x(0)=x^0$ *is solvable on* $0\leqslant t\leqslant\infty$; *the solution* $x(t)\in\Omega_0$ *for all* t; $x^*=\lim_{t\to\infty}x(t)$ *exists and* $F(x^*)=0$.

PROOF. Along $x(t), d\phi(x(t))/dt=-2\|F'(x(t))^*F(x(t))\|^2$. But

$$\|F(x)\|\leqslant\|[F'(x)^*]^{-1}\|\,\|F'(x)^*F(x)\|$$

$$\leqslant B\|F'(x)^*F(x)\|$$

so

$$\frac{d\phi(x(t))}{dt} \leqslant \frac{2}{B^2}\|F(x(t))\|^2 = -\frac{2}{B^2}\phi(x(t)).$$

Then

$$\phi(x(t)) \leqslant \phi(x^0)e^{-2t/B^2}$$

and

$$\|F(x(t))\| \leqslant \|F(x^0)\|e^{-t/B^2},$$

while

$$\|x'(t)\| \leqslant M\|F(x^0)\|e^{-t/B^2}.$$

Then $x(t)$ is of finite arc length, and $\|x(t)-x(0)\| \leqslant MB^2\|F(x^0)\|$ for all t. This implies the theorem. ■

Since these continuous processes cannot be realized numerically, it is necessary to 'solve' the differential equations approximately. For instance, if Euler's method with unit step size is applied to the Newton differential equation, then we obtain Newton's method. Boggs [1] has developed discrete methods that converge under the assumptions of Theorem 12.3.1, but more rapidly, using ideas from numerical integration theory for ordinary differential equations. (This paper also contains a thorough bibliography on such discretizations.)

EXERCISE TO SECTION 12.6

1. Let G be a bounded open set in a Banach space X; ∂G is the boundary of G. Let $F: X \to Y$, Y a Banach space. Let $F(x)$ possess Fréchet derivatives of first and second order on G, where $\|[F'(x)]^{-1}\| \leqslant B$ on G, while $F''(x)$ is bounded in x on any open ball in G. Suppose that there is a functional $\phi: Y \to R$ such that

   (a) $\phi(ty) = t\phi(y)$ for all $t > 0$;

   (b) $\phi(F(x)) \geqslant \phi(F(x^0)) > 0$ for all $x \in \partial G$. Then the Newton differential equation has a solution $x(t) \in G, 0 \leqslant t < \infty$, such that $\lim_{t\to\infty} x(t) = x^*$ exists in G and $F(x^*) = 0$. (Hint: $\phi(F(x(t))) = e^{-t}\phi(F(x^0))$.)

12.7 PROJECTION METHODS

If we try to realize Newton's method in a concrete numerical algorithm, we must eventually solve finite dimensional problems, but there are two obvious ways to get there. We may start by setting up the sequence of linear problems whose solutions converge to the root of $F(x)=0$, and then solve these problems by an approximate method, or else we may first construct a finite dimensional nonlinear approximate equation, and then apply Newton's method to it. These are equivalent, as a precise statement of the alternative procedures will make clear.

Suppose then that we have given Banach spaces X, Y, with $F(x): X \to Y$ the functional whose root we seek. We suppose $F(x)$ to be Fréchet differentiable, with invertible derivative $F'(x)$, and set up an iteration of the form

$$F'(x^n)(x^{n+1}-x^n)=F(x^n).$$

Then we find spaces $\hat{X} \subset X, \hat{Y} \subset Y$, and projections Q onto $\hat{X}, P$ onto $\hat{Y}$ which leave $\hat{X}$ and $\hat{Y}$ invariant, and set up the approximate Newton's method

$$PF'(\hat{x}^n)(\hat{x}^{n+1}-\hat{x}^n)=PF(\hat{x}^n),$$

where $\hat{x}^n \in R_Q, n=0,1,2,\ldots$.

Alternately we may replace $F(x)=0$ with the approximate equation $PF(QX)=0$, or $PF(x)=0$, for which Newton's method takes precisely the same form. This method is known as a *projection method* (Krasnosel'skii, Vainikko, *et al.* [1], 292–299). When $F(x)$ is linear, it coincides with the methods of Chapter 7. We assume, as in Chapter 7, that $X=Y$ and that $F(x)$ can be written in the form $x-T(x)$, where $T(x)$ is a Fréchet differentiable mapping on an open set $\Omega \subset X$. We assume further that there is a sequence of closed subspaces $\hat{X}_n \subset X$ and linear projections Ψ_n such that $\Psi_n x \subset \hat{X}_n$ if $x \in \Omega$, and $\Psi_n \hat{x}=\hat{x}$ if $\hat{x} \in \hat{X}_n$, where $\Psi_n T(x)$ is continuous on Ω. As an approximate equation we take

$$x^n=T_n(x^n)+S_n(x^n)$$

where $T_n(x)$ is some nonlinear operation in $\hat{X}_n$ that is continuous on $\Omega_n=\Omega \cap \hat{X}_n$, and $S_n(x)=T_n(x)-\Psi_n T(x)$.

We will use the following result relating the solvability of a nonlinear equation to the invertibility of its Fréchet derivative.

LEMMA 12.7.1. *Let $F(x)$ be continuous on some open set $\Omega \subset X$, a Banach space, and Fréchet differentiable on some ball $\Omega_r = \{x \mid \|x - x^0\| \leqslant r\} \subset \Omega$. Let $F'(x^0)$ be invertible, and let there exist two positive numbers $\delta < r$ and $q < 1$ such that*

1. $\sup_{\Omega_\delta} \|[F'(x^0)]^{-1}[F'(x) - F'(x^0)]\| \leqslant q$, *and*
2. $\alpha = \|[F'(x^0)]^{-1} F(x^0)\| \leqslant \delta(1-q)$.

Then on Ω_δ, $F(x) = 0$ has a unique solution x^*, and

$$\frac{\alpha}{1+q} \leqslant \|x^0 - x^*\| \leqslant \frac{\alpha}{1-q}.$$

PROOF. We consider the following equation whose solvability is obviously equivalent to the solvability of $F(x) = 0$:

$$G(x) = x^0 - \left[F'(x^0)\right]^{-1}\left[F(x^0) + F(x) - F(x^0) - F'(x^0)(x - x^0)\right] = x.$$

We will see that G is a contraction mapping on Ω_δ. First, for some $\lambda, 0 < \lambda < 1$,

$$\begin{aligned}
\|G(x) - x^0\| &\leqslant \|\left[F'(x^0)\right]^{-1} F(x^0)\| \\
&\quad + \|\left[F'(x^0)\right]^{-1}\left[F(x) - F(x^0) - F'(x^0)(x - x^0)\right]\| \\
&< \alpha + \|\left[F'(x^0)\right]^{-1}\left\{F'\left(x^0 + \lambda\left[x - x^0\right]\right) - F'(x^0)\right\}(x - x^0)\| \\
&\leqslant \alpha + q\|x - x^0\| \leqslant \delta(1-q) + \delta q = \delta,
\end{aligned}$$

if $x \in \Omega_\delta$. Second, if $x^1, x^2 \in \Omega_\delta$, then for some $\lambda, 0 < \lambda < 1$,

$$\begin{aligned}
\|G(x^1) - G(x^2)\| &\leqslant \|\left[F'(x^0)\right]^{-1}\left[F(x^1) - F(x^2) - F'(x^0)(x^1 - x^2)\right]\| \\
&\leqslant \|\left[F'(x^0)\right]^{-1}\left\{F'\left(x^1 + \lambda\left[x^2 - x^1\right]\right) - F'(x^0)\right\}\| \, \|x^1 - x^2\| \\
&\leqslant q\|x^1 - x^2\|.
\end{aligned}$$

Hence there is a unique $x^* \in \Omega_\delta$ such that $G(x^*) = x^*$, or $F(x^*) = 0$. Also,

$$\|G(x^*) - x^0\| = \|x^* - x^0\| \leqslant \alpha + q\|x^* - x^0\|,$$

so $\|x^* - x^0\| \leqslant \alpha/(1-q)$. On the other hand,

$$\begin{aligned}\|x^* - x^0\| &= \|G(x^*) - x^0\| \\ &\geqslant \alpha - q\|x^* - x^0\|,\end{aligned}$$

so $\|x^* - x^0\| \geqslant \alpha/(1+q)$. ■

The bounds on $\|x^0 - x^*\|$ found in this lemma are the best possible, as is seen from the case of a linear function of one real variable.

With the aid of this lemma we prove the following analog of Theorem 7.2.3:

THEOREM 12.7.1. *Let the operators $T(x)$ and $\Psi_n T(x)$ be Fréchet differentiable on an open set $\Omega \subset X$, a Banach space, and $T_n(x)$ be Fréchet differentiable on $\Omega_n = \Omega \cap \hat{X}_n$ where $\hat{X}_n$ is the range of the projection Ψ_n. Let $x^* = T(x^*) \in \Omega$, and let there exist $[I - T'(x^*)]^{-1} \in [X]$, while*

1. $\|x^* - \Psi_n x^*\| \to 0, n \to \infty$;
2. $\|S_n(\Psi_n x^*)\| \to 0, n \to \infty$;
3. $\|S_n'(\Psi_n x^*)\| \to 0, n \to \infty$;
4. $\|\Psi_n T(\Psi_n x^*) - T(x^*)\| \to 0, n \to \infty$;
5. $\|\Psi_n T'(\Psi_n x^*) - T'(x^*)\| \to 0, n \to \infty$.

Suppose that for any $\epsilon > 0$ there exists n_ϵ and $\delta = \delta(\epsilon)$ such that for all $n \geqslant n_\epsilon$,

$$\|T_n'(x) - T_n'(\Psi_n x^*)\| \leqslant \epsilon,$$

if $\|x - \Psi_n x^*\| < \delta$ and $x \in \Omega_n$. Then there exist N and p such that x^* is the only solution of $x = T(x)$ in the ball $B_p = \{x \mid \|x - x^0\| \leqslant p\}$, while for all $n \geqslant N, x = T_n(x) + S_n(x)$ has a unique solution $\hat{x}^n \in \Omega_n \cap B_p$, $\|\hat{x}^n - x^*\| \to 0, n \to \infty$, and for suitable constants r_1, r_2,

$$r_1\|x^* - \Psi_n x^* - S_n(\hat{x}^n)\| \leqslant \|\hat{x}^n - x^*\| \leqslant r_2\|x^* - \Psi_n x^* - S_n(\hat{x}^n)\|.$$

PROOF. Since $x^* \in \Omega$, assumption 1 implies $\Psi_n x^* \in \Omega$ for all $n \geqslant N^*$ for some N^*, and in fact that some ball $B_{\delta^*} = \{x \mid \|x - \Psi_n x^*\| \leqslant \delta^*\} \subset \Omega$ while $\Psi_n x^* \in B_{\delta^*} \cap \Omega_n$. Also $T_n(x)$ is Fréchet differentiable in $B_{\delta^*} \cap \Omega_n$ for $n \geqslant N^*$. Since $I - T'(x^*)$ is invertible, we may use assumption 5 to imply that $I - \Psi_n T'(\Psi_n x^*)$ is invertible and that for some M,

$$\|[I - \Psi_n T'(\Psi_n x^*)]^{-1}\| \leqslant M,$$

for all n not less than some $N^1 \geqslant N^*$. Last, using assumption 3 and $S_n' = T_n' - \Psi_n T'$, we conclude that for all n not less than some $N^2 \geqslant N^1$, and for some $M^1, I - T_n'(x^*)$ is invertible in $\hat{X}_n$ with

$$\|[I - T_n'(\Psi_n x^*)]^{-1}\| \leqslant M^1, \qquad n \geqslant N^2.$$

Similarly, there is an M^2 such that for some $N^3 \geqslant N^2$

$$\|I - T_n'(\Psi_n x^*)\| \leqslant M^2, \qquad n \geqslant N^3,$$

because of assumptions 5 and 3. We set $F_n(x) = \Psi_n x - T_n(\Psi_n x)$, and

$$\alpha_n = \|[F_n'(x^*)]^{-1} F_n(x^*)\|$$

as in Lemma 12.7.1. Then, for $n \geqslant N^3$,

$$\frac{\|F_n(x^*)\|}{M^2} \leqslant \alpha_n \leqslant M^1 \|F_n(x^*)\|,$$

and

$$\begin{aligned}
F_n(x^*) &= \Psi_n x^* - T_n(\Psi_n x^*) \\
&= \Psi_n T(x^*) - T_n(\Psi_n x^*) \\
&= \Psi_n T(x^*) - T(x^*) + T(x^*) - \Psi_n T(\Psi_n x^*) + \Psi_n T(\Psi_n x^*) - T_n(\Psi_n x^*) \\
&\to 0, \qquad n \to \infty,
\end{aligned}$$

using assumptions 1, 4, and 2.

We now pick $q, 0 < q < 1$, set $\epsilon = q/M^1$, and find $N^4 \geqslant N^3, \delta \leqslant \delta^*$ so that if $n \geqslant N^4$ and $\|x - \Psi_n x^*\| \leqslant \delta, x \in \Omega_n$, then

$$\|T_n'(x) - T_n'(\Psi_n x^*)\| \leqslant \epsilon.$$

This implies that condition 1 of Lemma 12.7.1 holds for $F_n(x)$, while we need only increase N^4 to N^5 to obtain condition 2. This shows that for $n \geqslant N^5, F_n(x) = 0$ has a unique solution $\hat{x}^n$ in Ω_n with $\|\hat{x}^n - \Psi_n x^*\|$ satisfying the last estimate of the theorem for $r_1 = 1/M^1(1+q), r_2 = M^2/(1-q)$; also we conclude,

$$\|\hat{x}^n - x^*\| \leqslant \|x^* - \Psi_n x^*\| + \|\Psi_n x^* - \hat{x}^n\| \to 0, \qquad n \to \infty.$$

The fact that x^* is an isolated root of $x = T(x)$ follows from the inverse function Exercise theorem 12.4.3. ■

As a converse we have

THEOREM 12.7.2. *Let T_n be Fréchet differentiable on $B_r^n = \{x \in \Omega_n | \|x - \hat{x}^n\| \leqslant r\}$ while $T, \Psi_n T$ are Fréchet differentiable on $B_r = \{x | \|x - \hat{x}^n\| \leqslant r\}$. Let $I - T_n'(\hat{x}^n)$ be invertible with*

$$\|[I - T_n'(\hat{x}^n)]^{-1}\| \leqslant M_n.$$

Let

$$p_n = (1 + M_n\|\Psi_n T'(\hat{x}^n)\|)\|T'(\hat{x}^n) - \Psi_n T'(\hat{x}^n)\| + M_n\|T_n'(\hat{x}^n) - \Psi_n T'(\hat{x}^n)\| < 1,$$

while for some positive δ_n and some $q_n, 0 < q_n < 1$,

1. $\sup_{\|x - \hat{x}^n\| \leqslant \delta_n} \|T'(x) - T'(x^n)\| \leqslant q_n / K_n$, *and*
2. $\|\hat{x}^n - T(\hat{x}^n)\| \leqslant \delta_n(1 - q_n)/K_n$,

where

$$K_n = \frac{1 + M_n\|\Psi_n T'(\hat{x}^n)\|}{1 - p_n}.$$

Then $I - T'(x^n)$ is invertible and $x = T(x)$ has a unique solution x^* in $\|x - x^n\| \leqslant \delta_n$, satisfying

$$\alpha_n(1 + q_n) \leqslant \|\hat{x}^n - x^*\| \leqslant \frac{\alpha_n}{1 - q_n},$$

with

$$\alpha_n = \|[I - T'(\hat{x}^n)]^{-1}[\hat{x}^n - T(\hat{x}^n)]\| \leqslant K_n\|\hat{x}^n - T(\hat{x}^n)\|.$$

PROOF. From the invertibility of $I - T_n'(x^n)$ in $\hat{X}_n$ we conclude the invertibility of $I - \Psi_n T'(\hat{x}^n)$ in $\hat{X}_n$, with

$$\|[I - \Psi_n T'(\hat{x}^n)]^{-1}\|_{\hat{X}_n} \leqslant \frac{M_n}{1 - M_n\|T_n'(\hat{x}^n) - \Psi_n T'(\hat{x}^n)\|},$$

which follows from Exercise 6.2.9 and the assumptions of the theorem.

Next note that

$$[I-\Psi_n T'(\hat{x}^n)]^{-1}=I+[I-\Psi_n T'(\hat{x}^n)]^{-1}\Psi_n T'(\hat{x}^n).$$

The right side of this equation is defined on X, and trivially has $I-\Psi_n T'(\hat{x}^n)$ as a left inverse. Also, we have the identity

$$[I-\Psi_n T'(\hat{x}^n)]^{-1}\Psi_n T'(\hat{x}^n)\Psi_n T'(\hat{x}^n)$$
$$=\{-I+[I-\Psi_n T'(\hat{x}^n)]^{-1}\}\Psi_n T'(\hat{x}^n),$$

which implies that $I-\Psi_n T'(\hat{x}^n)$ is also a right inverse, hence a two-sided inverse on X. Thus $[I-\Psi_n T'(\hat{x}^n)]^{-1}$ exists in $[X]$ and

$$\|[I-\Psi_n T'(\hat{x}^n)]^{-1}\|\leqslant\frac{1+M_n\|\Psi_n T'(\hat{x}^n)\|}{1-M_n\|T_n'(\hat{x}^n)-\Psi_n T'(\hat{x}^n)\|}.$$

We apply Exercise 6.2.9 to

$$[I-T'(\hat{x}^n)]+[T'(\hat{x}^n)-\Psi_n T'(\hat{x}^n)]$$

and obtain the existence of $[I-T'(\hat{x}^n)]^{-1}$ with

$$\|[I-T'(\hat{x}^n)]^{-1}\|\leqslant K_n.$$

If we turn to Lemma 12.7.1, then conditions 1 and 2 of our present theorem imply conditions 1 and 2 of the lemma. The application of the lemma here completes the proof. ∎

Vainikko [1], 28–38, has applied these theorems to the solution of nonlinear integral equations, generalizing the results that we have obtained in Section 8.3.

EXERCISES TO SECTION 12.7

1. Let $k(t,s,u)$ and $\partial k(t,s,u)/\partial u$ both be continuous on $\{0\leqslant t,s\leqslant 1\otimes(-\infty<u<\infty)\}$, and $x^*(s)$ be the unique solution in $[0,1]$ of

$$x(t)=\int_0^1 k(t,s,x(x))\,ds+f(t).$$

Suppose that the equation

$$y(t)=\int_0^1 \frac{\partial k(t,s,x^*(s))}{\partial u} y(s)\,ds$$

has no solution other than $y \equiv 0$. Apply the trapezoidal rule of numerical integration (as in Section 8.3) to obtain an approximate equation

$$x(t_{i,n})=\sum_{j=0}^{n} A_{j,n} k(t_{i,n}, t_{j,n}, x(t_{j,n}))+f(t_{i,n})$$

Establish the solvability and convergence of this procedure.

2. The boundary value problem $x''=f(t,x(t)), 0 \leqslant t \leqslant 1, x(0)=x(1)=0$ with solution $x^*(t)$ is transformed into an integral equation using the kernel of Exercise 11.7.4:

$$x(t)=-\int_0^1 k(t,s) f(s,x(s))\,ds.$$

Show that the trapezoidal rule applied to this equation is equivalent to the difference scheme

$$\frac{x(t_{i+1,n})-2x(t_{i,n})+-x(t_{i-1,n})}{(1/n)^2}=f(t_{i,n}, x(t_{in})),$$

$$x(t_{0,n})=x(t_{n,n})=0.$$

If $\partial f(t,u)/\partial u$ and $f(t,u)$ are both continuous on $\{[0,1]\otimes(-\infty<u<\infty)\}$ and the linear boundary value problem

$$y''=\frac{\partial f(t,x^*(t))}{\partial u} y, \qquad y(0)=y(1)=0,$$

has no solution other than $y=0$, show that the approximate equation has a unique solution, for all sufficient large n, which converges to x^*. (Hint: Show first that the linear system of equations

$$\frac{x_{i+1,n}-2x_{i,n}+x_{i-1,n}}{(1/n)^2}=a_{i,n}, \qquad i=1,\ldots,n-1$$

$$x_{0,n}=x_{n,n}=0$$

has the unique solution

$$x_{i,n}=-\frac{1}{n}\sum_{j=1}^{n} k(t_{i,n}, t_{j,n}) a_{j,n}, \qquad i=0,\ldots,n-1).$$

12.8 MONOTONICITY

Successive approximations and Newton's method (in its various forms) ignore the natural structure of the problem in order to produce a universal approach to solving functional equations. It is our intention here to show that this natural structure often has as much to say about convergence as universal methods. To do this, we select a very simple two-point boundary value problem:

$$\begin{cases} \dfrac{d^2x}{dt^2} = x^2, & 0<t<1; \\ x(0)=x_0, & x(1)=x_1. \end{cases}$$

Everything that we shall establish also holds true, with minor changes in terminology, for the Dirichlet problem in $\mathbf{E}^n$: $\triangle u = u^2$ on some region G with sufficiently smooth boundaries, $u=\phi$, a continuous function on ∂G, the boundary of G, and for more complicated partial differential operators than the Laplacian. (This problem is of interest in the theory of chemical reactions.) However, the analytic machinery needed in such cases clouds the simplicity of the points to be made. We adopt a notation that can be used in the more complicated situations, without confusing matters in the simple problem very much.

Thus Lx may be taken to be the differential operator $d^2x/dt^2, G$ the interior of the unit interval, ∂G the boundary $\{0,1\}$ and ϕ the function on G with $\phi(0)=x_0, \phi(1)=x_1$. (If L is a partial differential operator, a slight change in terminology is needed.) From Exercise 11.7.4 we conclude the equivalence of $(Lx)(t)=x^2(t)$ on $G, x=\phi$ on ∂G with an integral equation

$$x(t) = -\int_G k(t,s)x^2(s)\,ds + v(t)$$

where $v(t)$ is the unique solution of the linear boundary value problem

$$Lv=0 \text{ on } G, v=\phi \text{ on } \partial G,$$

and

$$k(t,s)=\begin{cases} t(1-s), & 0\leqslant t\leqslant s\leqslant 1, \\ s(1-t), & 0\leqslant s\leqslant t\leqslant 1, \end{cases}$$

is the kernel of the inverse integral operator (Green's function). We define

$$K_1 = \sup_{t\in G}\int_G k(t,s)\,dS,$$

and

$$B = \sup_{t \in G} |v(t)|.$$

For the one-dimensional case, $K_1 = 1$, and $B = \max(|x_0|, |x_1|)$. We take as our space $X = C[0,1]$, in which the mapping $[x]^2 : x(t) \to x^2(t)$ is defined everywhere. We use Kx to denote the linear integral operator with kernel $-k(t,s)$; then we may describe ordinary successive approximations in X by

$$\hat{x}^{n+1} = K[\hat{x}^n]^2 + v.$$

THEOREM 12.8.1. *If $4K_1 B < 1$, then $Lx = [x]^2$, on $G, x = \phi$ on ∂G possesses a solution x^*, unique in $\|x\| < t^{**} = [1 + (1 - 4K_1 B)^{1/2}]/2K_1$, reachable from any $\hat{x}^0$ with $\|\hat{x}^0\| \leqslant t^{**}$ by successive approximations, lying in $\|x\| \leqslant t^* = [1 - (1 - 4K_1 B)^{1/2}]/2K_1$,* and

$$\|\hat{x}^n - x^*\| \leqslant (t^*)^n \|\hat{x}^0 - x^*\|.$$

PROOF. We use the majorant method (Exercise 12.3.11) with $\hat{x}^0 = 0$. Set $\Psi(t) = K_1 t^2 + B, S(x) = K[x]^2 + v, t_0 = 0$. Then $\|S(\hat{x}^0) - \hat{x}^0\| \leqslant \Psi(t_0) - t_0$. Also

$$[S'(x)h](t) = \int_G k(t,s) 2x(s) h(s)\, dS$$

So $\|S'(x)\| \leqslant 2K_1 \|x\| \leqslant 2K_1 t$ on $\|x\| \leqslant t$. On $[0, t^{**}], \Psi(t) = t$ has a unique root t^*. If $4K_1 B < 1$, then $x = S(x)$ has a unique solution $x^* = S(x^*)$ on $\|x\| < t^{**}$ which can be reached by successive approximations starting from any $\hat{x}^0$ with $\|\hat{x}^0\| \leqslant t^{**}$, and

$$\begin{aligned}\|\hat{x}^{n+1} - \hat{x}^n\| &\leqslant \sup_{0 \leqslant \lambda \leqslant 1} \|S'(\hat{x}^{n-1} + \lambda(\hat{x}^n - \hat{x}^{n-1}))\| \, \|\hat{x}^n - \hat{x}^{n-1}\| \\ &\leqslant t^* \|\hat{x}^n - \hat{x}^{n-1}\|.\end{aligned}$$

Similarly $\|\hat{x}^n - \hat{x}^*\| \leqslant (t^*)^n \|\hat{x}^0 - \hat{x}^*\|$. ■

The constraint $4K_1 B < 1$ is achieved either by small boundary data, or a small domain G. The algorithm used in computation is

$$\begin{cases} \dfrac{d^2(\hat{x}^{n+1})}{dt^2} = [\hat{x}^n(t)]^2, \\ \hat{x}^{n+1}(0) = x_0, \hat{x}^{n+1}(1) = x_1. \end{cases}$$

The various Newton's methods are best applied to the functional equation

$$F(x)=x-K[x]^2-v=0.$$

Then

$$\{F'(x)h\}(t)=h(t)+2\int_G k(t,s)x(s)h(s)\,ds.$$

The Newton algorithm is

$$x^{n+1}(t)-x^n(t)+2\int_G k(t,s)x^n(s)[x^{n+1}(s)-x^n(s)]\,ds$$
$$+x^n(t)+\int_G k(s,t)[x^n(s)]^2\,ds-v(t)=0$$

which reduces to

$$x^{n+1}(t)+2\int_G k(t,s)x^n(s)x^{n+1}(s)\,ds-v(t)=0$$

or, equivalently,

$$(Lx^{n+1})(t)-2x^n(t)x^{n+1}(t)=0,$$
$$x^{n+1}(0)=x_0, x^{n+1}(1)=x_1.$$

The modified Newton's method with $\hat{x}^0=0$ is easily seen to reduce to successive approximations treated above.

THEOREM 12.8.2. *If* $4K_1B<1$ $(4K_1B=1)$, *then* $Lx=[x]^2$ *on* $G, x=\phi$ *on* ∂G, *possesses a solution* x^* *in* $\|x\|\leqslant t^*$ *which is unique in* $\|x\|<t^{**}$ $(\|x\|\leqslant t^{**})$. *The Newton iteration with* $x^0=0$ *converges to* x^* *and*

$$\|x^n-x^*\|\leqslant\frac{\left[1-(1-4K_1B)^{\frac{1}{2}}\right]^{2^n}}{2K_12^n}.$$

PROOF. We set $x^0=0$ and find $\|F(x^0)\|=\|v\|=B$. A simple calculation shows that $F'(x^0)=I$, so $\|F'(x^0)\|=1$, $\|[F'(x^0)]^{-1}F(x^0)\|=B$. Also $\|F'(x)-F'(y)\|\leqslant 2K_1\|x-y\|$ for all x,y. Then Theorems 12.3.1, 12.3.2, and 12.3.3 imply the results claimed. ■

If we start with $x^0 = v$, then we obtain improved convergence rates, but no increase in the class of problems whose solution is established. In fact the improvement over successive approximations in this regard is minimal.

The constraint on the product K_1B is an artifact of the method, not a natural consequence of the problem. We see this in a second look at the two iterations. The starting point is the observation that the operator L is associated with the following *maximum principles*:

1. *If $(Lx)(t) > 0$ on G, then $x(t)$ achieves its maximum on G if and only if $x(t)$ is a constant.*
2. *If $(Lx)(t) \geqslant 0$ on G, then*

$$\max_{t \in \partial G} x(t) = \sup_{t \in G} x(t)$$

3. *If $Lx = 0$, then, for all $t \in G$,*

$$\min_{t \in \partial G} x(t) < x(t) < \max_{t \in \partial G} x(t)$$

For $Lx = d^2x/dt^2$ these facts are obvious. Using the maximum principle we establish (Ablow and Perry [1]):

THEOREM 12.8.3 *Let $\phi \geqslant \alpha > 0$ and $Lv = 0$ on $G, v = \phi$ on G. Let $Lx^n = (x^{n-1})^2$ on $G, x^n = \phi$ on $\partial G, n = 1, 2, \ldots,$ and x^0 be any function continuous on $\bar{G}$. If $K_1B^2 < \alpha$ and $K_1B < \frac{1}{2}$ then x^i converges to a unique solution $x^*, 0 < x^* \leqslant v$ on G, and $\|x^n - x^*\| \leqslant (B + \|x^0\|)^2 (2K_1B)^{n-2}$. If $x^0 = v$, then $0 < x^1 < x^3 < \cdots < x^* < \cdots < x^4 < x^2 < v$.*

PROOF. From the maximum principle we have $v > \alpha$. If w is any continuous function on $\bar{G}$, and $Lx = [w]^2$ on $G, x = \phi$ on ∂G, then $L(x - v) \geqslant 0$ on G, so $x - v \leqslant 0$. If also $\|w\| \leqslant B$, then $x(t) \geqslant -K_1B^2 + \alpha$, using the integral equation representation of the problem. Thus $0 < x \leqslant v \leqslant B$. Successive approximations, starting anywhere, enter the set of functions $0 < x \leqslant v$ and stay there. Also if $n > m$,

$$\begin{aligned} |x^n(t) - x^m(t)| &\leqslant \int_G k(t,s) |x^{n-1}(s) - x^{m-1}(s)| \, |x^{n-1}(s) + x^{m-1}(s)| \, ds \\ &\leqslant 2K_1B \|x^{n-1} - x^{m-1}\| \\ &\leqslant (2K_1B)^{m-1} \|x^{n-m} - x^0\| \\ &\leqslant (B + \|x^0\|)^2 (2K_1B)^{m-2}. \end{aligned}$$

This establishes the existence of x^* and the convergence rate. The particular monotonicity of the odd and even iterates when $x^0=0, x^1=v$, is a consequence of the maximum principle and is left as an exercise. ■

The class of boundary data ϕ for which existence and convergence are established here is different from the class in Theorem 12.8.1, omitting some with zero values for the boundary data, adding others with larger maximum values on the boundary. For Newton's method, the change is even more striking. The results are based on the following *extended maximum principles*: *Let* $[c]w$ *denote the operation of pointwise multiplication*: $([c]w)(t)=c(t)w(t)$.

1. *If* $(Lw-[c]w)(t)\geqslant 0$ *on* $G, w(t)=0$ *on* $\partial G, c(t)>0$ *on* G, *then* $w(t)\leqslant 0$ *on* G.
2. *If* $(Lw-[c]w)(t)>0$ *on* $G, w(t)=0$ *on* ∂G *and* $c(t)\geqslant 0$ *on* G, *then* $w(t)<0$ *on* G.
3. *If* $(Lw-[c]w)(t)\leqslant 0$ (<0) *on* $G, w(t)=\phi(t)\geqslant 0$ *on* ∂G *and* $c(t)>0$ *on* G, *then* $w(t)\geqslant 0$ (>0) *on* G.

The proofs are left to the exercises. With these we can prove (Pohazaev [1]):

THEOREM 12.8.4. *Let* $\phi(t)\geqslant 0$ *but not identically zero, and* $Lv=0$ *on* $G, v(t)=\phi(t)$ *on* ∂G. *The Newton iterates* x^n *for* $Lx=[x]^2$ *on* $G, x(t)=\phi(t)$ *on* ∂G, *with* $x^0(t)=v(t)$ *satisfy* $0<\cdots<x^{n+1}(t)<x^n(t)\cdots<v(t)$, *and converge uniformly to an* x^*, *which is the unique nonnegative solution.*

PROOF. We set $x^0=v, w^n=x^{n+1}-x^n$ and consider the Newton iteration in terms of w^n:

$$\begin{cases} Lw^n-[2x^n]w^n=[x^n]^2-Lx^n \text{ on } G, \\ w^n(t)=0 \text{ on } \partial G. \end{cases}$$

Since $Lx^0=0, x^0(t)>0$ on G, and $w^1(t)<0$. That is, $x^1(t)<v(t)$ on G. At the same time x^1 satisfies

$$\begin{cases} Lx^1-[2x^0]x^1=-[x^0]^2 \text{ on } G, \\ x^1(t)=\phi(t) \text{ on } \partial G, \end{cases}$$

so $x^1(t)>0$ on G. Thus $0<x^1(t)<v(t)$ on G. Assume that we have already shown $0<x^n(t)<x^{n-1}(t)<\cdots<x^1(t)<v(t)$ on G. A straightforward

calculation shows that $Lw^n-[2x^n]w^n=[w^{n-1}]^2>0$ on G, so that $w^n(t)<0$ on G or $x^{n+1}(t)<x^n(t)$ there. At the same time, $Lx^{n+1}=[w^n+x^{n+1}]x^n$, so $Lx^{n+1}-[x^n]x^{n+1}=[x^n]w^n<0$ on G, which completes the argument for $0<x^{n+1}(t)<x^n(t)$ on G. We conclude that at each point $t\in G$, $x^n(t)\to x^*(t)$, $w^n(t)\to 0$, $n\to\infty$, and $(Lw^n)(t)\to 0$, $(Lx^n-[x^n]^2)(t)\to 0, n\to\infty$. At each $t, x^n(t)-x^*(t)\geqslant x^{n+p}(t)-x^*(t)\geqslant 0$, for $p=1,2,3,\ldots$. We wish to show that $\delta^n=\max_{t\in\bar{G}}|x^n(t)-x^*(t)|\to 0, n\to\infty$.

Suppose the contrary. Let $t_n\in G$ be the point at which $x^n(t_n)-x^*(t_n)$ equals the maximum. Observe that $\delta^n\geqslant\delta^{n+1}\geqslant 0$ for all n. Let $\delta^n\to\delta>0, n\to\infty$. The set of points t_n must have at least one convergent subsequence $t_{n_j}, j=1,2,\ldots$, in the compact set $\bar{G}$, with limit $\tilde{t}$. However, the family of functions $\{x^n(t)\}$ is equicontinuous in the set $\|x\|\leqslant v$. To see this, it is sufficient to analyze the integral equation formulation of Newton's method. Then

$$x^{n_j}(\tilde{t})-x^*(\tilde{t})=x^{n_j}(\tilde{t})-x^{n_j}(t_{n_j})+x^{n_j}(t_{n_j})$$
$$-x^*(t_{n_j})+x^*(t_{n_j})-x^*(\tilde{t})\to\delta, \qquad n\to\infty,$$

which contradicts the pointwise convergence of the sequence x^n. It now follows that x^n converges uniformly to x^*, that is, in the norm in X, and this implies that x^* solves our problem. The uniqueness among positive solutions follows from the maximum principle. ∎

It should be noted that this result does not give an estimate of the rate of convergence.

There are other monotone iterations for this problem (see Exercise 12.8.4). Maximum principles exist for differential operators other than those considered here, as well as for difference operators that approximate them (see Exercise 12.8.5). Other orderings may replace that induced by "less than or equal to." It should also be clear that the nonlinearity $[x]^2$, which lends itself easily to analysis, can be replaced by much more general nonlinear functionals of the form $f(t,x,x')$ (Collatz [1], 334–409.) A general survey of maximum principles may be found in Protter and Weinberger [1].

EXERCISES TO SECTION 12.8

1. In Theorem 12.4.2, starting at $x^0=v$, show that $h=2K_1^2B^2/(1-2K_1B)^2$ and $t^*=[1-(1-4K_1B)^{1/2}]/[1-2K_1B]$. Show that both h and t^* are not greater than the corresponding values obtained in the proof of Theorem

12.4.2. Start successive approximations at $\tilde{x}^0 = v$ and find results equivalent to Theorem 12.8.1.

2. Complete the proof of Theorem 12.8.3, using induction and the maximum principle.
3. Show that there is at most one nonnegative solution to $Lx=[x]^2$ on $G, x=\phi \geqslant 0$ on ∂G. (Hint: If there were two, say x^*, x^{**}, then on any set on which $[x^*]^2-[x^{**}]^2>0, L(x^*-x^{**})>0$. On the boundary of the largest such set, $x^*=x^{**}$.)
4. Show that the iteration $Lx^n=[x^{n-1}]x^n$ on $G, x^n=\phi$ on $\partial G, x^0=v$, also satisfies Theorem 12.8.4. (Cf. Ablow and Perry [1].)
5. Show that the difference operator $x(t_{i+1})-2x(t_i)-x(t_{i-1})$ of Exercise 12.7.2 satisfies both the maximum principle and the extended maximum principle.
6. Prove the extended maximum principle for $Lx=d^2x/dt^2$.

12.9 MINIMIZATION SUBJECT TO CONSTRAINTS

Let $f(x)$ be a real valued nonlinear functional on a Banach space X that achieves a minimum at $x^0 \in X$. Then (Exercises 12.1.14-15, and 12.2.3), if the appropriate derivatives exist, $f'(x^0)=0$ and $\langle h, f''(x^0)h \rangle \geqslant 0$ for all $h \in X$. If the domain of f is some proper subset of X, for instance some ball $\|x\| \leqslant \rho$, or some set that consists of the solutions of some other functional equation, then the arguments that lead to these results apply only if x^0 is an interior point of the domain. For the other case, in which x^0 is a boundary point, we need other results, the Lagrange multiplier rule and its consequences, to characterize the point x^0 at which a minimum exists.

In optimization theory, it is customary to use $\nabla f(x)$ to denote at least the Fréchet derivative of a functional $f(s)$, and $\nabla^2 f(x)$ to denote the second order Fréchet derivative. We will adopt these notations at this point. A discussion of the constrained minimization requires some geometric preliminaries about sets in a Banach space. (We follow Lusternik and Sobelev [1], 324–331.)

DEFINITION 12.9.1. Let X, Y be Banach spaces, $\phi(x)$ a nonlinear mapping from X to Y whose range contains the zero element in Y. Let $N=\{x \in X \mid \phi(x)=0\}$. At a point $x^0 \in N$ let $\phi(x)$ posses a Fréchet derivative $\nabla\phi(x^0) \in [X \to Y]$. The point x^0 is called a *regular* point of N if

$$R_{\nabla\phi(x^0)} = Y.$$

Let $T_0=\{h \in X \mid \nabla\phi(x^0)h=0\}$. The set $T_{x^0}=x^0+T_0=\{x \in X \mid x=x^0+h,$

$h \in T_0\}$ is called the *tangent manifold to N at* x^0. If there is another linear manifold $T_1 \subset X$ such that X is a direct sum of T_0 and T_1 (i.e., any x in X has a unique representation $x = h + g, h \in T_0, g \in T_1$), we say that T_0 *reduces* X.

The sense in which this is a tangent manifold is explained by the following two theorems.

THEOREM 12.9.1. *Suppose that T_0, the tangent manifold at x^0, reduces X. Then, for some $\eta > 0$, there is a one-to-one bicontinuous mapping Ψ between $N_\eta = \{x \mid \|x - x^0\| \leq \eta\} \cap N$ and $T_\eta = \{x \mid \|x - x^0\| \leq \eta\} \cap T_{x^0}$, such that $\|x - \Psi(x)\| = o(\|x - x^0\|) = o(\|\Psi(x) - x^0\|)$ for any $x \in N_\eta$, while $\|\Psi^{-1}(x) - x\| = o(\|x - x^0\|) = o(\|\Psi^{-1}(x) - x^0\|)$ for any $x \in T_\eta$.*

PROOF. For any $x \in X$, there is a unique representation $x = h + g$, with $h \in T_0, g \in T_1$. Then $\nabla\phi(x^0)x = y$; but $x = h + g$, where $\nabla\phi(x^0)x = \nabla\phi(x^0)h + \nabla\phi(x^0)g = \nabla\phi(x^0)g$. Let A denote the bounded linear operator from T_1 to Y given by $Ag = \nabla\phi(x^0)g, g \in T_1$. If $Ag_1 = Ag_2$, then $A(g_1 - g_2) = \nabla\phi(x^0)(g_1 - g_2) = 0$, so $g_1 - g_2 \in T_0$, or $g_1 - g_2 = 0$. Hence A is one-to-one, onto, and invertible. The closed graph theorem asserts that $A^{-1} \in [Y \to X]$.

Now, for any x with $\|x - x^0\| \leq \eta, x = x^0 + h + g$, where $h \in T_0, g \in T_1$. Then N is defined by $\Phi(h,g) = \phi(x^0 + h + g) = 0$ and $\Phi(0,0) = 0$. Also $\Phi_g(0,0) = A$. By the implicit function theorem, there is a function $g = g(h)$, so that, in N_η, we have $x = x^0 + h + g(h)$ with $h \in T_0, g(h) \in T_1$. The mapping $\Psi: x^0 + h \to x^0 + h + g(h)$ is one-to-one and bicontinuous between N_η and T_η, since the above is true for every $x \in N_\eta$, and differentiable at $h = 0$, with

$$g_h(0) = -\nabla\phi(x^0)^{-1}\nabla\phi(x).$$

Also

$$\Phi_h(0,0)h + \Phi_g(0,0)g_h(0)h = \Phi_h(0,0)h + Ag_h(0)h$$
$$= \nabla\phi(x^0)h + \nabla\phi(x^0)g_h(0)h = 0.$$

In addition,

$$g(h) = g(0) + g_h(0)h + o(\|h\|) = 0 - A^{-1}\nabla\phi(x^0)h + o(\|h\|)$$
$$= o(\|h\|).$$

Thus

$$\|g(h)\| = o(\|h\|) = o(\|h + g(h)\|). \quad \blacksquare$$

THEOREM 12.9.2. *If X is not reduced by T_0, then there exist two mappings, Ψ from N_η to T_η, χ from T_η to N_η that satisfy*

$$\|x-\Psi(x)\|=o(\|x-x^0\|)=o(\|\Psi(x)-x^0\|),$$

and

$$\|x-\chi(x)\|=o(\|x-x^0\|)=o(\|\chi(x)-x^0\|).$$

PROOF. Consider the space $X/T_0=\hat{X}$. $\hat{X}$ is a Banach space. If h_1, h_2 are in the same coset of X/T_0 (i.e., $h_1-h_0\in T_0$) then $\nabla\phi(x^0)(h_1-h_2)=0$, so $\nabla\phi(x^0)$ maps h_1,h_2 into the same element of Y. Conversely if $\nabla\phi(x^0)h_1=\nabla\phi(x^0)h_2$, then $\nabla\phi(x^0)(h_1-h_2)=0$, so $h_1-h_2\in T_0$, and h_1,h_2 belong to the same coset of X/T_0. Hence there exists a linear operator A mapping $\hat{X}=X/T_0$ into $Y: A\mathfrak{T}=\nabla\phi(x^0)h$ where $\mathfrak{T}\subset\hat{X}, h\in\mathfrak{T}$, and $A\mathfrak{T}$, the image does not depend upon the choice of $h\in\mathfrak{T}$. It is easy to see that $A\in[\hat{X}\to Y]$. Fix $y\in Y$. Since $R_{\nabla\phi(x^0)}=Y$, we can find $h\in X$ with $\nabla\phi(x^0)h=y$. But $h\in\mathfrak{T}\subset X$ for some $\mathfrak{T}$. Then A is one-to-one onto Y from all of $\hat{X}$; by the closed graph theorem, $A^{-1}\in[Y\to\hat{X}]$.

We must now establish the existence of a mapping χ from a neighborhood of $x^0\in T_0$ to a neighborhood of $x^0\in N$. (This is essentially an implicit function theorem; the existence of χ provides a local coordinate system in N from T_{x^0}, and the order of contact of this coordinate system at x^0 is the subject of our theorem.) We do this by showing, for any sufficiently small $h\in T_0$ the existence of $g(h)$ such that $x^0+h\in T_{x^0}$ and $\phi(x^0+h+g(h))=0$, so $x^0+h+g(h)\in N$.

Let $h\in\mathfrak{T}_0$. Set $g_0=0\in T_0$ and set up the iteration,

$$\mathfrak{T}_n=\mathfrak{T}_{n-1}-A^{-1}\phi(x^0+h+g_{n-1}),$$

where we select $g_n\in\mathfrak{T}_n$ such that $\|g_n-g_{n-1}\|\leqslant 2\|\mathfrak{T}_n-\mathfrak{T}_{n-1}\|$. This can be done because $\|\mathfrak{T}_n-\mathfrak{T}_{n-1}\|=\inf_{g_i\in\mathfrak{T}_i}\|g_1-g_2\|$. Further

$$A\mathfrak{T}_{n-1}=\nabla\phi(x^0)g_{n-1},$$

so

$$\mathfrak{T}_n=-A^{-1}[\phi(x^0+h+g_{n-1})-\nabla\phi(x^0)g_{n-1}]$$

and

$$\mathfrak{T}_{n-1}=-A^{-1}[\phi(x^0+h+g_{n-2})-\nabla\phi(x^0)g_{n-2}].$$

Then

$$\mathcal{T}_n - \mathcal{T}_{n-1} = -A^{-1}\big[\phi(x^0+h+g_{n-1}) - \phi(x^0+h+g_{n-2}) \\ - \nabla\phi(x^0)(g_{n-1}-g_{n-2})\big].$$

Also

$$\phi(x^0+h+g_{n-1}) - \phi(x^0+h+g_{n-2}) = \int_0^1 \nabla\phi(x^0+h+g(t))(g_{n-1}-g_{n-2})\,dt$$

where $g(t) = g_{n-2} + t(g_{n-1}-g_{n-2})$. Then

$$\mathcal{T}_n - \mathcal{T}_{n-1} = -A^{-1}\int_0^1 \big[\nabla\phi(x^0+h+g(t)) - \nabla\phi(x^0)\big]\,dt(g_{n-1}-g_{n-2}).$$

Let $\|h\| \leqslant \sigma, \|g_{n-2}\| \leqslant \sigma, \|g_{n-1}\| \leqslant \sigma$, so $\|g(t)\| \leqslant \sigma$ and $\|h+g(t)\| \leqslant 2\sigma$. Then, for any $\epsilon > 0$, there is $\sigma = \sigma_\epsilon$ with

$$\|\nabla\phi(x) - \nabla\phi(x^0)\| \leqslant \epsilon, \qquad \text{if } \|x - x^0\| \leqslant 2\sigma.$$

Hence

$$\|\mathcal{T}_n - \mathcal{T}_{n-1}\| \leqslant \|A^{-1}\| \int_0^1 \|\nabla\phi(x^0+h+g(t)) - \nabla\phi(x^0)\|\,dt\,\|g_{n-1}-g_{n-2}\|$$

$$\leqslant \epsilon\|A^{-1}\|\,\|g_{n-1}-g_{n-2}\|$$

and

$$\|g_n - g_{n-1}\| \leqslant 2\|A^{-1}\|\epsilon\|g_{n-1}-g_{n-2}\|.$$

Choose ϵ so small that $2\|A^{-1}\|\epsilon \leqslant \frac{1}{2}$. Then $\|g_n - g_{n-1}\| \leqslant \frac{1}{2}\|g_{n-1}-g_{n-2}\|$.

To complete an induction, suppose $\|h\| \leqslant \sigma, i = 1, 2, \ldots, 2^{n-1}$. Then

$$\|g_n - g_{n-1}\| \leqslant \frac{\|g_1\|}{2^{n-1}},$$

and

$$\|g_n\| = \|g_1 + (g_2 - g_1) + \cdots + (g_n - g_{n-1})\|$$

$$\leqslant \|g_1\|\left(1 + \tfrac{1}{2} + \tfrac{1}{4} + \cdots + \tfrac{1}{2^{n-1}}\right)$$

$$\leqslant 2\|g_1\|.$$

Since $g_0 = 0$, $\mathfrak{T}_1 = \mathfrak{T}_0 - A^{-1}\phi(x^0 + h)$ and $\|g_1\| \leqslant 2\|\mathfrak{T}_1 - \mathfrak{T}_0\| \leqslant 2\|A^{-1}\| \|(x^0+h)\|$, while

$$\phi(x^0+h)=\phi(x^0)+\nabla\phi(x^0)h+o(\|h\|)=o(\|h\|),$$

we have

$$\|g_1\| \leqslant 2\|A^{-1}\|o(\|h\|).$$

If σ is sufficiently small, and $\|\eta\| \leqslant \sigma$, we make the term $o(\|h\|) \leqslant \|h\|/4\|A^{-1}\|$. Then $\|g_1\| \leqslant \frac{1}{2}\|h\| \leqslant (\frac{1}{2})\sigma$, and $\|g_n\| \leqslant \sigma$.

This completes the induction: we know that $g_n \to g \in X$, and $\|g\| \leqslant 2\|g_1\| \leqslant 4\|A^{-1}\|o(\|h\|)$. Simultaneously $\mathfrak{T}_n \to \mathfrak{T} \in \hat{X}$ and $g \in \mathfrak{T}$. Then $\mathfrak{T} = \mathfrak{T} - A^{-1}\phi(x^0+h+g)$, so that $\phi(x^0+h+g)=0$. Then $x^0+h+g \in N$. Consider the mapping $\chi: x^0+h \in T_{x^0} \to x^0+h+g \in N$. Then $\|g\| = o(\|x^0+h-x^0\|)=o(\|h\|)$. Conversely, if $x=x^0+d \in N$, so $\phi(x^0+d)-\phi(x^0)=0$, then

$$\nabla\phi(x^0)d+o(\|d\|)=0.$$

Let $d \in \mathfrak{T} \in \hat{X}$. Then $\nabla\phi(x^0)d=A\mathfrak{T}=o(\|d\|)$, and $\mathfrak{T}=A^{-1}o(\|d\|)$, so $\|\mathfrak{T}\| \leqslant \|A^{-1}\|o(\|d\|)$. Choose $g \in \mathfrak{T}$ with

$$\|g\| \leqslant 2\|\mathfrak{T}\| \leqslant 2\|A^{-1}\|o(\|d\|).$$

Then $g \in \mathfrak{T}, d \in \mathfrak{T}$, so $g-d \in T_0$ and $x^0+d-g \in T_{x^0}$. Define Ψ as the mapping from x^0+d to x^0+d-g. Then $\|g\|=o(\|d\|)$. ■

We can now prove the *multiplier rule*:

THEOREM 12.9.3. *Let f map X to* **R**, *while $\phi(x)$ maps X to Y. Let $x^0 \in N=\{x|\phi(x)=0\}$ and for some $\eta>0$, let*

$$f(x) \geqslant f(x^0), \qquad x \in N, \|x-x^0\| \leqslant \eta.$$

If $\nabla f(x^0), \nabla\phi(x^0)$ exist, then there exists $y' \in Y'$ such that the functional

$$F(x)=f(x)-\langle\phi(x),y'\rangle$$

satisfies $\nabla F(x^0)=0$.

Suppose further than both $f(x)$ and $\phi(x)$ possess continuous second order Fréchet derivatives on $\|x-x^0\| \leqslant \eta$. Then

1. $\langle h, \nabla^2 F(x^0)h\rangle \geqslant 0$ *for all $h \in T_0$.*
2. *If in fact $\langle h, \nabla^2 F(x^0)h\rangle \geqslant \gamma^2\|h\|^2$ for all $h \in T_0$, then x^0 is a strict local relative minimum for f on N.*

PROOF. Let $h \in T_0$; suppose $df(x^0;h)=c \neq 0$. Then, by the preceeding theorem, we find a point $x^0+th+g \in N$ associated with x^0+th where $\|g\|=\|g(th)\|=o(\|th\|)$. Then

$$f(x^0+th+g)=f(x^0)+\langle th+g, \nabla f(x^0)\rangle + o(\|th+g\|)$$
$$=f(x^0)+ct+\langle g, \nabla f(x^0)\rangle + o(\|th+g\|).$$

Both g and $\langle g, \nabla f(x^0)\rangle$ are $o(t)$. Hence the sign of $f(x^0+th+g)-f(x^0)$ and of ct coincide for all sufficiently small t. This contradicts the assumption that x^0 is a relative minimum on N. Therefore $c=0$, and $\langle h, \nabla f(x^0)\rangle =0$ for all $h \in T_0$. This implies that if h_1, h_2 belong to the same coset $\mathfrak{T} \in \hat{X}$, then

$$\langle h_1, \nabla f(x^0)\rangle = \langle h_2, \nabla f(x^0)\rangle.$$

Define a linear functional τ' on $\hat{X}$ by

$$\langle \mathfrak{T}, \tau'\rangle = \langle h, \nabla f(x^0)\rangle$$

for any $h \in \mathfrak{T}$. Then for any $h \in \mathfrak{T}$

$$|\langle \mathfrak{T}, \tau'\rangle| = |\langle h, \nabla f(x^0)\rangle| \leqslant \|h\|\,\|\nabla f(x^0)\|,$$

so

$$|\langle \mathfrak{T}, \tau'\rangle| \leqslant \inf_{h \in \mathfrak{T}} \|h\|\,\|\nabla f(x^0)\| = \|\mathfrak{T}\|\,\|\nabla f(x^0)\|.$$

Then $\tau' \in \hat{\boldsymbol{X}}'$. But, if $\nabla\phi(x^0)h=y, y$ depends only upon $\mathfrak{T}$, not the representative $h \in \mathfrak{T}$, and $\mathfrak{T}=A^{-1}y$. This makes

$$\langle h, \nabla f(x^0)\rangle = \langle \mathfrak{T}, \tau'\rangle = \langle A^{-1}y, \tau'\rangle$$
$$= \langle y, [A^{-1}]'\tau'\rangle = \langle \nabla\phi(x^0)h, [A^{-1}]'\tau'\rangle$$

for all $h \in X$. Setting $[A^{-1}]'\tau'=y'$ and $F(x)=f(x)-\langle \phi(x), y'\rangle$ we see that $\langle h, \nabla f(x^0)-[\nabla\phi(x^0)]'y'\rangle=0$. It is obvious that this is the same statement as $\nabla F(x^0)=0$.

Using Taylor's theorem we can write both

$$F(x)-F(x^0)=\langle x-x^0, \nabla F(x^0)\rangle$$
$$+\langle x-x^0, \int_0^1 [\nabla^2 F(x^0)+\tau(x-x^0)](1-\tau)(x-x^0)\,d\tau\rangle$$

and

$$F(x)-F(x^0)=\langle x-x^0,\nabla F(x^0)\rangle+<(x-x^0),[\nabla^2F(x^0)](x-x^0)>$$
$$+\mathrm{o}(\|x-x\|^2).$$

Since $\nabla F(x^0)=0$, we conclude from the second form that, if $x\in N$,

$$\liminf_{\|x-x^0\|\to 0}\frac{F(x)-F(x^o)}{\|x-x^0\|^2}=\liminf_{\|x-x^0\|\to 0}\frac{f(x)-f(x^o)}{\|x-x^0\|^2}\geqslant 0.$$

Suppose now that we write $x\in T_{x^0}$ as x^0+d and let $g(d)=\mathrm{o}(\|d\|)$ be such that $x^0+d+g(d)\in N$. Then we must have

$$\liminf_{\|d+g(d)\|\to 0}\frac{\langle d+g(d),\nabla^2F(x^o)(d+g(d))\rangle}{\|d+g(d)\|^2}\geqslant 0.$$

But $\|d+g(d)\|/\|d\|\to 1$, as $\|d\|\to 0$. Hence

$$\liminf_{\|d\|\to 0}\frac{\langle d+g(d),\nabla^2F(x^0)[d+g(d)]\rangle}{\|d\|^2}\geqslant 0.$$

Since $\|g(d)\|/\|d\|\to 0, \|d\|\to 0$, this means

$$\liminf_{\|d\|\to 0}\frac{\langle d,[\nabla^2F(x^0)]d\rangle}{\|d\|^2}\geqslant 0.$$

Hence condition 1 is true. If condition 2 is true, then by continuity,

$$\langle h,[\nabla^2F(x^0+k)]h\rangle\geqslant\frac{\gamma^2}{2}\|h\|^2$$

for all k sufficiently small. From the first form of Taylor's theorem we see that for any $x\in N$, with $x-x^0$ sufficiently small,

$$F(x)-F(x^0)\geqslant\int_0^1(1-\tau)\,d\tau\frac{\gamma^2}{2}\|x-x^0\|^2$$
$$\geqslant\frac{\gamma^2}{2}\|x-x^0\|^2.$$

This makes

$$f(x)-f(x^0) \geqslant \frac{\gamma^2}{2}\|x-x^0\|^2,$$

and x^0 a strict local relative minimum. ■

We will apply Theorem 12.9.3 to obtain a first (analytic) form of the *Kuhn-Tucker conditions*(Kuhn and Tucker [1], Mangasarian [1], 94), which shows, among other facts, that the functional y' of Theorem 12.9.3 does in fact generalize *Lagrange multipliers* to function space. The Kuhn-Tucker conditions are necessary conditions that must be satisfied by any solution of the *nonlinear programming problem*:

Let $f(x): X \to \mathbf{R}$ be a real valued functional from a normed linear vector space X, let $g(x): X \to \mathbf{R}^m$ and let $\mathbf{R}^{m-}$ denote the orthant in $\mathbf{R}^m$ of vectors $(r_1, r_2, \ldots, r^m)$ with $r_i \leqslant 0$. Find $\bar{x} \in X$ such that

$$f(\bar{x}) = \min_{x \in \mathcal{F}} f(x)$$

where $\mathcal{F} = \{x \mid g(x) \in \mathbf{R}^{m-}\}$. The set $\mathcal{F}$ is known as the *feasible set*, $x \in \mathcal{F}$ is called a *feasible point*, and $\bar{x}$ is called the *minimum solution* and $f(\bar{x})$ the *minimum*. (Usually, $X = \mathbf{R}^n$.)

We introduce the space $\mathbf{X} = X \otimes \mathbf{R}^m$, the functional $\mathbf{F}(x,\chi): X \otimes \mathbf{R}^m \to \mathbf{R}$ and a mapping $\mathbf{G}(x;\chi): X \otimes \mathbf{R}^m \to \mathbf{R}^m$ by the definitions $\mathbf{F}(x,\chi) = f(x)$ and $\mathbf{G}(x,\chi) = (g_1(x)+\chi_1^2, g_2(x)+\chi_2^2, \cdots, g_m(x)+\chi_m^2)$, and consider the equivalent problem: Find $\bar{x},\chi$ such that

$$\mathbf{F}(\bar{x},\bar{\chi}) = \min_{x,\chi \in \mathfrak{F}} \mathbf{F}(x,\chi),$$

where $\mathfrak{F} = \{x,\chi \mid \mathbf{G}(x,\chi) = 0\}$. The new variables χ are known as *slack* or *dummy variables*. If f,g satisfy the various conditions of Theorem 12.9.3 then it is easy to see that $\mathbf{F}$ and $\mathbf{G}$ also satisfy them, and we can draw the appropriate conclusions. If we let ∇_x, ∇_χ denote the appropriate partial Fréchet derivatives, then we conclude that there exists a linear functional on $\mathbf{R}^m: \lambda' = (\lambda_1, \ldots, \lambda_m)$ such that

$$\begin{cases} \nabla_x \mathbf{F}(\bar{x},\bar{\chi}) - [\nabla_x \mathbf{G}(\bar{x},\bar{\chi})]'\lambda' = 0, \\ \nabla_\chi \mathbf{F}(\bar{x},\bar{\chi}) - [\nabla_\chi \mathbf{G}(\bar{x},\bar{\chi})]'\lambda' = 0, \end{cases}$$

or

$$\begin{cases} \nabla f(\bar{x}) - \sum_{i=1}^{m} \nabla g_i(\bar{x})\lambda_i = 0, \\ \qquad -2\chi_i\lambda_i = 0, \qquad i = 1, \ldots, m, \end{cases}$$

where we return to the notation ∇ for ∇_x. This must hold at any minimum solution that is a regular point of $g(x)=0$.

The condition $\chi_i\lambda_i=0$ is equivalent to $g_i(x)\lambda_i=0$, so that either $\lambda_i=0$ and the constraint $g_i(x)$ is *inactive*, or the constraint $g_i(x)$ is *active*$[g_i(x)=0]$ and λ_i need not be zero.

We may interpret the statements about $\nabla F, \nabla^2 F$ in Theorem 12.9.3 in terms of the Fréchet differentials, which is the form in which they are proven. Then, in addition to $d\mathbf{F}(x,\chi;h,\xi)=0$ for all h,ξ, which we translated above, we must also have $d^2\mathbf{F}(x,\chi;h,\xi)\geqslant 0$ for all h,ξ, which here translates into

$$\langle h,\{\nabla^2 f(\bar{x})\}h\rangle-\langle h,\{\nabla^2 g_i(\bar{x})h\}'\lambda'\rangle\geqslant 0,$$

$$-\lambda_i\geqslant 0, \qquad i=1,\ldots,m.$$

An illustration of the application of the Kuhn-Tucker conditions in the case where X is infinite dimensional is afforded by the following simple *optimal control problem*:

Let $x(t)$ denote the state of a system, $x(t):[0,1]\to\mathbf{R}^n$, and $u(t)$ denote a control of the system, $u(t):[0,1]\to\mathbf{R}^m$. Suppose that the dynamics of the system are given by $x'=Ax+Bu$ where A,B are real $n\times n$ and $n\times m$ matrices, respectively. Suppose that we wish to find, among all controls u that steer x from $x(0)=x^0$ to $x(1)=x^1$, a control that minimizes

$$f(u)=\int_0^1 (Ru(t),u(t))\,dt,$$

where R is a real $m\times m$ matrix and $(\cdot\,,\cdot)$ is the scalar product in $\mathbf{R}^m$.

From $x'=Ax+Bu$ we know that

$$x(t)=e^{At}x^0+\int_0^t e^{A(t-\tau)}Bu(\tau)\,d\tau$$

so

$$e^{-A}x^1-x^0-\int_0^1 e^{-A\tau}Bu(\tau)\,d\tau=0;$$

we call the left side of this equation $g(u)$. Any $u(t)$ that satisfies $g(u)=0$ will steer x from x^0 to x^1; call this the set of *admissible functions*, the manifold N. We compute the Fréchet differential of $g(u)$ to be

$$\nabla g(u)h=-\int_0^1 e^{-A\tau}Bh(\tau)\,d\tau.$$

If the range of this mapping is the whole of $\mathbf{R}^n$, the control u will be a regular point; this is known in optimal control theory as the condition of *controllability*. It is well known that this is equivalent to the statement that the rank of the $n \times nm$ matrix $[B, AB, A^2B, \ldots, A^{n-1}B]$ is equal to n (see, e.g., Padulo and Arbib [1], 537.) To complete the specification of the assumptions of Theorem 12.9.3, we take $u \in L_2^m(0, 1) = \{u(t) | u(t):[0,1] \to \mathbf{R}^m, u_i(t) \in L_2(0,1), i = 1, \ldots, m\}$. Then

$$\nabla f(u)h = \int_0^1 ((R^T + R)u(t), h(t))\,dt.$$

We conclude the existence of a linear functional $y \in \mathbf{R}^n$ such that, at a minimum,

$$\int_0^1 \left[((R^T + R)\bar{u}(t), h(t)) - (e^{-At}Bh(t), y)\right] dt$$

$$= \int_0^1 ((R^T + R)\bar{u}(t) - B^T e^{-A^T t} y, h(t))\,dt = 0$$

for all $h(t)$. This in turn, is easily seen to imply

$$(R^T + R)\bar{u}(t) - B^T e^{-A^T t} y (=) 0.$$

This result is not useful without a method of determining y; however it does give us a functional form of $\bar{u}$, which can be substituted into $f(u)$; this process leads to the minimization of a quadratic form to determine y. Alternatively, this functional form may be substituted in the admissibility equation $g(u) = 0$ to determine y and thus $\bar{u}$. If $R + R^T$ is nonsingular, $\bar{u}$ is uniquely determined by y; otherwise we must find some left inverses of $R + R^T$ to determine all candidates $\bar{u}$ for the minimum control. (Note that this process does not lead to the control that is usually of interest. For various reasons, a *feedback control* of the form $u = u(t, x)$ is the usual goal of optimal control theory. What we have considered is called an *open loop* control.) Because of the linearity of the dynamics of the system, and the fact that the amount of control is unrestricted, any terminal state x^1 is *reachable* from any initial state x^0. The general question of going from x^0 at time t_0 to x^1 at time t_1 is treated in the literature under the name of *controllability*.

EXERCISES TO SECTION 12.9

1. Let $X = \mathbf{E}^n, Y = \mathbf{E}^n$, in Theorem 12.9.3. Show that y' is the set of Lagrange multipliers.

2. Let X be a Hilbert space, A a bounded selfadjoint operator. Then $(Ax,x)\geqslant m(x,x)$ for some real m. Show that if for some $x\neq 0$, $(Ax,x)=m(x,x)$, then $Ax=mx$. (Hint: Consider the minimization problem for $f(x)=(Ax,x)$ subject to $\phi(x)=(x,x)-1=0$.)
3. Let $X=\mathbf{E}^n$ with the euclidean norm, and A be a selfadjoint ($n\times n$ matrix) operator. Show that $\lambda_1=\min_{\|x\|=1}(Ax,x)$ is an eigenvalue with eigenvector x^1 that can be taken to be of norm 1. Show that

$$\lambda_2=\min_{\|x\|=1,\,(x,x^1)=0}(Ax,x),$$

is an eigenvalue with eigenvector x^2 that can be taken to be of norm 1. Proceed inductively to construct n eigenvalues $\lambda_1\leqslant\lambda_2\leqslant\cdots\leqslant\lambda_n$ with an orthonormal set of eigenvectors $x^1,\ldots,x^n$.
4. Theorem 12.9.3 was established for Fréchet differentiable functionals. Show that it holds for Gâteaux differentials in the form $GDF(x^0;h)=0$ and $GD^2F(x^0;h,h)\geqslant 0$. Compute $GDF(x^0;h)$ and $GD^2F(x^0;h,h)$.
5. In the optimal control problem treated in the text, find the equivalent minimization problem for y, under the assumption that $R=R^T$ and is invertible.
6. As in Exercise 12.9.5 use the constraints $g(u)=0$ to determine y.
7. Let X be a real Banach space, $f(x)$ a real valued functional with a Gâteaux differential on a closed set $\Omega\subset X$. Suppose that $x^0\in\Omega$ and $f(x^0)=\inf_{x\in\Omega}f(x)$. Show that $d^+f(x^0;h)=\lim_{t\downarrow 0}[f(x^0+th)-f(x^0)]/t\geqslant 0$ for all h such that $x^0+th\in\Omega$ on some interval $0\leqslant t<\epsilon$. In particular, if $x^0\in\operatorname{int}\Omega$, show that this implies the existence of a Gâteaux differential that is zero. Show that if $f(x)$ is Fréchet differentiable, then $d^+f(x^0;h)=\langle h,\nabla f(x^0)\rangle$.

12.10 THE CALCULUS IN OPTIMAL CONTROL

In the preceding section we used the multiplier theorem in a simple version of the optimal control problem to characterize an optimal control. The multiplier theorem can be used directly in other, more general problems of optimal control theory. Here we will establish a multiplier theorem in which the multipliers are characterized in a different way, one that is unexpected from the analysis of Section 12.9.

One fairly general model for optimal control theory is the following: We suppose given a *state space* X of continuous functions $x(t):[t_0,t_1]\to\mathbf{R}^n$, a *control space* U of piecewise continuous functions $u(t):[t_0,t_1]\to\mathbf{R}^m$ (each $u_i(t)$ is continuous except at an at most finite number of points on $t_0\leqslant t\leqslant t_1$). Associated with these there is a *dynamic system* defined by the roots of a mapping $\Psi(x,u):X\otimes U\to X$ of the form

$$\Psi(x,u)(t)=x(t)-x(t_0)-\int_{t_0}^{t}f(x(s),u(s))\,ds, \qquad t_0\leqslant t\leqslant t_1;$$

that is, to each u in some set of *admissible controls*, and each $x(t_0)=a\in\mathbf{R}^n$, there is a unique *trajectory* x such that $\Psi(x,u)=0$. This will certainly be true if $f(x,u)$ is continuously differentiable in its $n+m$ arguments, in which case the trajectories $x(t)$ will be differentiable with piecewise continuous derivatives. We suppose also that we are given a function $F(x):X\to\mathbf{R}^n$ of the form $F(x)=g(x(t_0))$ and a functional $\eta(x):X\to\mathbf{R}$ of the form $\eta(x)=h(x(t_1))$, where $g:\mathbf{R}^n\to\mathbf{R}^n, h:\mathbf{R}^n\to\mathbf{R}$, are continuously differentiable in their n arguments. The control problem is that of finding a control $\bar{u}$ and its resulting trajectory $\bar{x}$, where $\Psi(\bar{x},\bar{u})=0$, and $F(x)=c\in\mathbf{R}^n$, such that

$$\eta(\bar{x})=\inf_{u\in\Omega}\eta(x)$$

where u ranges over all admissible controls, here taken to be some open set Ω. (When Ω is taken to contain boundary points, modification of the calculus is needed along the lines of Exercise 12.9.7.)

The mapping $F(x)$ is Fréchet differentiable with $F'(x)=g_x(x(t_0))$; $\eta(x)$ is also Fréchet differentiable, and

$$d\eta(x;y)=\langle y(t_1),\nabla h(x(t_1))\rangle_{\mathbf{R}^n}$$
$$=\sum_{i=1}^{n}\frac{\partial h(x(t_1))}{\partial x_i}y_i(t_1).$$

(Hereafter we will write $\langle\cdot,\cdot\rangle$ for $\langle\cdot,\cdot\rangle_{\mathbf{R}^n}$ where the meaning is clear.) Similarly $\Psi(x,u)$ is differentiable with

$$d\Psi(x,u;y,v)=y(t)-y(t_0)-\int_{t_0}^{t}[\,f_x(x(s),u(s))y(s)+f_u(x(u),u(s))v(s)\,]\,ds.$$

This is first shown to exist as a Gâteaux differential and then as a Fréchet differential.

The mapping from controls into trajectories we denote by $T:\mathbf{R}^n\otimes U\to X$. It is the mapping $T(a,u)$ such that $\Psi(T(a,u),u)=0$; that is $T(a,u)(t)=x(t)$ where

$$x(t)-a-\int_{t_0}^{t}f(x(s),u(s))\,ds=0,$$

or, at all continuity points of u,

$$\begin{cases}\dfrac{dx(t)}{dt}=f(x(t),u(t)),\\ x(t_0)=a.\end{cases}$$

Recalling our analysis of Section 12.4 we see that if $\Phi(t;t_0)$ is the fundamental matrix of the system of differential equations, (the variational equations,)

$$\frac{dy}{dt}=f_x(x(t),u(t))y, \qquad t_0\leqslant t\leqslant t_1,$$

(pieced together in a continuous fashion at the discontinuities of $u(t)$) then

$$dT(p,u;q,v)(t)=\Phi(t;t_0)\left[q+\int_{t_0}^{t}\Phi(t_0;s)f_u(x(s),u(s))v(s)\,ds\right].$$

We now form the composite functional $\eta(T(p,u))(t_1)=h(T(p,u)(t_1))$ and use the chain rule to see that

$$d\eta(T(p,u)(t_1);q,v)=\langle dT(p,u;q,v),\nabla h(x(t_1))\rangle.$$

The variational equations have an *adjoint* system

$$\frac{d\lambda}{dt}=-f_x^T(x(t),u(t))\lambda, \qquad t_0\leqslant t\leqslant t_1$$

to which we attach the initial condition

$$\lambda(t_1)=\nabla h(x(t_1)).$$

It is an easy calculation to show that $\Phi^T(t_1;t)$ is a fundamental matrix for the adjoint system and that, if continuous, $\lambda(t)=\Phi^T(t_1,t)\nabla h(x(t_1))$. If we evaluate $d\eta(T(p,u)(t_1);q,v)$ in terms of $dT(p,u;q,v)$ we obtain

$$\begin{aligned}
&\langle\Phi(t_1;t_0)\left[q+\int_{t_0}^{t_1}\Phi(t_0;t)f_u(x(t),u(t))v(t)\,dt\right],\nabla h(x(t_1))\rangle\\
&\quad=\langle q,\Phi^T(t_1;t_0)\nabla h(x(t_1))\rangle\\
&\qquad+\langle\int_{t_0}^{t_1}\Phi(t_1;t)f_u(x(t),u(t))v(t)\,dt,\nabla h(x(t_1))\rangle\\
&\quad=\langle q,\lambda(t_0)\rangle+\int_{t_0}^{t_1}\langle f_u(x(t),u(t))v(t),\Phi^T(t_1;t)\nabla h(x(t_1))\,dt\rangle\\
&\quad=\langle q,\lambda(t_0)\rangle+\int_{t_0}^{t_1}\langle f_u(x(t),u(t))v(t),\lambda(t)\rangle\,dt.
\end{aligned}$$

If we apply the multiplier theorem to the minimization of $\eta(x(t_1))$ subject to $g(x(t_0))=c$, then, *provided that* $R_{F'(x(t_0))}=\mathbf{R}^n$, we are led to the existence of a multiplier $\mu \in \mathbf{R}^n$ such that, when $\bar{u}(t)$ is admissible and optimal with corresponding initial condition $\bar{x}(t_0)$ and trajectory $\bar{x}(t)$,

$$-\langle g_x(\bar{x}(t_0))q, \mu\rangle + \langle q, \lambda(t_0)\rangle + \int_{t_0}^{t_1} \langle f_u(\bar{x}(t), \bar{u}(t))v(t), \lambda(t)\rangle dt = 0$$

for all q, v. (If the variation of v, q were to be restricted, then our condition must be modified in the light of Exercise 12.9.7.) Since q and v are independent in their variation we are quickly lead to the conditions that

$$[f_u(\bar{x}(t), \bar{u}(t))]'\lambda(t) = 0, \qquad t_0 \leqslant t \leqslant t_1,$$

or, equivalently,

$$f_u^T(\bar{x}(t), \bar{u}(t))\lambda(t) = 0, \qquad t_0 \leqslant t \leqslant t,$$

and

$$\lambda(t_0) = g_x^T(\bar{x}(t_0))\mu.$$

(Here we have used continuity to make the first equality hold for all t.) That is, there are two sets of multipliers; one set μ is connected with the constraint $g(x(t_0)) = c$, similar to the terminal constraint in Section 12.9, and another set λ which are continuous solutions of the adjoint system

$$\begin{cases} \dfrac{d\lambda}{dt} = -f_x^T(x(t), u(t))\lambda, & t_0 \leqslant t \leqslant t_1, \\ \lambda(t_1) = \nabla h(\bar{x}(t_1)). \end{cases}$$

The two sets are, of course, linked.

The functional $\langle f(x(t), u(t)), \lambda(t)\rangle$ is known as the *Hamiltonian* $\boldsymbol{H}$ of the problem. It satisfies the *Pontryagin minimum principle*: *In order for* $\bar{u}(t)$ *to minimize* η *among all admissible controls we must have*

$$\boldsymbol{H}(\bar{x}(t), \bar{u}(t)) = \min_{u \in \Omega} \boldsymbol{H}(\bar{x}(t), u), \qquad t_0 \leqslant t \leqslant t_1,$$

even if Ω is not a closed set. (For a sketch of a proof of this result see for instance Luenberger [1], Chapter 9 or Padulo and Arbib [1], 743–767.)

Note: The form of the control problem specified above is made simple to clarify the structure. The following devices, plus the use of slack variables, can convert many other problems to equivalent ones of this form. For instance, if we are given a problem of minimization of

$$\int_{t_0}^{t} L(t,x(t),u(t))\,dt$$

subject to

$$\frac{dx}{dt}=f(t,x,u), \qquad t_0 \leqslant t \leqslant t_1,$$

where $x \in \mathbf{R}^n$, we can augment x to $\mathbf{R}^{n+2}$ with the definitions

$$\frac{dx_0}{dt}=L(t,x(t),u(t)), x_0(t_0)=0,$$

and

$$\frac{dx_{n+1}}{dt}=1, x_{n+1}(t_0)=t_0,$$

and set $\eta(x(t_1))=x_0(t_1)$. The literature contains many other such devices.

The whole question of *reachability* of terminal states x^1 is untouched above. For nonlinear or constrained control situations, it is a nontrivial problem, treated in the literature of optimal control under the name of *controllability*.

EXERCISES TO SECTION 12.10

1. Consider the quadratic cost functional

$$\int_{t_0}^{t_1} \langle x, Qx \rangle_{\mathbf{R}^n} + \langle u, Ru \rangle_{\mathbf{R}^m}\,dt,$$

where Q is a nonnegative symmetric $n \times n$ matrix and R is a positive definite $m \times m$ matrix, and a dynamical system

$$\frac{dx}{dt}=Ax+Bu,$$

where A is an $n \times n$ real matrix and B is an $n \times m$ real matrix. Show that the optimal control is characterized by $\lambda(t_1)=0$, and find a G such that

$$u(t)=-R^{-1}G^T\lambda(t).$$

Show that the assumption $\lambda(t) = P(t)x(t)$ where $P(t)$ is an $n \times n$ matrix of functions leads to the matrix Riccati differential equation

$$\begin{cases} \dfrac{dP}{dt} = -PF - F^T P + PGR^{-1}GP - Q, \\ P(t_1) = 0. \end{cases}$$

Then (Reid[1]) there is a unique symmetric nonnegative matrix solution on $[t_0, t_1]$. Show that the resulting $\lambda(t)$ satisfies the adjoint problem, and $u(t) = -R^{-1}G^T P(t)x(t)$ furnishes a *feedback* control law. Show that it is optimal. (Hint: Padulo and Arbib [1], 611.)

13
CONVEXITY, OPTIMIZATION, AND CONTROL

13.0 INTRODUCTION

In minimizing quadratic functionals we associated to the problem of finding the minimum of a nonlinear functional $Q(x)=(Ax,x)-2\Re(x,f)$, the problem of finding roots of its Fréchet derivative. That this indeed was so follows from $Q(x+h)=Q(x)+(h,A^*x-f)+(Ax-f,h)+(Ah,h)$ which identifies $dQ(x;h)$ with $(h,A^*x-f)+(Ax-f,h)$. In the selfadjoint case for a real Hilbert space the Fréchet derivative is $Q'(x)=2(Ax-f)$, and the method of steepest descent involves moving an optimal distance h in the opposite direction. In a general setting, finding the direction h that minimizes the Gâteaux differential $GDF(x;h)$ of a nonlinear function $f(x)$ can also lead to similar descent methods of minimization. $GDF(x,h)=0$ for all h is necessary for a minimum. Thus root finding for the Gâteaux differential is a natural equivalent to minimization. Here we relate the sets Ω on which a minimum is achieved to some topology; thus we have to decide on the type of convergence to use. We show in this chapter that convexity of Ω and weak convergence are a natural association. All spaces considered will be real, for ease of exposition.

By studying convex sets, convex functionals, and generalizations of those ideas, we develop geometric methods for optimal control problems and constrained optimization. Last, we present a survey of algorithms for unconstrained and constrained optimization.

13.1 CONVEX SETS, HYPERPLANES, AND SEPARATION

For convenience, we shall restrict our attention to real normed linear vector spaces, although a similar theory can be developed over the complex field. We start with some notation.

DEFINITION 13.1.1. If A, B are any two sets in a linear vector space X, then their *sum* $A+B$ is the set

$$\{a+b \mid a \in A, b \in B\}.$$

If, further, A and B are linear subspaces whose intersection is the set $\{0\}$, then we call it a *direct sum* and write $A \oplus B$. If γ is any scalar, then $\gamma A = \gamma(A)$ is the set

$$\{\gamma a \mid a \in A\}.$$

If Γ is any subset of the scalar field, then ΓA is the set

$$\{\gamma a \mid \gamma \in \Gamma, a \in A\}.$$

DEFINITION 13.1.2. For any set $A \subset X$ and any x^0, the set $x^0 + A$ is called the *translation of A along x^0*. If A is a linear manifold, then $x^0 + A$ is called a *linear variety*.

In these terms, a set K is convex if $tK + (1-t)K \subset K$, $0 \leqslant t \leqslant 1$. Since a translation of a convex set is convex, we will frequently assume $0 \in \operatorname{int} K$, for technical reasons, in a proof, with no loss of generality.

DEFINITION 13.1.3. Let U be the closed unit sphere in the set of scalars, (i.e., $U = \{\mu \mid \mu \in \mathcal{C}, |\mu| \leqslant 1\}$.) Then a set $A \subset X$ is

1. *Balanced*, if $A = UA$;
2. *Symmetric*, if $A = (-1)A$;
3. *Absorbing*, if for any $x \in X$ there is an $\epsilon > 0$ such that $\alpha x \in A$ for all α, $0 < |\alpha| \leqslant \epsilon$. (Equivalently, A is *absorbing* if for each x there is an R with $x \in \alpha A$ for $|\alpha| \geqslant R$.)

DEFINITION 13.1.4. If A is a subset of X, then the *convex hull* of A, written $\operatorname{co} A$, is the intersection of all convex sets in X containing A. The closure of $\operatorname{co} A$ is written $\overline{\operatorname{co}} A$.

Obviously, $\operatorname{co} A$ is the smallest convex set containing A. Another description of $\operatorname{co} A$ is provided in the following:

THEOREM 13.1.1. *The convex hull of A coincides with* $B = \{x \mid x = \alpha_1 x^1 + \cdots + \alpha_n x^n, n = n_x, x^i \in A, \alpha_i \geqslant 0, \alpha_1 + \cdots + \alpha_n = 1\}$.

PROOF. Clearly B is convex and $B \supset A$ so $B \supset \operatorname{co} A$. To show $B \subset \operatorname{co} A$, we perform an induction on $n = n_x$. Let $K \supset A$, where K is convex. If $x = \alpha_1 x^1 \in B$, then $\alpha_1 = 1$, $x^1 = x \in A \subset K$. Assume: If $x = \alpha_1 x^1 + \cdots + \alpha_k x^k$, $x^i \in A$, $k \leqslant n-1$, $\alpha_1 + \cdots + \alpha_k = 1$, $\alpha_i \geqslant 0$, $i = 1, \ldots, k$, then $x \in K$. Let $x = \alpha_1 x^1 + \cdots + \alpha_n x^n \in B$. Assume: $1 > \alpha_n > 0$; otherwise $x \in K$ by the inductive hypothesis. Let $\beta = \alpha_1 + \cdots + \alpha_{n-1}$ and $\beta_k = \alpha_k / \beta$, $k = 1, \cdots, n-1$. Let $y = \beta_1 x^1 + \cdots + \beta_{n-1} x^{n-1}$. By the inductive hypothesis, $y \in K$ while $x^n \in A \subset K$. Since K is convex,

$$x = \beta y + (1-\beta) x^n \subset K.$$

Thus $B \subset K$; then $B \subset \operatorname{co} A$, which shows that $B = \operatorname{co} A$. ■

Convex sets may also be described in another more analytic way as follows:

DEFINITION 13.1.5. Let K be an absorbing convex set in a linear vector space X, and let $0 \in \operatorname{int} K$. For any $x \in X$, we define

$$\phi(x) = \inf_{\alpha} \{\alpha \mid \alpha > 0, x \in \alpha K\}.$$

The function $\phi(x)$ is known as the *gauge function* or *Minkowski functional* of K.

LEMMA 13.1.1. *Let K be a convex absorbing set in a linear vector space X. If $0 \in \operatorname{int} K$, then the gauge function of K satisfies*

1. $\phi(0) = 0$, $\phi(\lambda x) = \lambda \phi(x)$ if $\lambda > 0$;
2. $\phi(x+y) \leqslant \phi(x) + \phi(y)$;

further if, K is balanced, then

3. $\phi(\lambda x) = |\lambda| \phi(x)$.

PROOF. It is obvious that $\phi(0) = 0$. To prove condition 2 we note that, for positive α, β, if $x \in \alpha K, y \in \beta K$, then $x + y \in \alpha K + \beta K = (\alpha + \beta) K$, so $\phi(x+y) \leqslant \alpha + \beta$, or $\phi(x+y) \leqslant \inf\{\alpha + \beta \mid x \in \alpha K, y \in \beta K\} = \phi(x) + \beta$. Taking the infimum over β completes the proof of condition 2. If $x \in \alpha K$ and $\lambda > 0$, then $\lambda x \in \lambda \alpha K$, so $\phi(\lambda x) \leqslant \lambda \alpha$, or $\phi(\lambda x) \leqslant \lambda \phi(x)$. Now replace x by λx, λ by λ^{-1}, and read this as $\phi(x) \leqslant \lambda^{-1} \phi(\lambda x)$, so $\phi(\lambda x) = \lambda \phi(x)$. This completes the proof of condition 1. If K is balanced and $\lambda \neq 0$, let $x \in \alpha K$. Then $(\lambda / |\lambda| \alpha) x \in K$, so $\lambda x \in |\lambda| \alpha K$, $\phi(\lambda x) \leqslant |\lambda| \alpha$, and $\phi(\lambda x) \leqslant |\lambda| \phi(x)$. Replacing x by λx, and λ by λ^{-1}, reverses the inequality. ■

The following theorem characterizes convex sets with interior.

THEOREM 13.1.2. *Let X be a normed linear vector space. Let K be a convex absorbing set that contains 0. Let $\phi(x)$ be its gauge function. Let $K_1=\{x|\phi(x)<1\}$ and $K_2=\{x|\phi(x)\leqslant 1\}$. Then*

1. $\phi(x)\leqslant 1$ *whenever* $x\in\overline{K}$;
2. $\phi(x)<1$ *implies* $x\in\operatorname{int}K$;
3. $\operatorname{int}K\subset K_1\subset K\subset K_2\subset\overline{K}$;
4. $K=K_1$ *if* K *is open*;
5. $K=K_2$ *if* K *is closed*;
6. *If* $\phi(x)$ *is continuous then* $K_1=\operatorname{int}K, K_2=\overline{K}$;
7. $\phi(x)$ *is continuous if and only if* $0\in\operatorname{int}K$.

PROOF. From $x\in K$ we see that $\phi(x)\leqslant 1$. If $\phi(x)<1$, then there is an α, $0<a<1$, with $\alpha^{-1}x\in K$. Since $0\in K$, and K is convex, $\alpha(\alpha^{-1}x)+(1-\alpha)0\in K$. Suppose $x\in\operatorname{int}K$. Then for α near 1, $\alpha x\in K$, so choose $\alpha>1$ so that $\alpha x\in K$. Then $x\in\alpha^{-1}K$, $\phi(x)<\alpha^{-1}<1$ so $x\in K_1$. By the above, $\operatorname{int}K\subset K_1\subset K\subset K_2$. Now let $\phi(x)=1$. Then $\alpha x\in K_1\subset K$, $0<\alpha<1$, so $ax\to x\in\overline{K}$, $\alpha\to 1$. (This proves conditions 1–3.) The last remark shows $\phi(x)=1$ implies $x\notin\operatorname{int}K$; then K open implies $\phi(x)<1$ if $x\in K$. So $K=K_1$. Clearly K closed implies $K=K_2$.

Suppose now that $\phi(x)$ is continuous. Then the inverse image of an open set is open, so K_1 is open and $K_1=\operatorname{int}K$. Similarly K_2 is closed and $K_2=\overline{K}$. Since $0\in K_1$, $0\in\operatorname{int}K$. Next let $0\in\operatorname{int}K$ and U be a sphere about 0 in $\operatorname{int}K$. On U, $\phi(x)<1$, and on $\epsilon U, \phi(x)<\epsilon$, since $\epsilon^{-1}x\in U$, so $\phi(\epsilon^{-1}x)=\epsilon^{-1}\phi(x)<1$. Thus ϕ is continuous at $x=0$. Since ϕ satisfies the triangle inequality, it also satisfies

$$-\phi(y-x)\leqslant\phi(x)-\phi(y)\leqslant\phi(x-y).$$

Then ϕ is continuous everywhere.

Last, let K be bounded. If $x\neq 0$, there is a sphere U about 0 not containing x. Then for some β, $K\subset\beta U$. If $x\in\alpha K$, $x/\alpha\in\beta U$ or $x/\alpha\beta\in U$. If $\alpha\beta\leqslant 1$, $x=\alpha\beta(1/\alpha\beta)x\in U$, which is impossible. Thus $\alpha\beta>1$, $\alpha>1/\beta$, and $\phi(x)\geqslant 1/\beta>0$. Then $\phi(x)=0$ implies $x=0$. ■

If K is a symmetric convex set and $0\in\operatorname{int}K$, then its gauge function is a seminorm. If we are given a seminorm, then the unit ball it defines is a symmetric convex set with 0 in its interior, and it is the corresponding gauge function.

A final characterization of convex sets is obtained in terms of the generalization, to infinite dimensions, of the notion of a hyperplane.

DEFINITION 13.1.6. Let M be a subspace of a normed linear vector space X. If the factor space X/M is finite dimensional and of dimension k, $k=1,2,3,\ldots$, then we say that M is of *finite codimension* k and write $\operatorname{codim} M=\dim X/M=k$. If M is of codimension 1, then any translation of M is called a *hyperplane*.

LEMMA 13.1.2. *If M is of codimension k in a linear vector space X, then there exist $x^1,\ldots,x^k$ in X, such that $X=\mathbf{S}\{x^1,\ldots,x^k\}\oplus M$.*

The proof is left as an exercise. These algebraic concepts can be converted to analytic form as follows:

THEOREM 13.1.3. *Let X be a linear vector space, X^f the space of all linear functionals on X. The null space of any $x^f\in X^f$ is of codimension 1. Any subspace of codimension 1 is the null space of some $x^f\in X^f$.*

PROOF. First, let $\operatorname{codim} M=1$. Let $X=\mathbf{S}\{y,M\}$, $y\notin M$. Then $x\in X$ means $x=\lambda y+m$, $m\in M$. Define x^f by $\langle x,x^f\rangle=\lambda$. Then $\langle x,x^f\rangle=0$ if and only if $x\in M$. Conversely if x^f is given and $x^f\neq 0$, let y be such that $\langle y,x^f\rangle=\alpha\neq 0$. Then for any x, set $\beta=\langle x,x^f\rangle$ and $x=[x-(\beta/\alpha)y]+(\beta/\alpha)y$. This is a representation of x as a direct sum of a root of x^f and a fixed element not a root of x^f. Thus $X=M\oplus\mathbf{S}\{y\}$. ■

COROLLARY 1. *If M is of codimension k then there exist k vectors in X^f, $x^{1f},\ldots,x^{kf}$ with respective null spaces $M_1,\ldots,M_k$, such that $M=\cap_{i=1}^{k}M_i$.*

COROLLARY 2. *If $x^f\in X^f$, $x^f\neq 0$ then $\{x|\langle x,x^f\rangle=\alpha\}$ is a translation of a subspace of codimension 1. If M is a translation of a subspace of codimension 1 and $0\notin M$, then $M=\{x|\langle x,x^f\rangle=1\}$ for a unique $x^f\neq 0$.*

COROLLARY 3. *Let X be a normed linear vector space. A hyperplane in X is closed if and only if $x^f=x'\in X'$.*

(If X is a Banach space, this follows from the closed graph theorem. If a hyperplane is not closed, then it is everywhere dense in X. (See, e.g., Taylor [1], 139–140).)

Examples of hyperplanes that are not closed arise very naturally in the minimization of quadratic functionals whose linear operator A is not strictly positive (see Section 11.6).

If we consider closed hyperplanes, then they have an equation $\langle x,x'\rangle=\alpha$ for some fixed α. Any such hyperplane determines two open half spaces $\{x|\langle x,x'\rangle<\alpha\}$ and $\{x|\langle x,x'\rangle>\alpha\}$, and their closures $\{x|\langle x,x'\rangle\leqslant\alpha\}$ and $\{x|\langle x,x'\rangle\geqslant\alpha\}$.

DEFINITION 13.1.7. If a set A satisfies either $A\cap\{x|\langle x,x'\rangle<\alpha\}=\varnothing$, or $A\cap\{x|\langle x,x'\rangle>\alpha\}=\varnothing$, we say that A *lies to one side of the hyperplane* $\pi_{x'}=\{x|\langle x,x'\rangle=\alpha\}$. If, in addition, $A\cap\pi_{x'}=\varnothing$, we say that A *lies strictly to one side* of $\pi_{x'}$. If K_1 and K_2 lie(strictly) on opposite sides of a hyperplane $\pi_{x'}$, we say that $\pi_{x'}$*(strictly) separates* them. (If K is convex, then $K\cap\pi_{x'}=\varnothing$ is both necessary and sufficient in order that K lies strictly to one side of $\pi_{x'}$.) If K lies to one side of $\pi_{x'}$ and $\bar{K}\cap\pi_{x'}\neq\varnothing$, then $\pi_{x'}$ is called a *supporting hyperplane* of $\bar{K}$.

The relations between hyperplanes and convex sets are described by the following:

THEOREM 13.1.4. *Let X be a normed linear vector space.*

1. *Let B be a linear manifold, K an open convex set, and* $(B+x^0)\cap K=\varnothing$. *Then there is a hyperplane* $\pi_{x'}$ *containing* $B+x^0$ *such that* $\bar{K}$ *lies to one side of* $\pi_{x'}$ *and K lies strictly to one side of* $\pi_{x'}$.
2. *Through every boundary point of K there passes at least one supporting hyperplane of* $\bar{K}$.
3. $\bar{K}$ *is the intersection of all the closed half spaces containing K determined by supporting hyperplanes of* $\bar{K}$.

PROOF. (Condition 1) By a shift of the origin we can assume $0\in K$. Let $M=B+x^0$. Let $\phi(x)$ be the gauge function of K. Since $x\in K$ implies $\phi(x)<1$, choose αK as a neighborhood of $0\in K$. Then $x\in\alpha K$ implies $x/\alpha\in K$ so $\phi(x)<\alpha$ on αK. Let $B^1=\mathbf{S}\{B,x^0\}$; $\bar{x}\in B^1$ means $\bar{x}=b+\lambda x^0$, $b\in B$. Define f' by the equation $\langle b+\lambda x^0,f'\rangle=\lambda$. If $\lambda\neq 0$, $\bar{x}/\lambda=b/\lambda+x^0\in M$, so $\bar{x}/\lambda\notin K$, and $\phi(\bar{x}/\lambda)\geqslant 1$; this means $\phi(\bar{x})\geqslant\lambda$, so $\langle\bar{x},f'\rangle\leqslant\phi(\bar{x})$. By the Hahn-Banach theorem, we can extend f' to x' such that $\langle x,x'\rangle\leqslant\phi(x)$ for all x in X. Since $\phi(x)<1$ on K, $\langle x,x'\rangle<1$ on K. Since $\phi(x)\leqslant 1$ on $\bar{K}$, $\langle x,x'\rangle\leqslant\phi(x)\leqslant 1$, so K lies to one side of $\pi_{x'}$, as does $\bar{K}$. If we take $M=x^0$ where x^0 is a boundary point of K, we have condition 2. To prove condition 3, let $\hat{x}\notin\bar{K}$. Let $y\in\operatorname{int}K$, let z be the boundary point of K on the line segment connecting $\hat{x}$ and y. Then the hyperplane through z determined above excludes $\hat{x}$. Thus $\bar{K}$, and nothing else is in the intersection in question. ∎

An immediate translation of condition 1 is the following result often called the *geometric form of the Hahn-Banach theorem*.

COROLLARY 1. *Let K be an open convex set in a real normed linear vector space X and N a linear variety such that* $N\cap K=\varnothing$. *Then there is a closed hyperplane containing N, and disjoint from K.*

Another version of condition 1 is often called the *separation form of the Hahn-Banach theorem*.

COROLLARY 2. *If K_1, K_2 are two disjoint nonempty convex sets in a normed linear vector space X and at least one, say K_1, has a nonempty interior, then there exists a closed hyperplane that separates them.*

PROOF. Let $K = K_1 + (-1)K_2$. Then $0 \notin K$, which is nonempty and convex, with a nonempty interior (why?). We apply condition 1 to conclude the existence of $x' \in X'$ with $\langle x - y, x' \rangle \leqslant \langle 0, x' \rangle = 0$ for all $x \in K_1$, $y \in K_2$. That is for any $x \in K_1$, $y \in K_2$,

$$\langle x, x' \rangle \leqslant \langle y, x' \rangle.$$

Hence there exists α such that

$$\sup_{x \in K_1} \langle x, x' \rangle \leqslant \alpha \leqslant \inf_{y \in K_2} \langle y, x' \rangle.$$

Then $\pi_{x'} = \{x | \langle x, x' \rangle = \alpha\}$ separates K_1, K_2. ■

We can also rephrase condition 3 as follows:

COROLLARY 3. *If K is a convex set with interior, then K is the intersection of all the half spaces containing K determined by all the supporting hyperplanes of $\bar{K}$.*

PROOF. If $x \in \partial K$ but $x \notin K$, then the open half space determined by any supporting hyperplane through x also contains K. ■

From condition 3 we also conclude the following important result:

COROLLARY 4. *In a normed linear vector space, a convex set is closed in the weak topology if and only if it is closed in the norm topology.*

It follows that a minimization problem on a convex set may be solved by finding a weakly convergent minimizing sequence; if the convex set is weakly closed, then it is strongly closed, and the weak limit is a true solution of the problem. The corollary has the following consequence:

COROLLARY 5. *If $x^n \rightharpoonup x$, $n \to \infty$ in a closed convex set K, then in* $\operatorname{co}\mathbf{S}\{x^n\}$ *there is a sequence $\{y^n\}$ such that $y^n \to x, n \to \infty$.*

PROOF. The set $S = \mathbf{S}\{x^1, x^2, x^3, \ldots\}$ has a convex hull $\operatorname{co} S$; $x \in \operatorname{co} S$ since it is in the weak closure equals norm closure of $\operatorname{co} S$. Then we choose a sequence in $\operatorname{co} S$ which converges to x in norm. ■

There is another form of the Kuhn-Tucker conditions (obtained in Section 12.9 as a consequence of the multiplier theorem) which is a consequence of Corollary 2. This form is known as the *Kuhn-Tucker saddle point condition*. It requires the following:

DEFINITION 13.1.8. A functional $f(x)$: $X \to \mathbf{R}$, where X is a linear vector space, is called *convex* if $f(x)$ has a convex domain and

$$f(\lambda x+(1-\lambda)y) \leqslant \lambda f(x)+(1-\lambda)f(y),\ 0<\lambda<1.$$

If the inequality is strict, then $f(x)$ is *strictly convex*.

Let $f(x): X \to \mathbf{R}$ be a convex functional on a convex subset K of a normed linear vector space X, and $g(x)=(g_1(x),\ldots,g_m(x)): X \to \mathbf{R}^m$ where each $g_i(x)$ is convex. The (*constrained*) *convex programming problem* is that of finding an $\bar{x} \in K$ that minimizes $f(x)$ on K subject to $g(x) \in \mathbf{R}^{m-}$.

Suppose that for every nonzero $\lambda \in \mathbf{R}^{m-}$ *there is an* $x \in K$ *such that*

$$-\sum_{i=1}^{m} \lambda_i g_i(x) < 0.$$

Then there exists $\bar{\lambda} \in \mathbf{R}^{m-}$ *such that*

$$f(\bar{x}) - \sum_{i=1}^{m} \bar{\lambda}_i g_i(\bar{x}) = \min_{x \in K}\left[f(x) - \sum_{i=1}^{m} \bar{\lambda}_i g_i(x)\right].$$

Further, if $\mu \in \mathbf{R}^{m-}$, *then*

$$f(x) - \sum_{i=1}^{m} \bar{\lambda}_i g_i(x) \geqslant f(\bar{x}) - \sum_{i=1}^{m} \bar{\lambda}_i g_i(\bar{x}) \geqslant f(\bar{x}) - \sum_{i=1}^{m} \mu_i g_i(\bar{x}).$$

We show this by defining the following two obviously convex sets in $\mathbf{R}^{m+1}$:

$$K_1 = \{ y \in \mathbf{R}^{m+1} | f(x) \leqslant y_0, g_i(x) \leqslant y_i,\ i=1,\ldots,m, \text{ for some } x \in K_1\};$$

and

$$K_2 = \{ y \in \mathbf{R}^{m+1} | y_0 < f(\bar{x}), y_i < 0,\ i=1,\ldots,m\}.$$

Thus $K_1 \cap K_2 = \varnothing$, and K_2 is an open convex set with interior. Then there is a nonzero hyperplane in $\mathbf{R}^{m+1}$ which separates K_1 and K_2. Let λ_0,

$-\lambda_1, \ldots, -\lambda_m$, be the coefficients of that hyperplane, so that

$$\lambda_0 f(\bar{x}) - \sum_{i=1}^{m} \lambda_i(-|y_i|) \leq \inf_{x \in K} \left[\lambda_0 f(x) - \sum_{i=1}^{m} \lambda_i g_i(x) \right]$$

for all values of $|y_1|, \ldots, |y_m|$. If any λ_i, $i = 1, \ldots, m$, were positive, letting the corresponding $|y_i| \to \infty$ leads to a contradiction. Letting $|y_i| \to 0$, $i = 1, \ldots, m$, implies

$$\lambda_0 f(\bar{x}) \leq \lambda_0 f(\bar{x}) - \sum_{i=1}^{m} \lambda_i g_i(\bar{x}),$$

or $\sum_{i=1}^{m} \lambda_i g_i(\bar{x}) \leq 0$; but $\sum_{i=1}^{m} \lambda_i g_i(\bar{x}) \geq 0$ (why?), so $\sum_{i=1}^{m} \lambda_i g_i(\bar{x}) = 0$. If $(\lambda_1, \ldots, \lambda_m) = 0$, then $\lambda_0 \neq 0$, while $\underline{y} < f(\bar{x}) < \bar{y}$ implies $\lambda_0 \underline{y} \leq \lambda_0 f(\bar{x}) \leq \lambda_0 \bar{y}$; thus $\lambda_0 > 0$.

If $(\lambda_1, \ldots, \lambda_m) \neq 0$, then there is an $x \in k$ with $-\sum_{i=1}^{m} \lambda_i g_i(x) < 0$. Pick $\bar{y} \geq f(x)$. Then

$$\lambda_0 \bar{y} - \sum_{i=1}^{m} \lambda_i g_i(x) \geq \lambda_0 f(x) - \sum_{i=1}^{m} \lambda_i g_i(\bar{x}),$$

or, since the right side equals $\lambda_0 f(\bar{x})$,

$$\lambda_0 [\bar{y} - f(\bar{x})] \geq \sum_{i=1}^{m} \lambda_i g_i(x) > 0.$$

Again $\lambda_0 > 0$. Thus we can divide through by λ_0 and $\bar{\lambda} = (\lambda_1/\lambda_0, \ldots, \lambda_m/\lambda_0) \in \mathbf{R}^{m-}$ makes

$$f(x) - \sum_{i=1}^{m} \bar{\lambda}_i g_i(x) \geq f(\bar{x}) = f(\bar{x}) - \sum_{i=1}^{m} \bar{\lambda}_i g_i(\bar{x})$$

for all $x \in K$. If for some $\mu \in \mathbf{R}^{m-}$,

$$f(\bar{x}) - \sum_{i=1}^{m} \mu_i g_i(\bar{x}) > f(\bar{x}) - \sum_{i=1}^{m} \bar{\lambda}_i g_i(\bar{x}) = f(\bar{x}),$$

then $-\sum_{i=1}^{m} \mu_i g_i(\bar{x}) > 0$. But $\mu \in \mathbf{R}^{m-}$ and $g(\bar{x}) \in \mathbf{R}^{m-}$, so $\sum_{i=1}^{m} \mu_i g_i(x) \leq 0$, which is a contradiction.

No smoothness conditions are set for f, g. In the presence of differentiability, we clearly recover the results obtained in Section 12.9, provided that $\bar{x}$ is an interior point of K.

Observe that we have obtained a dual formulation of the convex programming problem: find

$$\sup_{\mu \in \mathbf{R}^{m-}} \left\{ \inf_{x \in K} \left[f(x) + \langle g(x), \mu \rangle \right] \right\}.$$

This dual problem is known as the *Lagrangian dual* (e.g., Bazaraa and Shetty [1], 167–176).

The restriction of the constraints to mappings $g(x): X \to \mathbf{R}^{m-}$ is connected with the problem of extending the notion of orthant, that is, of $\mathbf{R}^{m-}$, to an arbitrary Banach space. This extension is done in Section 13.7; in that context, a more general form of the saddle point conditions may be obtained.

EXERCISES TO SECTION 13.1

1. If A, B are convex, then show that $\alpha A + \beta B$ is convex. If A is convex, then show that $\alpha A + \beta A = (\alpha + \beta)A$ for all positive α, β.
2. For any scalars, show that $\operatorname{co}(\lambda A + \gamma B) = \lambda \operatorname{co} A + \gamma \operatorname{co} B$ and that $\operatorname{co}(A \cup B) = \cup_\alpha \{\alpha \operatorname{co} A + (1-\alpha) \operatorname{co} B \,|\, 0 \leqslant \alpha \leqslant 1\}$.
3. Show that $\overline{\operatorname{co}} A = \operatorname{co} \bar{A}$.
4. Show that if A is convex, then $\bar{A}$ is convex, as in $\operatorname{int} A$.
5. Show that if A is open then $\operatorname{co} A$ is open.
6. If $T \in [X \to Y]$ and $K \subset X$ is convex, show that its image $T(K)$ in Y is convex.
7. In a normed linear vector space X, show that $\|x\|$ is a gauge function for the unit ball. Is the unit sphere (a) convex; (b) absorbing; (c) symmetric; (d) balanced?
8. Prove Lemma 13.1.2.
9. Show that if K is convex, then $K \cap \pi_{x'} = \varnothing$ is both necessary and sufficient in order that K lie strictly to one side of $\pi_{x'}$.
10. Let A be a bounded positive selfadjoint operator on a real Hilbert space H and $Q(x) = (Ax, x) - (x, f) - (f, x)$. Show that if the set $E_{\boldsymbol{\alpha}} = \{x \,|\, Q(x) \leqslant \alpha\}$ contains a point $\hat{x}$ with $Q(\hat{x}) < \alpha$ then E_α is convex. What if E_α contains only points at which $Q(x) = \alpha$? (Hint: Consider the supporting hyperplane $x^0 + h$, where $x^0 \in \partial E_\alpha$ and $2(Ax^0 - f, h) = -\alpha$, and the quadratic form $Q(x^0 + h)$.)
11. Show that if K is a closed convex set in a normed linear vector space X, then any $x' \in X'$ such that

$$\inf_{x \in K} \langle x, x' \rangle > -\infty$$

achieves its minimum on K at points $\hat{x}$ such that x' is a supporting hyperplane of K at $\hat{x}$.

12. If $f: X \to \mathbf{R}$ is a continuous convex functional on a Banach space X and $x^n \rightharpoonup x$, $n \to \infty$, then

$$\liminf_{n \to \infty} f(x^n) \geqslant f(x).$$

(Hint: Use Corollary 6 of Theorem 13.1.4.)

13. Let $f(x): X \to \mathbf{R}$ be a convex functional on a normed linear vector space X, with domain K. Let $\bar{x} \in \operatorname{int} K$ and on $\{x \mid \|x - \bar{x}\| < \epsilon\}$, let $f(x) \leqslant \alpha < \infty$, for some $\epsilon > 0$. Show that f is continuous at $\bar{x}$. (Hint: Let $x = f(\bar{x}) = 0$. If $\|x\| \leqslant \delta$ then $-\epsilon\alpha \leqslant (1+\epsilon) f(0) - \epsilon f(-x/\epsilon) \leqslant f(x) \leqslant (1-\epsilon) f(0) + \epsilon f(x/\epsilon) < \epsilon\alpha$ provided that $\delta < \epsilon^2$.)
14. Show that the set K_1 used in the derivation of the Kuhn-Tucker conditions is convex.
15. Let $x = \mathbf{R}^2$ with the norms of l_p, $1 \leqslant p \leqslant \infty$. For each fixed p, find the support hyperplanes of the unit sphere $\|x\| = 1$.

13.2 FORMS OF CONVEXITY AND CONTINUITY OF FUNCTIONALS

We have seen that weak convergence is naturally associated with convexity. Unfortunately, such a simple set as $\{x \mid \|x\| \leqslant \alpha\}$, in a normed linear vector space X, is convex, but $\|x\|$ is *not* continuous with respect to weak convergence. It is easy to see this in the Hilbert space l_2: The sequence of vectors e^j, where $(e^j)_i = \delta_i^j$, are all of norm 1, but $e^j \rightharpoonup 0$, $j \to \infty$, which is of norm 0. On the other hand we can readily prove the following:

LEMMA 13.2.1. *If $x^n \rightharpoonup x$, $n \to \infty$ then $\|x\| \leqslant \liminf_{n \to \infty} \|x^n\|$.*

PROOF. There is an $x' \in X'$ with $\|x'\| = 1$ such that $\|x\| = \langle x, x' \rangle$. Then, for any $\epsilon > 0$, there is an N_ϵ such that $\|x\| = \langle x, x' \rangle \leqslant \langle x^n, x' \rangle + \epsilon$, $n \geqslant N_\epsilon$. So $\|x\| \leqslant \|x^n\| + \epsilon$, $n \geqslant N_\epsilon$, or $\|x\| \leqslant \liminf_{n \to \infty} \|x^n\| + \epsilon$ for all ϵ. Hence $\|x\| \leqslant \liminf_{n \to \infty} \|x^n\|$. ■

We can extend our definition of continuity to topologies other than the norm topology on a normed linear vector space. In a given topology (weak, weak*, norm, ...) we say that a functional $f: X \to \mathbf{R}$ is continuous if convergence of x^n to x *in that topology* implies $\lim_{n \to \infty} f(x^n)$ exists and equals $f(x)$. If instead, we only know that

$$f(x) \leqslant \liminf_{n \to \infty} f(x^n),$$

we call f *lower semicontinuous* in that topology: for example, *weakly lower semicontinuous*.

DEFINITION 13.2.1. In a normed linear vector space X, we call a set S *weakly (sequentially) compact* if every sequence $x^n \in S$, $n=1,2,\dots$, possesses a weakly convergent subsequence $\{z^{n_k}\}$ with a (weak) limit $x \in S$. (A similar definition holds for the weak* topology.)

From Section 9.4 we recall that a weakly (sequentially) compact set S is bounded. It is also *weakly sequentially closed*: every weakly convergent sequence in S has a limit in S. Boundedness and this closure are clearly necessary and sufficient for weak (sequential) compactness. For the weak* topology we show the following:

LEMMA 13.2.2. *If X is separable, every bounded sequence in X' contains a weak* convergent subsequence (that is, the closed bounded sets in X' are weak* (sequentially) compact).*

PROOF. Let $\{x'^n\} \in X'$ be a bounded sequence. Let $\{x^j\}$ be a dense denumerable set in X. For each fixed j, pick a subsequence of the last chosen sequence in X' such that $\langle x^j, x'^{n_k}\rangle$ converges. Then the diagonal sequence $\{x'^{n_n}\}$ converges at all x^j, hence for all $x \in X$. ■

If we translate Theorem 4.2.1 and its corollaries, noting that the proof of Theorem 4.2.1 utilizes the sequential form of compactness, we obtain:

THEOREM 13.2.1. *Let S be a weakly (sequentially) compact subset of a normed linear vector space X and $f: X \to \mathbf{R}$ a weakly lower semicontinuous functional. Then $\inf_{x \in S} f(x) > -\infty$ and there is an $s \in S$ with*

$$f(x) = \inf_{x \in S} f(x).$$

It is often convenient to assume that convex functions are defined everywhere. This can be accomplished by extending the defining equation

$$f(\lambda x + (1-\lambda)y) \leqslant \lambda f(x) + (1-\lambda) f(y), \qquad 0 \leqslant \lambda \leqslant 1,$$

to hold *even when the values $\pm\infty$ are admitted to the range of f*; since $\overline{\mathbf{R}} = \mathbf{R} \cup \{+\infty\} \cup \{-\infty\}$ this means $f: X \to \overline{\mathbf{R}}$, and that the defining equation is valid whenever the right side makes sense (i.e., is not $\infty - \infty$). Then any function f convex on a set $S \subset X$ is convex on X if we set $f(x) = \infty$, $x \notin S$. (Note that, if $f(x) = -\infty$ at any y, then $f(x) = \pm\infty$ everywhere. Thus if $f(x)$ takes on a finite value anywhere, the interesting case, it omits the value $-\infty$. With this convention, the indicator function of a set S,

$$\Phi_S(x)=\begin{cases}0, & x\in S,\\ \infty, & x\notin S,\end{cases}$$

is convex if and only if S is convex.) This extension of the range to $\overline{\mathbf{R}}$ will be used without regard to the convexity of f.

DEFINITION 13.2.2. If $f:X\to\overline{\mathbf{R}}$, then define the collection of level sets $L(f;\alpha)$, $\alpha\in\mathbf{R}$, by

$$L(f;\alpha)=\{x\,|\,f(x)\leqslant\alpha\},$$

the collection $L_\Omega(f;\alpha)$, $\alpha\in\mathbf{R}$, by

$$L_\Omega(f;\alpha)=\{x\in\Omega\,|\,f(x)\leqslant\alpha\},$$

and the collection

$$L_\Omega(f;x^0)=L_\Omega(f;f(x^0)).$$

The weak lower semicontinuous functionals are characterized by the following:

THEOREM 13.2.2. *Let $f:X\to\overline{\mathbf{R}}$ where X is a normed linear vector space. Then f is weakly lower semicontinuous if and only if, for every $\alpha\in\overline{\mathbf{R}}$, $L(f;\alpha)$ is weakly closed.*

PROOF. Suppose $L(f,\alpha)$ is weakly closed for every $\alpha\in\overline{\mathbf{R}}$, that $x^n\rightharpoonup x$, $n\to\infty$, yet that for $\epsilon>0$,

$$f(x)>\liminf_{n\to\infty}f(x^n)+\epsilon>-\infty.$$

From the definition of lim inf, there is a subsequence $\{x^{n_i}\}$ such that $f(x^{n_i})\leqslant\liminf_{n\to\infty}f(x^n)+\epsilon/2$, $i=1,2,3,\ldots$. But $x^{n_i}\rightharpoonup x$, $n_i\to\infty$, in a weakly closed set $L(f;\liminf_{n\to\infty}f(x^n)+\epsilon/2)$ so x is in that set, which is a contradiction. Thus $f(x)\leqslant\liminf_{n\to\infty}f(x^n)$. If $\liminf_{n\to\infty}f(x^n)=-\infty$, then there is a subsequence with $x^{n_i}\in L(f;-i)$; thus $f(x)\leqslant -i$ for all i.

Conversely, if $f(x)$ is weakly lower semicontinuous, then let $x^n\rightharpoonup x$, $n\to\infty$. Then $f(x)\leqslant\liminf_{n\to\infty}f(x^n)$. If $\{x^n\}\in L(f;\alpha)$, this implies that $x\in L(f;\alpha)$, so that $L(f;\alpha)$ is weakly closed. ■

LEMMA 13.2.3. *If $f(x)$ is lower semicontinuous, then for all $\alpha\in R$, $L(f;\alpha)$ is closed.*

PROOF. For any $\alpha \in R$, let $\{x^n\}$ be a Cauchy sequence in $L(f;\alpha)$ with limit x. If $f(x) > \alpha + \epsilon$, $\epsilon > 0$, then for some n, $f(x^n) \geqslant f(x) \geqslant \alpha + \epsilon/2$, which is impossible. Hence $x \in L(f;\alpha)$, which is thus a closed set. ■

COROLLARY 1. *If $f(x)$ is lower semicontinuous and, for all $\alpha \in \mathbf{R}$, $L(f;\alpha)$ is convex, then $f(x)$ is weakly lower semicontinuous.*

PROOF. In this case, $L(f;\alpha)$ is closed and convex, so weakly closed. This, with Theorem 13.2.2, implies the corollary. ■

In particular, *a convex lower semicontinuous functional is weakly lower semicontinuous*. But for instance, if $X = \mathbf{R}$ and $f(x) = x^3$, then $L(x^3;\alpha)$ is convex for all $\alpha \in \mathbf{R}$ but x^3 is not convex. A weakening of convexity is described in the following:

DEFINITION 13.2.3. $f: X \to \bar{R}$ is *quasiconvex* if $f(\lambda x + (1-\lambda)y) \leqslant \max[f(x), f(y)]$, $0 < \lambda < 1$. If $f(\lambda x + (1-\lambda)x) < \max[f(x), f(y)]$, $0 < \lambda < 1$, then $f(x)$ is *strictly quasiconvex*.

[Clearly, every (strictly) convex function is (strictly) quasiconvex.] A statement equivalent to this definition is $f: X \to \mathbf{R}$ is quasiconvex if and only if for all $\alpha \in R$, $L(f;\alpha)$ is convex. We can restate the corollary as:

THEOREM 13.2.3. (*Polyak*, [2],) *In a Banach space X, a quasiconvex lower semicontinuous functional is weakly lower semicontinuous.*

COROLLARY 1. *Every quasiconvex functional on a Banach space X attains a minimum on any weakly (sequentially) compact set $S \subset X$.*

COROLLARY 2. *If, in addition, X is reflexive, f achieves a minimum on an arbitrary convex closed bounded set.*

PROOF. In a reflexive space, a bounded closed set is weakly (sequentially) compact. ■

COROLLARY 3. *If f is strictly quasiconvex, then the minimum is unique.*

PROOF. If $f(x^1) = f(x^2) = \min f(x)$, then $f(x^1 + \alpha(x^2 - x^1)) < \max[f(x^1), f(x^2)]$, a contradiction. ■

Thus, if a functional is strictly quasiconvex, it achieves a unique minimum on every convex closed bounded set of a reflexive space. The converse is also true (Polyak [2]). Thus we are about as close to the correct class of functionals for convex optimization as we can get.

In the presence of a Gâteaux derivative weak lower semicontinuity and convexity may be guaranteed analytically.

THEOREM 13.2.4. *On a real normed linear vector space X, let $f(x): X \to \mathbf{R}$ possess a Gâteaux derivative $Df(x)$. If there is an $\epsilon > 0$ such that*

$$f(x+h) - f(x) - \langle h, Df(x) \rangle \geqslant 0$$

for all h with $\|h\| < \epsilon$, then $f(x)$ is weakly lower semicontinuous at x. If the same inequality holds for all x, $x+h$ in some open convex set $K \subset X$, then $f(x)$ is convex and weakly lower semicontinuous in K. If the inequality is strict for $h \neq 0$, then $f(x)$ is strictly convex.

PROOF. Let $\{h^n\}$ be a sequence with $h^n \rightharpoonup 0$, $n \to \infty$. Then $\langle h^n, Df(x) \rangle \to 0$, $n \to \infty$, so that

$$\liminf_{n\to\infty} [f(x+h^n) - f(x)] \geqslant \liminf_{n\to\infty} \langle h^n, Df(x) \rangle = 0$$

or

$$\liminf_{n\to\infty} f(x+h^n) \geqslant f(x).$$

Now let $x, \bar{x}, y \in K$. Then

$$\langle y - \bar{x}, Df(\bar{x}) \rangle + f(\bar{x}) \leqslant f(y),$$

and

$$\langle x - \bar{x}, Df(\bar{x}) \rangle + f(\bar{x}) \leqslant f(x),$$

so

$$\langle \lambda y + (1-\lambda)x - \bar{x}, Df(x) \rangle + f(x) \leqslant \lambda f(y) + (1-\lambda) f(x),$$

for $0 < \lambda < 1$. If $\bar{x} = y + (1-\lambda)x$, this becomes

$$\langle 0, Df(\bar{x}) \rangle + f(\bar{x}) \leqslant \lambda f(y) + (1-\lambda) f(x)$$

or

$$f(\lambda y + (1-\lambda)x) \leqslant \lambda f(y) + (1-\lambda) f(x), \qquad 0 < \lambda < 1. \quad \blacksquare$$

An analytic condition that guarantees the inequality of the theorem is given by:

COROLLARY 1. *If* $\langle x-\bar{x}, Df(x)-Df(\bar{x})\rangle \geqslant 0$, *for all* x, $\bar{x}\in K$, *then*

$$f(x)-f(\bar{x})-\langle x-\bar{x}, Df(\bar{x})\rangle \geqslant 0$$

for $\|x-\bar{x}\| \leqslant \epsilon$.

PROOF. Then, by the mean value theorem,

$$\begin{aligned} f(x)-f(\bar{x}) &= \langle x-\bar{x}, Df(\bar{x}+\tau(x-\bar{x}))\rangle \\ &= \langle x-\bar{x}, Df(\bar{x})\rangle + \langle x-\bar{x}, Df(\bar{x}+\tau(x-\bar{x}))-Df(\bar{x})\rangle, \end{aligned}$$

for some τ, $0<\tau<1$. But

$$\begin{aligned} \langle x-\bar{x}, Df(\bar{x}+\tau(x-\bar{x}))-Df(\bar{x})\rangle &= \frac{\langle \tau(x-\bar{x}), Df(\bar{x}+\tau(x-\bar{x}))-Df(\bar{x})\rangle}{\tau} \\ &\geqslant 0 \end{aligned}$$

so

$$f(x)-f(\bar{x})-\langle x-\bar{x}, Df(\bar{x})\rangle \geqslant 0. \quad ■$$

DEFINITION 13.2.4. If X is a normed linear vector space and $F(x): X \to X'$ satisfies

$$\langle x-y, F(x)-F(y)\rangle \geqslant 0$$

for all x,y in some set $S \subset X$ we say that F is a *monotone operator*. If the inequality is strict for $x \neq y$, we say that F is *strictly monotone*.

The corollary may be paraphrased: *If the Gâteaux derivative is a (strictly) monotone operator, then* $f(x)$ *is (strictly) convex and weakly lower semicontinuous*. If $f(x)$ is (strictly) convex and $Df(x)$exists, then $Df(x)$ is (strictly) monotone (Vainberg [2], 51). We have encountered monotone operators in studying the minimum of a quadratic functional. In a real Hilbert space X, the quadratic functional $Q(x)=\frac{1}{2}(Ax,x)-(f,x)$ has a Fréchet derivative $(Ax-f)$, which, when $A \geqslant 0$, satisfies

$$(Ax-f-[Ay-f], x-y) = (A[x-y], x-y) \geqslant 0,$$

so that $Q(x)$ is convex and A is monotone.

An inspection of the proof of Theorem 13.2.3 is instructive: nowhere was any use made of the fact that the linear functional $Df(x)$ was a Gâteaux derivative of $f(x)$. Any other element of X' would have suited the argument. We make the following:

DEFINITION 13.2.5. A linear functional $x' \in X'$, the dual space of a normed linear vector space X, is called a *subgradient* or a *support functional* at x if

$$f(x+h)-f(x) \geqslant \langle h, x' \rangle, \qquad h \in X.$$

The set of all subgradients of f at x is called the *subdifferential*, written $\partial f(x)$. (For a calculus of subdifferentials, with applications to optimization, see Pshenichnyi [1].)

We can restate Theorem 13.2.4 in the following form:

THEOREM 13.2.5. *Let $f(x): X \to \mathbf{R}$ be a functional on a real normed linear vector space X. If f has a subgradient at x, then f is weakly lower semicontinuous at x. If f has a subgradient at all x in some open convex set $K \subset X$, then f is convex and weakly lower semicontinuous in K.*

Conversely, we establish:

THEOREM 13.2.6. *Let $f(x): X \to \mathbf{R}$ be a convex-functional on an open convex set K in a real normed linear vector space X. Then at every $x \in K$ there exists at least one subgradient.*

PROOF. Fix x,h in X and consider the convex function of t: $\phi(t) = f(x+th)$. It is a simple calculation to show that if $t_1 < t_2$, then

$$\frac{\phi(t_1)-\phi(t)}{t_1-t} \leqslant \frac{\phi(t_2)-\phi(t)}{t_2-t}.$$

If $t_1 < t < t_2$, and we let $t_1 \uparrow t, t_2 \downarrow t$, it follows from the monotonicity and boundedness above (below) of the left (right) side of the inequality that the limit d^- (d^+) exists and $d^- \leqslant d^+$. Now $d^- = d^-(x;h) \leqslant d^+ = d^+(x;h)$, while $d^-(x;h) = -d^+(x;h)$ and $d^+(x;h) = -d^-(x;h)$. Further for any

h^1, h^2,

$$d^+(x;h^1+h^2)=\lim_{t\downarrow 0}\frac{f(x+(h^1+h^2)t/2)-f(x)}{(t/2)}$$

$$=\lim_{t\downarrow 0}\frac{2f((x+th^1)/2+(x+th^2)/2)-2f(x)}{t}$$

$$\leqslant \lim_{t\downarrow 0}\frac{f(x+th^1)-f(x)}{t}+\lim_{t\downarrow 0}\frac{f(x+th^2)-f(x)}{t}$$

$$\leqslant d^+(x,h^1)+d^+(x,h^2).$$

That is, d^+ is a sublinear functional. Fix $\bar{h}$ and define a linear functional z' on the line $t\bar{h}$, $-\infty<t<\infty$ by $\langle t\bar{h},z'\rangle = d^+(x,t\bar{h})$. Since $d^+(x;t\bar{h})= td^+(x;\bar{h})$ if $t\geqslant 0$ (why?), and d^+ is sublinear so $0=d^+(x;0)=d^+(x;\bar{h}-\bar{h}) \leqslant d^+(x;\bar{h})+d^+(x;-\bar{h})$, we see that, for $t<0$,

$$\langle t\bar{h},z'\rangle = -|t|d^+(x;\bar{h})\leqslant |t|d^+(x;-\bar{h})=d^+(x;t\bar{h}),$$

which means that z' satisfies the conditions of the Hahn-Banach theorem. Then we know that there exists an $x'\in X'$ with $\langle h,x'\rangle \leqslant d^+(x;h)$, $\langle \bar{h},x'\rangle = d^+(x;\bar{h})$. [We note that $\langle h,x'\rangle = -\langle -h,x'\rangle \geqslant -d^+(x;-h)=d^-(x;h)$, so that

$$d^-(x;h)\leqslant \langle h,x'\rangle \leqslant d^+(x;h).$$

Thus if f has a Gâteaux derivative $Df(x)$, it must coincide with x' and lie in $\partial f(x)$.]

Reverting to $\phi(t)$, we note that we must have

$$\lim_{t\downarrow 0}\frac{\phi(t)-\phi(0)}{t}\leqslant \frac{\phi(1)-\phi(0)}{1-0}.$$

Then for any h,

$$f(x+h)-f(x)\geqslant d^+(x;h)\geqslant \langle h,x'\rangle.$$

That is, x' is a subgradient of f at x. ■

EXERCISES TO SECTION 13.2

1. Show that a real functional $f(x)$ on a normed linear vector space X is quasiconvex

   (a) If and only if, for any $y \in X$, $L(x;y) = \{x | f(x) \leqslant f(y)\}$ is a convex set; or

   (b) If and only if $L(\mathrm{f}; \alpha)$ is convex for all $\alpha \in \mathbf{R}$.

2. Let $f(x)$ be a real valued functional on a normed linear vector space X. Define the *epigraph* of $f(x) = epi\,f(x) = \{(x, \alpha) \in X \otimes \mathbf{R} | f(x) \leqslant \alpha\}$. Show that $f(x) = \inf\{\alpha | (x, \alpha) \in epi\,f\}$.
3. Show that $f(x)$ is convex if and only if $epi\,f$ is convex. (Hint: f convex and $f(x) \leqslant \alpha$, $f(y) \leqslant \beta$ implies $f(\lambda x + (1-\lambda)y) \leqslant \lambda\alpha + (1-\lambda)\beta$ (why?).) Show that $f(x)$ is weakly lower semicontinuous if and only if $epi\,f$ is weakly closed.
4. Let $f_i(x): X \to \overline{\mathbf{R}}$ be a family of functionals where $i \in I$, some index set. If $g(x) = \sup_{i \in I} f_i(x)$, then show that

$$epi\,g(x) = \bigcap_{i \in I} epi\,f_i(x).$$

   Show that if all $f_i(x)$ are convex, then $g(x)$ is convex.

5. Show that the function

$$f(x) = \begin{cases} 0, & x \neq 0 \\ 1, & x = 0, \end{cases}$$

   is strictly quasiconvex but *not* quasiconvex.

6. Show that if $f(x)$ has a Gâteaux derivative, and a second Gâteaux differential satisfying

$$GD^2 f(x; h, h) \geqslant 0,\ h \in X,$$

   for $x \in K$, a convex set, then $f(y) - f(x) - \langle y - x, Df(x) \rangle \geqslant 0$ for all $x, y \in K$, f is convex and $Df(x)$ is monotone.

7. The mapping $g(t, x(t))$ of functions $x(t) \in L_p(0, 1)$, $1 \leqslant p < \infty$ to $L_q(0, 1)$, $1/p + 1/q = 1$, where g is continuous in both variables and, for some constant c, $|g(t, x)| \leqslant c + |x|^{p-1}$, is the Gâteaux derivative of

$$f(x) = D + \int_0^1 \int_0^{x(t)} g(x, t)\, ds\, dt$$

for any constant D. Show this and show that if g is monotone nondecreasing in x for each fixed t, then g is a monotone operator and f is convex.

8. The functional on $\mathbf{R} \to \mathbf{R}$, $|t|$, is obviously convex. Find all its subgradients at all finite points t.
9. Show that $\phi(t)$, in the proof of Theorem 13.2.6, is a convex function of t.
10. In the proof of Theorem 13.2.6 consider the case in which there exists an h such that $d^-(x;h) \neq d^+(x;h)$. Show that there exists more than one subgradient at x. (Hint: $-d^-(x;h) = d^+(x;-h) \neq d^+(x;h)$ is also a sublinear functional that can be used in the proof of Theorem 13.2.4.)
11. Riesz's Theorem 9.2.3 can be generalized to reflexive Banach spaces as follows: *If K is a nonempty closed convex set in a reflexive Banach space X, then there is at least one $\bar{x} \in K$ such that*
$$\|\bar{x}\| = \inf_{x \in K} \|x\|.$$
Prove this result. (Hint: Corollary 2 to Theorem 12.3.2.)
12. If $f: X \to \mathbf{R}$ has a subdifferential at $\bar{x}$ such that $0 \in \partial f(\bar{x})$, then show that $f(\bar{x}) = \min_{x \in X} f(x)$. Is the converse true?
13. If $f: X \to \mathbf{R}$ is bounded away from $-\infty$, convex, weakly lower semicontinuous and continuously Fréchet differentiable, then $f(\bar{x}) = \inf\{f(x) | x \in \mathbf{R}\}$ if and only if either (a) $\langle x - \bar{x}, \nabla f(\bar{x}) \rangle \geqslant 0$ for all $x \in X$, or (b) $\langle x - \bar{x}, \nabla f(x) \rangle \geqslant 0$ for all $x \in X$. (Hint: let $\lambda \downarrow 0$ in $f(x) - f(\bar{x}) \geqslant \{f([1-\lambda]\bar{x} + \lambda x) - f(\bar{x})\}/\lambda \geqslant 0$.)
14. Let $f(x) = \max[f_1(x), \ldots, f_n(x)]$, and let each $f_i(x)$ possess a subdifferential $f_i(x)$ at each point x of a convex set Ω. Then show that
$$\partial f(x) = \text{co}\{\partial f_i(x) | f(x) = f_i(x)\}.$$
15. (F. H. Clarke) Given the constrained optimization problem:
$$\min\{f_0(x) | f_1(x) \leqslant 0, \ldots, f_n(x) \leqslant 0\}$$
where each of $f_0, f_1, \ldots, f_n$ possesses a subdifferential at each x in a convex set Ω. Then there exist nonnegative $\lambda_0, \lambda_1, \ldots, \lambda_n$ with $\lambda_0 + \lambda_1 + \cdots + \lambda_n = 1$, such that
$$0 \in \lambda_0 \, \partial f_0(\bar{x}) + \lambda_1 \, \partial f_1(\bar{x}) + \cdots + \lambda_n \, \partial f_n(\bar{x})$$

at any local constrained minimum of $f_0(x)$ on Ω. From this derive the generalized multiplier rule. (Hint: Consider $f(x)=\max[f_0(x)-f(\bar{x}),f_1(x),\ldots,f_n(x)]$. Show that either $\lambda_i=0$ or $f_i(\bar{x})=0$.)

13.3 NORMS: CONVEXITY, DIFFERENTIABILITY, SMOOTHNESS, AND STRICT CONVEXITY

We take up the trail set in Exercise 12.1.10. In a Hilbert space both the inner product and the norm have Gâteaux derivatives. The extension of this property to Banach spaces is complicated. We start with the following:

LEMMA 13.3.1. *If X is a normed linear vector space whose norm has a Gâteaux differential, then, for $x\neq 0$, $\|x\|$ has a Gâteaux derivative.*

PROOF. Let $f(x)=\|x\|$. If $x\neq 0$ and h^1,h^2 are arbitrary, then

$$\frac{\|x+t(h^1+h^2)\|-\|x\|}{t}=\frac{\|(x+2th^1)+(x+2th^2)\|-2\|x\|}{2t}$$

$$\leqslant\frac{\|x+2th^1\|-\|x\|}{2t}+\frac{\|x+2th^2\|-\|x\|}{2t}.$$

If we let $t\to 0$, then the existence of the Gâteaux differential leads to

$$GDf(x;h^1+h^2)\leqslant GDf(x;h^1)+GDf(x;h^2);$$

replacing h^1,h^2 by $-h^1,-h^2$ gives

$$GDf(x;-h^1-h^2)\leqslant GDf(x;-h^1)+GDf(x;-h^2).$$

But the Gâteaux differential is homogeneous in h, so we must have

$$GDf(x,h^1+h^2)\geqslant GDf(x;h^1)+GDf(x;h^2).$$

That is, the Gâteaux differential is linear in h and, since

$$\frac{\|x+th\|-\|x\|}{t}\leqslant\frac{t}{|t|}\|h\|,$$

$|GDf(x;h)|\leqslant\|h\|$; thus there is a bounded linear operator $Df(x)$ such that $GDf(x,h)=Df(x)h$. ■

We write $\mathbf{D}\|x\|$ to denote this Gâteaux derivative. It satisfies:

LEMMA 13.3.2. *If the norm of a normed linear vector space X is Gâteaux differentiable, then*

1. $\|\mathbf{D}\|x\|\|=1$;
2. $\langle x,\mathbf{D}\|x\|\rangle=\|x\|$;
3. $\mathbf{D}\|\alpha x\|=(\alpha/|\alpha|)\mathbf{D}\|x\|$, $\alpha\neq 0$.

PROOF. In the proof of Lemma 13.3.1, we established $\|\mathbf{D}\|x\|\|\leqslant 1$. If we show condition 2 then it implies $\|x\|\leqslant\|x\|\,\|\mathbf{D}\|x\|\|$, or $\|\mathbf{D}\|x\|\|\geqslant 1$, which proves condition 1. But

$$\langle h,\mathbf{D}\|x\|\rangle=\lim_{t^2\to 0}\frac{\|x+t^2h\|-\|x\|}{t^2}.$$

Set $h=x$; then $\langle x,\mathbf{D}\|x\|\rangle=\lim_{t^2\to 0}\|x\|=\|x\|$, which is condition 2.

Last, if $x\neq 0$,

$$\begin{aligned}\langle h,\mathbf{D}\|\alpha x\|\rangle&=\lim_{t\to 0}\frac{\|\alpha x+th\|-\|\alpha x\|}{t}\\&=\lim_{t\to 0}\frac{|\alpha|(\|x+(t/\alpha)h\|-\|x\|)}{t}\\&=\frac{|\alpha|}{\alpha}\lim_{t/\alpha\to 0}\frac{\|x+(t/\alpha)h\|-\|x\|}{t/\alpha}\\&=\frac{\alpha}{|\alpha|}\langle h,\mathbf{D}\|x\|\rangle.\quad\blacksquare\end{aligned}$$

We can now characterize the Gâteaux differentiable norms in the following:

THEOREM 13.3.1. *The norm on a normed linear vector space X is Gâteaux differentiable at every $x\neq 0$ if and only if, at each x with $\|x\|=1$, there is a unique subgradient (equivalently, the unit sphere has a unique supporting hyperplane.)*

Since the norm is a convex functional, this theorem is a special case of the following:

THEOREM 13.3.2. *Let $f(x):X\to\overline{\mathbf{R}}$ be a convex functional on a normed linear vector space X; if $f(x)$ has a Gâteaux derivative at $x\in X$, then $f(x)$*

has a unique subgradient at x and $\partial f(x)=Df(x)$. Conversely, if $f(x)<\infty$ and the subdifferential $\partial f(x)$ reduces to a unique subgradient, then $f(x)$ has a Gâteaux derivative at x and $\partial f(x)=Df(x)$.

PROOF. If $f(x)$ has a Gâteaux derivative at x, and is convex, then we noted in the proof of Theorem 13.2.4 that $Df(x)\in\partial f(x)$. If $x'\in\partial f(x)$ and h is arbitrary, then, for all $\lambda>0$,

$$f(x+\lambda h)-f(x)\geqslant\lambda\langle h,x'\rangle.$$

Thus

$$\frac{f(x+\lambda h)-f(x)}{\lambda}\geqslant\langle h,x'\rangle,$$

or, letting $\lambda\downarrow 0$,

$$\langle h,Df(x)\rangle\geqslant\langle h,x'\rangle.$$

That is, $\langle h,Df(x)-x'\rangle\geqslant 0$ for all h, which is true if and only if $Df(x)-x'=0$. Suppose now that a convex functional $f(x)<\infty$ and possesses a unique subgradient x' at x. Then from Exercise 13.2.10 we see that $d^+(x;h)=d^-(x;h)$ for all h, so that $GDf(x;h)=\langle h,x'\rangle$ for the x' whose existence is assured by Theorem 13.2.4. ■

Corollary 1 to the Hahn-Banach Theorem 9.3.2 tells us that if $\|x\|=1$, then there is at least one $x'\in X'$ such that $\langle x,x'\rangle=1$; that is, *there is at least one supporting hyperplane at each point of the sphere* $\{x|\|x\|=1\}$.

Theorem 13.3.1 tells us that *the supporting hyperplane is unique if and only if the norm is Gâteaux differentiable*. We have seen that in a Hilbert space, the norm is Gâteaux differentiable. For the L_p spaces, we consider two cases:

1. $1<p\leqslant 2$. Here, on $0<\tau<1$,

$$\|x\|^p=\int_0^1|x(\tau)|^p\,d\tau,$$

and

$$\|x+th\|^p-\|x\|^p=\int_0^1\left[|x(\tau)+th(\tau)|^p-|x(\tau)|^p\right]d\tau.$$

Now, if $\alpha \neq 0$, then $d|\alpha|^p/d\alpha = \text{sign}[\alpha]p|\alpha|^{p-1}$, where

$$\text{sign}[\alpha] = \begin{cases} 1 & \text{if } \alpha > 0, \\ 0 & \text{if } \alpha = 0, \\ -1 & \text{if } \alpha < 0. \end{cases}$$

Thus our candidate for the Gâteaux differential is

$$p\int_0^1 h(\tau)|x(\tau)|^{p-1}\text{sign}[x(\tau)]\,d\tau.$$

(From $1/p + 1/q = 1$ it follows that $|x(\tau)|^{p-1} \in L_q(0,1)$, so the integral exists.)

If we consider the function $[|1+\alpha|^p - 1 - p\alpha]|\alpha|^{-p}$ it is immediate that it is bounded on $\overline{\mathbf{R}}$, so that there exist constants c_1, c_2 with

$$c_1|\alpha|^p \leqslant |1+\alpha|^p - 1 - p\alpha \leqslant c_2|\alpha|^p.$$

From this it follows that

$$c_1|th(\tau)|^p \leqslant |x(\tau) + th(\tau)|^p - |x(\tau)|^p - pth(\tau)|x(\tau)|^{p-1}\text{sign}[x(\tau)]$$
$$\leqslant c_2|th(\tau)|^p.$$

It now follows that if we set $g(x) = \|x\|^p$, divide by t and let $t \to 0$, that

$$GDg(x;h) = p\int_0^1 h(\tau)\big[|x(\tau)|^{p-1}\text{sign}[x(\tau)]\big]\,dt.$$

Set $\gamma(x) = \|x\|$; it follows that $GDg(x;h) = p[\gamma(x)]^{p-1}GDf(x;h)$, and

$$GD\gamma(x;h) = \|x\|^{1-p}\int_0^1 h(\tau)\big[|x(\tau)|^{p-1}\text{sign}[x(\tau)]\big]\,d\tau,$$

which is a linear operator on h, so $GD\gamma(x;h) = \langle h, D\gamma(x)\rangle$ where $D\gamma(x)$ is a Gâteaux derivative, and then by continuity a Fréchet derivative at $x \neq 0$.

2. $2 < p < \infty$. In this case we can repeat the argument of case 1 *except for the boundedness of* $[|1+\alpha|^p - 1 - p\alpha]|\alpha|^{-p}$ for $\alpha \in \overline{\mathbf{R}}$. We use instead the following function,

$$\frac{|1+\alpha|^p - 1 - p\alpha}{|\alpha|^p + \alpha^2},$$

which is bounded on $\overline{\mathbf{R}}$ and leads to the same result as in case 1.

It is clear that the calculation above extends to any L_p space, whatever the region of integration and the underlying measure.

It is obvious that the norm is Gâteaux differentiable at all $x \in \{x \mid \|x\| = 1\}$ if and only if it is Gâteaux differentiable for all $x \neq 0$. If the Banach space has such a Gâteaux differentiable norm, it is called *smooth*. (Cf., Day [1], 144 or Köthe [1], 346.) A closely related notion is contained in the following:

DEFINITION 13.3.1. If X is a normed linear vector space for which $\|x\| = \|y\| \leqslant 1$ implies $\|(x+y)/2\| < 1$, then X is called *strictly convex* (or *rotund*). (Cf., Day [1], 144, or Köthe [1], 342.)

A set of equivalent conditions is given in the following:

THEOREM 13.3.3. *A normed linear vector space X is strictly convex if and only if one of the following holds:*

1. *If $\|x+y\| = \|x\| + \|y\|$, then either $x = \lambda y$ or $y = \lambda x$ for some $\lambda \geqslant 0$;*
2. *$\|x\| \leqslant 1$ and $x = \lambda y + (1-\lambda)z$ with $y \neq z$, $\|y\| \leqslant 1$, $\|z\| \leqslant 1$ and $0 < \lambda < 1$ implies $\|x\| < 1$;*
3. *For any nonzero $x' \in X$, there is at most one $\bar{x} \in \{x \mid \|x\| = 1\}$ with $\langle \bar{x}, x' \rangle = \|x'\|$;*
4. *Supporting hyperplanes at distinct points of $\{x \mid \|x\| \leqslant 1\}$ are distinct;*
5. *For any convex set $K \subset X$ there is at most one $\bar{x} \in K$ such that*

$$\|\bar{x}\| = \inf_{x \in K} \|x\|.$$

PROOF. Assume X strictly convex and let $z = x + y$, where $x \neq 0$, $y \neq 0$, and $\|z\| = \|x\| + \|y\|$. Suppose that $x/\|x\| \neq z/\|z\|$. Then for some $\epsilon > 0$, $\|x/2\|x\| + z/2\|z\|\| < 1 - \epsilon$, while $\|y/2\|y\| + z/2\|z\|\| \leqslant 1$. Then

$$\begin{aligned}(2\|x\| + 2\|y\|)\|z\| &= (\|z\| + \|x\| + \|y\|)\|z\| \\ &= \|\|z\|x + \|x\|z + \|z\|y + \|y\|z\| \\ &< [2(1-\epsilon)\|x\| + 2\|y\|]\|z\|,\end{aligned}$$

so $\|x\| < (1-\epsilon)\|x\|$, which is impossible. Thus $x/\|x\| = z/\|z\|$, which implies condition 1.

Now assume condition 1. Suppose $\|x\| = 1$, where $x = \lambda y + (1-\lambda)z$, with $\|y\| \leqslant 1$, $\|z\| \leqslant 1$, $0 < \lambda < 1$. Then $1 \leqslant \lambda\|y\| + (1-\lambda)\|z\|$ implies $\|y\| = \|z\| = 1$, and $\|\lambda y + (1-\lambda)z\| = \|\lambda y\| + \|(1-\lambda)z\|$. Then for some $\alpha > 0$ either $\lambda y =$

$\alpha(1-\lambda)z$ or $\alpha\lambda y=(1-\lambda)z$. If the first holds, $\lambda=\alpha(1-\lambda)$ and $y=z$. If the second holds, $\alpha\lambda=1-\lambda$ with the same result. Thus $y\neq z$ implies $\|x\|<1$.

If condition 2 holds, and for some $x'\in X'$ there is a pair x,y, $\|x\|=\|y\|=1$, $x\neq y$, and $\langle x,x'\rangle=\langle y,x'\rangle=\|x'\|$, then set $\bar{x}=(x+y)/2$. Then $\|x'\|=\langle \bar{x},x'\rangle \leqslant \|\bar{x}\|\|x'\|$, so $\|\bar{x}\|=1$, which violates condition 2.

Now assume condition 3 holds. Let $\pi_{x'}$ be a supporting hyperplane to $\{x|\|x\|\leqslant 1\}$ at $\bar{x}$; then $\pi_{x'}$ has the equation $\langle x,x'\rangle=\|\bar{x}\|=1$, and we may assume that on $\{x|\|x\|\leqslant 1\}$, $\langle x,x'\rangle\leqslant 1$. Then

$$\left\langle \frac{\langle x,x'\rangle}{|\langle x,x'\rangle|}x,x'\right\rangle=|\langle x,x'\rangle|\leqslant 1$$

for $\|x\|\leqslant 1$, so $\|x'\|=1$. Then condition 3 implies condition 4.

If condition 4 holds and $0\in K$, then there is nothing to show. If $0\notin K$, then there is no loss in generality if we assume $1=\inf_{x\in K}\|x\|$. Now $\{x|\|x\|<1\}$ is an open convex set disjoint from K. Corollary 2 to Theorem 13.1.4 asserts that there is a separating hyperplane, $\pi_{x'}$. The set $\{x|x\in K, \|x\|=1\}$ which must lie in $\pi_{x'}$ can contain at most one element if condition 4 holds.

Last, if condition 5 holds, suppose that x,y satisfy $\|x\|\leqslant 1$, $\|y\|\leqslant 1$, and $\|(x+y)/2\|=1$. Then $2=\|x+y\|\leqslant\|x\|+\|y\|$, so $\|x\|=\|y\|=1$. If $z=x+\lambda(y-x)$, $w=y+\lambda(x-y)$, $0<\lambda<1$, then $\|z\|=\|w\|\leqslant 1$ and $\|(z+w)/2\|=\|(x+y)/2\|=1$, hence the closed line segment $\{x+\lambda(y-x)|0\leqslant\lambda\leqslant 1\}$ is a convex set on condition 5 is violated. Hence $\|(x+y)/2\|<1$. ■

DEFINITION 13.3.2. By an *extremal* of $x\in X$, $x\neq 0$ is meant an x' with $\|x'\|=1$ such that

$$\langle x,x'\rangle=\|x\|.$$

By an *extremal* of $x'\in X'$, $x'\neq 0$ is meant an x with $\|x\|=1$ such that

$$\langle x,x'\rangle=\|x'\|.$$

If $x\in X$, then the Hahn-Banach theorem asserts the existence of an extremal. Suppose $x''\in X''$; then there exists at most one x' with $\|x'\|=1$ and $\langle x',x''\rangle=\|x''\|$ *if and only if X' is strictly convex*. Since $JX\subset X''$, there exists at most one x' with $\|x'\|=1$, and $\langle x,x'\rangle=\|x\|=\langle x',Jx\rangle=\|Jx\|$ *if X' is strictly convex*. Then "*X is smooth*" is a weaker statement than "*X' is strictly convex*," which implies it. If X is norm reflexive, we can replace "*if X' is strictly convex*" by "*if and only if X is strictly convex*" in which case "*X is smooth*" is equivalent to "*X' is strictly convex*."

COROLLARY 1. *If X is norm reflexive, each $x \in X$ ($x' \in X'$) has at most one extremal if and only if X is smooth (X' is strictly convex). Then X is smooth (strictly convex) if and only if X' is strictly convex (smooth).*

COROLLARY 2. *If X is norm reflexive then every nonzero $x' \in X'$ ($x \in X$) has a unique extremal if and only if X is strictly convex (X' is smooth).*

If X is reflexive, smooth, and strictly convex, the mapping from a vector to its unique extremal: $(\cdot) \to (\cdot)^\wedge$ also satisfies

1. $(x)^{\wedge\wedge} = x/\|x\|$, if $x \in X$, $x \neq 0$;
2. $(x')^{\wedge\wedge} = x'/\|x'\|$, if $x' \in X$, $x \neq 0$;
3. $(\lambda x)^\wedge = (\lambda/|\lambda|)x^\wedge$, if $\lambda x \neq 0$.

Condition 2 of Theorem 13.33 has a name.

DEFINITION 13.3.3. Let $S \subset X$, a normed linear vector space, while $A \subset S$ has the property that if $x,y \in S$, $\alpha x + (1-\alpha)y \in A$, for $0 < \alpha < 1$, then both $x,y \in A$. A is called an *extreme subset* of A. If A consists of one point, that point is called an *extreme point* of S.

Note that S is always an extreme subset of itself. The extreme points, if any, of the set $\{x \mid \|x\| \leqslant 1\}$ are important in the study of extremals of $x' \in X'$. While the Hahn-Banach theorem asserts the existence of extremals of both x,x' when X is reflexive, and of x in general, it fails to guarantee the existence of extremals of x' in general. The extreme points of $\{x \mid \|x\| \leqslant 1\}$ turn out to be extremals of x', but the existence of such extreme points in general rests on the following Krein-Milman theorem, and through the proof of that theorem, on Zorn's lemma (compare, e.g., Rudin [1], 70–71). We state the theorem but will prove only the second, still interesting part of it. It should be remarked that compactness in any of the topologies (norm, weak, weak*) is acceptable for the theorem.

THEOREM 13.3.4 (*Krein-Milman*) *A nonempty compact set S in a normed linear vector space X has extreme points. If E is the set of all extreme points of S, then* $\operatorname{co} E = \operatorname{co} S$.

PROOF. Let E be the set of all extreme points in S. Since $S \supset E$, $\operatorname{co} S \supset \operatorname{co} E$. Suppose $z^0 \in S$, $z^0 \notin \operatorname{co} E$. Then there exists an $x' \in X'$ with

$$\sup\{\langle z,x'\rangle \mid z \in \operatorname{co} E\} < \langle z^0,x'\rangle.$$

To see this, cover co$E$ and z^0 with open spheres of radius ϵ such that the resultant open sets K_1, K_2 are disjoint nonempty convex sets in X. Let $K = K_1 - K_2$. K is nonempty, open, and convex, and $K_1 \cap K_2 = \varnothing$, so $0 \notin K = \operatorname{int} K$. Then by Theorem 13.1.4, there is an $x' \in X$ such that

$$\langle z, x' \rangle \leqslant \langle y, x' \rangle, \qquad z \in K_1, y \in K_2,$$

so

$$\sup\{\langle z, x'\rangle | z \in K_1\} \leqslant \alpha \leqslant \inf\{\langle t, x'\rangle | y \in K_2\}.$$

Since z^0, co$E$ are in the interior of K_1, K_2,

$$\sup\{\langle z, x'\rangle | z \in \operatorname{co} E\} < \langle z^0, x'\rangle.$$

Let $B \subset S$ be the set at which x' achieves its maximum on S. B is nonempty, compact, and has an extreme point b^0 while $B \cap \operatorname{co} E = \varnothing$, so $b^0 \notin \operatorname{co} E$ and is not an extreme point of S; then

$$b^0 = \alpha b^1 + (1-\alpha) b^2, \qquad \alpha > 0, b^1, b^2 \in S.$$

Now

$$\langle b^0, x'\rangle = \alpha\langle b^1, x'\rangle + (1-\alpha)\langle b^2, x'\rangle$$

so, by the maximality of x' on B, $b^1, b^2 \in B$, and b^0 is not extreme in B, which is a contradiction. Hence $z^0 \in \operatorname{co} E$, co$S \subset$ co$E$, so co$S$ = co$E$. ■

We can apply this result to the existence and location of extremals through:

THEOREM 13.3.5. *Let $x \neq 0$, $x \in X$, a normed linear vector space. At least one extremal of x is an extreme point of* $\{x' | \|x\| \leqslant 1\}$.

PROOF. Fix x. Consider $C^\wedge = \{x' | \|x'\| = 1, \langle x, x'\rangle = \|x\|\}$. It is nonempty, convex, closed in the weak* topology (Theorem 9.4.4) and is thus weak* (sequentially) compact. (This was shown in a separable space; in a nonseparable space we must use transfinite induction). By the Krein-Milman theorem, it has an extreme point $\bar{x}'$. If $\bar{x}' = \alpha y' + (1-\alpha) z'$ where $\|y'\| = \|z'\| \leqslant 1$ and $0 < \alpha < 1$, then

$$\|x\| = \langle x, \bar{x}'\rangle \leqslant \alpha|\langle x, y'\rangle| + (1-\alpha)|\langle x, z'\rangle|$$
$$\leqslant \alpha\|x\| + (1-\alpha)\|x\| = \|x\|.\|$$

Hence $y', z' \in C^\wedge$, and $\bar{x}'$ is an extreme point, so $\bar{x}' = y' = z'$. ■

COROLLARY 1. *If X is norm reflexive then each $x' \in X'$ achieves its maximum on $\{x \mid \|x\| \leqslant 1\}$ at an extreme point of that set.*

In the exercises, extremals (and extreme points) are found in various spaces. Generalizations of these results to vector valued functions exist and will be used with no restraint. Proofs of these generalizations are tedious, yet not informative (Porter [1], 423–431).

EXERCISES TO SECTION 13.3

1. In l_p, $1<p<\infty$, show that the norm is differentiable, with

$$<h, \boldsymbol{D}\|x\|> = \|x\|^{1-p} \sum_{i=1}^{\infty} h_i |x_i|^{p-1} \operatorname{sign}[x_i].$$

2. In any reflexive Banach space X, for any $x' \in X'$, show that there exists at least one $\bar{x} \in X$ such that

$$<\bar{x}, x'> = \|x'\|.$$

   It is a result of James [1] that the converse is true. That is: *X is reflexive if and only if every $x' \in X$ achieves its maximum on* $\{x \mid \|x\| = 1\}$.
3. On $\mathbf{R}^2$, consider $l_p, 1 \leqslant p \leqslant \infty$ and find the extreme points of the unit sphere. Describe the extremals of the unit sphere.
4. In $L_p(0,1), 1<p<\infty$, show that

$$<h, x^\wedge> = \|x\|^{1-p} \int_0^1 h(\tau) |x(\tau)|^{p-1} \operatorname{sign}[x(\tau)]\, d\tau,$$

   that is, that

$$x^\wedge(t) = \|x\|^{1-p} |x(t)|^{p-1} \operatorname{sign}[x(t)]$$

   is the unique extremal of $x(t)$ if $x(t) \in L_p(0,1)$. Here $x^\wedge \in L_q(0,1)$, $1/p + 1/q = 1$. Because of reflexivity, the extremal of an element $x' \in L_q(0,1)$ is given by

$$(x')^\wedge = \|x'\|^{1-q} |x'(t)|^{q-1} \operatorname{sign}[x'(t)].$$

5. Compute the extremals of $x \in l_p$, $1<p<\infty$.

6. Consider $\mathbf{R}^n$ with the l_p norm, $p=1,\infty$. These spaces are *not* strictly convex. What are the extreme points of their unit spheres. Use these extreme points to generate all the extremals of a given x.
7. Show that on $L_1(0,1)$, an extremal of x is any element $x'\in L_\infty(0,1)$ such that $x'(t)=\text{sign}[x(t)]$ if $x(t)\neq 0$ and $|x'(t)|\leqslant 1$ everywhere.
8. Show that in $L_\infty(0,1)$, the function $x'(t)=t$ has no extremal in $L_1(0,1)$. (Recall that $[L_1(0,1)]'=L_\infty(0,1)$.) On the other hand, if $\{t|x'(t)=\|x'\|\}$ has a nonzero measure μ, then

$$x(t)=\begin{cases} 1/\mu, & t\in\{t|x'(t)=\|x'\|\}, \\ 0, & \text{otherwise,}\end{cases}$$

is an extremal in $L_1(0,1)$ for $x'\in L_\infty(0,1)$.
9. In $C[0,1]$ show that $\{x|\,\|x\|=1\}$ has exactly two extreme points, $x\equiv+1$ and $x\equiv-1$. Also show that in $(C[0,1])'=BV[0,1]$, among the extremals of $x(t)\in C[0,1]$ are the functions

$$x'(t)=\begin{cases} 0, & 0\leqslant t<t^*, \\ \text{sign}[x(t^*)], & t^*\leqslant t\leqslant 1,\end{cases}$$

where t^* is any point at which $|x(t)|$ achieves its maximum. We can generate more extremals by convex combinations of these functions if there are multiple maxima of $|x(t)|$. If the set $\{t|x(t)=\|x\|\}$ is of positive measure, show that any L_1 function that is nonnegative there and zero elsewhere is an extremal.
10. Consider $BV[0,1]$ and the problem of finding its extremals in $C[0,1]$. Show that the functional

$$\int_0^1 x(t)\,dt$$

has a unique extremal in $C[0,1]$. On the other hand, for each a, $0<a<1$, the functional

$$\int_0^a x(t)\,dt-\int_a^1 x(t)\,dt$$

has norm 1, but no extremals.

13.4 SOME GEOMETRIC METHODS IN OPTIMAL CONTROL THEORY

We will apply the preceding study of norms and convexity to a model of a fairly wide class of optimal control problems, following Porter and Williams (Porter [1], 309–353 and further references cited there.) There is assumed to be a state space X which is a Banach space, as in Section 12.10, a control space U, *also a Banach space*, and a dynamical system that can be described by a linear mapping $T \in [U \to X]$. Associated with the control space is a concept of *effort*, which is (at first) taken to be $\|u\|, u \in U$. The *minimum effort problem* is: given a *target* $x \in X$, and a nonempty set of *admissible controls* such that $Tu = x$, find the admissible control of minimum effort (=norm):

$$\inf\{\|u\| \mid Tu = x\}.$$

(In Section 12.9 we considered a problem of precisely this form with $\|u\| = f(u)$, $x = x^1$, and $Tu = x$ equivalent to $g(u) = 0$. More general problems are discussed in the exercises.) It is instructive to consider the special case $X = \mathbf{R}$. Two interpretations of the dynamic equation $Tu = x$ are possible here.

First, assume that $U = V'$ for some Banach Space V, and that T corresponds to some fixed element $v_T \in V$. We now wish to minimize $\|v'\|$ subject to $\langle v_T, v' \rangle = x$, but Corollary 1 to the Hahn-Banach theorem assures us that there exists a v', $\|v'\| = 1$ such that $\langle v_T, v' \rangle = \|v_T\|$, while any solution of that equation satisfies $\|v_T\| \leqslant \|v_T\| \, \|v'\|$, so $\|v'\| \geqslant 1$. Then $xv'/\|v_T\|$ solves the minimal effort problem with target $x \in \mathbf{R}$.

Second, assume $T \in [U \to \mathbf{R}] = U'$. Then, for a fixed $u'_T \in U'$ we seek a u of minimum norm subject to $\langle u, u'_T \rangle = x$. In this case we know that norm reflexivity of U is the key to solvability of the minimal effort problem; if it is to be solved *for all* T, then it is known (James [1]) that reflexivity is necessary, while the argument of the preceding paragraph shows that it is sufficient. For a fixed T, the image of $\{u \mid \|u\| \leqslant 1\}$ under T is a convex subset of $\mathbf{R}$, thus either an open interval $(-\|T\|, \|T\|)$, or a closed interval $[-\|T\|, \|T\|]$. In the first case, no solution to the minimal effort problem can exist, in the second solutions do exist. That is, solutions exist if and only if the (convex) image of the unit ball in U under T is closed in $\mathbf{R}$.

We now return to the general case, with X an arbitrary Banach space. We fix our attention on the unit ball in U, $\Omega = \{u \mid \|u\| \leqslant 1\}$, its boundary $\partial\Omega = \{u \mid \|u\| = 1\}$, their images $T(\Omega) = \{x \mid x = Tu, u \in \Omega\}$ and $T(\partial\Omega) = \{x \mid x = Tu, u \in \partial\Omega\}$, and the boundary of $T(\Omega)$, $\partial T(\Omega)$. Always $R_T = X$. The fundamental properties of these sets are summarized in the following:

LEMMA (13.4.1) *Let $T \in [U \to X]$, U, X Banach spaces and $R_T = X$. Then*

1. *$T(\Omega)$ is convex, balanced, and contains an open sphere about* 0;
2. $T(\Omega) \cap \partial T(\Omega) \subset T(\partial\Omega)$;
3. *$T(\Omega)$ is closed if and only if $\partial T(\Omega) \subset T(\partial\Omega)$*,
4. *If $x \in \partial T(\Omega)$, then $Tu = x$ implies $\|u\| \geqslant 1$.*

PROOF. Since T is linear, $T(\Omega)$ is convex. If $u \in \Omega$, so is $-u$, and then $T(\Omega)$ is balanced. From $R_T = X$ and the continuity of T it follows that the image of $\{u \mid \|u\| < 1\}$ under T is an open set in X containing 0, and then some open sphere about 0.

If $x \in T(\Omega) \cap \partial T(\Omega)$, then there is at least one $u \in \Omega$ with $Tu = x$; at the same time, $x \in \partial T(\Omega)$ implies that λx is not in $T(\Omega)$ if $\lambda > 1$, so λu is not in Ω. This means $\|u\| \geqslant 1/\lambda$ for all $\lambda \geqslant 1$, so $\|u\| \geqslant 1$. Hence $u \in \partial\Omega$ and $x \in T(\partial\Omega)$.

If $T(\Omega)$ is closed, $\partial T(\Omega) \subset T(\Omega) \cap \partial T(\Omega) \subset T(\partial\Omega)$. Conversely, if $\partial T(\Omega) \subset T(\partial\Omega)$, then $T(\partial\Omega) \subset T(\Omega)$ implies that $T(\Omega)$ contains its boundary, so it is closed.

Last, suppose $x \in \partial T(\Omega)$. Since $T(\Omega)$ is convex, it has a supporting hyperplane $x' \in X'$ with $<y,x'> \leqslant <x,x'>$ for all $y \in T(\Omega)$. But $T(\Omega)$ is balanced, so that $|<y,x'>| \leqslant <x,x'>$ for all $y \in T(\Omega)$, while each $u \in \Omega$ is transformed by T into some $y \in T(\Omega)$. Then $<x,x'> \geqslant |<Tu,x'>| = |<u,T'x'>|$ for all $u \in \Omega$. This makes $\|T'x'\| \leqslant <x,x'>$. Now, if $Tu = x$, $<x,x'> = <u,T'x'> \leqslant \|u\|\,\|T'x'\|$, so that $\|u\| \geqslant 1$. ■

(Note: $R_T = X$ is a controllability condition.) We can now characterize solvability of the minimum effort problem in the following:

THEOREM 13.4.1. *If $x \in \partial T(\Omega)$, then there exists a solution of the minimum effort problem if and only if $x \in T(\Omega)$ [i.e., $x \in T(\Omega) \cap \partial T(\Omega)$].*

PROOF. The "if" statement is an immediate consequence of Lemma 13.4.1. Suppose now that $u(x)$ is the solution of the minimum norm problem for x; then $\alpha u(x)$ solves the same problem with target αx for every $\alpha > 0$. But, for $0 < \alpha < 1$, $\alpha x \in T(\Omega)$, so there is a u with $Tu = \alpha x$ and $\|u\| \leqslant 1$; hence, $\|\alpha u(x)\| \leqslant 1$ for all $\alpha < 1$. This means, that $\|u(x)\| \leqslant 1$ and, by the Lemma, $\|u(x)\| = 1$. That is, $u(x) \in \partial\Omega$ and $x = T(u(x)) \in T(\Omega)$. ■

The set $\{x \mid Tu = x\}$, known as the *preimage* of x, will be written as $T^{-1}(x)$; this is a *point-to-set mapping*, and no inverse of the operator T is implied to exist. In this notation we can extend our result, removing the normalization of the problem to Ω, in the following:

COROLLARY 1. *$T^{-1}(x)$ contains an element of minimal norm (i.e. a solution of the minimal effort problem) for every $x \in X$ if and only if $T(\Omega)$ is closed.*

PROOF. If the convex set $T(\Omega)$ is closed, and $\phi(x)$ is its gauge function, then, for any $x \in T(\Omega)$, $x/\phi(x) \in \partial T(\Omega) \subset \partial T(\Omega) \cap T(\Omega)$, and $T^{-1}[x/\phi(x)]$ contains a solution u to the minimal effort problem for target $x/\phi(x)$ by

Theorem 13.4.1. Then $\phi(x)u$ solves the minimal effort problem for target x. The converse is actually contained in Theorem 13.4.1. ■

Again we have the following corollaries, whose proofs are left as exercises.

COROLLARY 2. *The minimum effort problem is solvable for every* $T \in [U \to X]$ *if and only if* U *is norm reflexive.*

COROLLARY 3. *Let* $U = W'$, $X = V'$ *where* W', V' *are Banach spaces. Let* $T \in [U \to X]$ *be the adjoint of* $S \in [V \to W]$, *which is one-to-one, with closed range. Then the minimum effort problem for target* $x \in X$ *with dynamics* T *is solvable for all* x.

The proof of Corollary 1 assures us that it is sufficient to solve our problem for $x \in T(\Omega)$; the gauge function of $T(\Omega)$ then can be used to remove this normalization.

As in the case $X = \mathbf{R}$, the solution of the minimum effort problem is determined by extremals. We find it convenient to extend the notation $x^\wedge$.

DEFINITION 13.4.1. If $x \in X$, a normed linear vector space, then

$$x^\wedge = \{x' \mid \|x'\| = 1, \langle x, x' \rangle = \|x\|\}$$

is the set of all extremals of x. Similarly, if $x' \in X'$, then

$$(x')^\wedge = \{x \mid \|x\| = 1, \langle x, x' \rangle = \|x'\|\}.$$

We make a general observation: $x \in \partial T(\Omega)$ means that x is a boundary point of a convex set with interior, so that there is a support hyperplane to $T(\Omega)$ at x. Conversely, if there is a supporting hyperplane at x to $T(\Omega)$, then $x \in \partial T(\Omega)$. When also $x \in T(\Omega)$, the following result relates x and the set of supporting hyperplanes at x to $T(\Omega)$.

THEOREM 13.4.2 *A hyperplane* $x' \in X'$ *supports* $T(\Omega)$ *at* $x \in T(\Omega)$ *if and only if* $x \in T([T'x']^\wedge)$.

PROOF. Suppose that u is arbitrary in Ω and x' in X'. Then $\langle Tu, x' \rangle = \langle u, T'x' \rangle \leqslant \|T'x'\|$. If, however, $v \in [T'x']^\wedge$ then $\langle Tv, x' \rangle = \langle v, T'x' \rangle = \|T'x'\|$, and x' is a supporting hyperplane to $T(\Omega)$ at Tv, which lies in $T(\Omega) \cap \partial T(\Omega)$. Conversely, if x' is a supporting hyperplane to $T(\Omega)$ at x, then $x \in T(\Omega) \cap \partial T(\Omega)$, and there are solutions $u = u(x) \in T^{-1}(x)$

to the minimum effort problem with target x; now

$$\|T'x'\| = \sup_{u\in\Omega} <x, T'x'> = \sup_{u\in\Omega} <Tu, x'>$$

$$= \sup_{y\in T(\Omega)} <y, x'> \leqslant <x, x'>,$$

if x' is a supporting hyperplane at x, and $<y,x'> \leqslant <x,x'>$ for all $y \in T(\Omega)$.

On the other hand $<x,x'> = <Tu(x),x'> = <u(x), |T'x'> \leqslant \|T'x'\|$; hence $<x,x'> = <u(x), T'x'> = \|T'x'\|$. That is, every $u(x) \in [T'x']^\wedge$ and $x \in T([T'x']^\wedge)$. ■

COROLLARY 1. *$[T'x']^\wedge \neq \varnothing$ if and only if $T([T'x']^\wedge) \subset T(\Omega) \cap \partial T(\Omega)$; each $x \in T(\Omega) \cap \partial T(\Omega)$ is in $T([T'x']^\wedge)$ for some x'. Each element of $T([T'x']^\wedge)$ is a target for a solvable minimum effort problem.*

Thus, the solution of all solvable minimum effort problems is effected by finding all supporting hyperplanes x' to $T(\Omega)$ and then all extremals of the resulting set $T'x'$. Nonuniqueness in this process can occur as follows:

1. Given $x \in T(\Omega) \cap \partial T(\Omega)$, there may be more than one supporting hyperplane x' to $T(\Omega)$ at x with $[T'x']^\wedge \neq \varnothing$.
2. Given such an x', $[T'x']^\wedge$ may have more than one element. (Note: There may exist x'^1, x'^2, support hyperplanes to $T(\Omega)$ at x, with $x'^1 \neq x'^2$ yet $[T'x'^1]^\wedge \cap [T'x'^2]^\wedge \neq \varnothing$.)
3. There may be $u^1 \neq u^2, u^1, u^2 \in [T'x']^\wedge$, such that $Tu^1 = Tu^2$; equally plausible is $Tu^1 \neq Tu^2$.

These cannot occur *if U is strictly convex*. Then $T^{-1}(x)$ has a unique element $u = u(x)$ of minimum norm (Theorem 13.3.3). This $u(x) \in [T'x']^\wedge$ for some support hyperplane x'. In fact if more than one support hyperplane to $T(\Omega)$ exists at x, say x'^1, x'^2, then $[T'x'^1]^\wedge = [T'x'^2]^\wedge$. (Porter [1], 329–330). Under slightly stronger conditions we give an explicit solution to a uniquely solvable minimum effort problem.

THEOREM 13.4.3. *Let U be a smooth, strictly convex, norm reflexive Banach space. Let $T \in [U \to X]$, $R_T = X$, and $\phi(x)$ be the gauge function of $T(\Omega)$. To every target $x \in X$ there is a unique $x' = x'(x) \in X'$, $\|x'\| = 1$ such that*

$$u(x) = \phi(x)[T'x'(x)]^\wedge,$$

provides the unique solution of the minimum effort problem; $x'(x)$ *is uniquely determined by either*

1. $\|x'\|=1$, $<y,x'> \leqslant <x,x'>/\phi(x)$ *for all* $y\in T(\Omega)$; *or, equivalently,*
2. $\|x'\|=1, \|T'x'\|=<x,x'>/\phi(x)$.

PROOF. Let x'^1, x'^2 be support hyperplanes to $T(\Omega)$ at x. By the remark before the theorem, $u(x)=(T'x'^1)^\wedge=(T'x'^2)^\wedge$. Since U is smooth, $(T'x'^1)^\wedge=(T'x'^2)^{\wedge\wedge\wedge}$, so that $T'x'^1/\|T'x'^1\|=T'x'^2/\|T'x'^2\|$. Since $R_T=X$, T' is a one-to-one mapping. Hence $x'^1/\|T'x'^1\|=x'^2/\|T'x'^2\|$. But $\|x'^1\|=\|x'^2\|=1$, so $\|T'x'^1\|=\|T'x'^2\|$, and $x'^1=x'^2$. Thus $x'(x)$ exists, and both condition 1 and 2 assert that $x'(x)$ is a supporting hyperplane to $T(\Omega)$ at $x/\phi(x)$. ■

DEFINITION 13.4.2. The mapping $x\to\phi(x)[T'x'(x)]^\wedge$ is denoted by T^I and is called the *pseudoinverse* of T.

Note that T^I need not be a linear operator. The representation $T^I(x)=\phi(x)[T'x'(x)]^\wedge$ leaves something to be desired: a more direct representation in terms of x. To find this we first introduce the mapping $k(x')$: $X'\to X$ defined by

$$k(x')=\|x'\|(x')^\wedge$$

which maps x' into an x of the *same* norm along the extremal $(x')^\wedge$. Then

$$\begin{aligned}k(\lambda x')&=\|\lambda x'\|(\lambda x')^\wedge\\&=|\lambda|\frac{|\lambda|}{\lambda}\|x'\|(x')^\wedge=\lambda k(x'),\end{aligned}$$

and

$$\begin{aligned}\langle k(x'),x'\rangle&=\langle\|x'\|(x')^\wedge,x'\rangle\\&=\|x'\|\langle(x')^\wedge,x'\rangle\\&=\|x'\|^2.\end{aligned}$$

Define a (possibly nonlinear) operation from X' to $X, \Gamma(x')=Tk(T'x')$. Then $\Gamma(x')$ also equals $\|T'x'\|T(T'x')^\wedge$. Obviously $\|\Gamma(x')\|\leqslant\|T'\|\|T\|\|x'\|$, so Γ is bounded.

For real scalars $\alpha, \Gamma(\alpha x')=\alpha\Gamma(x')$. Also $R_\Gamma=X$; to see this, notice that Theorem 13.4.3 associates to each x, an x' with $\|x'\|=1$, with $x=$

$\phi(x)T(T'x')^{\wedge} = T(T'[\phi(x)x'])^{\wedge}$, so

$$x = \frac{\Gamma(\phi(x)x')}{\|T'\phi(x)x'\|}$$

$$= \Gamma\left(\frac{x'}{\|T'x'\|}\right).$$

Hence every x is in the range of Γ.

LEMMA 13.4.2. *$\Gamma(x')$ is monotone:*

$$0 \leqslant \langle \Gamma(x'^1) - \Gamma(x'^2), x'^1 - x'^2 \rangle$$

for any x'^1, x'^2 in X', and $\Gamma(x')$ is one-to-one.

PROOF. For any x'^1, x'^2,

$$\langle \Gamma(x'^1) - \Gamma(x'^2), x'^1 - x'^2 \rangle$$

$$= \langle \Gamma(x'^1), x'^1 \rangle - \langle \Gamma(x'^1), x'^2 \rangle - \langle \Gamma(x'^2), x'^1 \rangle + \langle \Gamma(x'^2), x'^2 \rangle$$

$$= \|T'x'^1\|^2 + \|T'x'^2\|^2 - \langle \|T'x'^1\| T(T'x'^1)^{\wedge}, x'^2 \rangle - \langle \|T'x'^2\| T(T - x'^2)^{\wedge}, x'^1 \rangle$$

$$= (\|T'x'^1\| - \|T'x'^2\|)^2 + \|T'x'^1\| \left[\|T'x'^2\| - \langle (T'x'^1)^{\wedge}, T'x'^2 \rangle \right]$$

$$+ \|T'x'^2\| \left[\|T'x'^1\| - \langle (T'x'^2)^{\wedge}, T'x'^1 \rangle \right].$$

But for $i,j = 1,2$, $|\langle (T'x'^i)^{\wedge}, T'x'^j \rangle| \leqslant \|T'x'^j\|$, so that $\Gamma(x')$ is monotone. $\Gamma(x')$ will be one-to-one if we show that $\langle \Gamma(x'^1) - \Gamma(x'^2), x'^1 - x'^2 \rangle$ vanishes if and only if $x'^1 = x'^2$. (Why?) But the "quadratic form" $\langle \Gamma(x'^1) - \Gamma(x'^2), x'^1 - x'^2 \rangle = 0$ if and only if simultaneously

$$\|T'x'^1\| = \|T'x'^2\|,$$

$$\|T'(x'^2)\| = \langle (T'x'^1)^{\wedge}, T'x'^2 \rangle,$$

$$\|T'(x'^1)\| = \langle (T'x'^2)^{\wedge}, T'x'^1 \rangle.$$

The last two conditions assert that $T'x'^1$ and $T'x'^2$ have a common extremal. Since U is reflexive, smooth, and strictly convex, extremals and supporting hyperplanes are unique. Then $T'x'^1 = T'x'^2$ and for $i,j = 1,2$, $\langle T(T'x'^i)^{\wedge}, x'^j \rangle = \langle T(T'x'^j)^{\wedge}, x'^i \rangle$. Then Theorem 13.4.3, condition 2 implies that $x'^1 = x'^2$. ■

That is, the (possibly nonlinear) operator $\Gamma(x')$ has an inverse $\Gamma^{-1}(x)$.

COROLLARY 1. $T^I = k(T'\Gamma^{-1}(x))$.

PROOF. For any $x \in X$ there is a unique $x' = x'(x) \in X'$ with

$$x = \phi(x) T(T'x')^{\wedge} = \phi(x) \| T'x' \|^{-1} \Gamma(x')$$
$$= \Gamma(\phi(x) \| T'x' \|^{-1} x').$$

Since $\Gamma(x')$ is one-to-one, the unique solution y' of $x = \Gamma(y')$ is a real scalar multiple of $x'(x)$, and $\Gamma^{-1}(x) = y' = \alpha x'(x)$ for $\alpha \in \mathbf{R}$. Thus $x = \Gamma(\Gamma^{-1}(x)) = Tk(T'\Gamma^{-1}(x)) = Tu(x)$ where $u(x)$ is the unique solution of the minimum effort problem with target x. Then $u(x) = k(T'\Gamma^{-1}(x)) = T^I(x)$. ■

COROLLARY 2. *If U is a Hilbert space, then $T^I = T^*(TT^*)^{-1}$.*

PROOF. For a real Hilbert space $T^* = T'$, since the mapping $\mathcal{G}$ of Section 10.3.1 reduces to the identity mapping. Also $k(\mathcal{G}x) = \|x\| (\mathcal{G}x)^{\wedge} = \|x\| x / \|x\| = x$. Then $T^I(x) = k(T'\Gamma^{-1}(x)) = T'\Gamma^{-1}(x)$. Also, $\Gamma(x') = Tk(T'x') = TT'x'$, so that TT' is invertible and $\Gamma^{-1}(x) = (TT')^{-1}$. Finally $T^I = T'(TT')^{-1} = T^*(TT^*)^{-1}$. ■

COROLLARY 3. *Under the conditions of Theorem* 13.4.3, $\|T^I(x)\| = \phi(x)$ *and is continuous.*

We show how this approach works in the problem first treated in Section 12.9. We assume that $x(t):[0,1] \to \mathbf{R}^n, u(t):[0,1] \to \mathbf{R}^m$ and

$$\frac{dx}{dt} = Ax(t) + Bu(t)$$

where A, B are real $n \times n$ and $n \times m$ matrices, respectively. We introduce a Hilbert space of admissible controls U with norm

$$\|u\|_R^2 = \int_0^1 \sum_{i=1}^m \sum_{j=1}^m (R)_{ij} u_i(t) u_j(t)\, dt$$
$$= \int_0^1 (Ru(t), u(t))_m\, dt.$$

In $\mathbf{R}^n$ we take the norm derived from $(x,x)_n = \sum_{i=1}^n x_i^2$. If $R + R^T$ is positive definite, or simpler, *if R is symmetric and positive definite*, then it is easy to see that U is a Hilbert space if the integral is a Lebesgue integral; (in fact if $u \in L_2(0,1)$, then it is easy to show that $\alpha \|u\|_2 \leqslant \|u\|_R \leqslant \beta \|u\|_2$ for suitable positive scalars α, β.) If x^0 is our initial state and x^1 is our target, remember that when $x=x^1-e^Ax^0$, x is a suitable target in $\mathbf{R}^n$ for the dynamic system

$$Tu=\int_0^1 e^{A(1-t)}Bu(t)\,dt.$$

(Why?) We assume that the matrices A, B satisfy the controllability condition of Section 12.9. According to our analysis, if $x\in\mathbf{R}^n$ is our target, then $u=T^Ix=T^*(TT^*)^{-1}x$.

Now

$$(Tu,x)_n=\left(\int_0^1 e^{A(1-t)}Bu(t)\,dt,x\right)_n$$
$$=\int_0^1 \left(Ru(t),R^{-1}B^Te^{A^T(1-t)}x\right)_m dt.$$

That is, $(T'x)(t)=R^{-1}B^Te^{A^T(1-t)}x$. Also $T'=T^*$. The mapping TT^* is easily seen to be given by

$$\left(\int_0^1 e^{A(1-t)}BR^{-1}B^Te^{A^T(1-t)}\,dt\right)x.$$

The coefficient of x is an $n\times n$ matrix that is invertible, and the optimal control is furnished by

$$u(x)(t)=R^{-1}B^Te^{A^T(1-t)}\left(\int_0^1 e^{A(1-t)}BR^{-1}B^Te^{A^T(1-t)}\,dt\right)^{-1}(x^1-e^Ax^0).$$

If U is norm reflexive, but not a Hilbert space, then the computation of T^I is more complicated, of course. Yet more complicated is the situation in which U is not reflexive. We sketch a development for the minimum effort problem with linear dynamics in a situation amenable to the use of Corollary 3 to Theorem 13.4.1. We describe the target and system dynamics in the following form:

$$Tu=\int_0^1 Me^{-At}Bu(t)\,dt,$$

$$x=M(e^{-A}x^1-x^0),$$

where M is an arbitrary $n\times n$ real matrix. (This formulation sometimes permits a somewhat simpler appearance in the results.) Again $X=\mathbf{R}^n$, but we take $U = L_\infty^m(0,1) = \{u(t) | u:[0,1] \to \mathbf{R}^m, u_i(t) \in L_\infty(0,1)\}$, with

$$\|u(t)\| = \operatorname*{ess\,sup}_{t\in[0,1]} \max_{1\leqslant i\leqslant m} |u_i(t)|.$$

It is a simple extension of Theorem 3.2.1 to show that U is a Banach space, and of Exercise 9.3.6 to show that it is the dual of $W = L_1^m[0,1] = \{w(t) | w:[0,1] \to \mathbf{R}^m, w_i(t) \in L_1(0,1)\}$ with

$$\|w\| = \int_0^1 \sum_{i=1}^m |w_i(t)|\, dt.$$

In $X = \mathbf{R}^n$ we use the usual inner product and norm, making it a Hilbert space. Then $V = V' = X = X' = \mathbf{R}^n$, and we find S such that $T = S'$ as follows: If v is arbitrary in V and u in U, then

$$\begin{aligned}(v, Tu)_n &= \sum_{i=1}^n v_i (Tu)_i \\ &= \int_0^1 (v, Me^{-At}Bu(t))_n\, dt \\ &= \int_0^1 (B^T e^{-A^T t} M^T v, u(t))_m\, dt.\end{aligned}$$

Then $(Sv)(t) = B^T e^{-A^T t} M^T v$. Now

$$R_S = \{y(t) | y:[0,1] \to \mathbf{R}^m, y_i(t) = [B^T e^{-A^T t} M^T v]_i\}$$

But the columns of $B^T e^{-A^T t} M^T$ are in $C[0,1]$, so that R_S is closed. Then $T(\Omega)$ is weak* closed, and convex; thus it is closed and $\partial T(\Omega) \subset T(\Omega)$. The extremals of Sv are easily determined:

$$(Sv, u)_m = \int_0^1 \sum_{i=1}^m u_i(t)(Sv)_i(t)\, dt.$$

Then

$$u_i(t) = (Sv)_i^\wedge(t) = \begin{cases} \operatorname{sign}(Sv)_i(t), & (Sv)_i(t) \neq 0, \\ \alpha, |\alpha| \leqslant 1, & (Sv)_i(t) = 0. \end{cases}$$

But the solutions of the minimum effort problem $u(x) = \phi(x)(Sv)^\wedge$, where ϕ is the gauge function of $T(\Omega)$ and v is a supporting hyperplane for $T(\Omega)$ at $x/\phi(x)$. *In the present case $\partial T(\Omega)$ can be determined by computing $T[(Sv)^\wedge]$ for all $v \in \mathbf{R}^n$*. Then the calculation of $u(x)$ for a given x is straightforward.

EXERCISES TO SECTION 13.4

1. In the minimal effort problem with $X=\mathbf{R}$, when is the solution unique?
2. Prove that U is norm reflexive if and only if the minimal effort problem is solvable for every target x and every $T\in[U\to X]$ with $R_T=X$. [Hint: $\{u|\|u\|\leqslant 1\}$ is weakly (sequentially) compact if U is norm reflexive (Taylor [1], 229–230).]
3. Prove Corollary 3 to Theorem 13.4.1. (Hint: In the weak* topology, the set $\{u'|\|u'\|\leqslant 1\}$ is weak* compact (Taylor [1], 228). Then, $T=S'$ is continuous in the weak* topology, $T(\Omega)$ is weak* compact, norm closed, and contains a multiple of $\{x|\|x\|\leqslant 1\}$ in X.)
4. Redo the text discussion of the optimal control problem of Section 12.9 for the Hilbert space case with dynamics

$$Tu=\int_0^1 Me^{-At}Bu(t)\,dt$$

and target

$$x=M(e^{-A}x^1-x^0)$$

for an arbitrary nonsingular $n\times n$ matrix M.

5. (Porter [1], 317–320.) In the second minimum effort problem of the text, take

$$A=\begin{bmatrix}\frac{1}{2} & -\frac{1}{2}\\ \frac{1}{2} & -\frac{1}{2}\end{bmatrix},\ B=\begin{bmatrix}1 & 1\\ 1 & -1\end{bmatrix}.$$

Show that

$$e^{At}=\begin{bmatrix}1+\frac{1}{2}t & -\frac{1}{2}t\\ \frac{1}{2}t & 1-\frac{1}{2}t\end{bmatrix}.$$

Let

$$M=\begin{bmatrix}\frac{1}{2} & \frac{1}{2}\\ \frac{1}{2} & -\frac{1}{2}\end{bmatrix}.$$

Compute

(a) The extremals of Sv,
(b) $\partial T(\Omega)$ and $T(\Omega)$, and
(c) Solutions of the minimum effort problem for $x=(1,5)$, for $x=(1,2)$ and for $x=(4.79,0.4)$.

(Hint: $T(\Omega)$ consists of portions of $x_2=\pm 1, 4x_1-6+(x_2+1)^2=0$.)

6. Suppose that we replace the minimum effort problem by one in which, for a given target x and a given $\bar{u}\in U$, we seek

$$\inf_{Tu=x} \|u-\bar{u}\|;$$

Show that if U is reflexive and strictly convex, then this problem has the unique solution

$$u(x)=T^I(x-T\bar{u})+\bar{u}.$$

(Hint: Show that $TT^I(x-T\bar{u})+T\bar{u}=x$; then $Tu=x$ implies $\|u-u\|\geqslant \|T^I(x-T\bar{u})\|=\|u(x)-\bar{u}\|$.) (Note the identity $TT^IT=T$.)

7. Let $F\in[U\to V], U, V$ both reflexive and strictly convex Banach spaces. Then consider the problem

$$\inf_{Tu=x} \|Fu\|_V$$

Show that if $\|Fu\|_V$ is a norm on U equivalent to the original norm (that is, if F is also invertible,) and $\phi_F(x)$ is the gauge function of $T_F(\Omega)=\{Tu \mid \|Fu\|\leqslant 1\}$, then this problem has the unique solution

$$u(x)=\phi_F(x)F^{-1}\big([F']^{-1}(T'x')\big)^{\wedge},$$

where x' is the unique element of X' with $\|x'\|=1$ and $\langle x,x'\rangle=\|[F']^{-1}T'x'\|$. (Hint: Consider the equivalent problem $\inf_{TF^{-1}v=x}\|v\|$.)

8. The *terminal controller problem* is

$$\inf_{u\in K} \|Tu-x\|$$

where $K\subset U$ is a closed convex set with interior, and U is reflexive, strictly convex, and smooth. Let $K_x=\{Tu-x \mid u\in K\}$. Show that $K_x\subset X$ is a closed convex set with interior and that *if X is reflexive and strictly convex, there is a unique $y\in K_x$ of minimum norm.*

9. In the terminal controller problem of Exercise 13.4.8, show that $\phi_K(T^I x) \leqslant 1$ implies $0 \in K_x$, where $\phi_K(u)$ is the gauge function of K, while the converse is true only if $\| T^I T u \| = u$ for any $u \in T^{-1}(x) \cap K$. If $\| T^I T u \| < \| u \|$, we can no longer conclude $\phi_K(T^I x) \leqslant 1$ from $Tu = x$, unless $\phi_K(u)$ is monotone in $\| u \|$, in which case we may as well take $K = \{ u \,|\, \| u \| \leqslant \alpha \}$ for some scalar α. In that case $0 \in K_x$ if and only if $\| T^I x \| \leqslant \alpha$. Show that if $\| T^I x \| < \alpha$, then the solution of the terminal controller problem is not unique.
10. Show that if $X = Y$ is a Hilbert space, then Corollary 1 to Lemma 13.4.2 can be restated as $T^I = T^*(TT^*)^{-1}$. (Hint: Show that $k(x) = x$, and $\Gamma(x) = TT^* x$.)

13.5 GEOMETRIC METHODS FOR GENERALIZED KUHN-TUCKER CONDITIONS

Following Guignard [1], we will establish a very general geometric form of both necessary and sufficient Kuhn-Tucker conditions for constrained optimization. X is a real Banach space throughout. To replace the reliance on the analytic tools used in Section 12.9, we need to generalize the notion of "tangent space." Some preliminary technical lemmas concern properties of cones.

DEFINITION 13.5.1. If $K \subset X$, and $x \in K$ implies $\alpha x \in K$ for all $\alpha > 0$, then K is called a *cone*. A similar definition holds for $K' \subset X'$. By K^-, K^+ we mean

$$K^- = \{ x' \in X' \,|\, \langle x, x' \rangle \leqslant 0 \text{ for all } x \in K \}$$

and

$$K^+ = \{ x' \in X' \,|\, \langle x, x' \rangle \geqslant 0 \text{ for all } x \in K \}.$$

K^- is usually called the *polar* cone of K, or simply the *polar*.

Since $0 \in \overline{K}$, each element of K^- (and of K^+) is a support hyperplane to $\overline{K}$ at 0, and every $x \in K$ lies to the "same side" of all $u' \in K^-$ (and of all $u' \in K^+$). If $K' \subset X'$ then we also set

$$K'^- = \{ x \in X \,|\, \langle x, x' \rangle \leqslant 0 \text{ for all } x' \in K' \}$$

and

$$K'^+ = \{ x \in X \,|\, \langle x, x' \rangle \geqslant 0 \text{ for all } x' \in K' \}.$$

It is obvious that K^-, K^+, K'^-, K'^+ are closed, convex cones. Less obvious are the following lemmas.

LEMMA 13.5.1. *Let K be a closed convex cone, $x \notin K$. Then there exists $x' \in X'$ such that $\langle k, x'\rangle \geqslant 0$, $k \in K$, and $\langle x, x'\rangle < 0$.*

PROOF. By Theorem 13.1.4 there is a real number α and a linear functional x' with $\langle x, x'\rangle < \alpha$ and $\langle k, x'\rangle \geqslant \alpha$, $k \in K$. Since $0 \in K$, $\langle x, x'\rangle < \alpha \leqslant 0$. But, if for some k, $\langle k, x'\rangle = \beta < 0$, then $\lambda > 0$, $\langle \lambda k, x'\rangle = \lambda\beta \to -\infty, \lambda \to \infty$ violates $\langle k, x'\rangle \geqslant \alpha$. Hence $\alpha = 0$ and $\langle k, x'\rangle \geqslant 0$. ■

LEMMA 13.5.2. *If K is a cone, then $K^{++} = \overline{\text{co}}\, K$.*

PROOF. K^{++} is a closed convex cone. If $k \in K$ then $\langle k, x'\rangle \geqslant 0$ for all $x' \in K^+$, so $k \in K^{++}$; that is, $K \subset K^{++}$. Then $\text{co}\, K \subset K^{++}$ and even $\overline{\text{co}}\, K \subset K^{++}$. If $x \in K^{++}$ but $x \notin \overline{\text{co}}\, K$ there is an $x' \in X'$ with $\langle k, x'\rangle \leqslant 0$, $k \in \overline{\text{co}}\, K$, since $\overline{\text{co}}\, K$ is easily seen to be a cone, while $\langle x, x'\rangle < 0$. Then $x' \in K^+$, so $x \notin K^{++}$, which is a contradiction. The lemma is now obvious. ■

LEMMA 13.5.3. $K^+ = (\overline{\text{co}}\, K)^+$.

PROOF. It is obvious that $K^+ \supset (\overline{\text{co}}\, K)^+$. If $x' \in K^+$, then $\langle k, x'\rangle \geqslant 0$, $k \in K$. If $g \in \text{co}\, K$, then $g = \lambda k^1 + (1-\lambda)k^2$ for some λ, $0 < \lambda < 1$ and $k^1, k^2 \in K$. Then $\langle g, x'\rangle \geqslant 0$. If $h \in \overline{\text{co}}\, K$, then $h = \lim_{n\to\infty} g^n, g^n \in \text{co}\, K$, so $\langle h, x'\rangle \geqslant 0$. Then $x' \in (\overline{\text{co}}\, K)^+$, so $(\overline{\text{co}}\, K)^+ \supset K^+$. ■

LEMMA 13.5.4. *If K_1, K_2 are closed convex cones, then*

$$(K_1 \cap K_2)^+ = \overline{K_1^+ + K_2^+}\,.$$

PROOF. From Lemma 13.5.2, $(K_1^+ + K_2^+)^{++} = \overline{(K_1^+ + K_2^+)} = K_1^+ + K_2^+$, since $K_1^+ + K_2^+$ is a closed convex cone. On the other hand $\text{co}(K_1^+ + K_2^+) = K_1^+ + K_2^+$. If we show that $K_1 \cap K_2 = (K^+ + K^+)^+$, then the lemma will be proven.

Let $x' \in K_1^+ + K_2^+$. Then $x' = x'^1 + x'^2, x'^1 \in K_1^+, x'^2 \in K_2^+$. Let $x \in K_1^{++} \cap K_2^{++} = K_1 \cap K_2$. Then $\langle x, x'\rangle \geqslant 0$, so $K_1 \cap K_2 \subset (K_1^+ + K_2^+)^+$. On the other hand, $0 \in K_1^+, K_2^+$ so $K_1^+ = K_1^+ + 0 \subset K_1^+ + K_2^+$ and $K_2^+ = K_2^+ + 0 \subset K_1^+ + K_2^+$. Thus

$$(K_1^+ + K_2^+)^+ \subset K_1^{++} = K_1$$

and

$$(K_1^+ + K_2^+)^+ \subset K_2^{++} = K_2.$$

Hence

$$(K_1^+ + K_2^+)^+ \subset K_1 \cap K_2. \quad \blacksquare$$

We generalize the notion of tangent as follows:

DEFINITION 13.5.2. Let $M \subset X$, $\bar{x} \in M$. We say y *is tangent to* M *at* $\bar{x}$ if there exist two sequences, $\{\lambda_k\}$ of *nonnegative* numbers and $\{x^k\}$ of elements of M, such that $\lambda_k(x^k - \bar{x}) \to y$, $k \to \infty$. The set $T(M,\bar{x})$ of all y tangent to M at $\bar{x}$ is called the *cone tangent to* M *at* $\bar{x}$.

Since $\alpha\lambda_k(x^k - \bar{x}) \to \alpha y$, $k \to \infty$, if $y \in T(M,\bar{x})$ then $\alpha y \subset T(M,\bar{x})$ for all $\alpha \geqslant 0$. Thus justifies calling $T(M,\bar{x})$ a cone. Definition 13.5.2 is related to the definition of $d^+(x^0;h)$ in Exercise 12.9.7 in an obvious way, but it avoids any questions of differentiability. $T(M,\bar{x})$ need not be convex; it is augmented to be so in the following:

DEFINITION 13.5.3. The set $P(M,\bar{x})$, the closed convex hull of $T(M,x)$, is called the *cone pseudotangent to* M *at* $\bar{x}$. That is

$$P(M,x) = \overline{\text{co}}\, T(M,\bar{x}).$$

LEMMA 13.5.5. *Let* $A_i \subset X$, $i \in I$, *for some index set* I *and let* $A = \cap_{i \in I} A_i \neq \varnothing$, *while* $\tilde{A} = \cup_{i \in I} A_i$. *Suppose that* $\bar{x} \in A$. *Then the following two properties hold*:

$$T_1: \begin{cases} T(A,\bar{x}) \subset \bigcap_{i \in I} T(A_i,\bar{x}), \\ P(A,\bar{x}) \subset \bigcap_{i \in I} P(A_i,\bar{x}); \end{cases}$$

$$T_2: \begin{cases} \bigcup_{i \in I} T(A_i,\bar{x}) \subset T(\tilde{A},\bar{x}) \\ \bigcup_{i \in I} P(A_i,\bar{x}) \subset P(\tilde{A},\bar{x}). \end{cases}$$

PROOF. (T_1) Since $A \subset A_i$ for all i, $T(A,\bar{x}) \subset T(A_i,\bar{x})$, so $T(A,\bar{x}) \subset \cap_{i \in I} T(A_i,\bar{x}) \subset \cap_{i \in I} P(A_i,\bar{x})$. But

$$\bigcap_{i \in I} P(A_i,\bar{x}) \supset \overline{\text{co}}\; T(A,\bar{x}).$$

(T_2) Since $A_i \subset \tilde{A}$ for all i, $T(A_i,x) \subset T(\tilde{A},x)$ so $\cup_{i\in I} T(A_i,\bar{x}) \subset T(\tilde{A},x)$. Further $T(A_i,\bar{x}) \subset T(\tilde{A},\bar{x}) \subset P(\tilde{A},\bar{x})$ while $P(\tilde{A},\bar{x})$ is a closed convex cone, so it must contain $P(A_i,\bar{x})$. Hence $\cup_{i\in I} P(A_i,\bar{x}) \subset P(\tilde{A},x)$. ■

The next definition relates to Theorem 13.1.4's characterization of a closed convex set as the intersection of all the closed half spaces containing it and determined by support hyperplanes. Although the definitions of tangent, $T(M,\bar{x})$, $P(M,\bar{x})$ make no reference to support hyperplanes, for a convex M there are obvious connections.

DEFINITION 13.5.4. If for all $x \in M$, $x-\bar{x} \in P(M,\bar{x})$, M is *pseudoconvex at* $\bar{x}$.

LEMMA 13.5.6. *If all* A_i, $i \in I$ *are pseudoconvex at* $\bar{x}$, *then* $\tilde{A}$ *is pseudoconvex at* $\bar{x}$.

PROOF. For any $i\in I$, $x-\bar{x} \in P(A_i,\bar{x}) \subset \cup_{i\in I} P(A_i,\bar{x})$. Then Lemma 13.5.5 implies $x-\bar{x} \subset P(\tilde{A},\bar{x})$. ■

LEMMA 13.5.7. *If* M *is convex, then* M *is pseudoconvex at all* $\bar{x} \in M$.

PROOF. Let $\bar{x} \in M$. Let $x \in M$ and set

$$x^k = \bar{x} + \frac{1}{k}(x-\bar{x}) = \left(1-\frac{1}{k}\right)\bar{x} + \frac{1}{k}x.$$

Then $x^k \to \bar{x}$, $k \to \infty$. But $x^k \in M$ by convexity, while

$$k(x^k-\bar{x}) = k\frac{1}{k}(x-\bar{x}) \equiv x-\bar{x}.$$

Then $x-\bar{x} \in T(M,\bar{x}) \subset P(M,\bar{x})$. Thus M is pseudoconvex at $\bar{x}$. ■

COROLLARY 1. *If* M_i *are convex,* $i \in I$, *and* $\cap_{i\in I} M_i \neq \emptyset$, *then* $\cup_{i\in I} M_i$ *is pseudoconvex.*

We turn now to minimizing a functional.

LEMMA 13.5.8. *If* ψ *is a real valued quasiconvex functional that is Fréchet differentiable at* $\bar{x}$, *then* $\langle x-\bar{x}, \nabla\psi(\bar{x})\rangle > 0$ *implies* $\psi(x) > \psi(\bar{x})$.

PROOF. The set $L(\psi;\bar{x})$ is convex and includes $\bar{x}$. If x is in that set, then $\psi(x)-\psi(\bar{x} \leqslant 0$. But $\psi(x)-\psi(\bar{x}) = \langle x-\bar{x}, \nabla\psi(\bar{x})\rangle + o(\|x-\bar{x}\|)$. Then set

$x(t)=\bar{x}+t(x-\bar{x})$, $0<t<1$. We have

$$0 \geqslant \psi(x(t))-\psi(\bar{x})=t\langle x-\bar{x}, \nabla\psi(\bar{x})\rangle+o(t\|x-\bar{x}\|).$$

For sufficiently small $t>0$ the right side takes the sign of $\langle x-\bar{x}, \nabla\psi(\bar{x})\rangle$, which is a contradiction, if $\langle x-\bar{x}, \nabla\psi(\bar{x})\rangle>0$. ■

DEFINITION 13.5.5. We say that ψ is *pseudoconvex over M at $\bar{x}$* if ψ has Fréchet derivative $\nabla\psi(\bar{x})$ at $\bar{x}$ and $x\in M$, $\langle x-\bar{x}, \nabla\psi(\bar{x})\rangle\geqslant 0$ implies $\psi(x)\geqslant\psi(\bar{x})$.

We relate pseudoconvexity, optimaliity, and the gradient in the following:

THEOREM 13.5.1. *Let $A\subset X$, where A is not empty. Let $\psi: A\to \mathbf{R}$ be Fréchet differentiable at $\bar{x}$. If*

1. $\psi(\bar{x})=\min_{x\in A}\psi(x)$, *then*
2. $\nabla\psi(\bar{x})\in[P(A,\bar{x})]^{+}$.

If ψ is pseudoconvex over A at $\bar{x}$ and A is pseudoconvex at $\bar{x}$ then condition 2 *implies condition* 1.

PROOF. Let $y\in T(A,\bar{x})$. Choose a sequence $\{x^k\}\in A$, $x^k\to\bar{x}$, and a sequence$\{\lambda_k\}$,$\lambda_k>0$, such that $\lambda_k(x^k-\bar{x})\to y, k\to\infty$. Let $\psi(x)\geqslant\psi(\bar{x})$ for all $x\in A$. Then $\psi(x^k)\geqslant\psi(\bar{x})$ for all k while

$$0\leqslant\psi(x^k)-\psi(\bar{x})=\langle x^k-\bar{x}, \nabla\psi(\bar{x})\rangle+\mathrm{o}(\|x^k-\bar{x}\|).$$

Then

$$\langle\lambda_k(x^k-\bar{x}), \nabla\psi(\bar{x})\rangle\geqslant-\frac{\mathrm{o}(\|x^k-\bar{x}\|)}{\|x^k-\bar{x}\|}(\lambda_k\|x^k-\bar{x}\|).$$

Let $k\to\infty$; we find $\langle y, \nabla\psi(\bar{x})\rangle\geqslant 0$. Thus $\nabla\psi(x)\in[T(A,\bar{x})]^{+}$. But $[T(A,\bar{x})]^{+}=[\mathrm{co}(T(A,\bar{x})]^{+}=[P(A,\bar{x})]^{+}$. This proves the first part of the theorem.

Now let $x\in A$. A is pseudoconvex at $\bar{x}$ and $x-\bar{x}\in P(A,\bar{x})$; if $\langle x-\bar{x}, \nabla\psi(\bar{x})\rangle\geqslant 0$, then $\psi(x)\geqslant\psi(\bar{x})$ and condition 2 implies condition 1. ■

We are now in a position to state and prove a generalization of the *first order optimality conditions* of Kuhn and Tucker.

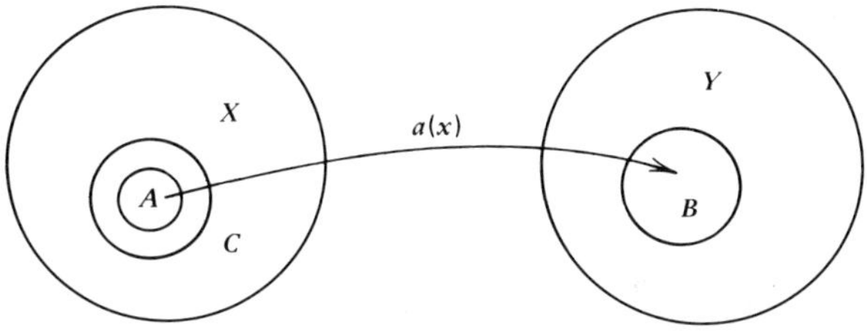

THEOREM 13.5.2. *Let X, Y be Banach spaces, and $B \subset Y, C \subset X$ be nonempty sets. Let $a(x): X \to Y$ be a mapping and let $A = \{x \in C | a(x) \in B\} \neq \varnothing$. Let*

$$\psi(\bar{x}) = \max\{\psi(x) | x \in A\} = \max\{\psi(x) | x \in C, a(x) \in B\}.$$

Let $a(x)$ posses a Fréchet derivative $\nabla a(\bar{x})$ at $\bar{x}$. Let

$$K = \{k \in X | \nabla a(\bar{x}) k \in P(B, a(\bar{x}))\}.$$

Let

$$H = \{h \in X' | h = \nabla a(\bar{x})' y', y' \in P(B, a(\bar{x}))^{+}\}.$$

Let $H = \overline{H}$; let G be a closed convex cone in X such that

$$K \cap G = P(A, \bar{x}),$$

and

$$K^{+} + G^{+} = \overline{K^{+} + G^{+}}.$$

If $\bar{x}$ exists satisfying the above conditions, then there exists a $y' \in P(B, a(x))^{-}$ such that

$$\nabla \psi(\bar{x}) + \nabla a(\bar{x})' y' \in G^{+}.$$

Conversely, if G is a closed convex cone in X such that $x - \bar{x} \in G$ for all $x \in A$, while

1. *Either A or $\Delta = \{x | a(x) \in B\}$ is pseudoconvex at $\bar{x}$, and*
2. *Either ψ is pseudoconvex over A at $\bar{x}$ or ψ is quasiconvex and $\nabla \psi(\bar{x}) \neq 0$,*

and there exists a $y' \in P(B, a(c))^-$ *such that*

$$\nabla \psi(\bar{x}) + \nabla a(x)' y' \in G^+,$$

then

$$\psi(\bar{x}) = \min\{\psi(x) | x \in A\}.$$

PROOF. Let $\bar{x}$ minimize ψ over A. Then $\nabla\psi(x) \in P(A, \bar{x})^+$ by Theorem 13.5.1. But, by Lemma 13.5.4

$$P(A, \bar{x})^+ = (K \cap G)^+ = K^+ + G^+,$$

since both K and G are closed convex cones. Thus $\nabla\psi(\bar{x}) - K^+ \subset G^+$, so there is a $u' \in K^- = -K^+$ with $\nabla\psi(\bar{x}) + u' \in G^+$. If we show that $u' = \nabla a(\bar{x})' y'$, $y' \in P(B, a(\bar{x}))^-$ then the first part of the theorem will be proven.

Suppose that $v \in H^+$, while y' is arbitrary in $P(B, a(\bar{x}))^+$; then

$$\langle \nabla a(\bar{x})' y, v' \rangle \geqslant 0$$

implies

$$\langle y', \nabla a(\bar{x}) v \rangle \geqslant 0.$$

Hence $\nabla a(\bar{x}) v \in P(B, a(\bar{x}))^{++} = P(B, a(\bar{x}))$; the last equality holds because $P(B, a(\bar{x}))$ is always a closed convex cone. Thus $v \in K$, so $H^+ \subset K$. Hence $H \supset K^+$. Then $u' \in K^- = -K^+$ implies $u' \in -H$ or $u' = \nabla a(\bar{x})' y'$ where $y' \in P(B, a(\bar{x}))^-$.

The second half of the theorem depends upon the following:

LEMMA 13.5.9. *If G is a closed convex cone in X with $x - \bar{x} \in G$ for all $x \in A$ and condition* 1 *of the theorem is true, then*

1. "*There exists $y' \in P(B, a(\bar{x}))^-$ with $\nabla\psi(\bar{x}) + \nabla a(\bar{x})' y' \in G^+$*" *implies*
2. "*For any $x \in A$, $\langle x - \bar{x}, \nabla\psi(\bar{x}) \rangle \geqslant 0$.*"

PROOF. We translate condition 1 of the lemma into the assertion: for all $x \in A$, $x - \bar{x} \in G$ and

$$\langle x - \bar{x}, \nabla\psi(\bar{x}) \rangle \geqslant -\langle x - \bar{x}, \nabla a(\bar{x})' y' \rangle.$$

Now if Δ is pseudoconvex at $\bar{x}$, $x - \bar{x} \in P(\Delta, \bar{x})$, while if A is pseudoconvex at $\bar{x}$, $x - \bar{x} \in P(A, \bar{x}) \subset P(\Delta, \bar{a})$ for all $x \in A \subset \Delta$. But if $z \in T(\Delta, \bar{x})$, then there are sequences $\{\lambda_k\}, \{x^k\}$ with $\lambda_k > 0$, $x^k \in \Delta$, and $\lambda_k(x^k - \bar{x}) \to z$, $k \to \infty$.

Also

$$a(x^k)-a(\bar{x})=\nabla a(\bar{x})(x^k-\bar{x})+o(\|x^k-\bar{x}\|),$$

or

$$\lambda_k\left[a(x^k)-a(\bar{x})\right]=\nabla a(\bar{x})z+o\left(\|\lambda_k(x^k-\bar{x})\|\right).$$

Suppose $\nabla a(\bar{x})z\notin P(B,a(\bar{x}))$. Since $P(B,a(\bar{x}))$ is a closed convex cone there is a linear functional $y'\in P(B,a(\bar{x}))^+$ such that

$$\langle\nabla a(\bar{x})z,y'\rangle<0,$$

while

$$\langle\lambda_k\left[a(x^k)-a(\bar{x})\right],y'\rangle\geqslant 0.$$

But as $k\to\infty$, the second inequality converges to $\langle\nabla a(\bar{x})z,y'\rangle\geqslant 0$, which is a contradiction. Hence, for all $z\in T(\Delta,\bar{x})$, $\nabla a(\bar{x})z\in P(B,a(\bar{x}))$. The same must be true for all $y\in P(\Delta,\bar{x})$, by continuity, also for all $y\in A\subset P(\Delta,\bar{x})$ if A or Δ is pseudoconvex. Then $-y'\in P(B,a(\bar{x}))^-$ implies

$$\langle x-\bar{x},\nabla\psi(\bar{x})\rangle\geqslant\langle x-\bar{x},\nabla a(\bar{x})'(-y')\rangle\geqslant 0$$

for all $x\in A$. ■

COROLLARY 1. *If in addition ψ is pseudoconvex over A at $\bar{x}$, then for all $x\in A$, $\psi(x)\geqslant\psi(\bar{x})$.*

If ψ is only quasiconvex, we proceed as follows: If $\nabla\psi(\bar{x})\neq 0$, then there is an $x^1\in X$ with $\langle x^1-\bar{x},\nabla\psi(\bar{x})\rangle>0$. (If we assume the contrary, i.e., $\langle z,\nabla\psi(\bar{x})\rangle\leqslant 0$ for all $z\in X$, then also $-\langle z,\nabla\psi(\bar{x})\rangle=\langle -z,\nabla\psi(\bar{x})\rangle\leqslant 0$, so $\nabla\psi(\bar{x})=0$.) Then we set

$$\begin{cases} x(t)=x+t(x^1-x), \\ \bar{x}(t)=\bar{x}+t(x^1-\bar{x}), \end{cases} \qquad 0\leqslant t\leqslant 1.$$

Hence

$$\langle\bar{x}(t)-\bar{x},\nabla\psi(\bar{x})\rangle=t\langle x^1-\bar{x},\nabla\psi(\bar{x})\rangle>0,\qquad 0<t\leqslant 1.$$

Also

$$\langle x(t)-\bar{x}(t),\nabla\psi(\bar{x})\rangle=(1-t)\langle x-\bar{x},\nabla\psi(\bar{x})\rangle\geqslant 0,\qquad 0\leqslant t<1$$

by Lemma 13.5.9. Also for any t, $0<t\leqslant 1$, $\langle x(t)-\bar{x}, \nabla\psi(\bar{x})\rangle>0$. Since ψ is quasiconvex, by Lemma 13.5.8, $\psi(x(t))<\psi(\bar{x})$ or, as $t\to 0$, $\psi(x)\geqslant\psi(\bar{x})$, for all $x\in A$.

It is possible to derive, in this way, second order optimality conditions that are an appropriately modified form of the second order conditions of Theorem 12.9.3, but only if X is a finite dimensional space (Guignard, [1]). Another generalization of the first order conditions, not involving differentiability of the functionals ψ and $a(x)$, is to be found in Pshenichnyi [1], 83–88.

EXERCISES TO SECTION 13.5

1. Show that if K is a closed cone, then $0\in K$.
2. Show that K^+ is a closed convex cone, for any set K. Show that $K_1^+ + K_2^+$ is a closed convex cone if K_1, K_2 are closed convex cones.
3. Show that $K^+ = -K^-[=(-1)K^-]$.
4. Show that $P(M,x)$ is a closed convex cone.
5. Let $X=\mathbf{R}^n$ and $Y=\mathbf{R}^m\subset\mathbf{R}^n$. Show that $Y^+=\{x|(y,x)=0, y\in Y\}$, i.e., $Y^+=Y^\perp$, the orthogonal complement of Y. Is this also true if X is an arbitrary Hilbert space?
6. If K_1, K_2 are convex cones, then show that $K_1\cap K_2$ and K_1+K_2 are convex cones.
7. If K_1, K_2 are nonempty cones, then show that $(K_1+K_2)^+ = K^+\cap K^+ = (K_1\cup K_2)^+$.
8. Let $X=\mathbf{R}^n$, $y=\mathbf{R}^m$; $A\in[X\to Y]$ is an $m\times n$ matrix. Show that
$$K=\{x|(Ax)_i\geqslant 0,\ i=1,\ldots,m\}$$
is a closed convex cone.
9. In Exercise 13.5.8, find K^-, K^+, K^{--}, and K^{++}.
10. As in Exercise 13.5.8, show that
$$C=\{y|y=Ax, x_i\geqslant 0,\ i=1,\ldots,n\}$$
is a closed convex cone. Find C^-, C^+, C^{--}, C^{++}. A set of the form of C is called a *polyhedral cone*. Show that a polyhedral cone satisfies $C=C^{--}=C^{++}$.
11. Let $g:\mathbf{R}^n\to\mathbf{R}^p$, $h:\mathbf{R}^n\to\mathbf{R}^q$ be differentiable and
$$S=\{x\in\mathbf{R}|\,g_i(x)\leqslant 0,\ i=1,\ldots,p, h_i(x)=0, i=1,\ldots,q\}.$$

Let $f:\mathbf{R}^n \to \mathbf{R}$. Let

$$f(\bar{x}) = \min_{x \in S} f(x).$$

Let

$$C = \{ x | \langle x, \nabla g_i(\bar{x}) \rangle \leqslant 0, i = 1, \dots, p, \langle x, \nabla h_i(\bar{x}) \rangle = 0, i = 1, \dots, q \}.$$

Show that if $C \subset P(S, \bar{x})$, then there exist scalars $\mu_1, \dots, \mu_p$, $\mu_i \geqslant 0$, and $\nu_1, \dots, \nu_q$ such that

$$\nabla f(\bar{x}) + \sum_{i=1}^{p} \mu_i \nabla g_i(\bar{x}) + \sum_{i=1}^{q} \nu_i \nabla h_i(\bar{x}) = 0.$$

(Hint: Apply Theorem 13.5.2 with $G = P(C, \bar{x})$. See Guignard [1] or Bazaraa and Shetty [1], 150–151.)

12. Show that in $L_p(0,1)$, $1 \leqslant p \leqslant \infty$, the functions $x(t)(\geqslant)0$ are a closed convex cone, K. What is K^-?
13. In $C[0,1]$ show that the functions $x(t) \geqslant 0$ are a closed convex cone K. What is K^-?
14. If $k(x,t)$ is such that

$$(Kx)(s) = \int_0^1 k(s,t) x(t) \, dt$$

maps $X \to Y$ where X, Y are dual L_p spaces, then is the set $C = \{ x | (Kx)(s) \geqslant 0 \}$ a closed convex cone? What if $X = Y = C[0,1]$? What is C^+?

13.6 CONVEX DUALITY AND SADDLE POINTS

In Section 13.1 we derived a Kuhn-Tucker saddle point condition for the solution of a restricted form of the constrained convex programming problem from the separation form of the Hahn-Banach Theorem. The proof involved the construction of a Lagrangian dual problem. Here we generalize the meaning of Lagrangian dual, following the ideas of Rockafellar [1],[2] and other references cited in [2]. (See also Ekeland and Temam [1], 1–70 or Céa [1], 193–216.) We start with a normed linear space X and a functional $f: X \to \overline{\mathbf{R}}$ whose minimum on X is the goal of our search. We recall (Exercises 13.2.2–4) the definition of the *epigraph* of $f(x)$:

$$epi f = \{ (x, \alpha) \in X \otimes \mathbf{R} | f(x) \leqslant \alpha \},$$

that $f(x)=\inf\{\alpha|(x,\alpha)\in epi f\}$, that f is convex if and only if $epi f$ is convex in $X\otimes\mathbf{R}$, and that f is weakly lower semicontinuous if and only if $epi f$ is weakly closed in $X\otimes\mathbf{R}$. The weak closure of $epi f$ is written $\widetilde{epi} f$. On $X\otimes\mathbf{R}$ a closed hyperplane takes the form $\langle x,x'\rangle+\alpha\beta=\gamma$ where $(x,\alpha)\in X\otimes\mathbf{R}$ and $x'\in X'$, $\beta,\gamma\in\mathbf{R}$ determine the hyperplane. A hyperplane that intersects the $\mathbf{R}$-axis in exactly one point is called a nonvertical hyperplane. This occurs if and only if $\beta\neq 0$, in which case we normalize $\beta=-1$.

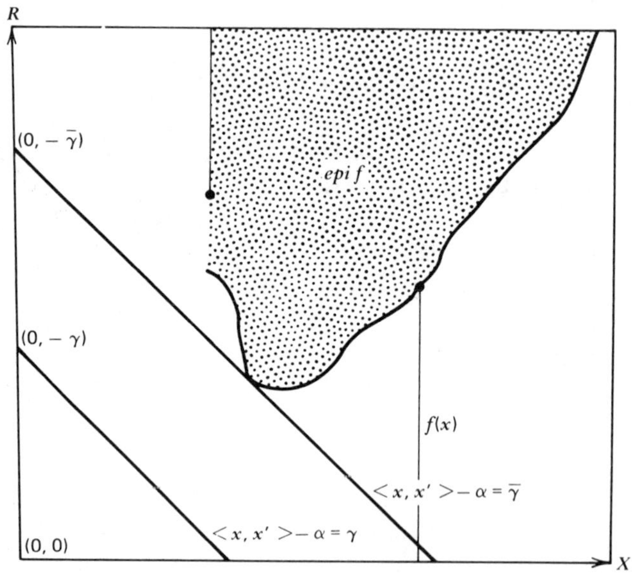

In this way we obtain a one parameter family of parallel nonvertical hyperplanes with fixed x' and parameter $-\gamma$, the $\mathbf{R}$-intercept. If even one such hyperplane does not intersect $epi f$, then $f(x)>-\infty$ for all $x\in X$, and $epi f$ lies in the corresponding half space $\langle x,x'\rangle-\alpha\leqslant\gamma$. If $f(x)$ is not everywhere $+\infty$, then we may decrease γ until at $\bar{\gamma}$ the hyperplane becomes a supporting hyperplane to $\widetilde{epi} f$. As x' varies over X' we generate the set of all nonvertical supporting hyperplanes of $\overline{epi f}$.

Now a convex set may be described either as the collection of its elements, or "dually," as the intersection of half spaces containing it. A nonconvex set S has a convex hull $\operatorname{co} S$ and its closure $\overline{\operatorname{co}} S$. For functions $f: X\to\overline{\mathbf{R}}$ we define the corresponding concepts.

DEFINITION 13.6.1. Let $f:X\to\overline{\mathbf{R}}$. The *convex hull* or *regularization* of $f(x)$, written $cof:X\to\overline{\mathbf{R}}$ is defined by

$$cof(x)=\sup_{x',\alpha}\{\langle x,x'\rangle-\alpha\,|\,\langle x,x'\rangle-\alpha\leqslant f(x)\text{ for all }x\in X\}.$$

The convex hull cof is the largest convex function that is everywhere not greater than $f(x)$. Thus if there is an $\bar{x}$ with $f(\bar{x})=-\infty$, then $cof(x)=-\infty$ everywhere. (Why?) We can make a dual definition.

DEFINITION 13.6.2. Let $f:X\to\overline{\mathbf{R}}$. The *conjugate function* or *Fenchel transform*: $f^*:X'\to\overline{\mathbf{R}}$, is

$$f^*(x')=\sup\{\langle x,x'\rangle-f(x)\,|\,x\in X\}.$$

Notice that $\langle x,x'\rangle-\alpha\leqslant f(x)$ if and only if $\langle x,x'\rangle-f(x)\leqslant\alpha$, so that $f^*(x')\leqslant\alpha$. Also if $(\bar{x},\bar{\alpha})\in epif$, $f(\bar{x})\leqslant\bar{\alpha}$, so $\langle\bar{x},x'\rangle-\bar{\alpha}\leqslant\langle\bar{x},x'\rangle-f(\bar{x})$. But $f(\bar{x})=\inf\{\alpha\,|\,(\bar{x},\alpha)\in epif\}$, so that

$$f^*(x')=\sup\{\langle x,x'\rangle-\alpha\,|\,(x,\alpha)\in epif\}.$$

Thus $f^*(x')$ is a convex function that is (weakly) lower semicontinuous (Exercise 13.2.1).

The operation of conjugation has taken us from X to X'. We return by way of the following:

DEFINITION 13.6.3. If $h:X'\to\overline{\mathbf{R}}$, then the *conjugate function* of $h(x')$ is

$$h^*(x)=\sup\{\langle x,x'\rangle-h(x')\,|\,x'\in X'\}.$$

If $h(x)=f^*(x)$ for some $f:X\to\overline{\mathbf{R}}$, then $h^*(x)=f^{**}(x)$ is called the *biconjugate function* of f.

In discussing f, cof, f^*, and f^{**}, much depends upon lower semicontinuity properties. We define two operators on functions that are useful in this context.

DEFINITION 13.6.4. To each $f:X\to\overline{\mathbf{R}}$ we associate its *lower semicontinuous hull*

$$lscf(x)=\liminf_{y\to x}f(y)$$

and its *closure*

$$clf(x)=\begin{cases} lscf(x), & \text{if } lscf(x)>-\infty,\ x\in X,\\ -\infty, & \text{if } lscf(x)=-\infty \text{ for some } x.\end{cases}$$

If $f(x)=clf(x)$, then $f(x)$ is called *closed.*

Clearly $lscf(x)\leqslant f(x)$ everywhere, but no weakly lower semicontinuous function $g(x)$ exists with $g(x)\leqslant f(x)$ and $g(\bar{x})>lscf(\bar{x})$ for some $\bar{x}\in X$. But $g(x)=\inf\{\alpha|(x,\alpha)\in\widetilde{epi f}\}$ is such a function unless $epi\ lscf=\widetilde{epi f}$, so these sets are the same. If f is convex then $epi f$ is convex, as is $\widetilde{epi f}$, and both $lscf$ and clf are also convex; we also establish the following:

LEMMA 13.6.1. *If $f:X\to\overline{\mathbf{R}}$ is convex and $lscf(x)$ is finite at some $x\in X$, then both $lscf(x)$ and $f(x)$ are bounded below and not identically $+\infty$, so that $lscf(x)=clf(x)$.*

PROOF. Suppose $x,\bar{x}$ are such that $lscf(\bar{x})=-\infty$ and $f(x)<\infty$. Let $(x,\alpha),(\bar{x},\beta)\in epi\ lscf$, which is a convex set. Then

$$lscf((1-\lambda)x+\lambda\bar{x})\leqslant(1-\lambda)\alpha+\lambda\beta,\qquad 0<\lambda\leqslant 1.$$

But $\inf\{\beta|(\bar{x},\beta)\in epi\ lscf\}=-\infty$, so $lscf((1-\lambda)x+\lambda\bar{x})=-\infty$, $0<\lambda\leqslant 1$. Let $\lambda\downarrow 0$ to get $lscf(x)=-\infty$. $lscf(\bar{x})$ finite implies $f(x)$ finite for some x near $\bar{x}$. The rest is obvious. ■

It is left as an exercise to show f^* is closed and convex and that f^{**} recovers $cl\ cof$, which is everywhere not greater than f.

There is a close relation between the conjugate and the subdifferential, which is summarized in the following:

LEMMA 13.6.2. *Let $f:X\to\overline{\mathbf{R}}$. Then, at any fixed $\bar{x}\in X$,*

1. *$f(\bar{x})+f^*(x')=\langle\bar{x},x'\rangle$ if and only if $x'\in\partial f(\bar{x})\neq\varnothing$;*
2. *If $\partial f(\bar{x})\neq\varnothing$, then $f(\bar{x})=f^{**}(\bar{x})$;*
3. *If $f(\bar{x})=f^{**}(\bar{x})$, then $\partial f(\bar{x})=\partial f^{**}(\bar{x})\neq\varnothing$;*
4. *$x'\in\partial f(\bar{x})$ implies $\bar{x}\in\partial f^*(x')$;*
5. *If $f(\bar{x})=f^{**}(\bar{x})$, then $x'\in\partial f(\bar{x})$ if and only if $\bar{x}\in\partial f^*(x')$.*

PROOF.

1. If $x'\in\partial f(\bar{x})\neq\varnothing$, then $f(\bar{x})\in\mathbf{R}$ and $f(x)\geqslant f(\bar{x})+\langle x-\bar{x},x'\rangle$, so that $\langle x,x'\rangle-f(x)\leqslant\langle\bar{x},x'\rangle-f(\bar{x})$ for all $x\in X$. Hence $f^*(x')=\langle\bar{x},x'\rangle$

$-f(\bar{x})$. Conversely, from $f^*(x') = \langle \bar{x}, x' \rangle - f(\bar{x}) \in \mathbf{R}$ we conclude $\langle \bar{x}, x' \rangle - f(\bar{x}) \geqslant \langle x, x' \rangle - f(x)$ for all $x \in X$, so $x' \in \partial f(\bar{x})$.

2. Next assume $x' \in \partial f(\bar{x})$; then $f(\bar{x}) + \langle x - \bar{x}, x' \rangle \leqslant f(x)$, so, for all x,

$$f(\bar{x}) + \langle x - \bar{x}, x' \rangle \leqslant co f(x) = f^{**}(x).$$

At $x = \bar{x}$ we find $f(\bar{x}) \leqslant f^{**}(\bar{x}) \leqslant f(\bar{x})$.

3. Let $f(\bar{x}) = f^{**}(\bar{x})$ and $x' \in \partial f(\bar{x})$. Then $f^{**}(\bar{x}) + \langle x - \bar{x}, x' \rangle = f(\bar{x}) + \langle x - \bar{x}, x' \rangle \leqslant f^{**}(x)$, so $x' \in \partial f^{**}(\bar{x})$. The converse is left as an exercise.
4. Since $f^{**}(x) \leqslant f(x)$, if $x' \in \partial f(\bar{x})$, then $f^{**}(\bar{x}) + f^*(x') \leqslant \langle \bar{x}, x' \rangle$. The reverse inequality is always true, so $f^{**}(\bar{x}) + f^*(x') = \langle \bar{x}, x' \rangle$. We now apply condition 1 to this equation.
5. If $f^{**}(\bar{x}) = f(\bar{x})$, then conditions 3 and 4 imply condition 5. ∎

We now turn to the *general optimization problem*: Given $f: X \to \overline{\mathbf{R}}$, find $\bar{x}$ such that

$$f(\bar{x}) = \inf\{ f(x) | x \in X \}.$$

We assume that this problem can be imbedded in an associated family of *perturbed problems* as follows:

Let there exist a second normed linear vector space U and a second functional $F: X \otimes U \to \overline{\mathbf{R}}$ such that

$$f(x) = F(x, 0).$$

We consider the family of perturbed problems

$$\inf\{ F(x, u) | x \in X \}$$

which reduce, at $u = 0$, to the original problem which we now name *the primal problem*. Dual to $X \otimes U$ is $X' \otimes U'$ with

$$\langle (x, u), (x', u') \rangle = \langle x, x' \rangle_X + \langle u, u' \rangle_U.$$

Using *this* duality, define the conjugate function

$$F^*(x', u') = \sup\{ \langle x, x' \rangle + \langle u, u' \rangle - F(x, u) | x \in X, u \in U \},$$

and the *dual problem*

$$\sup\{ -F^*(0, u') | u' \in U \}.$$

We let $\boldsymbol{P}$ denote the *value* of the primal problem, $\boldsymbol{P}^*$ the value of the dual problem. That is

$$\boldsymbol{P}=\inf\{f(x)|x\in X\}$$

and

$$\boldsymbol{P}^*=\sup\{-F^*(0,u')|u'\in U'\}.$$

They are related as follows:

LEMMA 13.6.3. $\boldsymbol{P}^*\leqslant\boldsymbol{P}$.

PROOF. From the definition, $F^*(0,u')=\sup\{\langle u,u'\rangle-F(x,u)|x\in X, u\in U\}\geqslant -F(x,0)$, so $-F^*(0,u')\leqslant f(x)$ for all u',x. ■

In this argument it is still possible that either or both values are infinite. We get $-\infty<\boldsymbol{P}^*$ $[\boldsymbol{P}<\infty]$ if we know that there is a $\bar{u}'$ [an $\bar{x}$] such that $F^*(0,\bar{u}')<\infty$ $[f(\bar{x})<\infty]$. A problem whose functional is somewhere less than $+\infty$ will be called *nontrivial*. We now restate matters.

DEFINITION 13.6.5. The function $\phi:U\to\overline{\mathbf{R}}$ defined as

$$\phi(u)=\inf\{F(x,u)|x\in X\}$$

is called the *optimal value function*.

The optimal value function has a close connection to $\boldsymbol{P}^*$.

THEOREM 13.6.1. *Let* $F:X\otimes U\to\overline{\mathbf{R}}$. *Then*

1. *If F is convex, then* $\phi(u)$ *is convex*;
2. $\phi^*(u')=F^*(0,u')$;
3. $\boldsymbol{P}^*=\sup\{-\phi^*(u')|u'\in U'\}=\phi^{**}(0)$.

PROOF. The convexity of ϕ is a direct consequence of the convexity of F. Using the definition of ϕ^* we have

$$\begin{aligned}\phi^*(u')&=\sup\{\langle u,u'\rangle-\inf\{F(x,u)|x\in X\}|u\in U\}\\&=\sup\{\sup\{\langle u,u'\rangle-F(x,u)|x\in X\}|u\in U\}\\&=F^*(0,u').\end{aligned}$$

This establishes condition 2 and the first equality of condition 3. The second equality follows from Exercise 13.6.8. ■

Now, for any function, ϕ, $\phi^{**}(0) \leqslant \phi(0)$, so that condition 3 above is another proof of $\boldsymbol{P}^* \leqslant \boldsymbol{P}$. Equality of the values of the primal and dual problems is equivalent to

$$\phi^{**}(0) = \phi(0).$$

But $\phi^{**}(u) = cl\ co\,\phi(u)$, and it seems plausible that we can restate $\boldsymbol{P}^* = \boldsymbol{P}$ in terms of ϕ alone. We do this in the convex case in the following:

THEOREM 13.6.2. *Let* $f: X \to \overline{\mathbf{R}}$ *and* $F: X \otimes U \to \overline{\mathbf{R}}$ *with* $F(x,0) = f(x)$. *Let* $F(x,u)$ *be convex. Let*

$$\boldsymbol{P} = \inf\{ f(x) | x \in X \},$$

and

$$\begin{aligned} \boldsymbol{P}^* &= -\inf\{ F^*(0,u') | u' \in U' \} \\ &= \sup\{ -F^*(0,u') | u' \in U' \}. \end{aligned}$$

Let

$$\phi(u) = \inf\{ F(x,u) | x \in X \}.$$

Then $\boldsymbol{P} = \boldsymbol{P}^*$ *and is finite if and only if*

1. $|\phi(0)| < \infty$, *and*
2. ϕ *is weakly lower semicontinuous at* $u = 0$.

PROOF. In the convex case, ϕ is convex, so $\phi(u) = co\,\phi(u)$. We always know that $\phi^{**}(u) < clco\,\phi(u) \leqslant \phi(u)$, so that $\phi^{***}(u') \geqslant (clco\,\phi)^*(u') \geqslant \phi^*(u')$. But $\phi^{***}(u') = \phi^*(u')$. Then $\phi^*(u') = (clco\,\phi)^*(u')$, and $\phi^{**}(u) = (clco\,\phi)^{**}(u) = clco\,\phi(u)$.

Now, assume conditions 1 and 2. From condition 1 we conclude $lsc\ co\,\phi(u) = cl\ co\,\phi(u) > -\infty$ for all u; since ϕ is convex, from condition 2 we conclude $lsc\ co\,\phi(0) = \phi(0)$. That is, $\phi(0) = \phi^{**}(0) = \boldsymbol{P} = \boldsymbol{P}^*$ and is finite. Conversely, if we know that $\boldsymbol{P} = \boldsymbol{P}^*$ and is finite, then $\phi^{**}(0) = \phi(0)$ and is finite. Convexity implies $\phi^{**}(0) = lsc\ co\,\phi(0) = lsc\,\phi(0)$. But $lsc\,\phi(0) = \phi(0)$ is condition 2. ■

Thus we have

$$\phi(0) = cl\ co\,\phi(0)$$

as the general criterion for $\boldsymbol{P}=\boldsymbol{P}^*$, and

$$\phi(0)=\liminf_{u\to 0}\phi(u)$$

as the criterion in the convex case. The equality of $\boldsymbol{P}$ and $\boldsymbol{P}^*$ does not imply the existence of solutions of either the primal or the dual problem (i.e., of an $\bar{x}$ such that $\boldsymbol{P}=f(\bar{x})$ or a $\bar{u}'$ such that $\boldsymbol{P}^*=-\phi^*(\bar{u}')$). The existence question can be linked to the existence of subgradients to ϕ at $u=0$. First we establish the following:

LEMMA 13.6.4. *If $F:X\otimes U\to\overline{\mathbf{R}}$, then*

$$\partial\phi^{**}(0)=\{u'|\boldsymbol{P}^*=-\phi^*(u')\}.$$

PROOF. The dual problem has a solution $\bar{u}'$ if and only if $-F^*(0,\bar{u}')\geqslant -F^*(0,u')$ for all $u'\in U'$. But this is just $-\phi^*(\bar{u}')\geqslant-\phi^*(u')$ for all $u\in U'$. Write this as

$$-\phi^*(\bar{u}')=\sup\{\langle 0,u'\rangle-\phi^*(u')|u'\in U'\}.$$

That is, $-\phi^*(\bar{u}')=\phi^{**}(0)$ if and only if $\bar{u}'$ solves the dual problem. On the other hand, $\bar{u}'\in\partial\phi^{**}(0)=\varnothing$ if and only if

$$\phi^{**}(0)+\phi^{***}(\bar{u}')=\langle 0,\bar{u}'\rangle=0,$$

by Lemma 13.6.2, condition 1. But $\phi^{***}(\bar{u}')=\phi^*(\bar{u}')$. ■

With convexity, the conditions of the lemma can be transferred to the following:

THEOREM 13.6.3. *Let $f:X\to\overline{\mathbf{R}}$, $F:X\otimes U\to\overline{\mathbf{R}}$ and $F(x,0)=f(x)$. If $F(x,u)$ is convex, then*

1. *$|\phi(0)|<\infty$ and $\partial\phi(0)\neq\varnothing$ if and only if*
2. *$\boldsymbol{P}=\boldsymbol{P}^*$ is finite and the dual problem is solvable.*

PROOF. Theorem 13.2.4 and condition 1 imply that $\phi(u)$ is weakly lower semicontinuous at $u=0$; we apply Theorem 13.6.2 to conclude that $\phi(0)=\phi^{**}(0)=\boldsymbol{P}=\boldsymbol{P}^*$ is finite. Lemma 13.6.2 implies $\partial\phi(0)=\partial\phi^{**}(0)=\{u'|\boldsymbol{P}^*=-\phi^*(u')\}\neq\varnothing$.

Conversely, condition 2 implies $\phi(0)=\phi^{**}(0)$; then Lemma 13.6.2 implies condition 1. (Note that convexity is not needed in this direction.) ■

If $F(x,u)=F^{**}(x,u)$ and is not identically $+\infty$ or $-\infty$, then $F(x,u)$ provides the dual to the primal problem for $F^*(x,u)$. (Why?) In that case it is easy to translate Theorem 13.6.3 into: $|\phi^*(0)|<\infty$ *and* $\partial\phi^*(0)\neq\varnothing$ *if and only if* $\boldsymbol{P}=\boldsymbol{P}^*$ *and the primal problem for* $f(x)$ *has solutions.*

A sufficient condition for 1 to hold is given in the following:

LEMMA 13.6.5. *Let* $f:X\to\overline{\mathbf{R}}$, $F:X\otimes U\to\overline{\mathbf{R}}$ *and* $F(x,0)=f(x)$. *Let* $F(x,u)$ *be convex. If there exists* $\bar{x}\in X$, *such that the mapping* $u\to F(\bar{x},u)$ *is finite and continuous at* $u=0$, *then* $\phi(0)$ *is finite and* $\partial\phi(0)\neq\varnothing$.

PROOF. ϕ is convex and $|\phi(0)|<\infty$. Further, as a function of u, $F(\bar{x},u)$ is convex and continuous at $u=0$, so bounded for all u in some ball $\|u\|\leqslant\rho$. Then $\phi(u)=\inf\{F(x,u|x\in X\}$ is also bounded on such a ball, so is continuous (Exercise 13.1.3). Theorem 13.2.4 implies $\partial\phi(0)\neq 0$. ■

To prepare for the Lagrangian saddle point property we need the following:

DEFINITION 13.6.6. A function $g:X\to\overline{\mathbf{R}}$, on a normed linear space X, is called *concave* and/or *upper semicontinuous* if $-g$ is convex and/or lower semicontinuous. The *upper semicontinuous hull* of g, *usc* g is given by

$$usc\,g(x)=\limsup_{y\to x} g(y)$$

and the *upper closure ucl* g is given by

$$ucl\,g=\begin{cases}+\infty, & \text{if } usc\,g(x)=+\infty \text{ for some } x,\\ usc\,g, & \text{otherwise.}\end{cases}$$

A function is *upper closed* if $g(x)\equiv ucl\,g(x)$. The *upper conjugate function* $g^{\vee}(x)$ to $g(x)$ is

$$g^{\vee}(x')=\inf\{\langle x,x'\rangle-g(x)|x\in X\}$$

and the *upper biconjugate function* $g^{\vee\vee}(x)$ is

$$g^{\vee\vee}(x)=\inf\{\langle x,x'\rangle-g^{\vee}(x')|x'\in X'\}.$$

Notice that if $g=-f$, then $g^{\vee}(x')=-f^*(-\mathbf{x}')$.

For a given perturbed family $F:X\otimes U\to\overline{\mathbf{R}}$, notice that we can write

$$\begin{aligned}F^*(x',u')&=\sup\{\langle x,x'\rangle+\langle u,u'\rangle-F(x,u)|x\in X,u\in U\}\\&=\sup\{\langle x,x'\rangle+\sup\{\langle u,u'\rangle-F(x,u)|u\in U\}|x\in X\}\\&=\sup\{\langle u,u'\rangle+\sup\{\langle x,x'\rangle-F(x,u)|x\in X\}|u\in U\}\end{aligned}$$

This represents the 'double' Fenchel transform $F \rightarrow F^*$ as a sequence of two 'almost' single Fenchel transforms. Pick the first representation and make the following:

DEFINITION 13.6.7. The *Lagrangian* of the primal problem with respect to the family of perturbations $F(x,u)$, $\mathfrak{L}: X \otimes U' \rightarrow \overline{\mathbf{R}}$ is

$$\mathfrak{L}(x,u') = -\sup\{\langle u,u'\rangle - F(x,u) | u \in U\}.$$

If we let $F_x(u) = F(x,u)$, and for each fixed x, let $F_x^*(u')$ be the conjugate of $F_x(u)$ over U, we can wriite

$$\mathfrak{L}(x,u') = -F_x^*(u') = F_x^{\vee}(-u').$$

Now from

$$F^*(x',u') = \sup\{\langle x,x'\rangle - \mathfrak{L}(x,u') | x \in X\}$$

we conclude that

$$-F^*(0,u') = \inf\{\mathfrak{L}(x,u') | x \in X\}.$$

Thus the dual problem to $\inf\{f(x)|x \in X\}$ is

$$\sup\{\inf\{\mathfrak{L}(x,u')|x \in X\}|u' \in U'\}.$$

If $F(x,u)$ is closed convex in u for each fixed x, so that $F_x^{**}(u) = F_x(u)$, then we may completely symmetrize the relation between the primal and dual problems. Now

$$\begin{aligned} F(x,u) &= F_x^{**}(u) \\ &= \sup\{\langle u,u'\rangle - F_x^*(u') | u' \in U'\} \\ &= \sup\{\langle u,u'\rangle + \mathfrak{L}(x,u') | u' \in U'\}, \end{aligned}$$

so that $F(x,0) = \sup\{\mathfrak{L}(x,u')|u' \in U'\}$, and the primal problem satisfies

$$\inf\{f(x)|x \in X\} = \inf\{\sup\{\mathfrak{L}(x,u')|u' \in U'\}|x \in X\}.$$

In particular this is true if $F(x,u)$ is closed convex on $X \otimes U$.

DEFINITION 13.6.8. We say that $(\bar{x},\bar{u}') \in X \otimes U'$ is a *saddle point* of the Lagrangian $\mathfrak{L}(x,u')$ if for all $x \in X, u' \in U'$,

$$\mathfrak{L}(\bar{x},u') \leqslant \mathfrak{L}(\bar{x},\bar{u}') \leqslant \mathfrak{L}(x,\bar{u}').$$

It is immediate from the definition of $\boldsymbol{P}$ and $\boldsymbol{P}^*$ that the values of the primal and dual problems satisfy

$$\boldsymbol{P}^* \leqslant \mathfrak{L}(\bar{x},\bar{u}') \leqslant \boldsymbol{P}.$$

For convex problems we have the following:

THEOREM 13.6.4. *Let the family of perturbations $F(x,u)$ be closed, convex, and not identically $+\infty$ or $-\infty$. Then $(\bar{x},\bar{u}')$ is a saddle point of the Lagrangian $\mathfrak{L}(x,u')$ if and only if $\bar{x}$ is a solution of the primary problem, $\bar{u}'$ is a solution of the dual problem, and $\boldsymbol{P}^*=\boldsymbol{P}$ is finite.*

PROOF. If $(\bar{x},\bar{u}')$ is a saddle point, then

$$\mathfrak{L}(\bar{x},\bar{u}') = -F^*(0,\bar{u}') = F(\bar{x},0)$$

follows immediately from the definition. (Why?) But $F(\bar{x},0)+F^*(0,\bar{u}')=0$ implies the first half of the theorem through Exercise 13.6.11.

For the converse, we use $F(\bar{x},0)+F^*(0,\bar{u}')=0$, and the definition of $\boldsymbol{P}^*$ and $\boldsymbol{P}$ to imply

$$\sup\{\mathfrak{L}(\bar{x},u')|u'\in U'\} \leqslant \mathfrak{L}(\bar{x},\bar{u}') \leqslant \inf\{\mathfrak{L}(x,\bar{u}')|x\in X\},$$

which in turn implies that $(\bar{x},\bar{u}')$ is a saddle point for $\mathfrak{L}(x,u')$. Notice that in this direction we do not use the assumption that F is closed and convex. ■

COROLLARY 1. *If $|\phi(0)|<\infty$ and $\partial\phi(0)\neq 0$ then $\bar{x}$ solves the primal problem if and only if there is a $\bar{u}'\in U'$ such that $(\bar{x},\bar{u}')$ is a saddle point of $\mathfrak{L}(x,u')$.*

EXERCISES TO SECTION 13.6

1. Let f be lower semicontinuous. Show that the hyperplane $\langle x,x'\rangle-\alpha=\gamma$ first intersects $\widetilde{epi f}$ as γ decreases at a point of $epi f$. Conclude that

$$f^*(x')=\sup\{\langle x,x'\rangle-\alpha|(x,\alpha)\in epi f\}.$$

2. Show that the regularization $co f$ of any f is convex and closed. (Hint: Exercises 13.2.3 and 13.2.4.)
3. Show that $epi\, co f=\widetilde{epi f}$. (Hint: $epi\, co f$ is closed. From $co f(x)\leqslant f(x)$ conclude $epi\, co f\supset \widetilde{epi f}$. Conversely, let $S(x)$ be the set $\{\alpha|(x,\alpha)\in$

$\widetilde{epi f}\}$. Let

$$g(x)=\begin{cases} +\infty, & S(x)=\varnothing \\ \bar{\alpha}, & S(x)=[\bar{\alpha},\infty). \end{cases}$$

Then $epi\, g=\widetilde{epi f}$, $g(x)\leqslant f(x)$ and $epi\, g\supset epi\, co f$.)

4. Show that f^* is closed and convex and that $f^{**}(x)=cl\, co f(x)$, while $lsc\, co f(y)=-\infty$ for one $x\in X$ implies $f^{**}(x)=-\infty$ for all x; thus if f is convex, closed, and bounded below, $f^{**}=f$, while for all $f, f^{***}=f^*$.
5. Let $\mathbf{\Phi}_S(x)$ be the indicator function of the set S, that is, $\mathbf{\Phi}_S(x)=0$, $x\in S$, and $\mathbf{\Phi}_S(x)=+\infty$, $x\notin S$. Show that $\mathbf{\Phi}_S^*$ is the *support function* of S:

$$\mathbf{\Phi}_S^*(x)=\sup\{\langle x,x'\rangle | x\in S\}.$$

If S is a cone, show that $\mathbf{\Phi}_S^*$ is the indicator function of S^-.

6. Let $\phi:\mathbf{R}\to\overline{\mathbf{R}}$ be closed, convex, and not identically $+\infty$ or $-\infty$. If $\phi(t)\geqslant 0$ for all $t\in\mathbf{R}$, $\phi(0)=0$, and X is any normed linear vector space, then show that the functions $\phi(\|x\|)$ and $\phi^*(\|x'\|)$ satisfy $(\phi(\|x\|))^*=\phi^*(\|x'\|)$ and $(\phi^*(\|x'\|))^*=\phi(\|x\|)$.
7. Let H be a real Hilbert space, A a bounded strictly positive selfadjoint operator, and $f\in H$. Let

$$Q(x)=\tfrac{1}{2}(Ax,x)-(x,f).$$

Show that

$$Q^*(x')=\tfrac{1}{2}(A^{-1}(x'-f),(x'-f)).$$

8. Let $f:X\to\overline{\mathbf{R}}$ and $g:X\to\overline{\mathbf{R}}$ where X is a normed linear vector space. Show that

   (a) $f^*(0)=-\inf\{f(x)|x\in X\}=\sup\{-f(x)|x\in X\}$;
   (b) If $f(x)\leqslant g(x)$, $x\in X$, then $f^*(x')\geqslant g^*(x')$, $x'\in X'$;
   (c) If λ is a positive scalar, then $(\lambda f)^*(x')=\lambda f^*(x'/\lambda)$;
   (d) If $\alpha\in\mathbf{R}$, then $(f+\alpha)^*(x')=f^*(x')-\alpha$;
   (e) For all $x\in X, x'\in X'$, $f(x)+f^*(x')\geqslant\langle x,x'\rangle$ (compare (e) with Lemma 13.6.2);
   (f) If $f_i(x)$ is any collection of functions: $X\to\overline{\mathbf{R}}$ with $i\in I$, an index set then $[\inf_{i\in I}f_i(x)]^*\leqslant\sup_{i\in I}f_i^*(x)$ and $[\sup_{i\in I}f_i(x)]^*\leqslant\inf_{i\in I}f_i^*(x)$.

9. In Lemma 13.6.1 show that when $|lsc f| = \infty$ everywhere, then $\{x \mid lsc f(x) = -\infty\} = \overline{\{x \mid f(x) < \infty\}}$.
10. Show that if $F(x,u)$ is convex, then so is the optimal value function $\phi(u)$. (Hint: If $\phi(u) \leqslant F(x^1,u) \leqslant a < \infty$ and $\phi(v) \leqslant F(x^2,v) \leqslant b < \infty$ then $\phi(\lambda u + (1-\lambda)v) \leqslant \inf F(x, \lambda u + (1-\lambda)v) \leqslant \lambda a + (1-\lambda)b$.)
11. Show that $\bar{x}$ solves the primal problem, $\bar{u}'$ the dual problem, and $\boldsymbol{P}^* = \boldsymbol{P}$ is finite if and only if

$$F(\bar{x},0) + F^*(0,u') = 0,$$

which is equivalent to

$$(0,u') \in \partial F(\bar{x},0).$$

(Hint: For any (x,u') we always have $F(x,0) + F^*(0,u') \geqslant \langle (x,0),(0,u') \rangle$.)
12. Show that if $g(x) = -f(x)$, then

$$g^{\vee}(x') = -f^*(-x').$$

13. Let $\mathfrak{L}_x(u') = \mathfrak{L}_{u'}(x) = \mathfrak{L}(x,u')$ define two one-parameter families of functions of u' and x respectively. Show that for each fixed x, $\mathfrak{L}_x(u')$ is a concave upper semicontinuous function of u' that is upper closed. Show that for each fixed u, if $F_x(u) = F(x,u)$ is closed convex, then $\mathfrak{L}_{u'}(x)$ is convex in x, and $-F_x(u) = \mathfrak{L}_x^{\vee}(u)$.

13.7 CONES, CONVEX DUALITY, AND KUHN-TUCKER POINTS

In the preceding section we have avoided specifying how to choose a family of perturbed problems for any specific problem, and thus any application of the ideas developed there. Here we use convex duality to treat a generalization of the constrained convex programming problem (Section 13.1): given $m+1$ real valued convex functions $f_0 : X \to \mathbf{R}$ and $f = (f_1, \ldots, f_m) : X \to \mathbf{R}^m$ on a normed linear vector space X, and a nonempty convex set $C \subset X$, consider the primal problem

$$\inf\{f_0(x) \mid x \in C; f_i(x) \leqslant 0, i = 1, \ldots, m\}.$$

The generalization we make is to add a linear vector space Z, in which there is a closed convex cone K. In Z we define a partial ordering of elements.

DEFINITION 13.7.1. Let Z be a normed linear vector space, K a closed convex cone in Z, and let $(-K)\cap K=\{0\}$ (the zero element in Z). We call an element $z\in Z$ *nonnegative*, written $z\geqslant 0$, if $z\in K$. We write $z^1\geqslant z^2$ if $z^1,z^2\in Z$ and $z^1-z^2\in K$. We call K *the cone of positive elements of* Z. In Z', the dual space, $K^+=\{z'\in Z'|\langle z,z'\rangle\geqslant 0,\ z\in K\}$ is called *the cone of positive elements of* Z' and defines a dual order in Z'.

With respect to *any such order*, we define a function ϕ from any linear vector space X to Z to be *convex* if for any $x,y\in X$,

$$\phi(\lambda x+(1-\lambda)y)\leqslant\lambda\phi(x)+(1-\lambda)\phi(y),\qquad 0<\lambda<1.$$

With respect to the ordering we can again define level sets of $\phi:X\to Z$ by

$$\hat{L}_\Omega(\phi;z)=\{x|x\in\Omega,\ \phi(x)\leqslant z\}$$

where Ω is a set in X, and the epigraph $\widehat{epi}_\Omega\phi\subset X\otimes Z$ by

$$\widehat{epi}_\Omega\phi=\{(x,z)|x\in\Omega,\ \phi(x)\leqslant z\}.$$

We need to generalize semicontinuity to obtain analogs of Theorem 13.2.2 and Exercise 13.2.3. What we really need are certain closure properties of convex sets; we find it sufficient to extend the definition as follows.

DEFINITION 13.7.2. A function $\phi:X\to Z$ is *weakly lower* (*upper*) *semicontinuous* if each of the functions $\langle\phi(u),z'\rangle$ is weakly lower (upper) semicontinuous, for all $z'\in Z'$.

We can now prove

LEMMA 13.7.1. *Let C be a convex set in X, and $\phi:X\to Z$ be a convex weakly lower semicontinuous function, where X,Z are normed linear vector spaces. Then, for all $z\in Z$, $\hat{L}_C(\phi;z)$ is a closed convex set in X, and $\widehat{epi}_C\phi$ is a closed convex set in $X\otimes Z$.*

PROOF. If $\phi(x)\leqslant z$, then $\langle\phi(x)-z,z'\rangle\leqslant 0$ for any $z'\in K^+$. Since ϕ is convex and weakly lower semicontinuous, so is $\langle\phi(x)-z,z'\rangle$ in the ordinary sense; this makes the set $\{x\in C|\langle\phi(x)-z,z'\rangle\leqslant 0\}$ closed and convex for each $z'\in K^+$. Hence

$$\hat{L}_C(\phi;z)=\bigcap_{z'\in K^+}\{x\in C|\langle\phi(x)-z,z'\rangle\leqslant 0\}$$

is closed and convex.

Next we note that as a function of x and z, $\langle \phi(x)-z, z' \rangle$ is convex and weakly lower semicontinuous. Again we take an intersection of all the resulting sets for $z' \in Z'$ to establish that $\widehat{epi}_C \phi$ is also closed and convex. ■

We are now prepared to state the *primal problem*. We suppose given two normed linear vector spaces X and U, a closed convex nonempty set $C \subset X$, a positive cone $K \subset U$, a convex weakly lower semicontinuous function $f_0 : X \to \mathbf{R}$, and a convex weakly lower semicontinuous function $g : X \to U$ such that $\hat{L}_C(g;0) = \{x | x \in C,\ g(x) \leqslant 0\} \neq \varnothing$. We seek

$$\inf\{ f_0(x) | x \in \hat{L}_C(g;0)\}.$$

Note: $g(x) \leqslant 0$ is equivalent to $g(x) \in -K$.

It is convenient to replace f_0 by an equivalent $f : X \to \overline{\mathbf{R}}$ where

$$f(x) = \begin{cases} f_0(x), & \text{if } x \in C \text{ and } g(x) \leqslant 0, \\ +\infty, & \text{otherwise,} \end{cases}$$

and to consider the primal problem $\inf\{ f(x) | x \in X\}$. We find $F(x,u)$ by replacing $g(x) \leqslant 0$ by $g(x) \leqslant u \in U$. To be specific we define $F : X \otimes U \to \overline{\mathbf{R}}$ by

$$F(x,u) = \begin{cases} f_0(x), & \text{if } x \in \hat{L}_C(g;u), \\ +\infty, & \text{otherwise.} \end{cases}$$

Another description of F is obtained if we let $\hat{\mathbf{\Phi}}(x;u)$ be the indicator function of $\widehat{epi}_C g$: if

$$\hat{f}(x) = \begin{cases} f(x), & \text{if } x \in C, \\ +\infty, & \text{otherwise,} \end{cases}$$

then $F(x,u) = \hat{f}(x) + \hat{\mathbf{\Phi}}(x;u)$. Both $\hat{f}$ and $\hat{\mathbf{\Phi}}$ are convex and weakly lower semicontinuous (why?), so that F is the same. Since $f_0 : X \to \mathbf{R}$ and $C \cap \{x | g(x) \in -K\} \neq \varnothing$, F is neither identically $+\infty$ nor identically $-\infty$, and the problem is not trivial. We calculate the dual problem from

$$\begin{aligned} F^*(0,u') &= \sup\{\langle u,u' \rangle - F(x,u) | x \in X,\ u \in U\} \\ &= \sup\{\langle u,u' \rangle - f_0(x) | x \in C,\ u \in U,\ g(x) \leqslant u\}. \end{aligned}$$

Let $u = g(x) + v$. Then $v \in K$, so

$$\begin{aligned} F^*(0,u') &= \sup\{\langle g(x),u' \rangle + \langle v,u' \rangle - f_0(x) | x \in C, v \in K\} \\ &= \sup\{\langle v,u' \rangle | v \in K\} + \sup\{\langle g(x),u' \rangle - f_0(x) | x \in C\}. \end{aligned}$$

Now $\sup\{\langle v,u'\rangle | v\in K\}=\sup\{\langle v,-u'\rangle | v\in -K\}=\mathbf{\Phi}_K^*(u')$, the support function of K, which is also equal to $\mathbf{\Phi}_{K^-}(u')$, the indicator function of $K^-=-K^+$, thus, finally, to $\mathbf{\Phi}_{K^+}(-u')$ (Exercise 13.6.5). Then

$$-F^*(0,u')=-\mathbf{\Phi}_{K^+}(-u')+\inf\{f_0(x)-\langle g(x),u'\rangle | x\in C\}$$

and the dual problem can be written as

$$\sup\{\inf\{f_0(x)-\langle g(x),u'\rangle | x\in C\} | u'\in K^-\},$$

since $-\mathbf{\Phi}_{K^+}(-u')=-\infty$ unless $u'\in K^-$. The Lagrangian is calculated as follows:

$$\begin{aligned}\mathfrak{L}(x,u')&=-\sup\{\langle u,u'\rangle-F(x,u) | u\in U\}\\ &=-\sup\{\langle u,u'\rangle-[\hat{f}(x)+\hat{\mathbf{\Phi}}(x;u)] | u\in U\}.\end{aligned}$$

If $x\notin C$, then $\hat{f}(x)=\hat{\mathbf{\Phi}}(x,u)=\infty$ and $\mathfrak{L}(x,u')=\infty$. If $x\in C$, then $\hat{f}(x)\in\mathbf{R}$ and

$$\begin{aligned}\mathfrak{L}(x,u')&=\hat{f}(x)-\sup\{\langle u,u'\rangle-\hat{\mathbf{\Phi}}(x;u) | u-g(x)\in K\}\\ &=\hat{f}(x)-\langle g(x),u'\rangle-\sup\{\langle u-g(x),u'\rangle-\hat{\mathbf{\Phi}}(x;u) | u-g(x)\in K\}\\ &=\hat{f}(x)-\langle g(x),u'\rangle-\sup\{\langle v,u'\rangle-\hat{\mathbf{\Phi}}(x;v+g(x)) | v\in K\}.\end{aligned}$$

But $\hat{\mathbf{\Phi}}(x;v+g(x))=\infty$ unless $v\in K$, in which case it is zero; that is $\hat{\mathbf{\Phi}}(x;v+g(x))=\mathbf{\Phi}_K(u)$, the indicator function of K, $\mathbf{\Phi}_K^*(u')=\mathbf{\Phi}_{K^-}(u')$ and

$$\mathfrak{L}(x,u')=\hat{f}(x)-\langle g(x),u'\rangle-\mathbf{\Phi}_{K^-}(u'); \qquad x\in C.$$

That is

$$\mathfrak{L}(x,u')=\begin{cases}\hat{f}(x)-\langle g(x),u'\rangle, & \text{if } x\in C, u'\in K^-,\\ -\infty, & \text{if } x\in C, u'\notin K^-,\\ +\infty, & \text{if } x\notin C.\end{cases}$$

It is now easy to write the saddle point condition for our problem:

$(\bar{x},\bar{u}')$ is a saddle point if and only if

$$\bar{x}\in C \text{ and } \bar{u}'\in K^-$$

and for all $x\in C, u'\in K^-$,

$$f_0(\bar{x})-\langle g(\bar{x}),u'\rangle\leqslant f_0(\bar{x})-\langle g(\bar{x}),\bar{u}'\rangle\leqslant f_0(x)-\langle g(x),\bar{u}'\rangle.$$

In the exercises other uses of duality, including Fenchel duality are explored. Connections to optimal control theory are considered there. Rockafellar [2] gives more applications and further references. Extensive applications to the theory of variational inequalities and the calculus of variations are found in Ekeland and Temam [1]. See also Céa [1], 192–217.

EXERCISES TO SECTION 13.7

1. Show that $F^*(x',u') = \Phi_{K^+}(-u') - \inf\{f_0(x) - \langle x, x'\rangle - \langle g(x), v'\rangle | x \in C\}$ for the constrained convex programming problem.
2. If $U = \mathbf{R}^m$ and $K = \mathbf{R}^{m+}$, show that our result for the constrained convex programming problem is identical with that of Section 13.1.
3. In the constrained convex programming problem, suppose that $U = \mathbf{R}^m$, $f_0(x) = \langle x, x'^0\rangle + \alpha_0$, and $g(x) = (g_1(x), g_2(x), \ldots, g_m(x))$, where each $g_i(x) = \langle x, x'^i\rangle + \alpha_i$ for some $x'^i \in X'$, $\alpha_i \in \mathbf{R}$, $i = 0, 1, \ldots, m$, and $C = X$. This is the *linear programming problem.* Compute the dual problem and the Lagrangian for this case. What is the Kuhn-Tucker saddle point condition in this case? Where does a constraint condition present itself?
4. Let $X = C[0,1]$ and $f(x) = \|x - [\sum_{i=1}^n a_i t^i]\| = \max\{|x(t) - \sum_{i=1}^n a_i t^i| | t \in [0,1]\}$. If $U = C[0,1]$, then we can set $F(x,u) = \|x - \sum_{i=1}^n a_i t^i - u\|$. Show that this is a convex family of perturbations. Find the dual problem and the Lagrangian. (This is the Chebyshev approximation problem.)
5. (Fenchel Duality) Let X, U be Banach spaces, $A \in [X \to U]$, $f_1 : X \to \overline{\mathbf{R}}$, $f_2 : U \to \overline{\mathbf{R}}$. Consider the primal problem

$$\inf\{f_1(x) + f_2(Ax) | x \in X\}.$$

Set $F(x,u) = f_1(x) + f_2(Ax - u)$. Compute $F^*(x',u')$, the dual problem and the Lagrangian. Suppose that f_1, f_2 are closed convex functions and neither is everywhere $+\infty$ or everywhere $-\infty$. Show that: if there is an $\bar{x} \in X$ such that $f_1(\bar{x}) < \infty, f_2(A\bar{x}) < \infty$ and f_2 is continuous at $A\bar{x}$, then $\boldsymbol{P}^* = \boldsymbol{P}$ is finite and the dual problem has solutions. Under the same conditions, find the results equivalent to Exercise 13.6.11. What is the saddle point condition?

6. If, in Exercise 13.7.5, A is only a closed operator with dense domain, find the dual problem and the Lagrangian.
7. (Fenchel Duality Theorem) On a normed linear vector space X, let $f_1 : X \to \mathbf{R}$ be convex on a closed convex set $C_1 \subset X$, $f_2 : X \to \mathbf{R}$ be convex on a closed convex set $C_2 \subset X$. Suppose that $C_1 \cap C_2$ contains points in the interior of C_i considered as a set in the normed linear

vector space $[C_i]$, $i=1,2$. Let

$$\hat{f}_i=\begin{cases} f_i(x), & x\in C_i,\\ +\infty, & x\notin C_i,\end{cases}$$

$i=1,2$. Suppose that $\operatorname{int} epi\,\hat{f}_1\cup \operatorname{int} epi\,\hat{f}_2\neq\varnothing$. Let the primal problem be $\inf\{f_1(x)+f_2(x)|x\in C_1\cap C_2\}$. Then $\boldsymbol{P}^*=\boldsymbol{P}$ is finite, and the dual problem is solvable. (Hint: Take $U=X$ above.)

8. Apply the Fenchel duality theorem to the optimal control problem of Section 12.1. (Hint: Take $f_2(x)\equiv 0$.)
9. Apply Fenchel duality to the problem of Exercise 12.10.1. Find the dual problem and the Lagrangian.
10. Treat the minimal effort problem of Section 13.4 by conjugate duality.
11. Use Fenchel duality for the problem of Exercise 13.4.7.
12. From the saddle point condition, derive as much as possible of Theorem 13.5.2 for the convex case.
13. (Quadratic Programming) In a real Hilbert space H_1 let A be a strictly positive bounded selfadjoint operator and K a positive cone defining an order in H_2, a second real Hilbert space. The primal problem is

$$\inf\{\tfrac{1}{2}(Ax,x)-(f,x)|Gx\leqslant y\}$$

where $G\in[H_1\to H_2]$ and $y\in H_2$. Letting $H_2=U$, show that the dual problem is

$$\sup\{\inf\{\tfrac{1}{2}(Ax,x)-(f,x)-(Gx-y,u')|x\in H_1\}|u'\in K^-\}$$

and that the unrestricted minimization in x occurs at $x=A^{-1}(f-G^*u')$. Find the dual problem that this substitution provides and note the simplicity of the constraint set.

14. Repeat Exercise 13.7.13 for the case in which G is only positive definite and then when G is only nonnegative.

13.8 ALGORITHMS FOR UNCONSTRAINED OPTIMIZATION

If f is a functional on a real Banach space X and we wish to solve the unconstrained optimization problem

$$\inf\{f(x)|x\in X\},$$

it is natural to follow the procedures that we first developed for quadratic functionals on a Hilbert space. (We follow the general path of Vainberg [2], 127–138, with some simplifications.) Thus, if f possesses a Gâteaux differential, then

$$f(x+th)=f(x)+tGDf(x;h)+o(t).$$

If $\bar{x}$ were to be a "local" solution of the problem, say $f(\bar{x}) \leqslant f(x)$ for all x in some ball $\|x-\bar{x}\| \leqslant \rho$, then $GDf(\bar{x};h) \equiv 0$. If $\bar{x}$ is not such a local minimum, then, in a method of steepest descent, we would first seek an $\bar{h}$ that solves

$$\sup\{GDf(\bar{x};h) \mid \|h\|=1\}=\sup\{GDf(\bar{x};h) \mid \|h\| \leqslant 1\},$$

and then a value of $t \in \mathbf{R}$ that solves $\min\{f(\bar{x}+t\bar{h} \mid t \in \mathbf{R}\}$. The norm is a convex function, so the sphere $\{x \mid \|x\| \leqslant 1\}$ is closed and bounded. In a separable norm reflexive space X, it is also weakly (sequentially) compact; then if $GDf(\bar{x};h)$ is even quasiconvex in h, there will be a unique optimal direction $\bar{h}$. This will certainly occur if $GDf(\bar{x};h)=\langle h,\bar{x}'\rangle$. Then we have the ingredients for an algorithm, if we solve $\min_t\{f(\bar{x}+t\bar{h}) \mid t \in \mathbf{R}\}$. Even if we cannot solve this, but instead set either an a priori sequence, or a procedure for calculation of a value to be used that yields $f(\bar{x}+t\bar{h})<f(\bar{x})$, we say that we have a *descent process*.

If we establish a descent process in the case of a Gâteaux differentiable functional $f(x)$, so that $GDf(\bar{x};h)=\langle h, \nabla f(\bar{x})\rangle$, then $|\langle h, \nabla f(\bar{x})\rangle| \leqslant \|h\| \, \|\nabla f(\bar{x})\|$ shows that we must choose $\bar{h}$ such that

$$\langle \bar{h}, \nabla f(\bar{x})\rangle = \|\nabla f(\bar{x})\|.$$

That is, $\bar{h}$ is an extremal of $\nabla f(\bar{x})$. If X is strictly convex, there is a unique extremal. If X is norm reflexive, there exist extremals. Using Lemma 13.3.2 we can solve for h under the condition that X is reflexive and that the norm in X' is Gâteaux differentiable. Then

$$\langle \mathbf{D}\|x'\|, x'\rangle = \|x'\| \qquad \text{and} \qquad \|\mathbf{D}\|x'\|\,\| = 1.$$

That is, we must take $\bar{h}=\mathbf{D}\|\nabla f(\bar{x})\|$ and determine $\bar{t}$ to yield

$$\inf\{f(\bar{x}+t\mathbf{D}\|\nabla f(\bar{x})\|) \mid t \in \mathbf{R}\}.$$

We call such a method a *gradient method*. (In fact, the computation of $\bar{t}$ is not often practical; usually some local minimum is used, and even that may not be available at low computational cost, even though we deal with a function of one real variable). We generalize the notion of a gradient method to any algorithm of the form

$$x^{n+1}=x^n-\epsilon_n\|\nabla f(x^n)\|\mathbf{D}\|\nabla f(x^n)\|.$$

Here we have normalized t, the multiplier, to be the product of $\|\nabla f(x^n)\|$, which is expected to tend to zero as $n\to\infty$, and an "arbitrary" factor ϵ_n. The choice of ϵ_n to guarantee $f(x^{n+1})<f(x^n)$, $n=1,2,3,\ldots$, is the subject of the following:

THEOREM 13.8.1. *Let X be a reflexive Banach space whose dual X' has a Gâteaux differentiable norm and $f: X\to\mathbf{R}$ a Gâteaux differentiable functional. Let $L(t)$ be any increasing function on $t\geqslant 0$. Let $L_n=\max\{1, L(\|x^n\|+\|\nabla f(x^n)\|)\}$ for the gradient method $x^{n+1}=x^n-\epsilon_n\|\nabla f(x^n)\|\mathbf{D}\|\nabla f(x^n)\|$. If*

1. $\|\nabla f(x+h)-\nabla f(x)\|\leqslant L(t)\|h\|$ *for all* $x, x+h\in\{x \mid \|x\|\leqslant t\}$, *and,*
2. *for some* $\alpha<1, 0<\epsilon_n\leqslant\alpha/L_n$,

then $f(x^{n+1})<f(x^n)$ *for all* $n=0,1,2,3,\ldots$.

PROOF. There is a τ, $0<\tau<1$ such that

$$\begin{aligned}
f(x^{n+1})-f(x^n) &= \langle x^{n+1}-x^n, \nabla f(x^{n+1}+\tau(x^n-x^{n+1}))\rangle \\
&= \langle x^{n+1}-x^n, \nabla f(x^n)\rangle \\
&\qquad + \langle x^{n+1}-x^n, \nabla f(x^{n+1}+\tau(x^n-x^{n+1}))-\nabla f(x^n)\rangle \\
&\leqslant -\epsilon_n\|\nabla f(x^n)\|^2 \\
&\qquad + \frac{1}{\tau}L(\max\{\|x^n\|, \|x^{n+1}\|\})\|\tau(x^{n+1}-x^n)\|^2 \\
&\leqslant -\epsilon_n\|\nabla f(x^n)\|^2+L(\|x^n\|+\|\nabla f(x^n)\|)\epsilon_n^2\|\nabla f(x^n)\|^2 \\
&= \|\nabla f(x^n)\|^2(\epsilon_n^2L_n-\epsilon_n) \\
&\leqslant (\alpha-1)\epsilon_n\|\nabla f(x^n)\|^2<0. \quad \blacksquare
\end{aligned}$$

If $\inf\{f(x) \mid x\in X\}>-\infty$, any descent process yields a sequence of values $f(x^n)$ which is strictly decreasing and bounded below; thus $\bar{f}=\lim_{n\to\infty}f(x^n)$ exists. If ϵ_n is bounded away from zero, this guarantees $\nabla f(x^n)\to 0$, $n\to\infty$.

THEOREM 13.8.2. *To the conditions of Theorem* 13.8.1, *add*:

1. $\inf\{f(x)|x \in X\} > -\infty$,
2. *$L(f;r)$ is bounded for some $r \in \mathbf{R}$, and*
3. *there is a $\beta > 0$ with $\beta/L_n \leqslant \epsilon_n \leqslant \alpha/L_n$.*

Then $\nabla f(x^n) \to 0$, $n \to \infty$, if $x^0 \in L(f;r)$.

PROOF. Since $f(x^n) \leqslant r$, $x^n \in L(f;r)$ for all n and so is a bounded sequence. It follows from the Lipschitz condition that $\|\nabla f(x^n)\|$ is bounded, and hence that L_n is bounded; say by $\bar{L}$. Then $\epsilon_n \geqslant \beta/\bar{L}$ and is bounded away from zero. ■

DEFINITION 13.8.1. A point $\bar{x}$ at which $\nabla f(\bar{x}) = 0$ is called a *critical point*. A sequence $\{x^n\}$ such that $\nabla f(x^n) \to 0$, $n \to \infty$ is called a *criticizing sequence*. If $\inf\{f(x)|x \in X\} > -\infty$, then any sequence $\{x^n\}$ such that $f(x^n) \to \inf\{f(x)|x \in X\}$ is called a *minimizing sequence*.

In terms of these definitions, we have found conditions that guarantee that $\{x^n\}$ is a criticizing sequence. We guarantee that it is a minimizing sequence in the following:

THEOREM 13.8.3. *Let X be a norm reflexive Banach space with dual X' that has a Gâteaux differentiable norm. Let $f: X \to \mathbf{R}$ be Gâteaux differentiable on X and let $\nabla f(x)$ be continuous in x. Let $L(t)$ be any increasing function on $t \geqslant 0$ and $L_n = \max\{1, L(\|x^n\| + \|\nabla f(x^n)\|\}$ for the gradient method $x^{n+1} = x^n - \epsilon_n \|\nabla f(x^n)\| \mathbf{D} \|\nabla f(x^n)\|$. Let*

1. *$\|\nabla f(x+h) - \nabla f(x)\| \leqslant L(t)\|h\|$ for all $x, x+h \in \{x| \, \|x\| \leqslant t\}$.*
2. *Let γ be a positive constant such that*

$$\langle h, \nabla f(x+h) - \nabla f(x) \rangle \geqslant \gamma \|h\|^2.$$

3. *Last, suppose there exist α, β with $0 < \beta < \alpha < 1$ and*

$$\frac{\beta}{L_n} \leqslant \epsilon_n \leqslant \frac{\alpha}{L_n}, \qquad n = 0, 1, 2, \ldots.$$

Then $f(x)$ has a unique global minimum at $\bar{x}$ and $x^n \to \bar{x}$, $n \to \infty$ (i.e., $\{x^n\}$ is a minimizing sequence).

PROOF. For some t, $L(f,t)$ is bounded, since

$$\begin{aligned}
f(x)-f(0) &= \int_0^1 \langle x, \nabla f(tx)\rangle\,dt \\
&= \langle x, \nabla f(0)\rangle + \int_0^1 \langle x, \nabla f(tx) - \nabla f(0)\rangle\,dt \\
&\geqslant -\|x\|\,\|\nabla f(0)\| + \gamma\|x\|^2 \int_0^1 t\,dt \\
&\geqslant \|x\|\left[\frac{\gamma}{2}\|x\| - \|\nabla f(0)\|\right] \\
&\to \infty, \quad \|x\| \to \infty,
\end{aligned}$$

while assumption 2 implies that $f(x)$ is strictly convex; this together with $f(x)\to\infty$, $\|x\|\to\infty$ shows that for all t, $L(f;t)$ is convex and bounded, and that $\inf\{f(x)|x\in X\} > -\infty$. Let $\bar{x}$ be the unique minimum point for $f(x)$. By Theorems 13.8.1 and 13.8.2, $f(x^n)$ is strictly decreasing and $\nabla f(x^n)\to 0$, $n\to\infty$. But $\nabla f(\bar{x})=0$ (why?) and

$$\begin{aligned}
\gamma\|x^n-\bar{x}\|^2 &\leqslant \left\langle \frac{x^n-\bar{x}}{\|x^n-\bar{x}\|}, \nabla f(x^n) - \nabla f(\bar{x})\right\rangle \\
&\leqslant \left\langle \frac{x^n-\bar{x}}{\|x^n-\bar{x}\|}, \nabla f(x^n)\right\rangle \\
&\leqslant \|\nabla f(x^n)\| \to 0, \qquad n\to\infty.
\end{aligned}$$

Thus $\{x^n\}$ is a minimizing sequence. ■

If $f(x)$ has a second Gâteaux derivative $H(x)$, and $\langle h, H(x)h\rangle \geqslant \gamma'\|h\|^2$ for some $\gamma'>0$, then obviously assumption 2 will be true.

In the case that X is a Hilbert space it is possible to give error estimates of the form $\|x^{n+1}-\bar{x}\| \leqslant [\epsilon + (M-m)/(M+m)]\|x^n-\bar{x}\|$ for any $\epsilon>0$, provided that $mI \leqslant H(x) \leqslant MI$. (Daniel [1], 93–94.)

A direct extension of steepest descent when X is a Banach space proceeds as follows. From Taylor's formula

$$f(\bar{x}+th) = f(\bar{x}) + t\langle h, \nabla f(\bar{x})\rangle + \tfrac{1}{2}t^2\langle h, H(\bar{x})h\rangle + o(t^2\|h\|^2)$$

with $\bar{h} = \mathbf{D}\|\nabla f(\bar{x})\|$, we choose $\bar{t}$ to minimize the quadratic form $f(\bar{x})+$

$t\langle\bar{h},\nabla f(\bar{x})\rangle+\frac{1}{2}t^2\langle\bar{h},H(\bar{x})\bar{h}\rangle$. Then

$$\bar{t}=\frac{-\|\nabla f(\bar{x})\|}{\langle\bar{h},H(\bar{x})\bar{h}\rangle}.$$

Using $\|\bar{h}\|=1$, and $\gamma'\leqslant\langle\bar{h},H(\bar{x})\bar{h}\rangle\leqslant\|H(\bar{x})\|$ establishes the bounds

$$\frac{1}{\|H(x)\|}\leqslant\frac{-\bar{t}}{\|\nabla f(\bar{x})\|}\leqslant\frac{1}{\gamma'}.$$

If $\inf\{f(x)|x\in X\}>-\infty$ and $L(f,\tau)$ is bounded for some τ, then it can be shown that this generates a criticizing sequence (Exercise 13.8.4), provided that $-\bar{t}/\|\nabla f(\bar{x})\|$ is bounded above by $\alpha/M(\|\bar{x}\|+\|\nabla f(\bar{x})\|)$, $\alpha<2$, where M is a positive nondecreasing function such that $\|H(x)\|\leqslant M(\|x\|)$. Further (Exercise 13.8.5), it can be shown that this is actually a minimizing sequence. Unfortunately, $-\bar{t}/\|\nabla f(\bar{x})\|\leqslant 1/\gamma'$ is not an equivalent condition, and so does not guarantee this convergence. Thus, the extra cost involved in this computation of $\bar{t}$ is not justified.

Various algorithms for a specific choice of ϵ_n may be found. Some references are Céa [1], 77–90 and Daniel [1], 76–92. See also, for a general theory and more references, Lyubich and Maistrovskii [1].

A slightly different approach to steepest descent based on Taylor's formula is more revealing. Suppose that $f(x)$ and $H(x)$ exist in the Fréchet sense on X, that $H(x)$ is continuous in x, and that for some γ, M, $0<\gamma<M<\infty$,

$$\gamma\|h\|^2\leqslant\langle h,H(\bar{x})h\rangle\leqslant M\|h\|^2.$$

Let $\bar{H}=H(\bar{x})$. Introduce on X the semi-inner product (Lumer, [1], see also Martin [1], 44–50)

$$[g,h]=\langle g,\bar{H}h\rangle.$$

That this is actually an inner product follows from Theorem 12.2.4 and $\langle h,\bar{H}h\rangle\geqslant\gamma\|h\|^2$. Then we can restate Taylor's formula as

$$f(\bar{x}+th)=f(\bar{x})+\tau[h,\bar{H}^{-1}\nabla f(\bar{x})]+\tfrac{1}{2}t^2[h,h]+o(t^2\|h\|^2),$$

provided that $\nabla f(\bar{x})\in R_{\bar{H}}$, since $\gamma\|h\|^2\leqslant\langle h,\bar{H}h\rangle\leqslant\|h\|\,\|\bar{H}h\|$ shows that $\bar{H}$ is invertible as a mapping from X to $R_{\bar{H}}$. We have a minimum in t for

$\bar{t}=-\left[h,\bar{H}^{-1}\nabla f(\bar{x})\right]/[h,h]$ and

$$f(\bar{x}+\bar{t}h)=f(\bar{x})-\tfrac{1}{2}\left[h,\bar{H}^{-1}\nabla f(\bar{x})\right]^2/\left[h,h\right]+o(\|h\|^2),$$

exactly as in the minimization of a quadratic functional on a Hilbert space. Then we should take $\bar{h}=\bar{H}^{-1}\nabla f(\bar{x})$, provided that this is not too large. If we establish an algorithm it would be

$$x^{n+1}=x^n-\epsilon_n\left[H(x^n)\right]^{-1}\nabla f(x^n)$$

where $\epsilon_n=[\bar{h},\bar{h}]/[\bar{h},\bar{h}]=1$! Put another way, we have been led to Newton's method for finding the roots of $\nabla f(x)=0$. As in the case of Newton's method, this generates the idea of avoiding the calculation of $\left[H(x^n)\right]^{-1}$. Thus we can generate a descent process by the following scheme.

Suppose that $\{H_n\}, H_n\in[X\to X']$, is a sequence which, for constants γ, M, $0<\gamma<M<\infty$, satisfies $\gamma\|h\|^2\leqslant\langle h,H_nh\rangle\leqslant M\|h\|^2$, $h\in X$. Suppose also that $R_{H_n}=X'$, for all n. Then we take as the direction of descent at each step, $h^n=H_n^{-1}\nabla f(x^n)$ and move a distance ϵ_n in that direction, so $x^{n+1}=x^n-\epsilon_nH_n^{-1}\nabla f(x^n)$. Then the Kantorovich theorem on modifications of Newton's method leads to convergence theorems here in a direct fashion. (For another approach see Céa [1], 73–77, 91–95.) From the development above it is easy to see why methods of this type are called *variable metric* methods. The usual computational practice is to compute H_n sequentially, with some minimization of computational effort, so as to approximate $H(x^n)$, or alternatively to compute H_n^{-1} so as to approximate $\left[H(x^n)\right]^{-1}$. The grandfather of this group of methods is found in Davidon [1], implemented by Fletcher and Powell [1].

It is left as an exercise to define conjugate gradients for general nonlinear problems.

We can also apply the Ritz method to nonlinear problems. Assume that X is a separable Banach space and that there is a sequence $\phi^1,\phi^2,\phi^3,\ldots$ in X that is total in X. We let $X^n=\mathbf{S}\{\phi^1,\ldots,\phi^n\}$. We consider a functional $f:X\to\mathbf{R}$ such that

$$\inf\{f(x)|x\in X\}>-\infty,$$

the sequence of approximate problems

$$\inf\{f(x)|x\in X^n\},\qquad n=1,2,3,\ldots,$$

and their solutions x^n:

$$f(x^n)=\inf\{f(x)|x\in X^n\}.$$

Since $X^n \subset X$, if f has a gradient $\nabla f(x)$ on X, it has the *same* gradient on X^n. (Why?) Furthermore, $\langle h, \nabla f(x^n)\rangle = df(x^n + th)/dt|_{t=0} = 0$ for all $h \in X^n$. For suitable scalars $\alpha_1,\ldots,\alpha_n$ and $\beta_1,\ldots,\beta_n$, $x^n = \alpha_1\phi^1 + \cdots + \alpha_n\phi^n$ and $h = \beta_1\phi^1 + \cdots + \beta_n\phi^n$ so that we must have

$$\sum_{i=1}^{n} \beta_i \langle \phi^i, \nabla f(\alpha_1\phi^1 + \cdots + \alpha_n\phi^n)\rangle = 0$$

for all $(\beta_1,\ldots,\beta_n) \in \mathbf{R}^n$. This leads to the Ritz system of equations

$$\langle \phi^i, \nabla f(\alpha_1\phi^1 + \cdots + \alpha_n\phi^n)\rangle = 0, \qquad i = 1,2,\ldots,n.$$

Thus minimization of f on X has been made equivalent to a Galerkin or projection method for solving $\nabla f(x) = 0$. In the linear case, the solvability of the Ritz system was inherited from the strict positivity of the selfadjoint operator, thus from the strict convexity of the quadratic functional. Here Theorem 13.2.3 and its corollaries may be restated as:

LEMMA 13.8.1. *If $f(x)$ is (strictly) quasiconvex and weakly lower semicontinuous, and if $L(f;t)$ is bounded for some $\tau < \infty$, then f achieves its (unique) minimum on every X^n and the Ritz system is (uniquely) solvable for every n.*

With a few assumptions the Ritz sequence $\{x^n\}$ (the solutions of the Ritz system) form a minimizing sequence. It is trivial that $f(x^n)$ is nonincreasing.

LEMMA 13.8.2. *Let $f: X \to \mathbf{R}$ and $\inf\{f(x)|x \in X\} > -\infty$. If $f(x)$ is continuous, and the Ritz system is solvable for all n, then the Ritz sequence is a minimizing sequence.*

PROOF. Choose any minimizing sequence, $y^1, y^2, y^3, \ldots$ such that

$$f(y^n) \leqslant \inf\{f(x)|x \in X\} + \frac{1}{n}.$$

Since $\{\phi^n\}$ is total, for each n there is an m such that $\inf\{\|y^n - z\| \,|\, z \in X^m\} = \|y^n - z^m\| \leqslant \delta$, for any $\delta > 0$. By continuity, if δ is sufficiently small, $\|f(y^n) - f(z^m)\| \leqslant 1/n$. Then $f(z^m) \to \inf\{f(x)|x \in X\}$, $n \to \infty$. On the other hand, $f(x^m) \leqslant f(z^m)$, by definition, so $f(x^n) \downarrow \inf\{f(x)|x \in X\}$, $n \to \infty$. ■

We can summarize the implications of this for convergence to the minimum in the following:

THEOREM 13.8.4. *If $f:X\to\mathbf{R}$ is a real continuous strictly quasiconvex functional on a reflexive, separable, real Banach space X, and $L(f,t)$ is bounded for some $t<\infty$, then the Ritz sequence exists for all n and converges weakly to the unique minimum of $f(x)$ on X.*

PROOF. Everything but the weak convergence is already contained in the preceding lemmas and Theorem 13.2.3 and its corollaries. Let $\{x^n\}$ be the (minimizing) Ritz sequence. It is bounded, since $L(f;t)$ is for some $t<\infty$. If $\bar{x}$ is the minimum, then either $x^n \rightharpoonup \bar{x}$, $n\to\infty$, or else there is a subsequence, say $\{y^n\}$, of $\{x^n\}$ such that $y^n \rightharpoonup \bar{y} \neq \bar{x}$. Since $f(x)$ is weakly lower semicontinuous (why?), $f(\bar{y}) \leqslant \liminf_{n\to\infty} f(y^n) = \lim_{n\to\infty} f(x^n) = f(\bar{x})$. Thus $\bar{y}$ is also a minimum, which violates the uniqueness of the minimum. ■

If f is twice Gâteaux differentiable and $\langle h, H(x)h\rangle \geqslant \gamma\|h\|^2$ for some $\gamma>0$, then it is possible to show that convergence takes place in the norm topology (Vainberg [2], 121–122 and Exercise 13.8.8.)

It should be noted that the problem $\inf\{f(x)|x\in\mathbf{R}^n\}$ is itself susceptible to gradient methods and variable metric methods of solution. An extensive literature on this problem exists, and Daniel [1], 142–179, treats this, as do Ortega and Rheinboldt [1], 473–519. Both give extensive references.

EXERCISES TO SECTION 13.8

1. Let $f(x)$ possess a second Gâteaux derivative $H(x)$ such that $\langle h, H(x)h\rangle \geqslant \gamma'\|h\|^2$. Find γ such that $\langle h, \nabla f(x+h) - \nabla f(x)\rangle \geqslant \gamma\|h\|^2$.
2. Let $f:X\to\mathbf{R}$ be convex and let $\inf\{f(x)|x\in X\} > -\infty$. Show that if $L(f;t)$ is bounded for some $t\in\mathbf{R}$ and $\{x^n\}$ is a criticizing sequence, then $\{x^n\}$ is a minimizing sequence (Daniel [1], 72).
3. Let X be a Hilbert space and A a strictly positive selfadjoint bounded linear operator. Let $f:X\to\mathbf{R}$ be twice Fréchet differentiable and

$$0 < m(x, A^{-1}x) \leqslant (x, f''(y)x) \leqslant M(x, A^{-1}x)$$

for constants m, M. Find a gradient method and show that it generates a minimizing sequence. (Hint: Set $h = -A\nabla f(x)$. See Daniel [1], 73–74.)

4. Suppose that, in Theorem 13.8.2, there is a nondecreasing function $M(t)$ with $0<M(t)<\infty$, $0\leqslant t<\infty$, such that

$$\langle g, H(x)h\rangle \leqslant M(\|x\|)\|g\|\,\|h\|, \qquad g,h\in X,$$

where $H(x)$ is the second Gâteaux derivative, continuous in x. If we use $M(t)$ in place of $L(t)$, show that

$$f(x^{n+1})-f(x^n)\leqslant -\epsilon_n\left(1-\tfrac{1}{2}\epsilon_n M_n\right)\|\nabla f(x^n)\|^2,$$

and that $0<\beta/M_n\leqslant\epsilon_n\leqslant\alpha/M_n$, $\alpha>2$, guarantees that $\{x^n\}$ is a criticizing sequence. (Hint: $\|\nabla f(x)\|\leqslant\|\nabla f(0)\|+\langle h, H(\tau x)x\rangle$ for some τ, if $\|\nabla f(x)-\nabla f(0)\|=\langle h,\nabla f(x)-\nabla f(0)\rangle$. Then $1\leqslant M_n\leqslant\overline{M}<\infty$.)

5. In Exercise 13.8.4 add the assumption that there is a $\gamma'>0$ with

$$\langle h, H(x)h\rangle \geqslant \gamma'\|h\|^2$$

and show that $\{x^n\}$ is a minimizing sequence. Note that $\epsilon_n\leqslant\alpha/M_n$ is not guaranteed by $\epsilon_n\leqslant 1/\gamma'$, which is what we would obtain from steepest descent.

6. If X is a Hilbert space, $f:X\to\mathbf{R}$ is twice Gâteaux differentiable and $mI\leqslant H(x)\leqslant MI$, show that $I-\gamma\nabla f(x)$ is a contraction for all γ, $0<\gamma<2m/M^2$, so that successive approximations $x^{n+1}=x^n-\gamma\nabla f(x^n)$ converge to the unique critical point (minimal point) of $f(x)$. Show that this coincides with the gradient method. Find an error estimate.

7. If X is a Hilbert space, show that the general gradient method takes the form $x^{n+1}=x^n-\epsilon_n\nabla f(x^n)$. (Hint: in this case $\mathbf{D}\|x\|=x/\|x\|$.)

8. Show that if $\langle h,H(x)h\rangle\geqslant\gamma\|h\|^2$ then the Ritz sequence for minimizing a bounded $f(x)$ is convergent in norm. (Hint: Then $\langle x-y,\nabla f(x)-\nabla f(y)\rangle\geqslant\gamma\|x-y\|^2$, so f satisfies $f((x+y)/2)\leqslant\frac{1}{2}f(x)+\frac{1}{2}f(y)-\gamma\|x-y\|^2$. Apply this to a minimizing sequence to show norm convergence.)

9. Suppose, in developing the Ritz method, that we have two total sequences $\{\phi^n\}$ and $\{\psi^n\}$ and that we expand $h=\beta_1\psi^1+\cdots+\beta_n\psi^n$ and $x^n=\alpha_1\phi^1+\cdots+\alpha_n\phi^n$. Develop as much of Lemmas 13.8.1 and 13.8.2 and Theorem 13.8.4 as seems correct. This is called a Galerkin-Petrov method in Hilbert space, a projection method in general, as applied to solving the equation $\nabla f(x)=0$.

10. (*Generalized Conjugate Gradients*) On a Hilbert space X let $f:X\to\mathbf{R}$ be

twice continuously Fréchet differentiable and let there exist γ, M such that $0<\gamma I \leqslant H(x) \leqslant MI < \infty$. Let

$$\begin{cases} p^0 = z^0 = -\nabla f(x^0), \\ x^{n+1} = x^n + a_n p^n, \\ z^{n+1} = -\nabla f(x^{n+1}), \\ b_n = \dfrac{-(z^{n+1}, H(x^{n+1})p^n)}{(p^n, H(x^{n+1})p^n)}, \\ p^{n+1} = z^{n+1} + b_n p^n, \end{cases}$$

and let a_n satisfy $f(x^n + a_n p^n) = \inf\{f(x^n + ap^n | a \in \mathbf{R}\}$, so that $(p^n, \nabla f(x^n + a_n p^n)) = 0$. Show that

(a) $(p^n, H(x^n)p^{n-1}) = (z^{n+1}, H(x^n)p^n) = 0$;
(b) $(z^n, p^n) = \|z^n\|^2$;
(c) $(p^n, H(x^n)z^n) = (p^n, H(x^n)p^n)$;
(d) $\|p^n\|^2 = \|z^n\|^2 + b_{n-1}^2 \|p^{n-1}\|^2$;
(e) $(z^n, H(x^n)z^n) = (p^n, H(x^n)p^n) + b_{n-1}^2 (p^{n-1}, H(x^n)p^{n-1})$;
(f) $\|z^n\|^2 \leqslant \|p^n\|^2 \leqslant (M/\gamma)\|z^n\|^2$.

11. In Exercise 13.8.10, show that for an arbitrary $x^0 \in H$, the sequence $\{x^n\}$ is a minimizing sequence converging in norm to the unique minimum $\bar{x}$, with $\|x^n - \bar{x}\| \leqslant \|\nabla f(x^n)\|/\gamma$. (Hint: Show that $a_n \geqslant \|z^n\|^2/\|p^n\|^2 \geqslant \gamma/M^2$, and from Taylor's formula that

$$f(x^{n+1}) \leqslant f\left(x^n + \frac{\gamma}{M^2} p^n\right) \leqslant f(x^n) - \frac{\gamma}{2M^2}\|p^n\|^2.$$

Also $\gamma\|x^{n+m} - x^m\|^2 \leqslant (x^{n+m} - x^n, \nabla f(x^{n+m}) - \nabla f(x^n))$. Daniel [1], 126–127.) (For an improved error estimate of the type found in the quadratic case see Daniel [1], 127–136.)

13.9 ALGORITHMS FOR CONSTRAINED OPTIMIZATION

When we restrict the set of values x at which a minimum of a functional f is of interest, all the approaches of Section 13.8 present conceptual difficulties. Suppose to fix ideas that our problem is posed as $\inf\{f(x) | x \in \Omega \subset X\}$ where X is a real Banach space, and that we wish to move from a specific $\bar{x}$ in some direction $\bar{h}$ (i.e., $\|\bar{h}\| = 1$) a distance $\bar{t}$, so as to decrease $f(x)$. Not only is there no guarantee that the steepest descent direction $-\nabla f(\bar{x})$ moves into Ω; that is, that $\bar{x}+\alpha[-\nabla f(\bar{x})]\in\Omega$ for all sufficiently small positive α, but, even if this is so, there is no guarantee that distances $\bar{t}$, determined as in Section 13.8 to lead to convergence, yield points $\bar{x}+\bar{t}\bar{h}\in\Omega$. Similarly, for the Ritz method and related direction methods, the condition that a certain linear combination y of n elements of a total sequence $\{\phi^n\}$ provides $\inf\{f(x)|x=\alpha_1\phi^1+\cdots+\alpha_n\phi^n\in\Omega\}$ may or may not be easily established. Thus, modifications are in order.

We may conceive of at least two distinct types of modification. In one we change the direction of descent so as not to leave the set Ω; this can be done either by modifying the gradient, or in an independent fashion. This has the advantage that we always will have a bound on $\inf\{f(x)|x\in\Omega\}$ available. In another direction, we may modify the set Ω with some procedure that guarantees that in the limit we are forced back to Ω.

We start with finding a replacement for the definition of critical point and criticizing sequence suited to the present situation. Since we no longer can use $\nabla f(\bar{x})=0$ at the minimum, we must look to the analysis of Section 13.5, in particular Lemma 13.5.8, and Theorem 13.5.1. Thus we expect $\langle x-\bar{x},\nabla f(\bar{x})\rangle\geqslant 0$ for all $x\in\Omega$, and for simplicity we limit ourselves to weakly (sequentially) compact convex sets. In addition we restrict ourselves to convex functionals f with continuous Fréchet derivatives.

From Taylor's formula

$$f(\bar{x}+h)=f(\bar{x})+\langle h,\nabla f(\bar{x})\rangle+o(\|h\|).$$

Proceeding as in the unconstrained case we would want

$$\begin{aligned} f(\bar{x}+\bar{h}) &= \inf\{f(\bar{x}+h)|\bar{x}+h\in\Omega\} \\ &= f(\bar{x})+\inf\{\langle h,\nabla f(\bar{x})\rangle|\bar{x}+h\in\Omega\}+o(\|h\|). \end{aligned}$$

If the value of the auxiliary problem

$$\inf\{\langle h,\nabla f(\bar{x})\rangle|\bar{x}+h\in\Omega\}$$

is not negative, we have a local constrained minimum. If it is negative, then, although we may not be able to solve it, we can always "approximately" solve it. The resulting method is called a *conditional gradient method*.

Suppose now that at $\bar{x}\in\Omega$

$$\sigma(\bar{x})=\inf\{\langle x-\bar{x},\nabla f(\bar{x})\rangle|x\in\Omega\}<0.$$

Then for any $\epsilon=\epsilon(\bar{x})>0$, it is possible to find, in a finite number of steps, an $\hat{x}$ such that

$$\langle\hat{x}-\bar{x},\nabla f(\bar{x})\rangle\leqslant\sigma(\bar{x})+\epsilon(\bar{x}).$$

We will use $\hat{x}-\bar{x}$ as the direction of descent.

LEMMA 13.9.1. *If $f:X\to\mathbf{R}$ is convex, Ω is weakly (sequentially) compact, $\inf\{f(x)|x\in\Omega\}>-\infty$, $\{\epsilon^n\}$ is a decreasing sequence in $\mathbf{R}$ with $\epsilon_n\to 0, n\to\infty$, and $\{x^n\},\{\hat{x}^n\}$ are sequences in Ω such that*

$$\langle\hat{x}^n-x^n,\nabla f(x^n)\rangle\leqslant\inf\{\langle x-x^n,\nabla f(x^n)\rangle|x\in\Omega\}+\epsilon_n,$$

and

$$\lim_{n\to\infty}\langle\hat{x}^n-x^n,\nabla f(x^n)\rangle=0,$$

then

$$\lim_{n\to\infty}f(x^n)=\inf\{f(x)|x\in\Omega\}.$$

(That is, $\{x^n\}$ is a minimizing sequence.)

PROOF. Let $\bar{x}$ be a minimizing point of f in Ω. Then

$$\begin{aligned}0&\geqslant f(\bar{x})-f(x^n)\geqslant\langle\bar{x}-x^n,\nabla f(x^n)\rangle\\&\geqslant\langle\bar{x}-x^n,\nabla f(x^n)\rangle+\langle\hat{x}^n-x^n,\nabla f(x^n)\rangle-\langle\hat{x}^n-x^n,\nabla f(x^n)\rangle\\&\geqslant\langle\hat{x}^n-x^n,\nabla f(x^n)\rangle-\epsilon_n\to 0,\qquad n\to\infty.\ \blacksquare\end{aligned}$$

Thus, a sequence of descent directions and distances that yields $\langle\hat{x}^n-x^n,\nabla f(x^n)\rangle\to 0, n\to\infty$ produces a minimizing sequence. We present one algorithm (Demyanov and Rubin [1], 114–119) in the following:

THEOREM 13.9.1. *Let $f:X\to\mathbf{R}$ be convex on Ω, a weakly (sequentially) compact convex set in a Banach space X. Let f be Fréchet differentiable or Ω and satisfy*

$$\|\nabla f(x)-\nabla f(y)\|\leqslant L\|x-y\|,\qquad x,y\in\Omega,$$

for some $L<\infty$. Let $\alpha_n\in(0,1]$ satisfying

$$f(x^n+\alpha_n[\hat{x}^n-x^n])\leqslant\tfrac{1}{2}[f(x^n)+\inf\{f(x^n+\tau[\hat{x}^n-x^n])|t\in[0,1]\}]$$

while for some sequence $\{\epsilon_n\}, \epsilon_n \downarrow 0, n \to \infty$, *and*

$$\langle \hat{x}^n - x^n, \nabla f(x^n)\rangle \leqslant \inf\{\langle x - x^n, \nabla f(x^n)\rangle | x \in \Omega\} + \epsilon_n.$$

Then

$$\lim_{n\to\infty} \langle \hat{x}^n - x^n, \nabla f(x^n)\rangle = 0.$$

PROOF. Note that if $\inf\{\langle x - x^n, \nabla f(x^n)\rangle | x \in \Omega\} \geqslant 0$, then we may set $\hat{x}^m = x^n$ for all $m \geqslant n$. In general

$$\begin{aligned} f(x+h) &= f(x) + \langle h, \nabla f(x)\rangle + \int_0^1 \langle h, \nabla f(x+\tau h) - \nabla f(x)\rangle\, d\tau \\ &\leqslant f(x) + \langle h, \nabla f(x)\rangle + L\|h\| \int_0^1 \tau \|h\|\, d\tau \\ &\leqslant f(x) + \langle h, \nabla f(x)\rangle + \tfrac{1}{2} L\|h\|^2. \end{aligned}$$

Thus, for $0 < \alpha < 1$,

$$f(x^n + \alpha[\hat{x}^n - x^n]) \leqslant f(x^n) + \alpha\langle \hat{x}^n - x^n, \nabla f(x^n)\rangle + \tfrac{1}{2}\alpha^2 L\|\hat{x}^n - x^n\|^2.$$

Suppose for some sequence $n_1, n_2, n_3, \ldots$ that

$$\langle \hat{x}^{n_k} - x^{n_k}, \nabla f(x^{n_k})\rangle \to -\rho < 0.$$

Ω is bounded. Let $D = \sup\{\|x - y\| \,|\, x \in \Omega, y \in \Omega\}$. Applying the estimate to the subsequence n_k, for $n_k \geqslant$ some N we get

$$f(x^{n_k} + \alpha[\hat{x}^{n_k} - x^{n_k}]) \leqslant f(x^{n_k}) + \tfrac{1}{2}\alpha(-\rho + \alpha L D^2).$$

By our construction $f(x^n)$ is a decreasing sequence bounded below. Let $f(x^n) \downarrow Q, n \to \infty$. Choose a ρ_1 such that for some $\alpha, 0 < \alpha < 1$

$$\tfrac{1}{2}\alpha(-\rho + \alpha L D^2) \leqslant -\rho_1 < 0.$$

Since $f(x^n) \downarrow Q, n \to \infty, f(x^{n_k}) = Q + \delta_{n_k}$, with $\delta_{n_k} \downarrow 0, k \to \infty$. For k large enough $\delta_{n_k} < \rho_1/2$ and

$$f(x^{n_k} + \alpha[\hat{x}^{n_k} - x^{n_k}]) \leqslant Q + \delta_{n_k} - \rho_1 < Q - \tfrac{1}{2}\rho_1.$$

Then

$$\inf\{f(x^{n_k} + \alpha[\hat{x}^{n_k} - x^{n_k}]) | 0 \leqslant \alpha \leqslant 1\} \leqslant Q - \tfrac{1}{2}\rho_1,$$

and by our choice of α_{n_k}

$$f(x_{n_k+1}) \leqslant Q + \tfrac{1}{2}\delta_{n_k} - \tfrac{1}{4}\rho_1 < Q,$$

which is impossible. Hence $\langle \hat{x}^n - x^n, \nabla f(x^n)\rangle \to 0, n \to \infty$. ■

COROLLARY 1. $\{x^n\}$ *is a minimizing sequence.*

Convergence rates are established in Levitin and Polyak [1], 23–27. If $X = \mathbf{R}^n$, then finding $\hat{x}^n$ and α_n are feasible operations. In general, we have an error estimate in terms of our computations of the form

$$f(x^n) - \inf\{f(x) | x \in \Omega\} \leqslant -\langle \hat{x}^n - x^n, \nabla f(x^n)\rangle$$

when $f(x)$ is convex. Other step size algorithms as well as further references can be found in Demyanov and Rubin [1], 122–128 and Daniel [1], 95–105. See also Levitin and Polyak [1] and Céa [1], 128–137.

In the conditional gradient method, at each step we (approximately or exactly) solve a linearized problem to obtain a new direction *into* Ω for the descent process. A general study of motions that are feasible (i.e., $\hat{x}^n - x^n \in \Omega$) is to be found in Zoutendijk [1].

If the space X is a Hilbert space, we can find another direction of rapid descent in (a closed bounded convex set) Ω if we use the projections onto Ω of motions in the gradient direction as the direction for the descent process. This is known as the *gradient projection method* (Rosen [1], [2], Levitin and Polyak [1].)

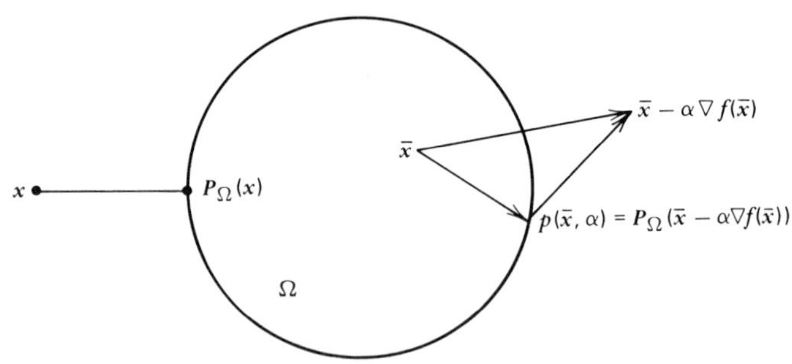

From Riesz's Theorem 9.2.3, there is a mapping $P_\Omega : X \to \Omega$ such that

$$\| P_\Omega(x) - x \| = \inf\{ \|y - x\| \,|\, y \in \Omega\}.$$

It is easy to see that it is equivalent to this minimization to write

$$(y-P_\Omega(x), x-P_\Omega(x)) \leqslant 0, y \in \Omega,$$

which says that the hyperplane $(z, x-P_\Omega(x))=0$ supports Ω at $z=P_\Omega(x)$, if $x \notin \Omega$. If $\bar{x} \in \Omega$ and we study all motions along the ray $\bar{x}-\alpha \nabla f(\bar{x})$, then, for α sufficiently large, we must replace this ray by $\bar{x}+\beta p(\bar{x},\alpha)$ where $p(\bar{x},\alpha) = P_\Omega(\bar{x}-\alpha\nabla f(\bar{x}))-\bar{x}$. Now $(y-p(\bar{x},\alpha), \bar{x}-\alpha\nabla f(\bar{x})-p(x,\alpha)) \leqslant 0$, so with $y=\bar{x}$ we find

$$-\alpha(\bar{x}-p(\bar{x},\alpha), \nabla f(\bar{x})) \leqslant -\|\bar{x}-p(\bar{x},\alpha)\|^2.$$

Thus for all $\alpha>0, (\bar{x}-p(\bar{x},\alpha), \nabla f(x)) \geqslant 0$, so the direction $p(\bar{x},\alpha)-\bar{x}$ is a direction of descent. We establish first the following:

LEMMA 13.9.2. *If $f: X \to \mathbf{R}$ is convex, Ω is weakly (sequentially) compact, $\inf\{f(x) | x \in \Omega\} > -\infty$, and $x^{n+1} = x^n + \beta_n p(x^n, \alpha_n) \in \Omega, n=1,2,3,\ldots,$ then*

$$\alpha_n \geqslant \underline{\alpha} > 0 \text{ and } (x^n - p(x^n,\alpha_n), \nabla f(x^n)) \to 0, \qquad n \to \infty.$$

implies

$$f(x^n) \downarrow \inf\{f(x) | x \in \Omega\}, \qquad n \to \infty.$$

(That is, $\{x^n\}$ is a minimizing sequence.)

PROOF. Under the hypotheses, there is an $\bar{x} \in \Omega$ with $f(\bar{x}) = \inf\{f(x) | x \in \Omega\}$. Let $p^n = p(x^n, \alpha_n)$. Then, by convexity

$$\begin{aligned} f(x^n)-f(\bar{x}) &\leqslant (x^n-\bar{x}, \nabla f(x^n)) \\ &\leqslant (x^n - p^n, \nabla f(x^n)) + \frac{(\bar{x}-p^n, -\alpha_n \nabla f(x^n))}{\alpha_n}. \end{aligned}$$

Now

$$\frac{(\bar{x}-p^n, -\alpha_n\nabla f(x^n))}{\alpha_n} = \frac{(\bar{x}-p^n, x^n-\alpha_n\nabla f(x^n)-p^n)}{\alpha_n} + \frac{(\bar{x}-p^n, p^n-x^n)}{\alpha_n},$$

so, if $\alpha_n > 0$,

$$f(x^n)-f(\bar{x}) \leqslant (x^n-p^n, \nabla f(x^n)) + \frac{\|\bar{x}-p^n\| \, \|p^n-x^n\|}{\alpha_n}.$$

As $n \to \infty$, the first term on the right tends to zero, while $\|\bar{x}-p^n\|$ is bounded, since Ω is. Also, reversing the inequality that shows that $p^n - x^n$ is a descent direction yields

$$\frac{\|p^n - x^n\|^2}{\alpha_n^2} \leqslant \frac{(x^n - p^n, \nabla f(x^n))}{\alpha_n},$$

which also tends to zero provided that $\alpha_n \geqslant \underline{\alpha} > 0$ for all n. Thus $\{x^n\}$ is a minimizing sequence. ■

We present one algorithm (Demyanov and Rubin [1], 148) in the following:

THEOREM 13.9.2. *Let $f: X \to \mathbf{R}$ be a convex functional on Ω, a weakly (sequentially) compact convex set in a Hilbert space X. Let f be Fréchet differentiable on Ω and satisfy*

$$\|\nabla f(x) - \nabla f(y)\| \leqslant L\|x-y\|, \qquad x,y \in \Omega,$$

for some $L < \infty$. Let $\underline{\alpha}$, $\bar{\alpha}$ and β be such that $0 \leqslant \underline{\alpha} < \bar{\alpha} < 2/L$ and $0 < \underline{\beta} \leqslant 1$. Let $\{\alpha_n\}, \{\beta_n\}$ be sequences satisfying $\underline{\alpha} < \alpha_n \leqslant \bar{\alpha}, \underline{\beta} \leqslant \beta_n \leqslant 1$. Let

$$x^{n+1} = x^n + \beta_n[p(x^n, \alpha_n) - x^n], \qquad n = 1,2,3,\ldots.$$

Then $\{f(x^n)\}$ is a monotone deceasing sequence and $\|x^n - p(x^n,\alpha_n)\| \to 0, n \to \infty$.

PROOF. For any β,

$$\begin{aligned} f(x^{n+1}) &= f(x^n + \beta_n[p(x^n,\alpha_n) - x^n]) \\ &\leqslant f(x^n) + \beta_n(p(x^n,\alpha_n) - x^n, \nabla f(x^n)) \\ &\quad + \tfrac{1}{2}\beta_n^2 L\|p(x^n,\alpha_n) - x^n\|^2 \\ &\leqslant f(x^n) - \left(\frac{\beta_n}{\alpha_n} - \tfrac{1}{2}\beta_n^2 L\right)\|p(x^n,\alpha_n) - x^n\|^2. \end{aligned}$$

Then

$$f(x^n) - f(x^{n+1}) \geqslant \underline{\beta}\left(\frac{1}{\bar{\alpha}} - \frac{L}{2}\right)\|p(x^n,\alpha_n) - x^n\|^2.$$

Thus $f(x^n)$ is a monotone decreasing sequence, bounded below, so convergent, since $1/\bar{\alpha} - L/2 > 0$. Hence $\|p(x^n,\alpha_n) - x^n\| \to 0, n \to \infty$. ■

COROLLARY 1. *$\{x^n\}$ is a minimizing sequence.*

PROOF. $\|\nabla f(x)\|$ is bounded on the bounded set Ω, so $(p(x^n,\alpha_n)-x^n, \nabla f(x^n))\to 0, n\to\infty$. ∎

For other convergence theorems and further references, see Demyanov and Rubin [1], 140–158. For convergence rates see Levitin and Polyak [1], 17–20, where asymptotic rates similar to unconstrained rates are found. As in the unconstrained case, it is possible to add the concept of a variable metric method to the conjugate gradient method. This is left as an exercise.

As in the unconstrained problem, if f is twice Fréchet differentiable on a Hilbert space X, then at each step we can consider the minimization of the quadratic part of Taylor's formula:

$$f(x^n+h)=f(x^n)+(h,\nabla f(x^n))+\tfrac{1}{2}(h,H(x^n)h)+o(\|h\|^2);$$

that is, we move in the direction of $\hat{x}^n-x^n$ where $\hat{x}^n$ solves

$$\inf\{(x-x^n,\nabla f(x^n))+\tfrac{1}{2}(x-x^n,H(x^n)[x-x^n])|x\in\Omega\}.$$

If

$$\phi_n(x)=(x-x^n,\nabla f(x^n))+\tfrac{1}{2}(x-x^n,H(x^n)[x-x^n])$$

then $\hat{x}^n$ satisfies

$$(x-\hat{x}^n,\nabla\phi_n(\hat{x}^n))=(x-\hat{x}^n,\nabla f(x^n)+H(x^n)[\hat{x}^n-x^n])\geqslant 0,\qquad x\in\Omega.$$

Then for $x=x^n$ we obtain

$$(\hat{x}^n-x^n,\nabla f(x^n))\leqslant -(\hat{x}^n-x^n,H(x^n)[\hat{x}^n-x^n]).$$

If $(h,H(x)h)\geqslant\gamma\|h\|^2$ for all h, this shows that x^n-x^n is a descent direction. That it can be used to establish a minimizing process is shown in the following:

THEOREM 13.9.3. *Let $f:X\to\mathbf{R}$ be a convex twice Fréchet differentiable functional on Ω, a closed bounded convex set in a Hilbert space X. Let there exist $\gamma,M,0<\gamma<M<\infty$, with*

$$\gamma(h,h)\leqslant(h,H(x)h)\leqslant M(h,h),\qquad x\in\Omega,h\in X.$$

Let $\{x^n\}$ be a sequence in Ω, let $\hat{x}^n$ minimize $\phi_n(x)$ on Ω and let $(\hat{x}^n-x^n, \nabla f(x^n))\to 0, n\to\infty$. Then $f(x^n)\downarrow f(\bar{x})$ where $\bar{x}$ minimizes $f(x)$ on Ω.

PROOF. As above, since $H(x^n)$ is a strictly positive bounded selfadjoint operator

$$(\hat{x}^n - x^n, \nabla f(x^n)) \leqslant -(\hat{x}^n - x^n, H(x^n)[\hat{x}^n - x^n])$$
$$\leqslant -\frac{(H(x^n)[\hat{x}^n - x^n], H(x^n)[\hat{x}^n - x^n])}{M}.$$

Also

$$(\hat{x}^n - x^n, \nabla f(x^n)) \leqslant -(\hat{x}^n - x^n, H(x^n)[\hat{x}^n - x^n]) \leqslant \gamma \|\hat{x}^n - x^n\|^2.$$

Then both $\|H(x^n)(\hat{x}^n - x^n)\|$ and $\|\hat{x}^n - x^n\| \to 0, n \to \infty$. Also

$$(\hat{x}^n - x, \nabla f(x^n)) = (\hat{x}^n - x^n, \nabla f(x^n)) + (x^n - x, \nabla f(x^n)).$$

Also

$$(\hat{x}^n - x, \nabla f(x^n)) = (\hat{x}^n - x, \nabla \phi_n(x^n)) - (\hat{x}^n - x, H(x^n)[\hat{x}^n - x^n])$$
$$\leqslant -(\hat{x}^n - x, H(x^n)[\hat{x}^n - x]).$$

Since Ω is bounded, $\|\hat{x}^n - x\|$ is bounded on Ω, while $\|H(x^n)(\hat{x}^n - x^n)\| \to 0, n \to \infty$. Thus

$$(\hat{x}^n - x^n, \nabla f(x^n)) \leqslant (x - x^n, \nabla f(x^n)) + \epsilon_n$$

where $\epsilon_n \to 0, n \to \infty$. We can now apply Lemma 13.9.1.

This Newton's method for constrained problems produces a minimizing sequence. In Levitin and Polyak [1], it is shown that the sequence is convergent in norm, at a geometric rate, provided that $H(x)$ satisfies a uniform Lipschitz condition in $x \in \Omega$.

It is also possible to use a Ritz or Galerkin procedure to produce finite dimensional approximating problems for which the computations may be effected by conditional gradient methods or variable metric gradient projection methods. The problem

$$\inf\{f(x) | x \in \Omega\}$$

is replaced by

$$\inf\{f(x^n) | x^n \in X^n \cap \Omega\}$$

where $X^n = \mathbf{S}\{\phi^1,\ldots,\phi^n\}$ and $\{\phi^n\}$ is total. The Ritz system becomes a system of *inequalities*. Thus, as in Section 13.8, if $x^n + h \in \Omega, 0 \leqslant t \leqslant \bar{t} < \infty$ for some $\bar{t}$, then $\langle h, \nabla f(x^n)\rangle = df(x^n + th)/dt_{t=0} \geqslant 0$, is necessary for x^n to be a local minimum; this is not a computationally advantageous approach. Analogs of Lemmas 13.8.1 and 13.8.2 and Theorem 13.8.4 are easily established, and the constrained minimization problem in X^n may be attacked by other methods. Examples of the use of this method are found in Levitin and Polyak [1], 32–34. It is important in this that the condition $\alpha_1\phi^1 + \cdots + \alpha_n\phi^n \in \Omega$ be computationally feasible.

Applications of these ideas to concrete problems may be found in Demyanov and Rubin [1], 173–248.

EXERCISES TO SECTION 13.9

1. For the conditional gradient method of Lemma 13.9.1 establish the error estimate

$$f(x^n) \leqslant \inf\{f(x) | x \in \Omega\} - \langle \hat{x}^n - x^n, \nabla f(x^n)\rangle$$

when f is convex.

2. Where was the assumption that f is convex used in Theorem 13.9.1?
3. Let $f(x)$ be a twice Fréchet differentiable functional on a Hilbert space X, with $\gamma I \leqslant H(x) \leqslant MI$ for $0 < \gamma < M < \infty$. Develop and prove equivalents to Lemma 13.9.2 and Theorem 13.9.2 for a variable metric gradient projection method. (Hint: Replace $\bar{x} - \alpha\nabla f(\bar{x})$ by $\bar{x} - \alpha[H(x)]^{-1}\nabla f(\bar{x})$.)
4. In the conditional gradient method, replace the definition of $\hat{x}^n$ as an (approximate) minimizer of the linearized problem $\langle x - x^n, \nabla f(x^n)\rangle$ on Ω by requiring that $\hat{x}^n$ minimize the quadratic

$$\langle x - x^n, \nabla f(x^n)\rangle + \tfrac{1}{2}\langle x - x^n, H(x^n)(x - x^n)\rangle$$

when $f(x)$ is twice Fréchet differentiable. Prove analogs to Lemma 13.9.1 and Theorem 13.9.1 under the assumptions that $f(x) \in R_{H(x)}$ for all $x \in \Omega$, and that X is a Hilbert space.

5. In Exercise 13.9.4, let H_n be any sequence in $[X]$ that satisfies $\gamma I \leqslant H_n \leqslant MI$ for some $0 < \gamma < M < \infty$ and $n = 1,2,3,\ldots$. Use this sequence to generate a *variable metric gradient projection method* (i.e., we use $P_\Omega(\bar{x} - \alpha H_n \nabla f(\bar{x})) - \bar{x}$ as a descent direction). Prove analogs of Lemma 13.9.2 and Theorem 13.9.2.

6. In the optimal control problem of Exercise 12.10.1, let the controls u be vector functions from $[t_0, t_1] \to \mathbf{R}^m$ that satisfy constraints of the form

$$|u_i(t)| \leqslant \mu, \qquad i=1,\dots,m, \qquad t_0 \leqslant t \leqslant t_1.$$

Apply the Ritz method in the following fashion. For each $k=1,2,3,\dots,$ define $\phi_k^j(t)$ as the characteristic function of the set $t_0+[(j-1)(t_1-t_0)/k, j(t_1-t_0)/k]$. Let $X^k=\mathbf{S}\{\phi_k^1,\dots,\phi_k^k\}$. Find the equivalent optimization problem in X^n, and its dual.

BIBLIOGRAPHY

ABLOW, C.M. and **PERRY, C.L.**

[1] Iterative solutions of the Dirichlet problem for $\Delta u = u^2$, *J. Soc. Ind. App. Math.*, **7** (1959), 459–467.

AKAIKE, H.

[1] On a successive transformation of probability distribution and its application to the analysis of the optimum gradient method, *Ann. Inst. Stat. Math. Tokyo*, **11** (1959), 1–16.

AL'BER, Ya.I.

[1] Continuous processes of the Newton Type, *Differ. Equations*, **7** (1971), 1461–1471, translated from *Differ. Uravn.*, **7** (1971), 1931–1943.

BANACH, STEFAN

[1] *Théorie des Opérations Linéaires*, Chelsea Publishing Company, New York, 1955.

BAZAARA, M.S. and **SHETTY, C.M.**

[1] *Foundations of Optimization*, Springer, New York, 1976.

BOGGS, PAUL T.

[1] The solution of nonlinear systems of equations by A-stable integration techniques, *SIAM J. Numer. Anal.*, **8** (1971), 767–785.

BREMERMAN, HANS

[1] *Distributions, Complex Variables and Fourier Transforms*, Addison-Wesley, Reading, Mass., 1965.

BROYDEN, C.

[1] Quasi-Newton methods, in *Numerical Methods for Unconstrained Optimization*, W. Murray, Ed., Academic Press, London, 1972, 87–106.

BRYSON, ARTHUR E. JR. and **HO, YU-CHI**

[1] *Applied Optimal Control*, Blaisdell Publishing Co., Waltham, Mass., 1969.

CÉA, JEAN

[1] *Optimization, théorie et algorithmes*, Dunod, Paris, 1971.

COLLATZ, LOTHAR

[1] *Functional Analysis and Numerical Mathematics*, Academic Press, New York, 1966.

DANIEL, JAMES W.

[1] *The Approximate Minimization of Functionals*, Prentice-Hall, Inc., Englewood Cliffs, N.J., 1971.

[2] The conjugate gradient method for linear and nonlinear operator equations, *SIAM J. Numer. Anal.*, **4** (1967), 10–26.

DAVIDENKO, D.F.

[1] On a new method of numerical solution of systems of nonlinear equations, *Dokl. Akad. Nauk. USSR (N.S.)*, **88** (1953), 601–602.

DAVIDON, W.C.

[1] Variable metric method for optimization, *AEC. Res. and Dev. Rep. ANL* 5990, Argonne Natl. Lab., 1959.

DAY, M.M.

[1] *Normed Linear Spaces*, Springer, New York, 1973.

DEMYANOV, V.F. and **RUBINOV, A.M.**

[1] *Approximate Methods in Optimization Problems*, American Elsevier Publishing Co., New York, 1970.

DENNIS, J.E. JR.

[1] Toward a unified convergence theory for Newton-like methods, in *Nonlinear Functional Analysis and Applications*, L.B. Rall, Ed., Academic Press, New York, 1971, 425–472.

EDWARDS, R.E.

[1] *Functional Analysis: Theory and Applications*, Holt, Rinehart & Winston, New York, 1965.

EKELAND, IVAR and **TEMAM, ROGER**

[1] *Analyse Convexe et Problemes Variationels*, Dunod, Paris, 1974.

FLETCHER, R. and **POWELL, M.J.D.**

[1] A rapidly convergent method for minimization, *Comp. J.*, **6** (1964), 163–169.

FORSYTHE, GEORGE E.

[1] On the asymptotic directions of the *s*-dimensional optimum gradient method, *Numer. Math.*, **11** (1968), 57–76.

FRANKLIN, PHILIP

[1] *A Treatise on Advanced Calculus*, Dover Publications Inc., New York, 1964.

FRIEDRICHS, KURT O.

[1] Spektraltheorie halbbeschänkter Operatoren und Anwendung auf die Spektralzerlegung von Differentialoperatoren, *Math. Ann.*, **109** (1934), 465–487.

GAVURIN, M.K.

[1] Nonlinear functional equations and continuous analogs of iterative methods, *Izv. Vyss. Ucebn. Zaved. Math.*, **6** (1958), 18–31, translated in *Tech. Rep.* 68–70, Univ. of Maryland, College Park, 1968.

GREENSPAN, H., KELBER, C.N. and **OKRENT, D.**

[1] *Computing Methods in Reactor Physics*, Gordon and Breach Science Publishers, New York, 1968.

GUIGNARD, MONIQUE

[1] Generalized Kuhn-Tucker conditions for mathematical programming in a Banach space, *SIAM J. Control*, **7** (1969), 232–241.

HARDY, G.H., LITTLEWOOD, J.E., and **POLYA, G.**

[1] *Inequalities*, Cambridge at the University Press, Cambridge, England, 1959.

HESTENES, M.R. and **STIEFEL, E.**

[1] Methods of conjugate gradients for solving linear systems, *J. Res. Natl. Bur. Stand.*, **49** (1952), 409–436.

HETRICK, DAVID L.

[1] *Dynamics of Nuclear Reactors*, University of Chicago Press, Chicago, 1971.

HILLE, EINAR

[1] *Methods in Classical and Functional Analysis*, Addison-Wesley, Reading, Mass., 1972.

HOLTZMAN, J.M.

[1] The use of the contraction mapping theorem with derivatives in a Banach Space, *Q. J. Appl. Math.*, **26** (1968), 462–465.

INTRILIGATOR, MICHAEL D.

[1] *Mathematical Optimization and Economic Theory*, Prentice-Hall, Inc., Englewood Cliffs, N.J., 1971.

ISAACSON, EUGENE and **KELLER, HERBERT B.**

[1] *Analysis of Numerical Methods*, John Wiley & Sons, New York, 1966.

JAMES, R.C.

[1] Characterizations of reflexivity, *Stud. Math.*, **23** (1946), 205–216.

KALMAN, R.E.

[1] Contributions to the theory of optimal control, *Bol. Soc. Mat. Mex.*, (2), **5** (1960), 102–119.

KANTOROVICH, L.V.

[1] Functional Analysis and Applied Mathematics, *Uspekhi*, **3** (1948), 89–185, translated in *Nat. Bur. Std. Rep.* 1509, 1952.

KANTOROVICH, L.V. and **AKILOV, G.P.**

[1] *Functional Analysis in Normed Spaces*, Pergamon Press, Oxford, England, 1964.

KANTOROVICH, L.V. and **KRYLOV, V.I.**

[1] *Approximate Methods of Higher Analysis*, Interscience Publishers Inc., New York, 1958.

KOLMOGOROV, A.N. and **FOMIN, S.V.**

[1] *Elements of the Theory of Functions and Functional Analysis*, Graylock Press, Rochester, N.Y., 1957.

KOROVKIN, P.P.

[1] *Linear Operators and Approximation Theory*, Hindustan Publishing Corp., Delhi, 1960.

KOTHE, G.

[1] *Topological Vector Spaces*, Vol. 1, Springer, New York, 1969.

KRASNOSEL'SKII, M.A. and **KREIN, S.G.**

[1] An iterative process with minimum residuals, *Mat. Sb.* **31** (1952), 73.

KRASNOSEL'SKII, M.A. and **VAINIKKO, G.M., ZABREIKO, P.P., RUTITSKII, Ya.B,** and **STETSENKO, V.Ya.**

[1] *Approximate Solution of Operator Equations*, Woltors-Noordhoff, Groningen, 1972, translated from *Priblizhennoe reshenie operatornykh uravnenii*, Moscow, 1969.

KUHN, H.W. and **TUCKER, A.W.**

[1] Nonlinear Programming, in *Proceedings of the Second Berkeley Symposium on Mathematical Statistics and Probability*, J. Neyman, Ed., University of California Press, Berkeley, 1951, 481–492.

KURATOWSKI, K.

[1] *Introduction to Set Theory and Topology*, Pergamon Press, Oxford, England, 1972.

LEVITIN, E.S. and **POLYAK, B.T.**

[1] Constrained minimization methods, *USSR Comp. Math. and Math. Phys.*, **6** (5) (1966), 1–50.

LUENBERGER, D.G.

[1] *Optimization by Vector Space Methods*, John Wiley & Sons, New York, 1969.

LUMER, G.

[1] Semi-inner-product spaces, *Trans. Am. Math. Soc.*, **100** (1961), 29–43.

LUSTERNIK, L.A. and **SOBOLEV, V.J.**

[1] *Elements of Functional Analysis*, John Wiley & Sons, New York, 1974.

LYUBICH, Yu.I. and **MAISTROVSKII, G.D.**

[1] The general theory of relaxation processes for convex functionals, *Uspekhi*, **25** (1970), translated in *Russ. Math. Surv.*, **25** (1970), 57–117.

MANGASARIAN, O.L.

[1] *Nonlinear Programming*, McGraw-Hill, New York, 1969.

MARTIN, ROBERT H. JR.

[1] *Nonlinear Operators and Differential Equations in Banach Spaces*, John Wiley & Sons, New York, 1976.

MIKHLIN, SOLOMON G.

[1] *The Problem of the Minimum of a Quadratic Functional*, Holden-Day, Inc., San Francisco, 1965.

[2] *Variational Methods in Mathematical Physics*, Pergamon Press, Oxford, England 1964.

[3] *The Numerical Performance of Variational Methods*, Wolters-Noordhoff, Groningen, Netherlands 1971.

MIKHLIN, SOLOMON G. and **SMOLITSKY, K.L.**

[1] *Approximate Methods for the Solution of Differential and Integral Equations*, American Elsevier Publishing Co., New York, 1967.

MORSE, PHILIP M.

[1] *Queues, Inventories and Maintenance*, John Wiley & Sons, New York, 1958.

NOBLE, B.

[1] Error analysis of collocation methods for solving Fredholm integral equations, in *Topics in Numerical Analysis*, Miller, J.J.H., Ed., Academic Press, New York, 1973, 211–232.

ORTEGA, J.M.

[1] The Newton-Kantorovich theorem, *Am. Math. Mon.*, **75** (1968), 658–660.

ORTEGA, J.M. and **RHEINBOLDT, W.C.**

[1] *Iterative Solution of Nonlinear Equations in Several Variables*, Academic Press, New York, 1970.

PADULO, L. and **ARBIB, M.A.**

[1] *System Theory*, W. B. Saunders Co., Philadelphia, 1974.

PARZEN, EMANUEL

[1] *Stochastic Processes*, Holden-Day, Inc., San Francisco, 1962.

POHAŽAEV, S.I.

[1] The Dirichlet problem for $\Delta u = u^2$, *Sov. Math.*, **1** (1960), 1143–1146.

POLYAK, B.T.

[1] The conjugate gradient method in extremal problems, *USSR Comp. Math. Math. Phys.*, **9** (4) (1969), 94–111.

[2] Existence theorems and the convergence of minimizing sequences in extremum problems with restrictions, *Sov. Math.*, **7** (1966), 72–75.

PORTER, WILLIAM A.

[1] *Modern Foundations of Systems Engineering*, Macmillan Co., New York, 1966.

PROTTER, M.H. and **WEINBERGER, H.F.**

[1] *Maximum Principles in Differential Equations*, Prentice-Hall, Englewood Cliffs, N.J., 1967.

PSHENICHNYI, B.N.

[1] *Necessary Conditions for an Extremum*, Dekker, Inc., New York, 1971.

RALL, LOUIS B.

[1] *Computational Solution of Nonlinear Operator Equations*, John Wiley & Sons, New York, 1969.

RALL, LOUIS B. Ed.

[2] *Nonlinear Functional Analysis and Applications*, Academic Press, New York, 1971.

REID, W.T.

[1] A matrix differential equation of Riccati type, *Am. J. Math.*, **68** (1946), 237–246.

RHEINBOLDT, W.C. and **VANDERGRAFT, J.S.**

[1] On the local convergence of update methods, *SIAM J. Numer. Anal.*, **11** (1974), 1069–1085.

RIESZ, FREDERIC

[1] *Les Systèmes d'Équations Linéaires a une Infinité d'Inconnues*, Gauthier-Villars, Paris, France 1913.

RIESZ, FREDERIC and **NAGY, BELA SZ.**

[1] *Functional Analysis*, Ungar, New York, 1955.

ROCKAFELLAR, R.T.

[1] *Convex Analysis*, Princeton University Press, Princeton, N.J., 1969.

[2] *Conjugate Duality and Optimization*, SIAM, Philadelphia, 1974.

ROSEN, J.B.

[1] The gradient projection method for nonlinear programming I: linear constraints, *J. Soc. Ind. Appl. Math.*, **8** (1960), 181–217.

[2] The gradient projection method for nonlinear programming II: nonlinear constraints, *J. Soc. Ind. Appl. Math.*, **9** (1961), 514–532.

RUDIN, W.

[1] *Functional Analysis*, McGraw-Hill, New York, 1973.

SOBELEV, S.L.

[1] *Applications of Functional Analysis in Mathematical Physics*, Am. Math. Soc., Providence, R.I., 1963.

STONE, MARSHALL H.

[1] *Linear Transformations in Hilbert Space*, Colloquium Publications Vol. 15, Am. Math. Soc., Providence, R.I., 1932.

TAPIA, RICHARD A.

[1] The Kantorovich theorem for Newton's method, *Am. Math. Mon.*, **78** (1971), 389–392.

[2] The weak Newton method and boundary value problems, *SIAM J. Numer. Anal.*, **6** (1969), 539–550.

TAYLOR, ANGUS E.

[1] *Introduction to Functional Analysis*, John Wiley & Sons, New York, 1958.

[2] *General Theory of Functions and Integration*, Blaisdell Publishing Co., New York, 1965.

VAINBERG, M.M.

[1] *Variational Methods for the Study of Nonlinear Operators*, Holden-Day, Inc., San Francisco, 1964.

[2] *Variational Method and the Method of Monotone Operators in the Theory of Nonlinear Equations*, John Wiley & Sons, New York, 1973.

VAINIKKO, G.M.

[1] Galerkin's perturbed method and the general theory of approximate methods for nonlinear equations, *Zh. Vychsl. Mat. Mat. Fiz.*, **7** (1967), 723–751, translated in *USSR Comp. Math. Math. Phys.*, **7** (4) (1967), 1–41.

[2] The compact approximation principle in the theory of approximation methods, *Zh. Vychsl. Mat. Mat. Fiz.*, **9** (1969), 739–761, translated in *USSR Comp. Math. Math. Phys.*, **9** (4) (1969) 1–32.

ZAANEN, ADRIAAN C.

[1] *Linear Analysis*, North-Holland Publishing Co., Amsterdam, Netherlands, 1960.

ZOUTENDIJK, G.

[1] *Method of Feasible Directions*, Elsevier Publishing Co., Amsterdam, Netherlands, 1960.

INDEX